# A FIRST UNDERGRADUATE COURSE IN ABSTRACT ALGEBRA

## THIRD EDITION



Abraham P. Hillman
*University of New Mexico*

Gerald L. Alexanderson
*University of Santa Clara*

*Wadsworth Publishing Company*
*Belmont, California*
*A division of Wadsworth, Inc.*

*To Josephine Hillman*

Mathematics Editor: Richard Jones
Production: Del Mar Associates
Cover designer: John Odam



Printed in the United States of America

3 4 5 6 7 8 9 10—87 86 85 84

**Library of Congress Cataloging in Publication Data**

Hillman, Abraham P.
A first undergraduate course in abstract algebra.

Bibliography: p.
Includes index.
1. Algebra, Abstract. I. Alexanderson, Gerald L.
II. Title.
QA162.H54 1983 512′.02 82-8484
ISBN 0-534-01195-0 AACR2

ISBN 0-534-01195-0

## *Contents*

## 4 *Rings and Fields 215*



## 5 *Polynomials 291*

# *Preface*

The basic structure of the earlier editions has been maintained. Among the larger additions are five new sections on discrete mathematics in Chapter 3, an expanded Section 4.7 on matrices, and a new Chapter 8 on applications to algebraic coding. Among the smaller changes are some new examples and explanatory remarks and increased use of the prominent display of definitions. The extensive problem sets have been augmented both for the new topics and to round out previous material.

No previous work in abstract algebra or linear algebra is assumed in this treatment of groups, rings, and fields.

Important concepts such as normal subgroups, quotient groups, and group homomorphisms and their analogues in ring theory are introduced gradually and developed thoroughly.

Many useful results from the Theory of Numbers and the Theory of Equations are obtained with little extra effort as by-products of the work with abstract structures. This greatly facilitates applications. For example, Lagrange's Theorem for finite groups leads easily to the well-known theorems of Euler and Fermat, and these are seen in Section 8.7 to be the basis for the application to public-key encryption. Also, the study of rings of polynomials provides a natural context for the investigation of polynomial equations.

This approach blends the motivation and insight available from concrete examples with the efficiency of abstract methods. The existence of infinite numbers of permutation groups, of modular rings, and of polynomial domains makes it possible for each student to have as many examples of the key concepts and techniques as are needed.

Groups are taken up before rings; thus group theory is available for application to the additive and multiplicative groups of rings. Also, concepts such as "homomorphism" are developed first for structures with one operation and then extended to structures with two operations with the aid of analogy and the experience with simpler material. Another advantage is that unfamiliar noncommutative structures (permutation groups) are studied at an early stage—when it is important to show the necessity of deducing properties from the axioms instead of taking them for granted on the basis of work with the special cases of elementary algebra.

The arrangement of topics in this text helps avoid tedious proofs of expected results and cuts down on unnecessary duplication and mere busy work. It thus permits effective teaching and retention of a greater amount of material.

In the following table, we illustrate that any section in this text can be reached effectively and covered in a one-quarter or one-semester course:

| *Goal* | *Sections Needed* |
|---|---|
| Automorphisms of Extension Fields | 1.2–1.6, 2.1–2.13, 3.1–3.3, 4.1–4.4, 4.6, 4.8, 5.1–5.4, 5.8, 5.9, 5.11, 5.12 |
| Euclidean Constructions | 1.2–1.6, 2.1–2.12, 3.1–3.3, 4.1–4.4, 4.6, 5.1–5.5, 6.1–6.3 |
| Number Theory through Quadratic Residues | 1.1–1.7, 2.1–2.12, 3.1–3.3, 4.1–4.4, 7.1, 7.3, 7.4 |
| Discrete Mathematics for Computer Science | 1.1–1.6, 2.1–2.12, 3.1–3.13 |
| Matrix Coding | 1.1–1.6, 2.1–2.12, 3.1–3.3, 4.1–4.4, 4.6, 4.7, 8.1–8.6 |
| Trapdoor Functions | 1.1–1.7, 2.1–2.12, 3.1–3.3, 4.1–4.4, 4.6, 8.7 |

Other objectives are even more readily accessible. However, achieving one's goal may require rapid progress through material that is easy or known to most students; this is especially true of Chapter 1. Also, one may have to resist the temptation to assign all the interesting problems or to delve into nonrequired sections. For this reason, Sections 3.10, 3.11, 3.12, 3.13, and 3.15 have been starred to indicate that they are not required for later chapters.

This forbearance is not necessary for those whose curricula afford the luxury of a year course in undergraduate abstract algebra; their students may reap maximum benefits from almost all of the sections and most of the problems. Others may be consoled by the fact that their students need not be limited to the assigned reading material and problems.

There is no "royal road" to effective understanding of abstract algebra (or any other branch of mathematics); effort is necessary to acquire knowledge and to develop the ability to apply that knowledge. Special care has therefore been used in the construction of the problem sets; this includes a number of aids for students willing to do the required work.

There are problems suitable for students with a very wide range of abilities in organized sets consisting of more than 1500 problems. In each set, problems with lower numbers tend to be easier and more concrete.

Some of the problems have hints such as references to previous results. Others are broken into parts to facilitate solution; these parts are numbered (i), (ii), (iii), etc. while parts that may be assigned separately are labeled (a), (b), (c), etc. There are problems that ask for a counterexample to show that a theorem is no longer true when part of the hypothesis is omitted (e.g., dropping "finiteness" or "commutativity").

The problems also provide a more gradual approach to new material through the foreshadowing of concepts to be introduced formally at a later stage.

Most of the odd-numbered problems have answers in the text. Some detailed solutions are given and thus are available for reference. If a problem is cited later in the text, this is indicated in a note at the beginning of the problem set. After the sections of each chapter, there is a special set of review problems which summarizes the topics and helps to integrate the material covered.

We also present some series of related problems leading the student through concrete examples to a more general or abstract formulation. Readers desiring greater tests of knowledge, ingenuity, and perseverance may consult the starred problems and the sets of supplementary and challenging problems at the ends of chapters.

In these ways, students gain considerable sophistication and frequently anticipate theorems or discover results new to them. They are helped to see mathematics as a living subject that is growing through the work of present-day mathematicians.

As a result of changes in high school curricula and in calculus sequences, there is a tendency toward an early introduction to the study of the basic number systems (the integers, rational numbers, real numbers, and complex numbers) as algebraic structures. To avoid unnecessary repetition, we assume some familiarity with these systems; however, many opportunities for learning their properties are provided in the context of studying generalizations and investigating examples.

The final sections in Chapters 2 to 5 are presented in order to motivate the study of other material and to show the connections between undergraduate and graduate courses in abstract algebra; these sections are starred, since they are built around results that have proofs too advanced for a text of this level.

To provide further evidence that mathematics is created by real people, we include historical and biographical notes on some of the great discoveries in algebra and the mathematicians who made them. We hope that reading about their lives while delving into consequences of their creative efforts will be more meaningful than reading their biographies alone.

We also furnish an annotated bibliography as a liaison with other published material. This is especially helpful in showing the many significant modern applications of groups, rings, and fields.

Previous editions of this text have been used successfully both for regular undergraduate courses and for summer institute courses, with different coverage and stress of topics. Students preparing for graduate work benefit from the opportunities for anticipation and generalization, the availability of supplementary material, and the emphasis on homomorphisms and quotient structures. Some of the advantages to prospective and in-service teachers are the large number of problems dealing with concrete algebraic structures, the material on constructions with straightedge and compass, the biographical and historical notes, and the references for applications.

The authors are grateful to Daniel M. Fendel, Marjorie Pickering Fitting, Richard M. Grassl, J. Myron Hood, Judith Q. Longyear, David G. Mead, Matthew Miller, Mervin E. Newton, Henrene E. Smoot, and Clifford N. Wagner for their many valuable comments and suggestions. We also express our appreciation to Josephine Hillman and to the editorial and production staffs of Wadsworth Publishing Company and Del Mar Associates, especially Rich Jones, Jerry Holloway, Sandra Craig, Marta Kongsle, and Nancy Sjoberg, for their invaluable assistance in the preparation of this edition.

*Abraham P. Hillman*
*Gerald L. Alexanderson*

# *Introduction*

*Strange as it may sound, the power of mathematics rests on its evasion of all unnecessary thought and on its wonderful saving of mental operations.*

*Ernest Mach*

Each of the natural numbers 1, 2, 3, ... is an abstraction; for example, the number 2 is something abstracted from all pairs of objects. An abstraction of higher order is the identity $(x + 1)(x - 1) = x^2 - 1$; in compact form this contains the infinite set of special cases in which $x$ is replaced by given numbers. We shall find that abstract algebra deals with abstractions of still higher order.

Algebra is concerned with sets of objects (such as numbers, polynomials, or matrices) and operations on these sets. In an axiomatic, or abstract, treatment of a given type of algebraic structure one assumes a small number of basic properties as axioms and then deduces many other properties from the axioms. Thus one deals simultaneously with all the structures satisfying a given set of axioms instead of with each structure individually.

Since this is for a first course in a branch of undergraduate mathematics, we try to help the reader gain maturity in the subject, instead of assuming this experience as a prerequisite. For example, each of the first few "if and only if" problems is restated as a proposition and its converse. There is a considerable amount of redundancy in the simultaneous use of phrases and their symbolic equivalents, such as "the quotient group $G/N$ of $G$ by its normal subgroup $N$." We hope that this helps some students to go on to courses in which this luxury can no longer be afforded and should no longer be expected.

It is assumed that the reader is acquainted with the rudiments of the underlying logic of proofs, for example, with the principle that things equal to the same thing are equal to each other.

This text includes some biographical and historical notes related to topics in abstract algebra. They generally appear at the ends of sections in which they are first mentioned and may also be located with the help of the index.

For ready reference, the inside front cover and its facing page have a description of set terminology and notation, the Greek alphabet, and a list of special symbols, each of which has fixed meaning when printed in boldface type.

# 1

## *Basic Properties of the Integers*

*All results of the profoundest mathematical investigation must ultimately be expressible in the simple form of properties of the integers.*

*Leopold Kronecker*

*God created the integers, all else is the work of man.*

*Leopold Kronecker*

Here we assume knowledge of the most elementary properties of the integers,

$$\mathbf{Z} = \{\ldots, -3, -2, -1, 0, 1, 2, 3, 4, \ldots\},$$

and treat some topics that are needed for our study of groups, rings, and fields. In later chapters, we shall return to the integers and discuss them further with the aid of techniques developed in our work with abstract algebraic structures.

### *1.1 Mathematical Induction*

*A mathematician, like a painter or a poet, is a maker of patterns. If his patterns are more permanent than theirs, it is because they are made with ideas.... The mathematician's patterns, like the painter's or the poet's, must be beautiful; the ideas, like the colours or the words, must fit together in a harmonious way. Beauty is the first test: there is no permanent place in the world for ugly mathematics.*

*G. H. Hardy*

First we consider a subset of the integers $\mathbf{Z}$, namely, the set

$$\mathbf{Z}^+ = \{1, 2, 3, 4, \ldots\}$$

of the positive integers. We assume that $\mathbf{Z}^+$ has the following property:

***Axiom 1 Well Ordering Principle***

*Every nonempty subset $S$ of $\mathbf{Z}^+$ has a least element.*

For example, the set $S = \{2, 4, 6, \ldots\}$ of the positive even integers is a nonempty subset of $\mathbf{Z}^+$ and we see that 2 is the smallest integer in $S$.

This property of $\mathbf{Z}^+$ is not possessed by the set $\mathbf{Z}$ of all the integers; for example, there is no smallest integer in the subset

$$\{\ldots, -4, -2, 0, 2, 4, 6, \ldots\}$$

of even integers in $\mathbf{Z}$.

A major application of the Well Ordering Principle is in the proof of the following:

***Theorem 1 Principle of Mathematical Induction***

*Let $S$ be a set of positive integers, that is, a subset of $\mathbf{Z}^+$, with the properties:*
(a) *The positive integer 1 is in $S$.*
(b) *If a positive integer $k$ is in $S$, so is $k + 1$.*
*Then every positive integer is in $S$; that is, $S = \mathbf{Z}^+$.*

***Proof***

Let $T$ be the set of all positive integers that are not in $S$. If $T$ is empty, $S = \mathbf{Z}^+$ and our task requires no more effort. Therefore, we assume that $T$ is nonempty. Then it follows from the Well Ordering Principle that there is a least integer $t$ in $T$.

By hypothesis (a), 1 is in $S$ and hence 1 is not in $T$. This means that $t$ is greater than 1. It follows that $t - 1$ is a positive integer. Since $t - 1$ is less than the least integer $t$ in $T$, $t - 1$ must be in $S$. Then

$$t = (t - 1) + 1$$

is in $S$ by hypothesis (b). Since $t$ is in $T$, this contradicts the definition of $T$ as the set of positive integers not in $S$. Thus the assumption that $T$ is nonempty is untenable; that is, $T$ is empty and $S = \mathbf{Z}^+$.

Next we illustrate the application of the Principle of Mathematical Induction to proofs of statements involving the positive integers.

We start by considering the Fibonacci sequence 1, 1, 2, 3, 5, 8, 13, 21, 34, 55, 89, 144, ..., which is defined by

$$F_1 = F_2 = 1, \qquad F_{n+2} = F_{n+1} + F_n \text{ for } n = 1, 2, 3, \ldots.$$

Which of the terms $F_m$ of this sequence are even? Among the first 12 terms, we see that $F_3 = 2$, $F_6 = 8$, $F_9 = 34$, and $F_{12} = 144$ are the only even integers.

It is natural to conjecture that $F_m$ is even if and only if $m$ is an integral multiple of 3. But verifying this conjecture in 12 or one million cases will not prove it true for the infinite number of terms. However, with the help of mathematical induction we now prove the conjecture true for all positive integers.

***Example 1***

Prove that $F_{3n-2}$ and $F_{3n-1}$ are odd and $F_{3n}$ is even for all positive integers $n$.

***Proof***

Let $S$ be the set of all positive integers $n$ for which the statement is true. We wish to prove that every positive integer is in $S$; that is, $S = \mathbf{Z}^+$.

When $n = 1$, we have $F_{3n-2} = F_1 = 1$, $F_{3n-1} = F_2 = 1$, and $F_{3n} = F_3 = 2$. Since 1 is odd and 2 is even, the statement holds for $n = 1$; that is, the integer 1 is in $S$ and condition (a) of the Principle of Mathematical Induction is satisfied.

Next we let $k$ be a positive integer in $S$; in other words, we assume that $F_{3k-2}$ and $F_{3k-1}$ are odd and that $F_{3k}$ is even. Now

$$F_{3(k+1)-2} = F_{3k+1} = F_{3k} + F_{3k-1},$$

using the definition of the Fibonacci sequence. Then $F_{3k+1}$ is odd since it is the sum of an even integer and an odd integer. For the same reason,

$$F_{3(k+1)-1} = F_{3k+2} = F_{3k+1} + F_{3k}$$

is odd. Also

$$F_{3(k+1)} = F_{3k+3} = F_{3k+2} + F_{3k+1}$$

is even, since it is the sum of two odd integers.

This shows that the assumption that $k$ is in $S$ implies that $k+1$ is in $S$; thus, condition (b) of Theorem 1 is satisfied. Since both conditions (a) and (b) hold, it follows from Theorem 1 that $S = \mathbf{Z}^+$. This completes the proof.

In proofs by mathematical induction, one generally streamlines the procedure of Example 1 by omitting explicit mention of the set $S$ of integers $n$ for which the statement is true and instead proving that a proposition $P(n)$ is true for all positive integers $n$ by showing both of the following:

(A) The proposition holds for $n = 1$; that is, $P(1)$ is true.

(B) Whenever $k$ is a positive integer such that $P(k)$ is true, then $P(k + 1)$ is true; that is, $P(k)$ implies $P(k + 1)$.

We next illustrate this streamlined technique.

*Example 2*

Let $F_n$ be the $n$th Fibonacci number. Prove that

$$(F_{n+1})^2 - F_n F_{n+2} = (-1)^n$$

for all positive integers $n$.

*Proof*

When $n = 1$, $(F_{n+1})^2 - F_n F_{n+2} = (F_2)^2 - F_1 F_3 = 1^2 - 1 \cdot 2 = -1$. Since $(-1)^n$ is also $-1$ when $n = 1$, the statement is true for $n = 1$. This establishes part (A).

Next we assume that $k$ is a positive integer for which the statement is true. We introduce the notation $F_k = a$, $F_{k+1} = b$, $F_{k+2} = c$, $F_{k+3} = d$. Our assumption concerning $k$ is that $b^2 - ac = (-1)^k$. We want to use this assumption to show that the statement is true for $k + 1$; that is, $c^2 - bd = (-1)^{k+1}$.

By definition of the Fibonacci sequence, $d = c + b$ and $c = b + a$. Hence

$$\begin{aligned} c^2 - bd &= c^2 - b(c + b) \\ &= c(c - b) - b^2 \\ &= ca - b^2 \\ &= -(b^2 - ac) \\ &= -(-1)^k \\ &= (-1)^{k+1}. \end{aligned}$$

Since this is the desired result for $k + 1$, part (B) is accomplished. Together, parts (A) and (B) prove the statement true for all positive integers $n$.

*Example 3*

Conjecture a simple formula for the sum

$$A_n = \frac{1}{1 \cdot 3} + \frac{1}{3 \cdot 5} + \frac{1}{5 \cdot 7} + \cdots + \frac{1}{(2n-1)(2n+1)},$$

and prove that the formula holds for all positive integers $n$.

***Solution***

$A_1 = 1/(1 \cdot 3) = 1/3$; $A_2 = (1/3) + (1/15) = 6/15 = 2/5$; $A_3 = (1/3) + (1/15) + (1/35) = (2/5) + (1/35) = 15/35 = 3/7$; $A_4 = (3/7) + (1/63) = 28/63 = 4/9$; and $A_5 = (4/9) + (1/99) = 45/99 = 5/11$. We assemble these data in the table:

| $n$ | 1 | 2 | 3 | 4 | 5 |
|---|---|---|---|---|---|
| $A_n$ | $\frac{1}{3}$ | $\frac{2}{5}$ | $\frac{3}{7}$ | $\frac{4}{9}$ | $\frac{5}{11}$ |

This table leads us to conjecture that $A_n = n/(2n+1)$. Let us see if mathematical induction can be used to make this conjecture into a theorem, that is, to prove the formula correct for all positive integers $n$.

We have already seen that the formula holds for $n = 1$; hence part (A) is done. Now we assume the formula correct for $n = k$; specifically, we assume that $A_k = k/(2k+1)$. Then

$$A_{k+1} = A_k + \frac{1}{[2(k+1)-1][2(k+1)+1]} = \frac{k}{2k+1} + \frac{1}{(2k+1)(2k+3)}$$

$$= \frac{k(2k+3)+1}{(2k+1)(2k+3)} = \frac{2k^2+3k+1}{(2k+1)(2k+3)}$$

$$= \frac{(2k+1)(k+1)}{(2k+1)(2k+3)} = \frac{k+1}{2k+3}.$$

We see that $A_{k+1} = (k+1)/[2(k+1)+1]$, which is the formula $A_n = n/(2n+1)$ for $n = k+1$. This shows that if the formula is true for $k$, it is also true for $k+1$. This finishes part (B); together (A) and (B) prove the theorem.

It is frequently convenient to have the following generalizations of the Well Ordering Principle and the Principle of Mathematical Induction:

***Theorem 2 Generalized Well Ordering Principle***

*Let a be any integer (positive, negative, or zero). Let X be the set of all integers x with $x \geq a$. Then every nonempty subset S of X has a least integer.*

***Theorem 3*** ***Generalized Mathematical Induction***

*Let $a$ be an integer and let $X$ consist of all integers $x$ with $x \geq a$. Let $S$ be a subset of $X$ such that $a$ is in $S$ and whenever an integer $k$ is in $S$ so is $k + 1$. Then $S = X$.*

Theorems 2 and 3 can be deduced from Axiom 1 (the original Well Ordering Principle) and Theorem 1, respectively, by a change of variable. The proofs are left to the reader.

Here we present a fairly small problem set on mathematical induction. There is a larger, and perhaps more interesting, collection in Section 7.2.

## *Problems*

1. Let $F_1, F_2, F_3, \ldots$ be the Fibonacci sequence. Prove that

$$F_{n+1}F_{n+2} - F_n F_{n+3} = (-1)^n$$

for all positive integers $n$.

2. The Lucas sequence 1, 3, 4, 7, 11, 18, 29, ... is defined by

$$L_1 = 1, \qquad L_2 = 3, \qquad L_{n+2} = L_{n+1} + L_n \text{ for } n = 1, 2, \ldots.$$

(i) Calculate $(L_{n+1})^2 - L_n L_{n+2}$ for $n = 1, 2, 3, 4$, and 5.
(ii) Make a conjecture based on the data of part (i). Test your conjecture for $n = 6$ and $n = 7$.
(iii) Prove your conjecture true for all positive integers $n$ by mathematical induction.

3. Let the Lucas sequence 1, 3, 4, 7, ... be as in Problem 2. Make a conjecture concerning the subscripts $m$ for which $L_m$ is even or odd and prove it true for all the terms of this sequence by mathematical induction.
4. Do as in Problem 2, with $(L_{n+1})^2 - L_n L_{n+2}$ replaced by $L_{n+1} L_{n+2} - L_n L_{n+3}$.
5. Let $T_2, T_3, T_4, \ldots$ be defined by

$$T_n = \left(1 - \frac{1}{2^2}\right)\left(1 - \frac{1}{3^2}\right)\left(1 - \frac{1}{4^2}\right)\cdots\left(1 - \frac{1}{n^2}\right).$$

(i) Note that $T_2 = 3/4$ and $T_3 = (3/4)(8/9) = 2/3$. Find $T_4$, $T_5$, $T_6$, $T_7$, and $T_8$.

(ii) Conjecture a simple formula for $T_n$ based on the data for $T_2$, $T_4$, $T_6$, and $T_8$. Is it true for $T_{10}$? Is it true for $T_3$, $T_5$, $T_7$?

(iii) Prove the formula true for all integers $n \geq 2$ by mathematical induction.

6. For every positive integer $s$, let $c_s = \cos(x/2^s)$. Let

$$P_n = c_1 c_2 c_3 \cdots c_n.$$

Prove that $P_n = (\sin x)/[2^n \sin(x/2^n)]$ for all positive integers $n$. [*Hint:* The case $n = 1$ follows from the double angle formula $\sin(2\theta) = 2 \sin \theta \cos \theta$.]

7. Let $A_n = 1 + 3 + 5 + \cdots + (2n - 1)$; that is, let $A_n$ be the sum of the first $n$ positive odd integers. Note that $A_1 = 1$ and $A_2 = 1 + 3 = 4$.
   (i) Find $A_3$, $A_4$, and $A_5$.
   (ii) Conjecture a simple formula for $A_n$.
   (iii) Prove the formula for all positive integers $n$.

8. Prove that, for all positive integers $n$,

$$1^3 + 2^3 + 3^3 + \cdots + n^3 = [n(n + 1)/2]^2.$$

9. Let $u_0$, $u_1$, $u_2$, ... be the sequence

$$0, 1, \frac{1}{2}, \frac{3}{4}, \frac{5}{8}, \frac{11}{16}, \ldots$$

in which the average of two consecutive terms is the term that follows them; that is, $u_{s+2} = (u_{s+1} + u_s)/2$. Let $v_s = u_s - u_{s-1}$. Do the following:
   (i) Calculate $v_1$, $v_2$, $v_3$, $v_4$, and $v_5$ and conjecture a simple formula for $v_n$.
   (ii) Prove the formula for all positive integers $n$.
   (iii) Show that $u_n = v_1 + v_2 + \cdots + v_n$.
   (iv) Use (iii) to find a simple formula for $u_n$.

10. Let $x_n = (n^3 + 5n)/6$ and $y_n = x_n - x_{n-1}$. Prove the following for all positive integers $n$:
   (i) $y_n$ is an integer. [*Hint:* $n(n + 1)$ is an even integer.]
   (ii) $x_n$ is an integer. [*Hint:* Use mathematical induction and part (i).]

11. Conjecture a simple formula for the following sum and prove it for all positive integers $n$ by mathematical induction:

$$B_n = 1(1!) + 2(2!) + 3(3!) + \cdots + n(n!).$$

[*Note:* $n! = 1 \cdot 2 \cdot 3 \cdots n$ for positive integers $n$. Also $B_1 = 1$, $B_2 = 5$, and $B_3 = 23$.]

12. Do as in Problem 11 for the sum

$$C_n = \frac{1}{2!} + \frac{2}{3!} + \frac{3}{4!} + \cdots + \frac{n}{(n + 1)!}.$$

13. Let $a_n$ be the number of representations of the positive integer $n$ as a sum of 1's and 2's, taking order into account. For example, $a_4 = 5$ since the representations of 4 are

$$2+2, \quad 2+1+1, \quad 1+2+1, \quad 1+1+2, \quad 1+1+1+1.$$

Prove that $a_n$ has a simple expression in terms of the Fibonacci numbers.

*14. Prove that every positive integer in $\{1, 2, 3, \ldots, 2^n - 1\}$ is expressible in one and only one way in the form

$$c_0 + 2c_1 + 2^2c_2 + 2^3c_3 + \cdots + 2^{n-1}c_{n-1}, \qquad \text{(R)}$$

with each $c_i$ in $\{0, 1\}$. [The representation (R) is the binary representation in which the base is 2 instead of the base 10 of the decimal system.]

---

***Leopold Kronecker*** *(1823–1891) Kronecker is a counterexample to the often stated belief on the part of some that good mathematicians, though very creative in mathematics, are not very adept at other human pursuits. Kronecker, while still a young man, fell heir to a sizable fortune and, although he already had pursued mathematical studies, went on to manage his business interests with great success. For eleven years he was a businessman and did almost no mathematics. Then he entered academic life and eventually succeeded E. E. Kummer as professor at the University of Berlin. His scholarly interests also ranged beyond mathematics to include Greek, Latin, Hebrew, and philosophy.*

*He was a good friend of Karl Weierstrass but carried on a long standing dispute with him concerning the proper direction of mathematics. Weierstrass was one of the foremost proponents of analysis during that period while Kronecker advocated the "arithmetization" of mathematics, that is, the treatment of mathematics in terms of the natural numbers. It was this view that led him to make his famous comment: "Die ganze Zahl schuf die liebe Gott, alles Übrige ist Menschenwerk" (God created the integers, all else is the work of man).*

*Kronecker was one of the first mathematicians to investigate thoroughly the work of Galois and he wrote on Galois theory in a way that made the subject accessible to many. He gave the first axiomatic formulation of abstract groups.*

---

***Fibonacci** (**or Leonardo of Pisa**)* *The Fibonacci numbers are named for a thirteenth century mathematician, Leonardo of Pisa. This is not as unlikely as it sounds since he was also called Fibonacci, literally "son of Bonaccio." His principal work,* **Liber Abaci**, *published in 1202, contained much algebra known to the Arabs, in addition to original work; it marked the renaissance of mathematical studies in Europe following the Dark Ages.*

***François Édouard Anatole Lucas*** *(1842–1891)* *Although the Fibonacci numbers are named after Leonardo of Pisa, the great exponent of this and similar sequences with recursion relations was the prolific French number theorist, Lucas. He summarized this aspect of his research in two articles in the first volume of the first mathematical journal published in the United States: "Théorie des fonctions numériques simplement périodiques,"* **American Journal of Mathematics,** *1 (1878), 184–240, 289–321.*

## *1.2 Multiples and Divisors, Primes in* **Z**

Let $a$, $b$, and $c$ be integers, that is, elements of

$$\mathbf{Z} = \{\ldots, -2, -1, 0, 1, 2, 3, \ldots\}.$$

If $a = bc$, we say that $b$ is an *integral divisor* of $a$ or that $b$ is a *divisor* of $a$ in **Z**; this is often denoted by $b \mid a$.

If $b \mid a$, we also say that $a$ is an *integral multiple* of $b$ or that $a$ is a *multiple* of $b$ in **Z**. Another phrase for $b \mid a$ is $b$ is a *factor* of $a$ in **Z**.

For example, $14 \mid 42$ and $(-15) \mid 45$. Also, the positive integral divisors of 30 (or of $-30$) are

$$1, 2, 3, 5, 6, 10, 15, \text{ and } 30, \tag{1}$$

and the other divisors of $\pm 30$ in **Z** are the negatives of the integers of (1). However, 7 is not an integral divisor of 30 since 30/7 is not an integer.

*Example 1*

Let $a$, $b$, $d$, $h$, and $k$ be integers with $d|a$ and $d|b$. Prove that $d|(ha + kb)$.

*Proof*

By definition, $d|a$ and $d|b$ imply that there are integers $u$ and $v$ such that $a = du$ and $b = dv$. Then

$$ha + kb = hdu + kdv = d(hu + kv),$$

which shows that $d|(ha + kb)$.

The only expressions for 1 as a product $bc$ of integers are

$$1 = 1 \cdot 1 = (-1)(-1);$$

this tells us that each of 1 and $-1$ is its own reciprocal and that the only integers with reciprocals in **Z** are 1 and $-1$. We use this fact in the following example.

*Example 2*

Prove that $s|t$ and $t|s$ together imply that $s = \pm t$.

*Proof*

The hypothesis tells us that there are integers $b$ and $c$ such that $t = bs$ and $s = ct$. Substituting from one of these equations into the other, we have $t = bct$. This implies that either $1 = bc$ or $t = 0$.

If $1 = bc$, then $b = c = \pm 1$ and $s = ct = \pm t$. If $t = 0$, then $s = ct = c \cdot 0 = 0$. In either case, the hypothesis implies that $s = \pm t$, as desired.

Every integer $a$ has at least the factorizations $a = a \cdot 1$ and $a = (-a)(-1)$; such a factorization is said to be *trivial*. That is, $a = bc$ is a trivial factorization if either $b$ or $c$ is in the set $\{1, -1\}$.

***Definition 1 Prime in Z***

*An integer $p$ that is not in $\{1, -1\}$ and has only the trivial factorizations $p = p \cdot 1 = (-p)(-1)$ is a* **prime** *in* **Z**.

It follows from this definition that an integer $p$ is a prime if and only if $p$ does not have a reciprocal in **Z** and any equation $p = bc$, with $b$ and $c$

integers, implies that both $b$ and $c$ are in the set $\{-p, -1, 1, p\}$. One easily sees that the first ten positive primes in **Z** are

$$2, 3, 5, 7, 11, 13, 17, 19, 23, \text{ and } 29.$$

### *Definition 2 Composite in Z*

*A nonzero integer that has a nontrivial factorization is a* **composite integer**.

For example, 15 is composite since $15 \neq 0$ and $15 = 5 \cdot 3$ is a nontrivial factorization. Other composite integers are $\pm 4$, $\pm 6$, $\pm 8$, $\pm 9$, $\pm 10$, $\pm 12$, and $\pm 14$. We note that 0, 1, and $-1$ are neither prime nor composite.

The following definition helps us to characterize some important subsets of **Z**.

### *Definition 3 Set Closed under Subtraction*

*A set $S$ is* **closed under subtraction** *if $a - b$ is in $S$ whenever $a$ and $b$ are in $S$. (The elements $a$ and $b$ need not be distinct.)*

#### *Example 3*

Let 10 and 14 be in a set $S$ of integers that is closed under subtraction. Show that 0 and 2 must be in $S$.

#### *Solution*

We are given that $a - b$ is in $S$ when $a$ and $b$ are in $S$. We can let $a = 10$ and $b = 10$ and thus find that $10 - 10 = 0$ is in $S$. Letting $a = 14$ and $b = 10$ shows that $14 - 10 = 4$ is in $S$. Similarly, $10 - 4 = 6$ and $6 - 4 = 2$ are in $S$.

## *Problems*

1. Restate the result in Example 1 of this section (without the proof), replacing each use of the symbolism "$x|y$" by "$x$ is an integral divisor of $y$."
2. Restate the result in Example 1 (without the proof) replacing each use of "$x|y$" by "$y$ is an integral multiple of $x$."
3. Let $a$, $b$, $c$, and $q$ be integers with $a = qb + c$. Show both of the following:

(a) If $d|a$ and $d|b$, then $d|c$.
(b) If $d|b$ and $d|c$, then $d|a$.

4. Let $a$ be an integer and let $S$ be the set of all integers $x$ which are integral multiples of $a$. Use Example 1 to show that $S$ is closed under subtraction.

5. (a) Find all integers $a$ such that $a|1$.
(b) Which integers are integral divisors of 0?
(c) How many integral divisors are there of 3?
(d) How many integers $a$ are there with $a|25$?
(e) Of which integers $n$ is 6 an integral multiple?

6. (a) How many integral divisors are there of a prime $p$?
(b) What is the least number of integral divisors that a composite integer can have? [See Problem 5(d).]

7. The integers 21 and 27 are in a set $T$ that is closed under subtraction.
(a) Which of the integers in the set $U = \{-20, -19, -18, \ldots, 19, 20\}$ must be in $T$?
(b) Can all the integers of $U$ be in $T$? Explain.

8. The integer 4 is in a set $R$ that is closed under subtraction.
(i) Show that 0 must be in $R$.
(ii) Show that $-4$ must be in $R$.
(iii) Show that 8 must be in $R$.
(iv) Which of the integers between $-25$ and 25 must be in $R$?

9. Which of the integers 0, 1, 2, 3, 4, 5, and 6 is expressible in the form $10x + 14y$, with $x$ and $y$ integers? Justify your answer.

10. Do the previous problem with $10x + 14y$ replaced by $15x - 9y$.

11. Given that $a|b$ and $b|c$, prove that $a|c$.

12. Let $a$, $b$, and $c$ be integers and let $a|b$. Prove that $(ac)|(bc)$.

13. Explain why the negative $-p$ of a prime $p$ is also a prime.

14. Explain why $-c$ is composite whenever $c$ is composite.

15. Let $a$, $b$, $c$, $d$, $h$, $k$, and $q$ be integers with

$$d = bh + ck \quad \text{and} \quad a = qb + c.$$

Find integers $x$ and $y$ (in terms of $h$, $k$, and $q$) such that $d = ax + by$.

16. Given that each of $R$ and $S$ is a set of integers closed under subtraction, prove that their intersection $T = R \cap S$ is also closed under subtraction. ($T$ consists of the integers that are in both $R$ and $S$.)

17. For each of the following integers $d$, find the smallest positive integer $n$ such that $d|(10^n - 1)$.
(a) $d = 7$; (b) $d = 11$; (c) $d = 13$; (d) $d = 37$.

18. Do as in Problem 17 for the following values of $d$.
(a) $d = 77$; (b) $d = 91$; (c) $d = 407$.

19. Find all solutions in integers $x$ and $y$ of the equation

$$xy + 5x - 8y = 79.$$

[It may be helpful to show that the given equation implies that
(i) $y = [39/(x - 8)] - 5$;
(ii) $(x - 8) | 39$;
(iii) $x \in \{-31, -5, 5, 7, 9, 11, 21, 47\}$.]

20. Tabulate the solutions in integers $x$ and $y$ of the equation

$$xy - 10x + 21y = 228.$$

## 1.3 *The Division Algorithm*

Multiplication of positive integers is a form of addition; for example, $4 \cdot 5$ may be thought of as

$$5 + 5 + 5 + 5 \quad \text{or} \quad 4 + 4 + 4 + 4 + 4.$$

Similarly, division of positive integers can be accomplished by repeated subtraction. Thus one tests whether or not 8 is an integral divisor of 48 by repeatedly subtracting 8's and seeing if 0 is obtained after some number of steps. In this case, 0 results after 6 subtractions of 8 from 48; hence $48 = 6 \cdot 8$ and $8 | 48$.

If we start with 53 instead of 48, a positive remainder that is less than 8 is reached after 6 subtractions of 8 and we find that

$$53 = 6 \cdot 8 + 5.$$

Further subtractions of 8 would give negative results, not 0. Hence we see that 8 is not a divisor of 53 in **Z**.

This motivates the following result.

### *Theorem 1 Division Algorithm*

*If $a$ and $b$ are integers and $b$ is positive, there exist unique integers $q$ and $r$ such that*

$$a = qb + r, \qquad 0 \le r < b.$$

*Also, $b \mid a$ if and only if $r = 0$. (One says that $q$ is the* **quotient** *and $r$ is the* **remainder** *in the division of $a$ by $b$.)*

***Proof***

Let $X$ be the set $\{0, 1, 2, \ldots\}$ of the nonnegative integers. Let $S$ be the subset of $X$ consisting of all the nonnegative integers that are expressible in the form $a - qb$, with $q$ an integer. We next show that the set $S$ is not empty.

Since $b$ is a positive integer, we have $b \ge 1$. Now

$$|a| \cdot b \ge |a| \cdot 1 = |a| \ge -a;$$

that is, $|a| \cdot b \ge -a$. It follows that $a + |a| \cdot b \ge a - a = 0$ and hence that

$$a - (-|a|)b \ge 0.$$

This inequality tells us that one of the values of $q$, for which $a - qb$ is in $S$, is $q = -|a|$; hence $S$ is nonempty. Then it follows from the Generalized Well Ordering Principle (Theorem 2 of Section 1.1) that the nonempty set $S$ has a least integer $r$. Associated with this $r$ is an integer $q$ such that $a - qb = r$.

By definition of $S$, its least integer $r$ satisfies $0 \le r$. We show that $r$ also satisfies $r < b$ by assuming $r \ge b$ and obtaining a contradiction. Let $r' = r - b$ and $q' = q + 1$. If $r \ge b$, then $r' \ge 0$ and

$$r' = r - b = (a - qb) - b = a - (q + 1)b = a - q'b.$$

Together $r' \ge 0$ and $r' = a - q'b$ imply that $r'$ is in $S$. Since $r' = r - b < r$, this contradicts the fact that $r$ is the least integer in $S$. Hence $r < b$; that is, $0 \le r < b$.

We next prove that $r$ and $q$ are unique. We start by assuming that

$$a = qb + r = q_1 b + r_1, \qquad 0 \le r < b, \qquad 0 \le r_1 < b.$$

Without loss of generality, one may assume that $r \ge r_1$. Then

$$0 \le r - r_1 < b, \qquad r - r_1 = (q_1 - q)b. \tag{1}$$

Since there is no multiple of $b$ between 0 and $b$, it follows from (1) that $r - r_1 = 0$. Then $r = r_1$, $qb = q_1 b$, and finally $q = q_1$ since $b \ne 0$. Thus $r$ and $q$ are uniquely determined by $a$ and $b$.

If $r = 0$, then $a = qb + r = qb$ and we have $b \mid a$. Conversely, if $b \mid a$, then there is an integer $m$ such that $a = mb$ and hence the unique $q$ and $r$ are $m$ and 0, respectively. We see that $b \mid a$ if and only if $r = 0$; this completes the proof.

***Corollary Extended Division Algorithm***

*If a and c are in* $\mathbf{Z}$ *and* $c \neq 0$, *there exist unique q and r in* $\mathbf{Z}$, *such that*

$$a = qc + r \quad \text{and} \quad 0 \leq r < |c|.$$

This corollary is proved by letting $b = |c|$ and then applying the theorem.

If the divisor $b$ in the division algorithm is chosen as 2, the remainder $r$ must be 0 or 1. Hence every integer $a$ is of the form $2q$ or $2q + 1$, with $q$ an integer.

***Definition 1 Parity of Integers***

*Integers m and n that are either both even or both odd have the same* **parity.**

That is, $m$ and $n$ have the same parity if and only if they have the same remainder when divided by 2.

The division algorithm has many applications; one of these is the following result.

***Theorem 2 Closure under Subtraction***

*Let S be a set of integers closed under subtraction. Then,*

(*a*) *If S is nonempty,* 0 *is in S.*

(*b*) *If a is in S, so is* $-a$.

(*c*) *If a and b are in S and q is an integer,* $a - qb$ *is in S.*

(*d*) *S consists of all the integral multiples of some b in* $\mathbf{Z}$ *or S is empty.*

(*e*) *If there is at least one nonzero integer in S, then S consists of all the integral multiples of the smallest positive integer in S.*

***Proof***

If $S$ is nonempty, there is an integer $a$ in $S$ and $a - a = 0$ is in $S$ by closure under subtraction. Then $0 - a = -a$ is in $S$. Hence we have proved parts (a) and (b).

Now let $a$ and $b$ be in $S$. Then $a - qb$ is the result of subtracting $q$ $b$'s from $a$ if $q$ is positive or of subtracting $-b$ a total of $-q$ times if $q$ is negative. In either case, $a - qb$ is in $S$ by closure under subtraction. This proves part (c).

If $S$ is the single element set $\{0\}$, $S$ consists of all the integral multiples of 0. Hence we now assume that there is a nonzero element $c$ in $S$. Then one of $c$ and $-c$ is positive and is in $S$. It follows from the Well Ordering Principle that the nonempty set $T$ of the positive integers in $S$ has a least integer $b$.

Let $a$ be any integer in $S$. The division algorithm gives us integers $q$ and $r$ such that

$$a = qb + r, \qquad 0 \leq r < b.$$

Then $r = a - qb$ is in $S$ by part (c).

Since $r$ is smaller than the least positive integer $b$ in $S$, $r$ cannot be positive. This and the condition $0 \leq r$ imply that $r = 0$. Hence $a = qb$; that is, every $a$ in $S$ is an integral multiple of the least positive integer $b$ in $S$.

Conversely, if $a$ is an integral multiple $qb$ of the smallest positive integer $b$ in $S$, then $a = 0 - (-q)b$ is in $S$ by parts (a) and (c).

We will generalize the concept of sets of integers closed under subtraction when we deal with *subgroups* (see especially Theorem 1 of Section 2.5) and when *ideals in rings* are introduced (in Section 4.2).

Problems related to the material of this section are among those at the end of Section 1.4.

## *1.4 Common Divisors*

Let $a$ and $b$ be nonzero integers. A common integral divisor $t$ of $a$ and $b$, that is, an integer $t$ such that both $t|a$ and $t|b$, satisfies the inequalities

$$-|a| \leq t \leq |a|, \qquad -|b| \leq t \leq |b|.$$

Hence the set $T$ of such common integral divisors is finite. Since 1 is in $T$, there is at least one positive integer in $T$. It follows from the last two statements that there is a largest positive integer in $T$.

### *Definition 1 Greatest Common Divisor*

*If $a \neq 0$ and $b \neq 0$, the largest positive integer $d$, such that both $d|a$ and $d|b$, is called the* **greatest common divisor** (*gcd*) *of $a$ and $b$ and is denoted by* $(a, b)$.

If $c \neq 0$, the largest positive integer $d$, such that both $d|c$ and $d|0$, is $d = |c|$. Hence we extend Definition 1 so that

$$(c, 0) = |c| = (0, c).$$

The gcd of 0 and 0 is not defined.

If one wants to avoid confusion with other uses of the notation $(a, b)$, a more explicit notation is $\gcd(a, b)$.

Some specific examples of greatest common divisors are

$$(15, 6) = 3, \quad (26, -10) = 2, \quad (0, -7) = 7, \quad (42, 14) = 14.$$

If $a$ and $b$ are integers, one can readily show the following:

(1) $(a, b) = (b, a)$ if $a \neq 0$ or $b \neq 0$.

(2) $(a, b) = (a, -b) = (-a, b) = (-a, -b)$.

(3) $(1, b) = 1$.

(4) If $b|a$ and $b \neq 0$, then $(a, b) = |b|$.

Terminology for an important special case, in which only 1 and $-1$ are common integral divisors of $a$ and $b$, is given in the following definition.

***Definition 2 Relatively Prime Integers***

*If* $gcd(r, s) = 1$, *the integers* $r$ *and* $s$ *are* **relatively prime** (*or* **coprime**).

For example, 22 and $-15$ are relatively prime since $(22, -15) = 1$. Also, 7 and 24 are relatively prime. We note that $r$ and $s$ may be relatively prime even though one (or each) of $r$ and $s$ is composite.

The examples $(17, 51) = 17$ and $(19, -19) = 19$ show that $r$ and $s$ are not necessarily relatively prime even when one (or each) of $r$ and $s$ is a prime.

Next we state the main results of this section.

***Theorem 1 Linear Combinations***

*Let $a$ and $b$ be in* $\mathbf{Z}$ *with* $a \neq 0$ *or* $b \neq 0$. *Let*

$$L = \{n : n = ax + by, x \in \mathbf{Z}, y \in \mathbf{Z}\};$$

*that is, let $L$ be the set of all $ax + by$ with $x$ and $y$ integers. Then there is a smallest positive integer $t$ in $L$ and $L$ consists of all the integral multiples of $t$.*

***Proof***

For definiteness, let $a \neq 0$. Let $n_1 = ax_1 + by_1$ and $n_2 = ax_2 + by_2$ be any two integers in $L$. Then

$$n_1 - n_2 = a(x_1 - x_2) + b(y_1 - y_2)$$

is in $L$ since the differences $x_1 - x_2$ and $y_1 - y_2$ of integers are also integers. Hence $L$ is closed under subtraction.

Using $x = 1$ and $y = 0$ in $ax + by$, we see that $a \cdot 1 + b \cdot 0 = a$ is in $L$. Similarly, $-a$ and $b$ are in $L$.

Since $a \neq 0$, either $a$ or $-a$ is positive; that is, $L$ has at least one positive integer in it. Then, among the positive integers in $L$ there is a smallest positive integer $t$ by the Well Ordering Principle. Also, $L$ consists of all the integral multiples of $t$ by Theorem 2 of Section 1.3.

We are now able to give several other characterizations of the gcd which are frequently used as alternative definitions of this concept. In fact, they will furnish a model for our definition of the gcd of two polynomials in Section 5.8.

### *Theorem 2 Characterizations of the GCD*

*Let $a$ and $b$ be in $\mathbf{Z}$ with $a \neq 0$ or $b \neq 0$. Then:*
*(a) $\gcd(a, b)$ is the smallest positive integer in the set*

$$L = \{n : n = ax + by, x \in \mathbf{Z}, y \in \mathbf{Z}\}.$$

*(b) $\gcd(a, b)$ is the unique positive common divisor $d$ of $a$ and $b$ such that $c \mid d$ for all common divisors $c$ of $a$ and $b$.*

### *Proof*

In Theorem 1, we saw that there is a smallest positive integer $t$ in $L$, that $t \mid n$ for all $n$ in $L$, and that $a$ and $b$ are in $L$. Hence $t \mid a$ and $t \mid b$. Since $t$ is a common divisor of $a$ and $b$ while $d$ is the greatest common divisor, we have $t \leq d$.

Since $t \in L$, we have $t = ah + bk$ with $h$ and $k$ in $\mathbf{Z}$. If both $c \mid a$ and $c \mid b$, then $c \mid t$ by Example 1 of Section 1.2. One of the things this tells us is that $d \mid t$. Since $d$ and $t$ are positive, it follows that $d \leq t$. This and the previously shown $t \leq d$ imply that $d = t$.

Having proved that $d = t$, we now know that $d$ is a positive common divisor of $a$ and $b$ with the property that $c \mid d$ for every common divisor $c$ of $a$ and $b$. If $d'$ is another such integer, then both $d \mid d'$ and $d' \mid d$; and it follows that the positive integers $d$ and $d'$ are equal. This establishes the uniqueness in part (b).

### *Corollary Linear Combinations of Relatively Prime Integers*

*(a) Let $a$ and $b$ be fixed integers. Then every integer $c$ is a linear combination $c = ax + by$, with $x$ and $y$ integers, if and only if $a$ and $b$ are relatively prime.*
*(b) Integers $a$ and $b$ are relatively prime if and only if there exist integers $h$ and $k$ such that $1 = ah + bk$.*

The proof of the corollary is left to the reader in Problem 16 below.

*Theorem 3 Multiples of the Greatest Common Divisor*

*Let a and b be in Z with $a \neq 0$ or $b \neq 0$. Let $d = \gcd(a, b)$. Then an integer n is of the form $ax + by$, with x and y in $\mathbf{Z}$, if and only if $d \mid n$.*

*Proof*

Theorem 2 tells us that $d$ is the smallest positive integer in the set $L = \{n : n = ax + by, x \in \mathbf{Z}, y \in \mathbf{Z}\}$; that is, $d$ is the $t$ of Theorem 1. Then Theorem 1 states that $d \mid n$ for all $n$ in $L$.

## Problems

1. For each of the following choices of $a$ and $b$, find the quotient $q$ and the remainder $r$ in the division of $a$ by $b$:
   (a) $a = 203, b = 17$; (b) $a = -203, b = 17$; (c) $a = 0, b = 17$.
2. Do as in Problem 1 for the following:
   (a) $a = 1000, b = 13$; (b) $a = 1001, b = 13$; (c) $a = 1002, b = 13$.
3. (a) Find $d = \gcd(11, 99)$, $e = \gcd(-21, 14)$, $f = \gcd(14, 15)$, and $g = \gcd(1, 0)$.
   (b) Which integers in the set $\{1, 2, 3, 4, 5\}$ are relatively prime to 5?
   (c) Which integers in the set $\{1, 2, 3, 4, 5, 6\}$ are relatively prime to 6?
4. (a) Find $d = \gcd(18, 498)$, $e = \gcd(154, 35035)$, $f = \gcd(30, 1001)$, and $g = \gcd(0, -1)$.
   (b) Which integers in the set $\{1, 2, 3, 4, 5, 6, 7\}$ are relatively prime to 7?
   (c) Which integers in the set $\{1, 2, 3, 4, 5, 6, 7, 8\}$ are relatively prime to 8?
5. Given that $\gcd(a, 0) = 1$, what are the possibilities for $a$?
6. Given that $b$ is a positive integer with $\gcd(6, b) = 2$, what are the possibilities for the remainder when $b$ is divided by 6?
7. (i) What is the least positive integer $c$ of the form $c = 22x + 55y$, with $x$ and $y$ integers? Explain.
   (ii) Let $c$ be the answer to part (i) and find integers $x$ and $y$ such that $c = 22x + 55y$ and $0 < x < 5$.
8. Describe the set of all integers $c$ that are expressible in the form $c = 1470x + 119y$, with $x$ and $y$ integers. (*Hint*: See the Linear Combinations Theorem and Theorem 3.)
9. Let $72 = ah + bk$ with $a$, $b$, $h$, and $k$ in $\mathbf{Z}$. What are the possibilities for the value of $\gcd(a, b)$?
10. Let $72 = 99h + bk$ with $b$, $h$, and $k$ in $\mathbf{Z}$. What are the possibilities for the value of $\gcd(99, b)$?

11. (i) What are the positive integral divisors of 19?
    (ii) Let $a$ be an integer. Explain why either $\gcd(a, 19) = 1$ or $19|a$, but not both.
12. Let $p$ be a positive prime and $q$ a positive integer. Find $\gcd(p, q)$ under each of the following assumptions:
    (a) $1 \leq q < p$.
    (b) $p|q$.
    (c) $q = mp + r$ with $0 < r < p$.
    (d) $q$ is a prime and $q \neq p$.
13. Let $a$ and $p$ be in **Z** with $p$ a positive prime and $p|a$. Explain why $\gcd(a, p) = p$.
14. Let $a$ and $p$ be in **Z** with $p$ a positive prime and $a$ not an integral multiple of $p$. Explain why $\gcd(a, p) = 1$, that is, why $a$ and $p$ are relatively prime.
15. Let there be at least two integers in a set $S$ of integers that is closed under subtraction. Explain why $S$ consists of the integral multiples of the least positive integer in $S$.
16. Let $a$ and $b$ be in **Z**. Show that:
    (a) If $1 = ah + bk$ with $h$ and $k$ in **Z**, then $(a, b) = 1$.
    (b) If $(a, b) = 1$, then for every $n$ in **Z** there exist $x$ and $y$ in **Z** such that $n = ax + by$.
    (It is easily seen that the corollary to Theorem 2 follows from the parts of this problem.)
17. Let $t$, $u$, and $v$ be nonzero integers. Let $a = tu$, $b = tv$, and $d = (a, b)$. Show that $u$ and $v$ are relatively prime if and only if $t = \pm d$; that is, show both of the following:
    (i) If $u$ and $v$ are relatively prime, then $t = \pm d$.
    (ii) If $t = \pm d$, then $u$ and $v$ are relatively prime.
    [Part (ii) implies that a rational number $a/b$ always has an equivalent form $u/v$ in which the numerator and denominator are relatively prime.]
18. (i) Explain why the division algorithm implies that every integer $a$ can be expressed in one and only one of the forms

$$3q, \qquad 3q + 1, \qquad 3q + 2,$$

    with $q$ an integer.
    (ii) Let $a$ and $q$ be integers. Show that 3 is not an integral divisor of $(3q + 1)^2$ or of $(3q + 2)^2$ and hence that $3|(a^2)$ if and only if $3|a$.
19. Prove that $\sqrt{3}$ is not rational by doing the following parts:
    (i) Assume that $a/b = \sqrt{3}$ with $a$ and $b$ integers. Let $d = (a, b)$, $a = du$, and $b = dv$. Explain why $u/v = \sqrt{3}$, with $u$ and $v$ relatively prime.
    (ii) Show that $u^2 = 3v^2$, $3|u^2$, $3|u$, and hence $u = 3w$ with $w$ an integer.
    (iii) Show that $3w^2 = v^2$ and hence $3|v$.

(iv) Note that $3|v$, together with $3|u$, contradicts the fact that $u$ and $v$ are relatively prime. This contradiction completes the proof that $\sqrt{3}$ is not a rational number $a/b$.

20. (a) Prove that $\sqrt{2}$ is irrational
    (b) Prove that $\sqrt{5}$ is irrational.
21. Let 170 and $-102$ both be in a set $S$ that is closed under subtraction.
    (a) Explain why $S$ must contain all the integral multiples of 34.
    (b) Can the integer 2 be in $S$? Explain.
22. Describe all the possibilities for the set $S$ of Problem 21.
23. (i) Let $a$ and $b$ be in $\mathbf{Z}$. Explain why $a$ and $a + 2b$ always have the same parity.
    (ii) Explain why $m - n$ and $m + n$ have the same parity for any $m$ and $n$ in $\mathbf{Z}$.
    (iii) Find all the solutions of the equation

$$x^2 - y^2 = 88$$

    in positive integers $x$ and $y$.
24. Find all solutions of $x^2 - y^2 = 77$ in positive integers $x$ and $y$.
25. Explain why there are no positive integers $x$ and $y$ such that $x^2 - y^2 = 34$.
26. Let $n = a^2 - b^2$ with $a$ and $b$ integers. Explain why $n$ must be either odd or an integral multiple of 4.
27. Explain why $n$ and $n + 1$ are relatively prime for all integers $n$.
28. Explain why the following are true.
    (i) If $n$ is an even integer, $\gcd(n, n + 2) = 2$.
    (ii) If $n$ is an odd integer, $\gcd(n, n + 2) = 1$.
29. What are the possible values of $\gcd(n, n + 10)$, where $n$ is an integer?
30. Let $n$ be an integer. Give the set of integers that might be values of $\gcd(n, n + 30)$.
31. Show that $\gcd(n - 1, n^2 + n + 1)$ is either 1 or 3 for all $n \in \mathbf{Z}$.
32. Show that $\gcd(n + 1, n^2 - n + 1)|3$ for all $n \in \mathbf{Z}$.

## 1.5 The Euclidean Algorithm

If $a$, $b$, and $c$ are fixed integers, it follows from Theorem 3 of Section 1.4 that the equation

$$ax + by = c$$

has integer solutions for $x$ and $y$ if and only if $c$ is of the form $kd$, with $d = \gcd(a, b)$ and $k$ in $\mathbf{Z}$. This section contains a constructive method for solving such an equation.

We start with the special case in which $k = 1$ and $a \geq b > 0$; in this case the euclidean algorithm described below shows how to find integers $x$ and $y$ such that

$$ax + by = \gcd(a, b).$$

Then in Example 2 we illustrate the solution of more general equations $ax + by = kd$, in which $k$ need not be 1, with the help of the euclidean algorithm.

Let $a$ and $b$ be positive integers with $a \geq b$. It follows from the division algorithm that there exist integers $q$ and $r$ with

$$a = qb + r \tag{1}$$

and $0 \leq r < b$. If $r = 0$, we have $a = qb$ and hence $b = \gcd(a, b)$. If $r > 0$, it follows from equation (1) and Example 1 of Section 1.2 that any common divisor of $b$ and $r$ is a divisor of $a$ and thus a common divisor of $a$ and $b$. Also, equation (1) can be rewritten as

$$r = a - qb,$$

and this similarly implies that every common divisor of $a$ and $b$ is a common divisor of $b$ and $r$. Hence it follows from (1) that

$$\gcd(a, b) = \gcd(b, r). \tag{2}$$

Equation (2) reduces the problem of finding $\gcd(a, b)$ to the same problem with smaller positive integers. Then the Well Ordering Principle helps us show that $\gcd(a, b)$ is found after a finite number of such reductions.

We next look at an example and then give the details of the general technique, which is called the *euclidean algorithm.* (See the biographical note on Euclid after this section.)

***Example 1***

Find gcd(803, 154) and integers $x$ and $y$ such that

$$\gcd(803, 154) = 803x + 154y.$$

***Solution***

Since $803 = 5 \cdot 154 + 33$, we have $(803, 154) = (154, 33)$. Similarly, $154 = 4 \cdot 33 + 22$ implies that $(154, 33) = (33, 22)$ and $33 = 1 \cdot 22 + 11$ implies that $(33, 22) = (22, 11)$. Finally, $22 = 2 \cdot 11$ gives us

$$11 = (22, 11) = (33, 22) = (154, 33) = (803, 154);$$

that is, the last nonzero remainder 11 is the desired gcd(803, 154). Next we seek $x$ and $y$.

We rearrange the above data as follows:

$$803 = 5 \cdot 154 + 33, \tag{a}$$

$$154 = 4 \cdot 33 + 22, \tag{b}$$

$$33 = 1 \cdot 22 + 11, \tag{c}$$

$$22 = 2 \cdot 11.$$

Solving the next to the last equation (c) for the gcd, 11, we have

$$11 = 33 - 22. \tag{d}$$

Then solving (b) for the previous remainder 22 and substituting in (d) gives us

$$11 = 33 - (154 - 4 \cdot 33)$$

$$11 = 5 \cdot 33 - 154. \tag{e}$$

Finally, we use (a) to solve for the first remainder 33 and substitute into (e), thus obtaining

$$11 = 5(803 - 5 \cdot 154) - 154$$

$$11 = 5 \cdot 803 - 26 \cdot 154. \tag{f}$$

Equation (f) shows that $x = 5$ and $y = -26$ is one solution for this problem. (A description of all the solutions is given in Problem 15 at the end of Section 1.6.)

### *Description of the Euclidean Algorithm*

Let $a$ and $b$ be integers with $0 < b \leq a$. Then it follows from the division algorithm that there are $q_1$ and $r_1$ in $\mathbf{Z}$ with

$$a = q_1 b + r_1, \quad 0 \leq r_1 < b.$$

If $r_1 > 0$, there are $q_2$ and $r_2$ in $\mathbf{Z}$ with

$$b = q_2 r_1 + r_2, \quad 0 \leq r_2 < r_1.$$

If $r_2 > 0$, there are integers $q_3$ and $r_3$ such that

$$r_1 = q_3 r_2 + r_3, \quad 0 \leq r_3 < r_2.$$

In this process, smaller remainders occur at each step. Hence we must ultimately have a remainder $r_n$ that is 0 since otherwise there would be a set $\{r_1, r_2, r_3, \ldots\}$ of positive integers without a least element, thus contradicting the Well Ordering Principle.

Letting $r_n$ be the first zero remainder, we have

$$\begin{aligned}
a &= q_1 b + r_1, \\
b &= q_2 r_1 + r_2, \\
r_1 &= q_3 r_2 + r_3, \\
r_2 &= q_4 r_3 + r_4, \\
&\cdots \\
r_{n-3} &= q_{n-1} r_{n-2} + r_{n-1}, \\
r_{n-2} &= q_n r_{n-1}.
\end{aligned}$$

Since $r_n = 0$, we have $r_{n-1} \mid r_{n-2}$, and it follows that

$$\begin{aligned}
r_{n-1} &= (r_{n-2}, r_{n-1}) = (r_{n-3}, r_{n-2}) = \cdots = (r_1, r_2) \\
&= (b, r_1) = (a, b).
\end{aligned}$$

Thus we obtain gcd$(a, b)$ as the last nonzero remainder in this chain of divisions.

In the next example we show how to use the euclidean algorithm to find integers $x$ and $y$ satisfying

$$ax + by = kd,$$

where $a$, $b$, and $k$ are integers and $d = \gcd(a, b)$. Note that here we do not require that $a$ and $b$ are positive or that $k = 1$.

***Example 2***

Find integers $x$ and $y$ such that $803x - 154y = 33$.

***Solution***

Using the euclidean algorithm on the positive integers 803 and 154, one finds (see the results of Example 1) that $\gcd(803, 154) = 11$ and

$$803 \cdot 5 + 154(-26) = 11.$$

Multiplying both sides by 3, one obtains

$$803 \cdot 15 + 154(-78) = 33 \quad \text{or} \quad 803 \cdot 15 - 154 \cdot 78 = 33.$$

Hence one solution is $x = 15$ and $y = 78$.

Clearly, the technique of Example 2 was successful only because the equation is of the form $ax + by = c$, with $c$ an integral multiple of $\gcd(a, b)$; but Theorem 3 of Section 1.4 tells us that $c$ must be of this form if the equation can be solved for integers $x$ and $y$.

Problems related to the material of this section are among those at the close of Section 1.6.

---

***Euclid of Alexandria*** (365?–275? B.C.) *Although Euclid is known to be the author of a number of mathematical treatises and it is known that the most famous of these, the* **Elements**, *was written about 300* B.C., *little is known of his life. The dates of his birth and death, his birthplace, and even his nationality are not known.*

*The* **Elements** *has certainly been one of the most influential books in Western history, a surprising distinction for something that is essentially a textbook of elementary mathematics. Over a thousand editions have been issued, the first printed version having appeared in Venice (printed by E. Ratdolt) in 1482, not many years after the Gutenburg Bible.*

*The subject matter of the* **Elements** *is usually thought to be geometry; however, other topics of elementary mathematics are treated, albeit from a geometric point of view. Books VII–IX treat the so-called arithmetic, which is, in fact, what the English call "higher arithmetic" and the Americans call "theory of numbers." The euclidean algorithm appears at the beginning of Book VII. Book IX contains the proof that there exists an infinite number of primes, a formula for the sum of a geometric progression, and the formula for even perfect numbers.*

*It is doubtful that Euclid discovered many of the results the* **Elements** *contains. Mainly he is credited with the organization of existing knowledge, although some of the proofs are thought to be his. The style and organization of the* **Elements** *has been the model for much mathematical writing and is still influential today.*

---

## 1.6 Common Multiples

Let $a$ and $b$ be nonzero integers and let $M$ be the set of all common integral multiples of $a$ and $b$; that is, let $M$ consist of the integers $m$ such that both $a \mid m$ and $b \mid m$. Then $|ab|$ is in $M$, hence there is at least one positive integer in $M$.

It follows from Example 1 of Section 1.2 that $M$ is closed under linear combinations and therefore is closed under subtraction. Then Theorem 2(e) of Section 1.3 tells us that $M$ consists of all the multiples of the smallest positive integer $m$ of $M$.

### *Definition 1 Least Common Multiple*

*If $a$ and $b$ are nonzero integers, their least positive common integral multiple is called their* **least common multiple** *and is denoted by $[a, b]$ or lcm$[a, b]$. Also, $[a, 0] = 0 = [0, b]$ for all integers $a$ and $b$.*

Some examples are $[6, 15] = 30$, $[7, 21] = 21$, $[10, 27] = 270$, and $[23, 0] = 0$. It is also easily seen that for all integers $a$, $b$, and $c$ we have

$$[a, b] = [a, -b] = [-a, b] = [-a, -b],$$
$$[a, ac] = |ac|.$$

Now let $a|c$, $b|c$, and $m = \text{lcm}[a, b]$. Then it follows from Definition 1 and the first two paragraphs of this section that $m|c$. That is, any common integral multiple of $a$ and $b$ is an integral multiple of the least common multiple of $a$ and $b$.

### *Theorem 1 LCM of Relatively Prime Integers*

*Let $u \neq 0$ and $v \neq 0$. Then lcm$[u, v] = |uv|$ if and only if gcd$(u, v) = 1$.*

*Proof*

Let $\gcd(u, v) = 1$. Then the Corollary to Theorem 2 of Section 1.4 tells us that there exist integers $h$ and $k$ such that

$$1 = hu + kv.$$

Let $m = \text{lcm}[u, v]$. Then $m = su = tv$ with $s$ and $t$ in **Z**. We see that

$$s = s \cdot 1 = s(hu + kv) = h(su) + skv = htv + skv = (ht + sk)v, \text{ i.e., } v|s.$$

Then $v|s$ implies that $(uv)|(su)$, that is, $(uv)|m$. Clearly, $uv$ is a common integral multiple of $u$ and $v$. Also, the least common multiple $m$ is an integral divisor of all common multiples and hence $m|(uv)$. Together, $(uv)|m$ and $m|(uv)$ imply that $m = \pm uv$. Since $m > 0$, this tells us that $m = |uv|$.

Conversely, let lcm $[u, v] = |uv|$. Let $\gcd(u, v) = d$. We wish to show that $d = 1$. By definition of $d$ there exist $e$ and $f$ in **Z** such that $u = ed$ and $v = fd$. The hypotheses $u \neq 0$ and $v \neq 0$ imply that $e \neq 0$ and $f \neq 0$. Now

$$efd = uf = ve$$

shows that $efd$ is a common integral multiple of $u$ and $v$. This implies that $|uv| \leq |efd|$, since the hypothesis is that $|uv|$ is the least positive common multiple. Using $uv = edfd = efd^2$, we have

$$|uv| = |efd^2| \le |efd|.$$

Since $e \neq 0$ and $f \neq 0$, it follows that the positive integer $d$ is 1; that is, $\gcd(u, v) = 1$.

## Problems

Problems 9 and 10 below are cited in Section 5.5.

1. For each of the following pairs $c$ and $d$ find

$$(c, d), [c, d], (c, d)[c, d], \text{ and } cd.$$

   (a) $c = 8$ and $d = 12$;
   (b) $c = 7$ and $d = 35$;
   (c) $c = 3$ and $d = 14$.

2. Do as in Problem 1 for each of the following pairs of integers.
   (a) $c = 12$ and $d = 0$;
   (b) $c = 12$ and $d = 1$;
   (c) $c = 12$ and $d = -24$.

3. Let $a = 864371$ and $b = 735577$. Use the euclidean algorithm to find $d = (a, b)$ and integers $x$ and $y$ such that $d = by - ax$ and $0 < x < 400$.

4. Let $a = 980051$ and $b = 926213$. Find $d = (a, b)$ and positive integers $x$ and $y$ such that $d = by - ax$ and $0 < y < 100$.

5. Let $a$, $b$, and $c$ be integers such that $a = 33b + c$ and $(a, b) = 14$. What is $(b, c)$?

6. Find $(b, c)$ given that $a$, $b$, and $c$ are integers with $a = 47b - c$ and $(a, b) = 15$.

7. Let $a$, $b$, and $c$ be integers with $(a, c) = 1 = (b, c)$. Show that $c$ and $ab$ are relatively prime. [*Hint*: Use the existence of integers $e$, $f$, $g$, and $h$ such that $ae + cf = 1$ and $bg + ch = 1$, and show that there exist integers $x$ and $y$ with $abx + cy = 1$.]

8. Let $a$, $b$, and $c$ be integers with $a|c$, $b|c$, and $(a, b) = 1$. Show that $(ab)|c$. [*Hint*: Explain why there exist integers $h$, $k$, $u$, and $v$ such that $ah + bk = 1$, $c = ua$, and $c = vb$. Then find an integer $m$, in terms of $h$, $k$, $u$, and $v$, such that $c = mab$.]

9. Given that $a$, $b$, and $c$ are integers with $(a, b) = 1$ and $a|(bc)$, show that $a|c$.

10. Let $a$ and $b$ be relatively prime integers and let $n$ be a positive integer. Prove by mathematical induction that $a$ and $b^n$ are relatively prime.

11. Let $t$, $u$, and $v$ be integers. Show that $t[u, v]$ is a common multiple of $tu$ and $tv$.

*12. Given that $t$ is a positive integer, show that $t[u, v] = [tu, tv]$.

*13. Show that $[a, b](a, b) = |ab|$. [*Hint*: Use Problem 12 above, Theorem 1 of Section 1.6, and Problem 16(a) of Section 1.4.]

14. Let $a$, $b$, $c$, $m$, $q$, and $v$ be positive integers with $[b, c] = m$, $a = qb + c$, and $m = vc$. Show the following:
    (i) There is an integer $u$ such that $vc = ub$.
    (ii) $b|(va)$ and hence $va$ is a common multiple of $a$ and $b$.
    (iii) $[a, b]|(va)$.
    *(iv) $[a, b] = va$.

*15. (i) Show that $u$ and $v$ are integers such that $803u + 154v = 0$ if and only if $u = 14t$ and $v = -73t$, with $t$ an integer.
    (ii) Show that $x$ and $y$ are integers such that $803x + 154y = 11$ if and only if $x = 5 + 14t$ and $y = -26 - 73t$, with $t$ an integer.

## 1.7 Unique Factorization in Z

*(Arithmetic) is one of the oldest branches, perhaps the very oldest branch, of human knowledge; and yet some of its most abstruse secrets lie close to its tritest truths.*

*H. J. S. Smith*

One way in which the number 1 is special is that it is the only positive integer whose reciprocal is an integer. The other positive integers fall into one of two categories—the primes,

$$2, 3, 5, 7, 11, 13, 17, 19, 23, 29, 31, \ldots, \tag{P}$$

and the composites,

$$4 = 2 \cdot 2, \quad 6 = 2 \cdot 3, \quad 8 = 2 \cdot 2 \cdot 2, \quad 9 = 3 \cdot 3,$$
$$10 = 2 \cdot 5, \quad 12 = 2 \cdot 2 \cdot 3, \quad 14 = 2 \cdot 7, \ldots. \tag{C}$$

Below we prove that every integer greater than 1 is either a prime or the product of a finite number of positive primes and that this factorization is unique, except for rearrangement of the factors. The factorizations in (C) are examples of such representations.

We begin the proof with some preliminary results.

***Lemma 1 Euclid's Lemma***

*Let a, b, and p be integers with p a prime. If* $p|(ab)$, *then either* $p|a$ *or* $p|b$.

***Proof***

Our approach is to assume that $p$ is an integral divisor of $ab$ but not of $a$ and to show that these assumptions imply $p|b$.

The only integral divisors of the prime $p$ are $-p$, $-1$, 1, and $p$. With the assumption that $p$ is not an integral divisor of $a$, it follows that the only common divisors of $a$ and $p$ are 1 and $-1$.

Hence $\gcd(a, p) = 1$ and it follows from Corollary (b) to Theorem 2 of Section 1.4 that $ah + pk = 1$ with $h$ and $k$ in **Z**. The hypothesis $p|(ab)$ tells us that $ab = pt$ with $t$ in **Z**. Then

$$b = b \cdot 1 = b(ah + pk) = (ba)h + bpk = pth + bpk = p(th + bk).$$

This shows that $p|b$, as desired.

***Lemma 2 Generalized Euclid's Lemma***

*Let* $p, a_1, a_2, \ldots, a_n$ *be integers with p a prime. If* $p|(a_1 a_2 \cdots a_n)$, *then* $p|a_i$ *for at least one i in* $\{1, 2, \ldots, n\}$.

The proof follows readily for $n \geq 2$ by mathematical induction and is left to the reader. (Note that Lemma 1 is the case $n = 2$ of Lemma 2.)

***Lemma 3***

*Let* $p, u_1, u_2, \ldots, u_t$ *be positive primes and let*

$$p|(u_1 u_2 \cdots u_t). \tag{1}$$

*Then p is equal to at least one of the* $u_i$.

***Proof***

It follows from the hypothesis (1) and Lemma 2 that $p|u_i$ for some $i$. But the only positive integral divisors of a positive prime $u_i$ are 1 and $u_i$. Since $p \neq 1$, we have $p = u_i$.

The results in the two following theorems are frequently called the ***Fundamental Theorem of Arithmetic***.

**Theorem 1 Factorization into Primes**

*Every integer $n > 1$ is expressible as*

$$n = p_1 p_2 \cdots p_r \tag{2}$$

*with the $p_i$ positive primes in $\mathbf{Z}$.*

***Proof***

Let $S$ be the set of all integers $n > 1$ such that $n$ is not expressible in the form (2). We show that $S$ is the empty set by assuming that it is not empty and obtaining a contradiction.

A nonempty set $S$ of positive integers has a least integer $m$ by the Well Ordering Principle. If $m$ is a prime, it is of the form (2) with $r = 1$. Hence $m$ must be composite and there exist integers $u$ and $v$ such that

$$m = uv, \qquad 1 < u < m, \quad 1 < v < m.$$

Owing to the minimal nature of $m$, neither $u$ nor $v$ is in $S$. Hence both $u$ and $v$ are of the form (2); that is,

$$u = q_1 q_2 \cdots q_s, \qquad v = q_1' q_2' \cdots q_t'$$

with the $q_i$ and $q_i'$ positive primes. Then we have

$$m = uv = q_1 q_2 \cdots q_s q_1' q_2' \cdots q_t'.$$

This contradicts the assumption that $m$ is not of the form (2). Hence $S$ is empty and the result is proved.

**Theorem 2 Unique Factorization**

*Let $n$ have the representations*

$$p_1 p_2 \cdots p_r = q_1 q_2 \cdots q_s \tag{3}$$

*with the $p_i$ and $q_j$ positive primes. Then $r = s$ and the $p_i$ are the same as the $q_j$ except, possibly, for the order of appearance of the primes.*

***Proof***

Let us assume that the result is not true and use this assumption to obtain a contradiction. Then there is a prime that appears more times as a $p_i$ than as a $q_j$, or vice versa. Let $p$ be such a prime and let it appear $a$ times as a $p_i$ and $b$ times as a $q_j$. Since $a \neq b$, either $a > b$ or $a < b$. For definiteness, let $a > b$. (We may have $b = 0$.)

Now we cancel $b$ of the factors $p$ from each side of (3) and let $u_1 u_2 \cdots u_t$ be the product of the $q_j$ remaining on the right side of (3) after this cancellation. Since $a > b$, there is at least one $p$ on the new left side. This means that $p | (u_1 u_2 \cdots u_t)$. It then follows from

Lemma 3 that $p$ is one of the $u$'s and this contradicts the fact that all the $q$'s that are equal to $p$ have been removed when we obtained the $u$'s. This completes the proof of the Fundamental Theorem of Arithmetic.

Next we present a unique representation for integers $n > 1$ as products of powers of distinct primes. Let

$$n = q_1 q_2 \cdots q_u, \tag{4}$$

where the $q$'s are positive primes. Let $p_1, p_2, \ldots, p_v$ be the distinct primes of (4) in increasing order (that is, with $p_1 < p_2 < \cdots < p_v$). Collecting the multiple appearance of a $p$ among the $q$'s of (4), we have

$$n = p_1^{e_1} p_2^{e_2} \cdots p_v^{e_v}, \tag{S}$$

with each $e_i$ a positive integer. It follows from Theorem 2 that the primes $p$ and their exponents in (S) are unique for a given integer $n > 1$.

***Definition 1 Standard Factorization***

*Let $n$ have the representation (S) where the $p_j$ are primes such that $0 < p_1 < p_2 < \cdots < p_v$ and the $e_j$ are positive integers. Then we call the right side of (S) the* **standard factorization** *of $n$.*

For example, the standard factorization of 720 is $2^4 \cdot 3^2 \cdot 5$ and the positive prime 23 is its own standard factorization. Note that this representation only applies to integers $n > 1$.

One could apply the above factorization to a negative integer $m < -1$ by using the fact that its negative $-m$ is a positive integer with $-m > 1$.

## *Problems*

Problem 13(b) below is cited in Section 6.3.

1. Give the standard factorization for each of the following.
   (i) 5040; (ii) $5040^2$; (iii) $5040^3$.
2. Give the standard factorization for each of the following.
   (i) 2042040; (ii) $2042040^2$; (iii) $2042040^3$.
3. List all the positive integral divisors of each of the following.
   (i) 8; (ii) 16; (iii) 32; (iv) 64.
4. Give the number of positive integral divisors of each of the following.
   (i) $3^3$; (ii) $3^4$; (iii) $3^5$; (iv) $3^6$.
5. Let $2^a 3^b$ be the standard factorization for $n$. Explain why the following are true.

(i) Let $d$ be a positive integer. Then $d|n$ if and only if $d = 2^h 3^k$ with $h$ and $k$ integers satisfying $0 \le h \le a$ and $0 \le k \le b$.
(ii) There are $(a + 1)(b + 1)$ positive integral divisors of $n$.

6. How many positive integral divisors are there of $72 = 2^3 \cdot 3^2$?

7. Let $a$, $b$, and $c$ be positive integers. How many positive integral divisors are there of $2^a \cdot 5^b \cdot 11^c$?

8. Let $p_1^{e_1} p_2^{e_2} \cdots p_v^{e_v}$ be the standard factorization of $n$. Give the number of positive integral divisors of each of the following:
(i) $n$; (ii) $n^2$; (iii) $n^3$.

9. (a) Show that $a|b$ implies $a^2|b^2$.
(b) Show that $a^2|b^2$ implies $a|b$.
(c) Show that $a|b$ if and only if $a^3|b^3$.

10. Let $x, y, z \in \mathbf{Z}^+$ and $x^2 + y^2 = z^2$.
(a) Show that $\gcd(x, y) = 1$ if and only if $\gcd(x, z) = 1$.
*(b) Does $\gcd(x, y) = \gcd(x, z)$? Explain.

11. State a necessary and sufficient condition on the exponents $e_1, \ldots, e_v$ in the standard factorization (S) of $n$ for the integer $n \ge 2$ to be the square of an integer $m$. [See Problems 1(ii) and 2(ii).]

12. Do the analogue of Problem 11 concerning cubes. [See Problems 1(iii) and 2(iii).]

13. Use unique factorization into primes to prove each of the following:
(a) There do not exist integers $a$ and $b$ with $a^2 = 30b^2$ and hence $\sqrt{30}$ is irrational.
(b) There are no integers $c$ and $d$ with $c^3 = 2d^3$ and hence $\sqrt[3]{2}$ is irrational.

14. Use unique factorization to prove the following:
(a) $\sqrt{68}$ is irrational; (b) $\sqrt[3]{24}$ is irrational.

15. Let $m \in \mathbf{Z}^+$. Prove that $\sqrt{m}$ is rational if and only if it is an integer.

*16. Let $m, n \in \mathbf{Z}^+$. Prove that $\sqrt[n]{m}$ is rational if and only if it is an integer.

17. Let $a$ and $b$ be positive integers. Explain why $\gcd(a, b) = 1$ if and only if there is no positive prime $p$ that appears in the standard factorizations of both $a$ and $b$.

18. Let $s$ and $t$ be any integers. Explain why $\gcd(s, t) = 1$ if and only if there is no prime $p$ with both $p|s$ and $p|t$.

19. Show that $\gcd(a, bc) = 1$ if and only if

$$\gcd(a, b) = 1 = \gcd(a, c).$$

20. Given that $\gcd(a, b) = 1$, use Problem 18 to explain why $\gcd(a^m, b^n) = 1$ for all positive integers $m$ and $n$.

21. Let $a = 2^{30} \cdot 5^{21} \cdot 19 \cdot 23^3$ and $b = 2^6 \cdot 3 \cdot 7^4 \cdot 11^2 \cdot 19^5 \cdot 23^7$. Give the standard factorization for $\gcd(a, b)$ and $\text{lcm}[a, b]$.

22. (a) Let $e = 2^{10} \cdot 3^7 \cdot 13^6 \cdot 29^8$ and $f = 2^9 \cdot 5^3 \cdot 11^4 \cdot 13^8$. Give the standard factorization for $\gcd(e, f)$ and $\text{lcm}[e, f]$.

    (b) Let each of $a$ and $b$ be a positive integer greater than 1. Explain how to obtain the standard factorization for $\gcd(a, b)$ and $\text{lcm}[a, b]$ from those for $a$ and $b$.

23. Let $m$ and $n$ be positive integers. Use unique factorization to explain why

$$\gcd(m, n) \cdot \text{lcm}[m, n] = mn.$$

24. Use the preceding problem to explain why $\gcd(a, b) \cdot \text{lcm}[a, b] = |ab|$ for all integers $a$ and $b$.

25. Given that 307 is a prime and that $n$ is an integer such that

$$(1 \cdot 2 \cdot 3 \cdots 99)n = 307 \cdot 306 \cdot 305 \cdots 209,$$

use unique factorization to show that $307 | n$.

26. The binomial coefficients in the expansion

$$(x + 1)^n = \binom{n}{n}x^n + \binom{n}{n-1}x^{n-1} + \cdots + \binom{n}{1}x + \binom{n}{0}$$

are known to be positive integers and to be given by

$$\binom{n}{k} = \frac{n(n-1)(n-2)\cdots(n-k+1)}{1 \cdot 2 \cdot 3 \cdots k}.$$

Prove that $p \Big| \binom{p}{k}$ for $0 < k < p$, when $p$ is a prime.

27. How many pairs $\{s, t\}$ of positive integers are there satisfying both $st = 55440$ and $\gcd(s, t) = 1$?

28. Let the standard factorization of $n$ be

$$(p_1)^{e_1}(p_2)^{e_2} \cdots (p_r)^{e_r}.$$

How many pairs $\{s, t\}$ of positive integers are there satisfying both $st = n$ and $\gcd(s, t) = 1$? (The answer depends only on the number $r$ of positive prime divisors of $n$.)

## *Review Problems*

1. (i) Find $d = \gcd(8051, 8633)$, using the euclidean algorithm.
   (ii) Check that the $d$ of part (i) is a common divisor of 8051 and 8633.
   (iii) Find integers $x$ and $y$ such that $-20 < x < 0$ and

$$d = 8051x + 8633y.$$

2. Find (751903, 800881) and check that it is a common divisor.
3. Explain why $(n, n+4) \in \{1, 2, 4\}$ for all $n \in \mathbf{Z}$.
4. Let $30 = 84h + bk$ with $b$, $h$, and $k$ integers. What are the possible values of $\gcd(84, b)$?
5. Let $S$ be the set of all common integral multiples of 4 and 6.
   (i) Show that $S$ is closed under subtraction.
   (ii) What is the smallest positive integer $t$ in $S$?
   (iii) Does $S$ consist of all the integral multiples of the $t$ of part (ii)? Explain.
6. Give an example of positive integers $a$, $b$, and $c$ such that $a \mid (bc)$ but neither $a \mid b$ nor $a \mid c$.
7. Let $a$ and $b$ be integers such that $23 \mid (ab)$. Use the fact that 23 is a prime to show that either $23^2 \mid a^2$ or $23^2 \mid b^2$.
8. Let $n \in \mathbf{Z}^+$ and $r = \sqrt[3]{n}$. Use unique factorization into primes to show that $r$ is rational if and only if $r$ is an integer.

## *Supplementary and Challenging Problems*

1. Let $a$ and $b$ be integers with $b$ odd and $\gcd(a, b) = 1$. Prove that either $\gcd(a, 2b) = 1$ or $\gcd(a + b, 2b) = 1$.
2. Show that for every integer $n$ there exist unique integers $q$ and $r$ such that $n = 15q + r$ and $|r| \leq 7$.
3. Let $a$ and $b$ be integers with $b \neq 0$. Show that there exist unique integers $q$ and $r$ with $a = qb + r$ and $2|r| \leq |b|$ or give an example of lack of uniqueness.
4. Let $a$, $b$, $c$, and $q$ be integers with $c \neq 0$ and $a = bq + c$. Let $m = \text{lcm}[b, c] = hb = kc$. Prove that $ka = \text{lcm}[a, b]$.
5. With the help of Problem 4, explain how the division algorithm could be used to calculate the least common multiple of two integers.
6. Let $a$, $b$, and $c$ be integers. Let $d_1 = \gcd(a, b)$ and $d_2 = \gcd(b, c)$. Prove that $\gcd(d_1, c) = \gcd(a, d_2)$.
7. Generalize the result in Problem 6.
8. Do the problem analogous to Problem 6 in which gcd is replaced by lcm.
9. Generalize the result in Problem 8.
10. Let $p_1, p_2, \ldots, p_r$ be $r$ distinct positive primes.
    (i) Explain why the standard factorization for

$$1 + p_1 p_2 \cdots p_r$$

    involves only primes which are different from all these $p_j$.
    (ii) Explain why there must be an infinite number of primes.

11. Let $(p_1)^{e_1}(p_2)^{e_2} \cdots (p_r)^{e_r}$ be the standard factorization of the positive integer $n$. Express each of the following in terms of the $e_i$:
   (a) the number of ordered pairs of positive integers $(x, y)$ such that $\operatorname{lcm}[x, y] = n$.
   (b) the number of ordered pairs of positive integers $(x, y)$ such that
$$\frac{1}{x} + \frac{1}{y} = \frac{1}{n}.$$

12. Prove a relationship between the answers for the preceding problem and that for problem 8(ii) of Section 1.7.

# 2

# *Groups*

*Wherever groups disclosed themselves,*
*or could be introduced, simplicity*
*crystallized out of comparative chaos.*

*E. T. Bell*

Groups are involved in mathematics and its applications in much the same way that prose is involved in communication; that is, one tends to use groups extensively before seeing an exact definition of the term. In this chapter, the experiences of past study together with new concrete examples are used to abstract a powerful concept for further study and application. (Some of the many descriptions of applications of group theory are listed in the Applications part of our Bibliography.)

Our first topic is groups of permutations on a finite set. We shall see that this topic has the advantage of being both concrete and general.

## *2.1 Permutations on a Finite Domain*

A one-to-one function $\theta$ whose domain and range are the same finite set $X = \{x_1, x_2, \ldots, x_n\}$ is a *permutation* on $X$. This means that a permutation $\theta$ on the set $X$ is a rule $y = \theta(x)$ that assigns to each $a$ in $X$ exactly one $b$ in $X$ in such a way that for each $b$ in $X$ there is a unique $a$ with $\theta(a) = b$. In this chapter, $X$ will always be a set $\mathbf{X_n} = \{1, 2, \ldots, n\}$ of positive integers.

An example with $n = 6$ is the permutation $\theta$ given by the following table:

| $x$ | 1 | 2 | 3 | 4 | 5 | 6 |
|---|---|---|---|---|---|---|
| $y = \theta(x)$ | 3 | 2 | 6 | 5 | 4 | 1 |

(1)

The permutation $\theta$ characterized by table (1) is not changed if we replace the letters $x$ and $y$ by some other letters. A form of expressing $\theta$ that does not use letters for the variables is

$$\theta = \begin{pmatrix} 1 & 2 & 3 & 4 & 5 & 6 \\ 3 & 2 & 6 & 5 & 4 & 1 \end{pmatrix}. \tag{2}$$

The function $\theta$ is also unaltered by any rearrangement of the columns in representation (2). For example, this permutation may be designated by

$$\theta = \begin{pmatrix} 6 & 2 & 1 & 5 & 4 & 3 \\ 1 & 2 & 3 & 4 & 5 & 6 \end{pmatrix},$$

which indicates the same data for $\theta$ as table (1) above, namely, that

$$\theta(1) = 3, \quad \theta(2) = 2, \quad \theta(3) = 6, \quad \theta(4) = 5, \quad \theta(5) = 4, \quad \theta(6) = 1. \tag{3}$$

Another expression for the permutation $\theta$ is

$$\theta\colon 1 \mapsto 3, 2 \mapsto 2, 3 \mapsto 6, 4 \mapsto 5, 5 \mapsto 4, 6 \mapsto 1. \tag{4}$$

One may read this, or any of the other forms, as "$\theta$ sends 1 to 3, 2 to itself, 3 to 6, 4 to 5, 5 to 4, and 6 to 1."

Since a permutation is a one-to-one function,

$$\theta\colon 1 \mapsto a_1, 2 \mapsto a_2, \ldots, n \mapsto a_n \tag{5}$$

represents a permutation on $\mathbf{X}_n = \{1, 2, \ldots, n\}$ if and only if each integer in $\mathbf{X}_n$ appears exactly once among

$$a_1, a_2, \ldots, a_n. \tag{6}$$

Permutations $\alpha$ and $\beta$ on $\mathbf{X}_n = \{1, 2, \ldots, n\}$ are equal when they are the same function; that is, $\alpha = \beta$ if and only if $\alpha(x) = \beta(x)$ for all $x$ in $\mathbf{X}_n$. Hence $\alpha \neq \beta$ if and only if $\alpha(c) \neq \beta(c)$ for at least one $c$ in $\mathbf{X}_n$.

In courses dealing with "permutations and combinations," form (6) is generally used to denote a permutation. However, form (6) is not convenient in group theory and will not be used in this text.

## *Problems*

Problem 7 below is cited in Section 2.2.

1. Express each of the two permutations on $\{1, 2\}$ in the 2-row form

$$\begin{pmatrix} 1 & 2 \\ a & b \end{pmatrix}.$$

2. Express each of the six permutations on $\{1, 2, 3\}$ in the arrow form

$$1 \mapsto a, \quad 2 \mapsto b, \quad 3 \mapsto c.$$

3. How many permutations are there on each of the following sets?

$$\text{(a) } \{1, 2, 3, 4\} \qquad \text{(b) } \{1, 2, 3, 4, 5\}.$$

4. How many permutations are there on $\{1, 2, \ldots, n\}$?
5. How many permutations $\theta$ on $\{1, 2, 3, 4\}$ have $\theta(4) = 4$?
6. How many permutations $\theta$ on $\{1, 2, 3, 4, 5\}$ have $\theta(1) = 1$?
7. Let $\alpha$ and $\beta$ be the permutations on $\{1, 2, 3\}$ given by

$$\alpha\colon 1 \mapsto 2,\ 2 \mapsto 1,\ 3 \mapsto 3;$$
$$\beta\colon 1 \mapsto 2,\ 2 \mapsto 3,\ 3 \mapsto 1.$$

   (i) Find the permutation $\gamma$ on $\{1, 2, 3\}$ such that $\gamma(x) = z$ whenever $\alpha(x) = y$ and $\beta(y) = z$.
   (ii) Find the permutation $\delta$ on $\{1, 2, 3\}$ such that $\delta(x) = z$ whenever $\beta(x) = y$ and $\alpha(y) = z$.

8. Let $\alpha$ and $\beta$ be the permutations on $\{1, 2, 3, 4, 5\}$ given by:

$$\alpha\colon 1 \mapsto 2,\ 2 \mapsto 3,\ 3 \mapsto 1,\ 4 \mapsto 5,\ 5 \mapsto 4;$$
$$\beta\colon 1 \mapsto 3,\ 2 \mapsto 1,\ 3 \mapsto 2,\ 4 \mapsto 4,\ 5 \mapsto 5.$$

   Find $r$, $s$, $t$, $u$, and $v$, so that

$$\gamma\colon 1 \mapsto r,\ 2 \mapsto s,\ 3 \mapsto t,\ 4 \mapsto u,\ 5 \mapsto v$$

   has $\gamma(x) = z$ whenever $\alpha(x) = y$ and $\beta(y) = z$.

9. Let

$$\alpha = \begin{pmatrix} 1 & 2 & 3 & 4 & 5 & 6 \\ 6 & 4 & 1 & 2 & 5 & 3 \end{pmatrix}.$$

   Rewrite $\alpha$ with the columns rearranged so that

$$\alpha = \begin{pmatrix} 1 & a & b & 2 & c & d \\ a & b & 1 & c & 2 & d \end{pmatrix}.$$

10. Let

$$\beta = \begin{pmatrix} 1 & 2 & 3 & 4 & 5 & 6 & 7 \\ 5 & 3 & 7 & 6 & 2 & 1 & 4 \end{pmatrix}.$$

   Rewrite $\beta$ in the form

$$\beta = \begin{pmatrix} 1 & a & b & c & d & e & f \\ a & b & c & d & e & f & 1 \end{pmatrix}.$$

11. Let $\alpha$ be the permutation of Problem 9 and let the sequence $u_1, u_2, u_3, \ldots$ be determined by

$$u_1 = 1, \quad u_2 = \alpha(u_1), \quad u_3 = \alpha(u_2), \quad u_4 = \alpha(u_3), \ldots.$$

(i) Which integer in $\mathbf{X}_6 = \{1, 2, 3, 4, 5, 6\}$ is the first to be repeated in the sequence $u_1, u_2, \ldots$?
(ii) Which integers in $\mathbf{X}_6$ appear in the sequence $u_1, u_2, \ldots$?
(iii) Which integers appear in the sequence $v_1, v_2, \ldots$ with $v_1 = 2$ and $v_{j+1} = \alpha(v_j)$ for $j = 1, 2, 3, \ldots$?

12. Let $\beta$ be the permutation of Problem 10. Let the sequence $w_1, w_2, w_3, \ldots$ be such that $w_1$ is one of the integers in $\{1, 2, \ldots, 7\}$ and $w_{j+1} = \beta(w_j)$ for $j = 1, 2, 3, \ldots$.
(i) If $w_1 = 1$, which integer is the first to be repeated in $w_1, w_2, \ldots$?
(ii) If $w_1 = 1$, which integers appear among the $w$'s?

13. Write in the form $1 \mapsto a, 2 \mapsto b, 3 \mapsto c, 4 \mapsto d$ each of the 10 permutations $\alpha$ on $\{1, 2, 3, 4\}$ such that $\alpha(x) = y$ implies that $\alpha(y) = x$, that is, $\alpha[\alpha(x)] = x$ for $x = 1, 2, 3, 4$.

14. How many permutations $\beta$ are there on $\mathbf{X}_5 = \{1, 2, 3, 4, 5\}$ such that $\beta(x) = y$ implies that $\beta(y) = x$; i.e., $\beta[\beta(x)] = x$ for all $x$ in $\mathbf{X}_5$?

---

***Évariste Galois*** *(1811–1832)* *The story of the life of Galois is certainly one of the most dramatic, tragic, and often told in the history of mathematics. He died at the age of twenty after being wounded in a duel. Fortunately for mathematics, the night before the duel he wrote down his principal mathematical results in a letter to a friend. His work was edited by Liouville and published in 1846 and only then did Galois' ideas become known to the mathematical community at large.*

*Prior to his death, Galois' life was a long series of tragedies, disappointments, and frustrations. He was rejected by the École Polytechnique twice, probably because of unorthodox answers to the entrance examination questions. When he presented a preliminary version of some important results to Cauchy for presentation to the Académie des Sciences, Cauchy lost the paper. Galois' father committed suicide. Shortly after giving a copy of his results to Fourier in an attempt to win the mathematics prize of the Académie, Fourier died and Galois' paper was lost. Another paper submitted to Poisson was returned since Poisson could not understand it. After gaining admission to the École Normale, Galois was expelled in his second year for writing a letter attacking the king. He finally became an active rebel against the monarchy, was jailed twice, and then was challenged and killed in a duel that was probably politically motivated. His life, though short, was eventful.*

*Galois followed up the work of Lagrange of 1770 and developed what is known as Galois Theory. He showed that the solvability of an algebraic equation by rational operations and extraction of roots is equivalent to the solvability of an associated group of permutations on its roots. Although the insolvability of the quintic had already been settled by Abel, nevertheless the methods of Galois provided a whole new look at algebraic structure and the stimulus for the subsequent development of abstract algebra in the late nineteenth and twentieth centuries.*

## 2.2 Multiplication of Permutations

Next we use composition of functions to define the product of permutations $\alpha$ and $\beta$ on $\mathbf{X}_n = \{1, 2, \ldots, n\}$ and then prove the basic properties of this operation. In later sections, we will derive many other properties from those presented here.

Let $\alpha$ and $\beta$ be permutations on $\mathbf{X}_n = \{1, 2, \ldots, n\}$. We write

$$a \overset{\alpha}{\mapsto} b$$

to indicate that $\alpha(a) = b$ and compress

$$a \overset{\alpha}{\mapsto} b \quad \text{and} \quad b \overset{\beta}{\mapsto} c$$

into

$$a \overset{\alpha}{\mapsto} b \overset{\beta}{\mapsto} c.$$

### *Definition 1 Multiplication of Permutations*

*If $\alpha$ and $\beta$ are permutations on $\mathbf{X}_n = \{1, 2, \ldots, n\}$, the* **product** *$\alpha\beta$ is defined to be the function $\gamma$ on $\mathbf{X}_n$ such that $\gamma(a) = c$ when $\alpha(a) = b$ and $\beta(b) = c$.*

This definition tells us that the product (or composite) $\alpha\beta$ is the function $\gamma$ such that

$$a \overset{\alpha}{\mapsto} b \overset{\beta}{\mapsto} c \quad \text{implies} \quad a \overset{\gamma}{\mapsto} c.$$

The product $\beta\alpha$ (taken in the other order) need not equal $\alpha\beta$. (See Problem 7 of Section 2.1.)

***Definition 2*** ***Set of Permutations on*** $\{1, 2, \ldots, n\}$

*The set of all permutations on* $\{1, 2, \ldots, n\}$ *is denoted by* $\mathbf{S_n}$.

***Theorem 1*** $\mathbf{S_n}$ ***Is Closed under Multiplication***

*If* $\alpha$ *and* $\beta$ *are permutations on* $\mathbf{X_n} = \{1, 2, \ldots, n\}$, *then so is their product* $\alpha\beta$.

***Proof***

Let

$$\alpha = \begin{pmatrix} 1 & 2 & \cdots & n \\ a_1 & a_2 & \cdots & a_n \end{pmatrix}.$$

By definition of a permutation, every one of the numbers 1, 2, ..., $n$ appears exactly once in the bottom row $a_1, a_2, \ldots, a_n$. We can therefore arrange the columns in the representation for $\beta$ so that

$$\beta = \begin{pmatrix} a_1 & a_2 & \cdots & a_n \\ b_1 & b_2 & \cdots & b_n \end{pmatrix}.$$

In this form, the top row for $\beta$ is the bottom row of $\alpha$. Then we have

$$k \overset{\alpha}{\longmapsto} a_k \overset{\beta}{\longmapsto} b_k \quad \text{or} \quad k \overset{\alpha\beta}{\longmapsto} b_k$$

for each $k$ in $\mathbf{X_n}$ and hence

$$\alpha\beta = \begin{pmatrix} 1 & 2 & \cdots & n \\ b_1 & b_2 & \cdots & b_n \end{pmatrix}.$$

Every number in $\mathbf{X_n}$ occurs exactly once among the $b$'s since $\beta$ is a permutation; this tells us that the product $\alpha\beta$ is also a permutation.

Using the technique of the above theorem, we easily establish the following result.

***Lemma 1***

*If*

$$\alpha = \begin{pmatrix} u_1 & u_2 & \cdots & u_n \\ v_1 & v_2 & \cdots & v_n \end{pmatrix} \text{ and } \beta = \begin{pmatrix} v_1 & v_2 & \cdots & v_n \\ w_1 & w_2 & \cdots & w_n \end{pmatrix}$$

*are permutations on* $\{1, 2, \ldots, n\}$, *their product, in that order, is*

$$\alpha\beta = \begin{pmatrix} u_1 & \cdots & u_n \\ v_1 & \cdots & v_n \end{pmatrix}\begin{pmatrix} v_1 & \cdots & v_n \\ w_1 & \cdots & w_n \end{pmatrix} = \begin{pmatrix} u_1 & u_2 & \cdots & u_n \\ w_1 & w_2 & \cdots & w_n \end{pmatrix}.$$

***Example 1***

Find $\alpha\beta$ and $\beta\alpha$ given that

$$\alpha = \begin{pmatrix} 1 & 2 & 3 \\ 1 & 3 & 2 \end{pmatrix} \quad \text{and} \quad \beta = \begin{pmatrix} 1 & 2 & 3 \\ 3 & 1 & 2 \end{pmatrix}.$$

***Solution***

Rewriting $\beta$ so that its top row is the same as the bottom row of $\alpha$ and using the lemma we have

$$\alpha\beta = \begin{pmatrix} 1 & 2 & 3 \\ 1 & 3 & 2 \end{pmatrix}\begin{pmatrix} 1 & 3 & 2 \\ 3 & 2 & 1 \end{pmatrix} = \begin{pmatrix} 1 & 2 & 3 \\ 3 & 2 & 1 \end{pmatrix}.$$

Similarly,

$$\beta\alpha = \begin{pmatrix} 1 & 2 & 3 \\ 3 & 1 & 2 \end{pmatrix}\begin{pmatrix} 3 & 1 & 2 \\ 2 & 1 & 3 \end{pmatrix} = \begin{pmatrix} 1 & 2 & 3 \\ 2 & 1 & 3 \end{pmatrix}.$$

We note that $\alpha\beta \neq \beta\alpha$ in Example 1.

***Theorem 2 Identity Permutation***

Let $\boldsymbol{\varepsilon} = \begin{pmatrix} 1 & 2 & \cdots & n \\ 1 & 2 & \cdots & n \end{pmatrix}$. *Then* $\alpha\boldsymbol{\varepsilon} = \alpha = \boldsymbol{\varepsilon}\alpha$ *for all permutations* $\alpha$ *on* $\mathbf{X_n} = \{1, 2, \ldots, n\}$.

***Proof***

Let $\alpha = \begin{pmatrix} 1 & 2 & \cdots & n \\ a_1 & a_2 & \cdots & a_n \end{pmatrix}$. It follows immediately from Lemma 1 that $\boldsymbol{\varepsilon}\alpha = \alpha$. Rearranging the columns of $\boldsymbol{\varepsilon}$ so that its top row is the bottom row of $\alpha$ and using the lemma, we also find that

$$\alpha\boldsymbol{\varepsilon} = \begin{pmatrix} 1 & 2 & \cdots & n \\ a_1 & a_2 & \cdots & a_n \end{pmatrix}\begin{pmatrix} a_1 & a_2 & \cdots & a_n \\ a_1 & a_2 & \cdots & a_n \end{pmatrix} = \begin{pmatrix} 1 & 2 & \cdots & n \\ a_1 & a_2 & \cdots & a_n \end{pmatrix} = \alpha.$$

This completes the proof.

The $\boldsymbol{\varepsilon}$ of Theorem 2 is called the ***identity permutation*** on $\mathbf{X_n}$.

**Theorem 3 Inverse of a Permutation**

*For any permutation*

$$\alpha = \begin{pmatrix} 1 & 2 & \cdots & n \\ a_1 & a_2 & \cdots & a_n \end{pmatrix}$$

*the permutation*

$$\beta = \begin{pmatrix} a_1 & a_2 & \cdots & a_n \\ 1 & 2 & \cdots & n \end{pmatrix}$$

*is such that* $\alpha\beta = \boldsymbol{\varepsilon} = \beta\alpha$, *where* $\boldsymbol{\varepsilon}$ *is the identity permutation on* $\{1, 2, \ldots, n\}$.

This follows readily from Lemma 1; the proof is left to the reader.

The permutation $\beta$ of Theorem 3 is denoted as $\alpha^{-1}$ and is called the ***inverse permutation*** of $\alpha$.

**Theorem 4 Associativity**

*For all permutations* $\alpha$, $\beta$, *and* $\gamma$ *on* $\{1, 2, \ldots, n\}$,

$$(\alpha\beta)\gamma = \alpha(\beta\gamma).$$

***Proof***

Let

$$\alpha = \begin{pmatrix} 1 & 2 & \cdots & n \\ a_1 & a_2 & \cdots & a_n \end{pmatrix}.$$

Then we can write $\beta$ and $\gamma$ as

$$\beta = \begin{pmatrix} a_1 & a_2 & \cdots & a_n \\ b_1 & b_2 & \cdots & b_n \end{pmatrix}, \qquad \gamma = \begin{pmatrix} b_1 & b_2 & \cdots & b_n \\ c_1 & c_2 & \cdots & c_n \end{pmatrix}.$$

Using Lemma 1, we have

$$\alpha\beta = \begin{pmatrix} 1 & 2 & \cdots & n \\ b_1 & b_2 & \cdots & b_n \end{pmatrix} \quad \text{and then} \quad (\alpha\beta)\gamma = \begin{pmatrix} 1 & 2 & \cdots & n \\ c_1 & c_2 & \cdots & c_n \end{pmatrix}.$$

Similarly,

$$\beta\gamma = \begin{pmatrix} a_1 & a_2 & \cdots & a_n \\ c_1 & c_2 & \cdots & c_n \end{pmatrix} \quad \text{and} \quad \alpha(\beta\gamma) = \begin{pmatrix} 1 & 2 & \cdots & n \\ c_1 & c_2 & \cdots & c_n \end{pmatrix}.$$

Hence $(\alpha\beta)\gamma = \alpha(\beta\gamma)$, as desired.

### *Problems*

Problems 1 and 3 below are cited in Section 2.3.

1. Let $\boldsymbol{\varepsilon}$ and $\theta$ be the permutations on $\{1, 2\}$ with

$$\boldsymbol{\varepsilon}: 1 \mapsto 1, 2 \mapsto 2 \quad \text{and} \quad \theta: 1 \mapsto 2, 2 \mapsto 1.$$

Find each of the products $\boldsymbol{\varepsilon\varepsilon}$, $\boldsymbol{\varepsilon}\theta$, $\theta\boldsymbol{\varepsilon}$, and $\theta\theta$ in the form $1 \mapsto a$, $2 \mapsto b$.

2. Let $\boldsymbol{\varepsilon} = \begin{pmatrix} 1 & 2 & 3 & 4 \\ 1 & 2 & 3 & 4 \end{pmatrix}$, $\alpha = \begin{pmatrix} 1 & 2 & 3 & 4 \\ 2 & 1 & 4 & 3 \end{pmatrix}$, $\beta = \begin{pmatrix} 1 & 2 & 3 & 4 \\ 3 & 4 & 1 & 2 \end{pmatrix}$, and $\gamma = \alpha\beta$.

Complete the following multiplication table of products of permutations in $\{\boldsymbol{\varepsilon}, \alpha, \beta, \gamma\}$; the value of the product $xy$ should be the entry on the row for $x$ and column for $y$:

| | $\boldsymbol{\varepsilon}$ | $\alpha$ | $\beta$ | $\gamma$ |
|---|---|---|---|---|
| $\boldsymbol{\varepsilon}$ | $\boldsymbol{\varepsilon}$ | $\alpha$ | $\beta$ | $\gamma$ |
| $\alpha$ | $\alpha$ | $\boldsymbol{\varepsilon}$ | $\gamma$ | $\beta$ |
| $\beta$ | | | | |
| $\gamma$ | | | | |

3. Let $\boldsymbol{\varepsilon} = \begin{pmatrix} 1 & 2 & 3 \\ 1 & 2 & 3 \end{pmatrix}$, $\rho = \begin{pmatrix} 1 & 2 & 3 \\ 2 & 3 & 1 \end{pmatrix}$, $\phi = \begin{pmatrix} 1 & 2 & 3 \\ 1 & 3 & 2 \end{pmatrix}$.
   (i) Find $\rho^2 = \rho\rho$.
   (ii) Show that $\phi\rho = \rho^2\phi$.
   (iii) Show that $\{\boldsymbol{\varepsilon}, \rho, \rho^2, \phi, \rho\phi, \rho^2\phi\}$ is the set $\mathbf{S}_3$ of all permutations on $\{1, 2, 3\}$.
   (iv) Complete the following multiplication table for $\mathbf{S}_3$:

| | $\boldsymbol{\varepsilon}$ | $\rho$ | $\rho^2$ | $\phi$ | $\rho\phi$ | $\rho^2\phi$ |
|---|---|---|---|---|---|---|
| $\boldsymbol{\varepsilon}$ | $\boldsymbol{\varepsilon}$ | $\rho$ | $\rho^2$ | $\phi$ | $\rho\phi$ | $\rho^2\phi$ |
| $\rho$ | $\rho$ | $\rho^2$ | $\boldsymbol{\varepsilon}$ | $\rho\phi$ | $\rho^2\phi$ | $\phi$ |
| $\rho^2$ | $\rho^2$ | | | $\rho^2\phi$ | | |
| $\phi$ | | | | | | |
| $\rho\phi$ | | | | | | |
| $\rho^2\phi$ | | | | | | |

4. (i) Does $\alpha\beta = \beta\alpha$ for all permutations $\alpha$ and $\beta$ on $\{1, 2\}$? Explain. (See Problem 1.)
   (ii) Does $\alpha\beta = \beta\alpha$ for all permutations $\alpha$ and $\beta$ on $\{1, 2, 3\}$? Explain. (See Problem 3.)

5. Let $\alpha$ be the permutation on $\{1, 2, 3, 4, 5, 6\}$ with

$$\alpha: 1 \mapsto 6, 2 \mapsto 4, 3 \mapsto 1, 4 \mapsto 2, 5 \mapsto 5, 6 \mapsto 3.$$

Express each of the following in this arrow form.
(i) $\alpha^2 = \alpha\alpha$.
(ii) $\alpha^2\alpha$.
(iii) $\alpha\alpha^2$.
(iv) The inverse $\alpha^{-1}$ of the permutation $\alpha$.

6. Let $\beta$ be the permutation on $\{1, 2, 3, 4, 5, 6, 7\}$ with

$$\beta: 1 \mapsto 5, 2 \mapsto 3, 3 \mapsto 7, 4 \mapsto 6, 5 \mapsto 2, 6 \mapsto 1, 7 \mapsto 4.$$

Find each of the following in the arrow form.
(a) $\beta^2 = \beta\beta$.
(b) The inverse $\beta^{-1}$ of the permutation $\beta$.

*7. Let $y = \theta(x)$ determine a permutation $\theta$ on $\{1, 2, \ldots, n\}$. In the sequence $u_1, u_2, \ldots$, let $u_2 = \theta(u_1)$, $u_3 = \theta(u_2)$, $\ldots$ . Let $r$ be the least positive integer such that $u_{r+1}$ is one of the previous terms $u_1, \ldots, u_r$. Explain why $u_{r+1}$ must be $u_1$.

*8. Let $\theta$ and $u_1, u_2, \ldots, u_r$ be as in the previous problem. Let $v_1$ be an element of $\{1, 2, \ldots, n\}$ that is not in $A = \{u_1, \ldots, u_r\}$. Let $v_2 = \theta(v_1)$, $v_3 = \theta(v_2)$, $\ldots$ . Explain why no one of the $v$'s is in $A$.

## 2.3 Abstract Groups

The four theorems of the preceding section state the basic properties of the operation of multiplication (i.e., composition) of permutations on $\{1, 2, \ldots, n\}$. The definition which follows helps us to deduce many other properties in a way that allows us to apply the results to any one of the algebraic structures that also has these four basic properties. This illustrates the power of abstract methods.

### *Definition 1 Group*

*A* **group** *G is a set $\hat{G}$ with an operation having the following four properties:*

$G_1$ **Closure** *If a and b are elements (not necessarily distinct) in $\hat{G}$, there is a unique product ab in $\hat{G}$.*

$G_2$ **Identity** *There is an element* **e** *in $\hat{G}$ such that*

$$a\mathbf{e} = a = \mathbf{e}a$$

*for all a in $\hat{G}$.*

$G_3$ **Inverses** *For each a in $\hat{G}$ there is an h in $\hat{G}$ such that*

$$ah = \mathbf{e} = ha.$$

$G_4$ **Associativity** *If a, b, and c are elements (not necessarily distinct) in $\hat{G}$, then*

$$(ab)c = a(bc).$$

$G_1$, $G_2$, $G_3$, and $G_4$ are called the ***group axioms***; hence a group is a set with an operation satisfying all four group axioms. In other texts the group axioms may be stated in a somewhat different, but equivalent, form.

***Example 1***

Let $\hat{G}$ be the set $\{1, -1, \mathbf{i}, -\mathbf{i}\}$ of four complex numbers. Show that the operation of complex multiplication makes $\hat{G}$ into a group.

***Solution***

The multiplication table for $\hat{G}$ is:

| | 1 | $-1$ | $\mathbf{i}$ | $-\mathbf{i}$ |
|---|---|---|---|---|
| 1 | 1 | $-1$ | $\mathbf{i}$ | $-\mathbf{i}$ |
| $-1$ | $-1$ | 1 | $-\mathbf{i}$ | $\mathbf{i}$ |
| $\mathbf{i}$ | $\mathbf{i}$ | $-\mathbf{i}$ | $-1$ | 1 |
| $-\mathbf{i}$ | $-\mathbf{i}$ | $\mathbf{i}$ | 1 | $-1$ |

$\hat{G}$ is closed under the operation since each entry in the table is in $\hat{G}$. One sees that 1 is an identity. Also, each of 1 and $-1$ is its own inverse while $\mathbf{i}$ and $-\mathbf{i}$ are inverses of each other. The operation for $\hat{G}$ is associative since multiplication of any complex numbers is associative. Hence $\hat{G}$ and its operation form a group $G$.

The two following definitions are means of classifying groups.

***Definition 2 Order of a Group***

*If the set $\hat{G}$ of a group G is finite, the number of elements in $\hat{G}$ is called the* **order** *of G and is sometimes denoted as ord G. If $\hat{G}$ is infinite, G is a group of* **infinite order**.

### *Definition 3 Abelian Group*

*If $xy = yx$ for all $x$ and $y$ in $\hat{G}$, the group $G$ is said to be* **abelian**, *or* **commutative**.

Abelian groups are named after Niels Henrik Abel. (See the biographical note on Abel after this section.)

Clearly, the group of Example 1 above is an abelian group of order 4. The set of positive real numbers, with the operation of multiplication of real numbers, is an infinite abelian group.

A group is more than a set in that a group is a set with an operation satisfying the group axioms. Although it is important that one appreciate the distinction between a group $G$ and its set of elements $\hat{G}$, consistently stressing the difference results in notational awkwardness. Hence we shall frequently write $G$ for $\hat{G}$. For example, we shall refer to "an element of the set $\hat{G}$ of the group $G$" as "an element of the group $G$." When this is done, the context will indicate whether $G$ represents a group or its set.

Theorems 1–4 of Section 2.2 show that the $n!$ permutations on $\{1, 2, \ldots, n\}$ form a group under the operation of composition of functions.

A function $f(x_1, x_2, \ldots, x_n)$ is said to be symmetric in $x_1, x_2, \ldots, x_n$ if its value is unchanged by all permutations of its $n$ variables. This concept, which will be discussed further in Section 2.8, is responsible for the use of the word "symmetric" in the definition given next.

### *Definition 4 The Symmetric Groups* $\mathbf{S_n}$

*For every positive integer $n$, the group $\mathbf{S_n}$ of all the permutations on $\mathbf{X_n} = \{1, 2, \ldots, n\}$ is called the* **symmetric group** *on* $\mathbf{X_n}$.

We note that $\text{ord}\,\mathbf{S_1} = 1$, $\text{ord}\,\mathbf{S_2} = 2$, $\text{ord}\,\mathbf{S_3} = 6$, $\text{ord}\,\mathbf{S_4} = 24$, $\text{ord}\,\mathbf{S_5} = 120$, and in general

$$\text{ord}\,\mathbf{S_n} = n! = 1 \cdot 2 \cdot 3 \cdots n.$$

The symmetric group $\mathbf{S_2}$ is abelian. (See Problem 1 of Section 2.2.) The symmetric group $\mathbf{S_3}$ is not abelian since there is at least one pair of permutations $\alpha$ and $\beta$ on $\{1, 2, 3\}$ such that $\alpha\beta \neq \beta\alpha$. (See Problem 3 of Section 2.2.)

Even in a nonabelian group $G$, there are some pairs of elements that commute; for example, the definition of the identity $\mathbf{e}$ of $G$ implies that $g\mathbf{e} = \mathbf{e}g$ for all $g$ in $G$.

Our treatment of group theory includes results concerning specific groups, such as a symmetric group $\mathbf{S_n}$ or the group of Example 1. We shall also deal with theorems about groups $G$ in general, that is, with theorems which are consequences of the group axioms and whose proofs depend in no other way on the nature of the elements of the set $G$ or its operation. Such theorems are part of ***abstract group theory***.

As an aid in distinguishing permutation groups from other groups, we generally will use lowercase Greek letters for permutations and lowercase Latin letters for elements of an abstract group.

Some abstract results on groups follow:

***Lemma 1***

*There is only one element $e$ in a group $G$ such that*

$$ae = a = ea$$

*for all $a$ in $G$.*

***Proof***

Axiom $G_2$ tells us that there is at least one element with this property. Let each of $e_1$ and $e_2$ designate such an element, that is, let

$$ge_1 = g = e_1 g, \qquad ge_2 = g = e_2 g$$

for all $g$ in $G$. Then we can replace $g$ with $e_2$ in $g = e_1 g$ and replace $g$ with $e_1$ in $ge_2 = g$, thus obtaining

$$e_2 = e_1 e_2 \quad \text{and} \quad e_1 e_2 = e_1.$$

Since each of $e_2$ and $e_1$ is equal to $e_1 e_2$, they are equal to each other. This proves the stated uniqueness.

Now we can say "the identity **e**" instead of "an identity **e**."

***Theorem 1 Cancellation***

*If either $ab = ac$ or $ba = ca$ in a group $G$, then $b = c$.*

***Proof***

We show left cancellation, namely that $ab = ac$ implies $b = c$, and leave right cancellation to the reader as Problem 16 of this section.

Let **e** be the identity of $G$. Axiom $G_3$ tells us that there is an element $h$ of $G$ such that $ah = \mathbf{e} = ha$. Substituting the equal $ac$ for $ab$ in $h(ab)$, we obtain

$$h(ab) = h(ac).$$

Then it follows from associativity and properties of the identity that

$$(ha)b = (ha)c, \qquad \mathbf{e}b = \mathbf{e}c, \qquad b = c.$$

Note that Theorem 1 deals with either left or right cancellation but not with "mixed" cancellation. That is, an equation $ab = ca$ need not imply $b = c$. (See Problem 15 of this section.)

We are now able to strengthen the result in Lemma 1 above.

***Theorem 2 Uniqueness of the Identity***

*Let $u$ be an element of a group $G$. If there exists at least one element $v$ of $G$ such that either*

$$uv = v \quad \text{or} \quad vu = v,$$

*then $u$ is the identity $\mathbf{e}$ of $G$.*

***Proof***

Let $uv = v$. Since $\mathbf{e}v = v$, we then have $uv = \mathbf{e}v$. This and right cancellation give us $u = \mathbf{e}$. Similarly, $vu = v$ implies $u = \mathbf{e}$.

***Theorem 3 Uniqueness of the Inverse***

*Let $a$ be an element of a group $G$. Then there is only one element $h$ in $G$ such that $ah = \mathbf{e}$ or $ha = \mathbf{e}$.*

***Proof***

Axiom $G_3$ tells us that there is an $h_1$ in $G$ such that

$$ah_1 = \mathbf{e} = h_1 a.$$

If there also is an element $h_2$ in $G$ with $ah_2 = \mathbf{e}$, then $ah_2 = ah_1$ and left cancellation gives us $h_2 = h_1$. Similarly, $h_3 a = \mathbf{e}$ implies that $h_3 = h_1$.

***Notation Inverse of an Element of a Group***

*The unique inverse of an element $a$ of a group is designated as $a^{-1}$.*

If $g$ is an element of a group $G$, closure of $G$ under its operation implies that the product of $g$ by itself is also in $G$; we use the notation $g^2$ of ordinary algebra for this product $gg$. Similarly, we let $g^2g = g^3$, $g^3g = g^4$, and so on. To complete the process of defining $g^n$ for all integers $n$, we let

$$g^0 = \mathbf{e}, \quad g^1 = g, \quad \text{and} \quad g^{-m} = (g^m)^{-1} \text{ for } m = 1, 2, 3, \ldots .$$

Using associativity, one sees that

$$gg^2 = g(gg) = (gg)g = g^2g = g^3.$$

Associativity can be used together with mathematical induction to prove the general rules

$$g^m g^n = g^{m+n}, \qquad (g^m)^n = g^{mn} \tag{E}$$

for any element $g$ of a group $G$ and any integers $m$ and $n$. It is also clear that $\mathbf{e}^n = \mathbf{e}$ for all integers $n$. We assume these rules without proof.

The exponent rules (E) apply to commutative and to noncommutative groups since one can use associativity as a substitute for commutativity in dealing with powers having a fixed base. However, $(ab)^n$ need not equal $a^n b^n$ in a noncommutative group. For example, $(ab)^2 = abab$ and this can be proved to be different from $a^2b^2 = aabb$ when $a$ and $b$ do not commute. (See Problem 6 of this section.)

The following result facilitates construction of tables for operations of finite groups.

### *Theorem 4 Operation Table for a Group*

*Each element occurs exactly once as an entry on each row and on each column of the table for the operation of a finite group $G = \{g_1, \ldots, g_m\}$.*

#### *Proof*

The entries on the row for an element $x$ are $xg_1, xg_2, \ldots, xg_m$. Any element $y$ of $G$ appears as the entry $xg$ with $g = x^{-1}y$. No element can appear more than once, since that would mean that $xg_i = xg_j$ with $g_i \neq g_j$; but this is impossible because of left cancellation. Similarly, one shows that each element of $G$ appears exactly once on any given column.

We next show that the multiplication table for a group of order 3 is completely determined once one knows which element is the identity.

#### *Example 2*

Assume that there exists a group $G = \{\mathbf{e}, b, c\}$ of order 3 with $\mathbf{e}$ as the identity. Show that $bc = \mathbf{e} = cb$, $b^2 = c$, $c^2 = b$, and $b^3 = \mathbf{e} = c^3$ in such a group. Also tabulate the operation of such a group.

*Solution*

Since $b \neq \mathbf{e}$ and $c \neq \mathbf{e}$, it follows from Uniqueness of the Identity (Theorem 2) that $bc \neq b$ and $bc \neq c$. But $bc$ is some element of $G$ by closure of a group, and hence $bc = \mathbf{e}$.

The three products $b\mathbf{e}$, $bb$, and $bc$ are different from each other since equality of two of them and left cancellation would imply equality of two of the three distinct elements $\mathbf{e}$, $b$, and $c$ of $G$. We note that $b\mathbf{e} = b$ and $bc = \mathbf{e}$. Since $b^2 = bb$ is different from the other two products, we must have $b^2 = c$. Then

$$b^3 = bb^2 = bc = \mathbf{e}.$$

Similarly, $cb = \mathbf{e}$, $c^2 = b$, and $c^3 = \mathbf{e}$. Now we can use these facts to give the table for the operation in $G$.

| | $\mathbf{e}$ | $b$ | $c$ |
|---|---|---|---|
| $\mathbf{e}$ | $\mathbf{e}$ | $b$ | $c$ |
| $b$ | $b$ | $c$ | $\mathbf{e}$ |
| $c$ | $c$ | $\mathbf{e}$ | $b$ |

We see that no element is repeated among the entries in a given row or column; this is as promised by Theorem 4.

Example 2 does not prove that groups of order 3 exist. It only shows that there is just one possibility for the operation of a group with three elements. One way to show the existence of such a group is to verify that all the axioms of Definition 1 are satisfied by the operation tabulated in Example 2. This would be somewhat tedious since 27 cases are involved in checking associativity. We prefer to produce a group $\{1, \beta, \gamma\}$ of complex numbers whose multiplication table is that of Example 2 with $\mathbf{e}$, $b$, and $c$ replaced by 1, $\beta$, and $\gamma$, respectively. (See Problem 9 of this section.) Then it follows from associativity of multiplication of complex numbers that the operation given by the table in Example 2 is associative. Alternatively, one can show that $\{\boldsymbol{\varepsilon}, \rho, \rho^2\}$, with $\boldsymbol{\varepsilon}$ and $\rho$ given by

$$\boldsymbol{\varepsilon}: 1 \mapsto 1, 2 \mapsto 2, 3 \mapsto 3;$$
$$\rho: 1 \mapsto 2, 2 \mapsto 3, 3 \mapsto 1,$$

is a group of permutations whose multiplication table is that of Example 2 with $\mathbf{e}$, $b$, and $c$ replaced by $\boldsymbol{\varepsilon}$, $\rho$, and $\rho^2$, respectively; and then use the fact that multiplication of permutations is associative.

## *Problems*

Problem 11 below is cited in Section 4.1.

1. For each of the following permutations $\theta$, find $\theta^{-1}$, $\theta^2$, $\theta^3$, $\theta^4$, $\theta^{100}$, and $\theta^{101}$.
   (a) $\theta: 1 \mapsto 2, 2 \mapsto 1.$
   (b) $\theta: 1 \mapsto 2, 2 \mapsto 3, 3 \mapsto 1.$
   (c) $\theta: 1 \mapsto 3, 2 \mapsto 4, 3 \mapsto 1, 4 \mapsto 2.$

2. Do as in Problem 1 for each of the following permutations.
   (a) $\theta\colon 1 \mapsto 1, 2 \mapsto 3, 3 \mapsto 2.$
   (b) $\theta\colon 1 \mapsto 3, 2 \mapsto 1, 3 \mapsto 2, 4 \mapsto 4.$
   (c) $\theta\colon 1 \mapsto 2, 2 \mapsto 3, 3 \mapsto 4, 4 \mapsto 1.$
3. Show that $(\alpha\beta)^2 \neq \alpha^2\beta^2$ if

$$\alpha\colon 1 \mapsto 2, 2 \mapsto 3, 3 \mapsto 1;$$
$$\beta\colon 1 \mapsto 2, 2 \mapsto 1, 3 \mapsto 3.$$

4. Show that $(\alpha\beta)^2 = \alpha^2\beta^2$ if $\alpha$ is as in Problem 3 and

$$\beta\colon 1 \mapsto 3, 2 \mapsto 1, 3 \mapsto 2.$$

5. Prove that if $ab = ba$ in a group $G$, then $(ab)^2 = a^2b^2$.
6. Prove that if $(ab)^2 = a^2b^2$ in a group $G$, then $ab = ba$.
7. Prove that the subset $\{1, -1\}$ of $\mathbf{Z}$ is a group under the operation of multiplication of integers.
8. Show that multiplication in a group $\{\mathbf{e}, a\}$ of order 2 is completely determined when one knows that $\mathbf{e}$ is the identity.
9. Let $\beta = (-1 + \mathbf{i}\sqrt{3})/2$ and $\gamma = (-1 - \mathbf{i}\sqrt{3})/2$.
   (i) Show that $\{1, \beta, \gamma\}$ is a group (of order 3) under multiplication of complex numbers.
   (ii) Explain why the multiplication table for this group must be the table of Example 2 with $\mathbf{e}$, $b$, and $c$ replaced by 1, $\beta$, and $\gamma$, respectively.
10. (i) Assume that a group $G = \{\mathbf{e}, u, v, w\}$ of order 4 exists with $\mathbf{e}$ the identity, $u^2 = v$, and $v^2 = \mathbf{e}$. Construct the table for the operation of such a group.
    (ii) Does there exist a group $G$ with the properties given in (i)? Explain.
11. Given that $G = \{\mathbf{e}, b, c, d\}$ is a group of order 4 with

$$\mathbf{e}^2 = b^2 = c^2 = d^2 = \mathbf{e},$$

construct its multiplication table. (In Problem 12 of Section 2.5, we shall see that such a group exists.)

12. Given that $G = \{\mathbf{e}, a, a^2, a^3, a^4\}$ is a group of order 5 with $a^5 = \mathbf{e}$, construct its multiplication table.
13. Is a group of order 2 necessarily commutative? Explain.
14. Is a group of order 3 necessarily abelian? Explain.
15. Find permutations $\alpha$, $\beta$, and $\gamma$ in the symmetric group $\mathbf{S}_3$ such that $\alpha\beta = \gamma\alpha$ and $\beta \neq \gamma$ (and thus show that "mixed cancellation" is not always valid).
16. Prove the "right cancellation" part of Theorem 1 that was left to the reader by showing that $ba = ca$ implies $b = c$ in a group.

17. Let $a$, $b$, and $c$ be elements of a group. Show the following:
    (i) $(a^{-1})^{-1} = a$
    (ii) $(ab)^{-1} = b^{-1}a^{-1}$; that is, $(ab)(b^{-1}a^{-1}) = \mathbf{e} = (b^{-1}a^{-1})(ab)$.
    (iii) $(abc)^{-1} = c^{-1}b^{-1}a^{-1}$.
18. Let $a_1, a_2, \ldots, a_s$ be elements of a group.
    (i) Express $(a_1a_2a_3a_4)^{-1}$ in terms of $a_1^{-1}$, $a_2^{-1}$, $a_3^{-1}$, and $a_4^{-1}$.
    (ii) Express $(a_1a_2 \cdots a_s)^{-1}$ in terms of the $a_i^{-1}$.
19. How can one tell, by looking at the multiplication table of a group $G$, whether or not $G$ is commutative?
20. Let $a$ and $b$ be elements of a group. Show that $(ab)^{-1} = a^{-1}b^{-1}$ if and only if $ab = ba$; that is, show both of the following:
    (i) If $ab = ba$, then $(ab)^{-1} = a^{-1}b^{-1}$.
    (ii) If $(ab)^{-1} = a^{-1}b^{-1}$, then $ab = ba$.
21. Let $a$ be an element of a group $G$. Let $a$ and $a^2$ be different from the identity $\mathbf{e}$ but let $a^3 = \mathbf{e}$. Let $m$ and $n$ be integers. Show the following:
    (i) $a^m = \mathbf{e}$ if and only if $3|m$.
    (ii) $a^m = a^n$ if and only if $3|(m - n)$.
    (iii) There are exactly 3 different elements among the powers
    $$\ldots, a^{-2}, a^{-1}, \mathbf{e}, a, a^2, a^3, a^4, \ldots .$$
22. Let $a$ be an element of a group $G$ and let $H$ be the subset of $G$ consisting of all $a^n$ with $n$ an integer. Explain why the following are true:
    (i) If $h$ and $h'$ are in $H$, so is $hh'$.
    (ii) The identity $\mathbf{e}$ of $G$ is in $H$.
    (iii) If $h$ is in $H$, so is $h^{-1}$.
23. Let $G$ be a finite group. Explain why there are an even number of elements $x$ of $G$ such that $x^2 \neq \mathbf{e}$.
24. Show that a finite group $G$ of even order has at least one element $y$ with $y^2 = \mathbf{e}$ and $y \neq \mathbf{e}$ by showing that $G$ has an odd number of such elements.
25. Let $g^2 = \mathbf{e}$ for all $g$ in a group $G$. Show that $G$ is abelian.
26. Given that $abc = \mathbf{e}$ in a group $G$, show that $bca = \mathbf{e} = cab$.
27. Show that $abcd = \mathbf{e}$ implies $bcda = \mathbf{e}$ in a group $G$.
28. Generalize on Problems 26 and 27.
29. Tell why $S = \{a, b, c, d\}$ is not a group under the operation of the following table:

| | $a$ | $b$ | $c$ | $d$ |
|---|---|---|---|---|
| $a$ | $c$ | $a$ | $d$ | $b$ |
| $b$ | $a$ | $b$ | $c$ | $d$ |
| $c$ | $d$ | $c$ | $b$ | $c$ |
| $d$ | $b$ | $d$ | $c$ | $a$ |

---

***Niels Henrik Abel*** *(1802–1829) Abel was born in a small town in Norway, one of the seven children of the local pastor. The responsibility of supporting the family fell largely on his shoulders at the age of eighteen, when his father died. One year later, he proved the insolvability of the quintic equation by rational operations and extraction of roots; thus he succeeded where the greatest mathematicians of the previous 300 years had all failed. He also did highly original work in such fields as elliptic functions, abelian functions, and integration in finite terms.*

*Like Galois, Abel received little recognition during his short lifetime. In Paris he submitted a memoir to Cauchy and had the same misfortune that Galois experienced—Cauchy mislaid it! It was not found until after Abel's death and not published until 1841. Abel sought an academic post without success. He died of tuberculosis in Norway in 1829. Two days later he was appointed to a professorship at the University of Berlin, where they had not yet heard of his death.*

*A statue of Abel by Vigeland stands in the Royal Park in Oslo. His pose is heroic and it appears that he is standing on two prone and vanquished figures. It is rumored that one represents the general quintic, the other elliptic functions.*

---

## 2.4 Cycle Notation

This section presents the last of the notations commonly used for permutations. It is a convenient notation and one that brings out many important facts about permutations.

Let $\{a_1, a_2, \ldots, a_s\}$ be a subset of $\mathbf{X_n} = \{1, 2, \ldots, n\}$ and let $\gamma$ be the permutation on $\mathbf{X_n}$ with

$$\gamma(a_1) = a_2, \quad \gamma(a_2) = a_3, \ldots, \quad \gamma(a_{s-1}) = a_s, \quad \gamma(a_s) = a_1,$$

and $\gamma(x) = x$ if $x$ is not one of the $a_i$. This permutation $\gamma$ is called a ***cycle of length s***, or an ***s-cycle***, and is denoted by $(a_1 a_2 \ldots a_s)$. In the cycle notation, the value of $n$ is not given explicitly. For example, $\gamma = (123)$ designates

$$\begin{pmatrix} 1 & 2 & 3 \\ 2 & 3 & 1 \end{pmatrix}, \quad \begin{pmatrix} 1 & 2 & 3 & 4 \\ 2 & 3 & 1 & 4 \end{pmatrix}, \quad \ldots, \quad \begin{pmatrix} 1 & 2 & 3 & 4 & 5 & \cdots & n \\ 2 & 3 & 1 & 4 & 5 & \cdots & n \end{pmatrix}$$

as an element of $\mathbf{S_3}, \mathbf{S_4}, \ldots, \mathbf{S_n}$, respectively.

Every cycle of length one represents the identity $\boldsymbol{\varepsilon}$. A cycle of length two is also called a ***transposition***.

Clearly the cycle notation saves space. We shall see other advantages later.

***Example 1***

In $\mathbf{S_3}$ let $\rho = (123)$ and $\phi = (23)$. Then represent $\rho\phi$, $\rho^2$, and $\rho^2\phi$ in the cycle notation.

***Solution***

We see that

$$1 \overset{\rho}{\mapsto} 2 \overset{\phi}{\mapsto} 3 \quad \text{or} \quad 1 \overset{\rho\phi}{\mapsto} 3,$$

$$3 \overset{\rho}{\mapsto} 1 \overset{\phi}{\mapsto} 1 \quad \text{or} \quad 3 \overset{\rho\phi}{\mapsto} 1,$$

$$2 \overset{\rho}{\mapsto} 3 \overset{\phi}{\mapsto} 2 \quad \text{or} \quad 2 \overset{\rho\phi}{\mapsto} 2.$$

This shows that $\rho\phi = (13)$. Similarly, one finds that $\rho^2 = (132)$ and $\rho^2\phi = (12)$. Thus the six permutations of the symmetric group $\mathbf{S_3}$ are

$$\boldsymbol{\varepsilon} = (1), \qquad \rho = (123), \qquad \rho^2 = (132),$$

$$\phi = (23), \qquad \rho\phi = (13), \qquad \rho^2\phi = (12).$$

Juxtaposition of cycles indicates multiplication. We next illustrate some techniques for finding the product of cycles.

***Example 2***

For all $n \geq 3$, show that $(12)(13) = (123)$ in $\mathbf{S_n}$.

***Solution***

$$(12)(13) = \begin{pmatrix} 1 & 2 & 3 & \cdots & n \\ 2 & 1 & 3 & \cdots & n \end{pmatrix}\begin{pmatrix} 2 & 1 & 3 & 4 & \cdots & n \\ 2 & 3 & 1 & 4 & \cdots & n \end{pmatrix} = \begin{pmatrix} 1 & 2 & 3 & 4 & \cdots & n \\ 2 & 3 & 1 & 4 & \cdots & n \end{pmatrix} = (123).$$

***Example 3***

In any $\mathbf{S_n}$ with $n \geq 5$, let $\alpha = (12345)$ and $\beta = \alpha^2$. Show that $\beta$ is a 5-cycle.

***Solution***

We see that

$$1 \overset{\alpha}{\mapsto} 2 \overset{\alpha}{\mapsto} 3 \quad \text{or} \quad 1 \overset{\beta}{\mapsto} 3,$$

$$3 \overset{\alpha}{\mapsto} 4 \overset{\alpha}{\mapsto} 5 \quad \text{or} \quad 3 \overset{\beta}{\mapsto} 5,$$

$$5 \overset{\alpha}{\mapsto} 1 \overset{\alpha}{\mapsto} 2 \quad \text{or} \quad 5 \overset{\beta}{\mapsto} 2,$$

$$2 \overset{\alpha}{\mapsto} 3 \overset{\alpha}{\mapsto} 4 \quad \text{or} \quad 2 \overset{\beta}{\mapsto} 4,$$

$$4 \overset{\alpha}{\mapsto} 5 \overset{\alpha}{\mapsto} 1 \quad \text{or} \quad 4 \overset{\beta}{\mapsto} 1.$$

Also, if $x$ is in $\{1, 2, \ldots, n\}$ but not in $\{1, 2, 3, 4, 5\}$, then

$$x \overset{\alpha}{\mapsto} x \overset{\alpha}{\mapsto} x \quad \text{or} \quad x \overset{\beta}{\mapsto} x.$$

Hence $\beta$ is the 5-cycle (13524).

***Example 4***

Show that (123)(234) = (13)(24) in $\mathbf{S_n}$ for $n \geq 4$.

***Solution***

Let $\alpha = (123)$, $\beta = (234)$, $\gamma = (13)$, $\delta = (24)$. Then

$$1 \overset{\alpha}{\mapsto} 2 \overset{\beta}{\mapsto} 3 \quad \text{or} \quad 1 \overset{\alpha\beta}{\mapsto} 3; \quad 1 \overset{\gamma}{\mapsto} 3 \overset{\delta}{\mapsto} 3 \quad \text{or} \quad 1 \overset{\gamma\delta}{\mapsto} 3.$$

Hence $\alpha\beta$ and $\gamma\delta$ both send 1 to 3. Similarly, we see that $\alpha\beta$ and $\gamma\delta$ have the same value at every $x$ in $\{1, 2, \ldots, n\}$ and hence $\alpha\beta = \gamma\delta$.

A set of cycles is said to be ***disjoint*** if no element of $\mathbf{X_n} = \{1, 2, \ldots, n\}$ is involved in more than one of the cycles. For example,

$$\{(13), (24), (567)\}$$

is a set of three disjoint cycles. However, (143) and (2356) are not disjoint since 3 is present in both of them.

The permutations that are cycles do not have unique representations in the cycle notation; for example,

$$(1) = (2) = (3), \quad (12) = (21), \quad (123) = (231) = (312).$$

To avoid confusion and to be able to write all the permutations in a given $\mathbf{S_n}$ without the danger of duplication, we shall present conventions that pick out just one way of representing a given permutation after proving the following result:

***Theorem 1 Product of Disjoint Cycles***

*Every $\theta$ in $\mathbf{S_n}$ is either a cycle or a product*

$$(a_1 a_2 \ldots a_r)(b_1 b_2 \ldots b_s) \cdots (m_1 m_2 \ldots m_v)$$

*of disjoint cycles.*

***Proof***

Let $a_1 = 1$, $a_2 = \theta(a_1)$, $a_3 = \theta(a_2)$, etc. Since $\mathbf{X_n} = \{1, 2, \ldots, n\}$ is finite, there must be repetitions among these $a$'s. Let $r$ be the smallest positive integer such that $a_{r+1}$ is one of the previous terms $a_1, \ldots, a_r$.

Then $a_1, \ldots, a_r$ are distinct terms and it follows from the one-to-one nature of a permutation $\theta$ that $\theta(a_1), \ldots, \theta(a_r)$ are distinct. Since $\theta(a_i) = a_{i+1}$, it follows that $a_2, \ldots, a_{r+1}$

are distinct. Therefore $a_{r+1} = a_1$, since this is the only remaining way in which $a_{r+1}$ can equal an earlier term.

Let $A$ be the subset $\{a_1, \ldots, a_r\}$ of $\mathbf{X_n}$. If $r < n$, let $b_1$ be the least number in $\mathbf{X_n}$ that is not in $A$. Then let $\theta(b_1) = b_2$, $\theta(b_2) = b_3$, etc. As was true with the $a$'s, the first repetition among the $b$'s is a reappearance of $b_1$ as some $b_{s+1}$. Let $B = \{b_1, \ldots, b_s\}$.

We next show that no integer in $B$ is in $A$ by assuming that $b_k = a_j$ for some $j$ and $k$ and obtaining a contradiction. First we note that $k$ cannot be 1 since $b_1$ is not in $A$.

If $j \geq 2$, it follows from $b_k = \theta(b_{k-1})$, $a_j = \theta(a_{j-1})$, $b_k = a_j$, and the fact that $\theta$ is a one-to-one function that $b_{k-1} = a_{j-1}$. If $j = 1$, it follows similarly that $b_{k-1} = a_r$. In either case, the assumption that $b_k$ is in $A$ implies that $b_{k-1}$ is in $A$. Then this implies that $b_{k-2}$ is in $A$. Continuing in this manner, we find that the assumption that $b_k$ is in $A$ leads to the contradiction that $b_1$ is in $A$. Hence $A$ and $B$ have no integer in common.

If there are integers in $\mathbf{X_n}$ that are not in $A$ or $B$, we let $c_1$ be the smallest such integer and continue the process; that is, we let $\theta(c_1) = c_2$, $\theta(c_2) = c_3$, etc.

We stop this process when $\mathbf{X_n}$ is exhausted. Then $\theta$ can be written in the form

$$\begin{pmatrix} a_1 & a_2 & \ldots & a_{r-1} & a_r & b_1 & b_2 & \ldots & b_{s-1} & b_s & \ldots & m_1 & \ldots & m_{v-1} & m_v \\ a_2 & a_3 & \ldots & a_r & a_1 & b_2 & b_3 & \ldots & b_s & b_1 & \ldots & m_2 & \ldots & m_v & m_1 \end{pmatrix}$$

and hence $\theta$ is the product of disjoint cycles

$$\theta = (a_1 a_2 \ldots a_r)(b_1 b_2 \ldots b_s)(c_1 c_2 \ldots c_t) \cdots (m_1 m_2 \ldots m_v). \tag{P}$$

Now we use the representation (P) to define the ***standard form*** for a permutation $\theta$. The representation (P) for the identity of $\mathbf{S_n}$ is $(1)(2) \cdots (n)$; we delete the unnecessary 1-cycles after the first one and let (1) be the standard form for the identity. If $\theta$ is not the identity, we obtain the standard form for $\theta$ by deleting any 1-cycles that might be present in the representation (P).

Every cycle $(d_1 d_2 \cdots d_u)$ that appears in a standard form must be such that $d_1$ is the smallest integer in the set $\{d_1, d_2, \ldots, d_u\}$. Also, in a standard form

$$(a_1 a_2 \ldots a_r)(b_1 b_2 \ldots b_s)(c_1 c_2 \ldots c_t) \cdots (m_1 m_2 \ldots m_v)$$

one has $a_1 < b_1 < c_1 < \ldots < m_1$.

The product (123)(234) is not in standard form since (123) and (234) are not disjoint. Example 4 above shows that (123)(234) = (13)(24). Since (13)(24) meets the conditions for a standard form, it is the standard form for (123)(234).

The product (543)(17) of disjoint cycles is not in standard form since 5 is not the least integer in $\{5, 4, 3\}$. We note that

$$(543)(17) = (354)(17) = (17)(354)$$

and that (17)(354) is the standard form for (543)(17).

These conditions are imposed on standard forms so that a given permutation should have one and only one standard form.

If $\theta$ is in $\mathbf{S_n}$, the sum of the lengths of the cycles in its standard form cannot exceed $n$. For example, in $\mathbf{S_4}$ the sum of the lengths is at most 4. Therefore each of the 24 permutations in $\mathbf{S_4}$ has a standard form of one of the following types:

$$(1),\ (ab),\ (abc),\ (abcd),\ (ab)(cd).$$

In our discussion above, we have noted the equality (354)(17) = (17)(354), which is a special case of the following result.

***Theorem 2 Disjoint Cycles Commute***

*Let $\alpha = (a_1 a_2 \ldots a_r)$ and $\beta = (b_1 b_2 \ldots b_s)$ be disjoint cycles. Then $\alpha\beta = \beta\alpha$.*

***Proof***

Since no $a_j$ is a $b_k$, $\alpha(b_k) = b_k$ and $\beta(a_j) = a_j$ for all $j$ and $k$. Hence

$$a_1 \overset{\alpha}{\mapsto} a_2 \overset{\beta}{\mapsto} a_2 \quad \text{or} \quad a_1 \overset{\alpha\beta}{\mapsto} a_2,$$

$$a_1 \overset{\beta}{\mapsto} a_1 \overset{\alpha}{\mapsto} a_2 \quad \text{or} \quad a_1 \overset{\beta\alpha}{\mapsto} a_2.$$

This shows that $a_1$ is sent to $a_2$ by both $\alpha\beta$ and $\beta\alpha$. Similarly, one shows that each of $\alpha\beta$ and $\beta\alpha$ sends $a_2$ to $a_3, \ldots, a_{r-1}$ to $a_r$, $a_r$ to $a_1$, $b_1$ to $b_2, \ldots, b_{r-1}$ to $b_r$, and $b_r$ to $b_1$. Also, each of $\alpha\beta$ and $\beta\alpha$ sends $x$ to $x$ if $x$ is not an $a_i$ or a $b_j$. Thus $\alpha\beta = \beta\alpha$.

The following result is very important in the study of permutation groups.

***Theorem 3 Product of 2-Cycles***

*For $n > 1$, every permutation in $\mathbf{S_n}$ is a transposition or a product of transpositions (which need not be disjoint).*

***Proof***

Let $\theta$ be in $\mathbf{S_n}$. If $\theta = (1)$, an expression for $\theta$ as a product of transpositions is $(1) = (12)(12)$. If $\theta \neq (1)$, one obtains a representation of the desired form by replacing each cycle with length greater than 2 in the standard form of $\theta$ with a product of transpositions using

$$(d_1 d_2 d_3 \ldots d_s) = (d_1 d_s)(d_2 d_s) \cdots (d_{s-1} d_s).$$

## Problems

1. (a) Eight of the permutations of $\mathbf{S_4}$ have standard forms of the type $(abc)$. Write these eight standard forms.
   (b) Write the three standard forms of the type $(ab)(cd)$ for permutations of $\mathbf{S_4}$.
2. (a) Write the six standard forms of the type $(ab)$ for permutations of $\mathbf{S_4}$.
   (b) Write the six standard forms of the type $(abcd)$ for permutations of $\mathbf{S_4}$.
3. For each of the following cycles $\gamma$, express each of $\gamma^{-1}$, $\gamma^2$, $\gamma^3$, $\gamma^4$, $\gamma^5$, and $\gamma^6$ in the standard form:
   (a) (123); (b) (1234); (c) (12345); (d) (123456).
4. Let $\gamma = (123 \ldots s)$. Express $\gamma^{-1}$ and $\gamma^{s-1}$ in standard form.
5. Express each of the following products in standard form.
   (i) (13)(23); (ii) (14)(24)(34);
   (iii) (15)(25)(35)(45); (iv) (16)(26)(36)(46)(56).
6. Express each of the following products in standard form.
   (i) (12)(13); (ii) (12)(13)(14);
   (iii) (12)(13)(14)(15); (iv) (12345)(16).
7. Express $(123 \ldots s)$ as a product of transpositions.
8. Express $(a_1 a_2 \ldots a_s)$ as a product of transpositions.
9. (i) Express $(abc)^2$ as a 3-cycle.
   (ii) Show that $(xyz)$ is the square of a 3-cycle.
   (iii) Express $(abc)^5$ as a 3-cycle.
   (iv) Show that $(xyz)$ is the fifth power of a 3-cycle.
10. (i) Express $(abcde)^2$ as a 5-cycle.
    (ii) Show that $(vwxyz)$ is the square of a 5-cycle.
    (iii) Express $(abcde)^3$ as a 5-cycle.
    (iv) Show that $(vwxyz)$ is the cube of a 5-cycle.
11. Let $\alpha = (a_1 a_2 a_3 \ldots a_{2r})$. Write $\alpha^2$ as the product of two $r$-cycles.
12. Let $\beta = (b_1 b_2 \ldots b_{3s})$. Write $\beta^3$ as the product of three $s$-cycles.
13. Prove that the symmetric group $\mathbf{S_n}$ is nonabelian for $n \geq 3$.
14. (a) Which elements $\alpha$ of $\mathbf{S_3}$ are of the form $\alpha = \beta^2$, with $\beta$ in $\mathbf{S_3}$?
    (b) Which elements $\alpha$ of $\mathbf{S_3}$ are of the form $\alpha = \gamma^3$, with $\gamma$ in $\mathbf{S_3}$?
15. Let $\mathbf{A_3}$ be the subset $\{(1), (123), (132)\}$ of $\mathbf{S_3}$. Show that:
    (i) Each $\alpha$ of $\mathbf{A_3}$ is expressible as $\alpha = \beta\gamma$, with $\beta$ and $\gamma$ transpositions. ($\beta$ and $\gamma$ cannot be disjoint.)
    (ii) $\mathbf{A_3}$ is closed under composition of permutations.
    (iii) $\alpha^{-1}$ is in $\mathbf{A_3}$ for every $\alpha$ in $\mathbf{A_3}$.

16. Express each of 12 of the 24 permutations of $\mathbf{S}_4$ as a product $\beta\gamma$ of two (not necessarily disjoint) transpositions.
17. Let $\alpha = \gamma_1\gamma_2 \cdots \gamma_{2h}$ and $\beta = \delta_1\delta_2 \cdots \delta_{2k}$, where the $\gamma_i$ and $\delta_j$ are (not necessarily disjoint) transpositions. Is $\alpha\beta$ expressible as the product of an even number of transpositions? Explain. (Here $h$ and $k$ are positive integers.)
18. Is every transposition $\gamma = (ab)$ its own inverse? Explain.
19. Let $\alpha = \gamma_1\gamma_2 \cdots \gamma_{2h}$, where each $\gamma_i$ is a transposition. Is $\alpha^{-1}$ the product of an even number of transpositions? Explain.
20. Let $\theta$ be the $s$-cycle $(a_1 a_2 \ldots a_s)$. What is the smallest positive integer $m$ such that $\theta^m = (1)$?
21. Let $\theta = (12345)$. Show that the subset

$$H = \{(1), \theta, \theta^2, \theta^3, \theta^4\}$$

of $\mathbf{S}_5$ is closed under multiplication by making a multiplication table for $H$.
22. Let $\theta$ be the $s$-cycle $(123 \ldots s)$. How many distinct permutations are there among the powers

$$\ldots, \theta^{-3}, \theta^{-2}, \theta^{-1}, \theta^0, \theta, \theta^2, \theta^3, \ldots ?$$

23. Let $\theta = (12)(345)$. Find the smallest positive integer $r$ such that $\theta^r = (1)$.
24. Let $\theta = \alpha\beta$, where $\alpha$ and $\beta$ are disjoint cycles of lengths 9 and 6, respectively. Find the smallest positive integer $s$ such that $\theta^s = (1)$.
25. Let $\theta = \alpha\beta$, where $\alpha$ and $\beta$ are disjoint cycles of lengths $r$ and $s$, respectively. Let $t$ be an integer. Prove that $\theta^t = (1)$ if and only if $(\text{lcm}[r, s]) \mid t$.
26. Let $\theta = (12345)(567)$. Find the smallest positive integer $q$ such that $\theta^q = (1)$. (*Hint*: First find the standard form for $\theta$.)
27. Find the smallest positive integer $m$ such that $\theta^m = (1)$ for all $\theta$ in $\mathbf{S}_4$.
28. Find the smallest positive integer $n$ such that $\theta^n = (1)$ for all $\theta$ in $\mathbf{S}_5$.
29. A permutation $\theta$ with $\theta^2 = (1)$ is called an ***involution***.
    (i) Explain why the standard form for an involution $\theta$ involves no cycle with length greater than 2.
    *(ii) Let $T(n)$ be the number of involutions in $\mathbf{S}_n$. Prove that

$$T(n + 1) = T(n) + nT(n - 1) \text{ for } n > 1.$$

## 2.5 Subgroups in a Group

Using the cycle notation, the multiplication table for the symmetric group $\mathbf{S}_3$ is shown in Table 2.1.

*Table 2.1 The Symmetric Group* $S_3$

| | (1) | (123) | (132) | (23) | (13) | (12) |
|---|---|---|---|---|---|---|
| (1) | (1) | (123) | (132) | (23) | (13) | (12) |
| (123) | (123) | (132) | (1) | (13) | (12) | (23) |
| (132) | (132) | (1) | (123) | (12) | (23) | (13) |
| (23) | (23) | (12) | (13) | (1) | (132) | (123) |
| (13) | (13) | (23) | (12) | (123) | (1) | (132) |
| (12) | (12) | (13) | (23) | (132) | (123) | (1) |

Let us consider the subset $T = \{(1), (23)\}$ of $\mathbf{S}_3$. If we strike out of the table all the rows and columns except those for (1) and (23), what remains is the following multiplication table for $T$:

| | (1) | (23) |
|---|---|---|
| (1) | (1) | (23) |
| (23) | (23) | (1) |

Does the operation of this table make $T$ into a group? Let us check the axioms. Since each entry in this smaller table is (1) or (23), the set $T$ is closed under multiplication. The permutation (1) is an identity of $T$ since it is an identity in the larger set $\mathbf{S}_3$. The small table also shows that each element of $T$ is its own inverse. Finally, the elements of the subset $T$ obey the associative law since they are elements of the group $\mathbf{S}_3$. These facts show that $T$ is also a group under the operation of composition of permutations.

We generalize on this example as follows:

***Definition 1 Subgroup in a Group***

*Let $H$ consist of some or all of the elements of a group $G$. Let the product $ab$ of elements of $H$ be the same as when $a$ and $b$ are thought of as elements of $G$. If $H$ is a group under this operation, $H$ is called a* **subgroup** *in $G$.*

***Lemma 1 Identity and Inverses in a Subgroup***

*Let $H$ be a subgroup in $G$. Then the identity of $H$ is the identity of $G$ and the inverse of an $h$ of $H$ is the inverse of $h$ as an element of $G$.*

***Proof***

Let $\mathbf{e}$ be the identity of $H$. Then $\mathbf{ee} = \mathbf{e}$ in $H$ and also in $G$, since multiplication in $H$ is the same as in $G$. It then follows from Uniqueness of the Identity (Theorem 2 of Section 2.3) that $\mathbf{e}$ is the identity of $G$.

Similarly, Uniqueness of the Inverse (Theorem 3 of Section 2.3) shows that $h^{-1}$ in $H$ is the same as in $G$.

### *Lemma 2 Sufficient Conditions for a Subgroup*

*Let $H$ be a nonempty subset closed under the operation of a group $G$. For every $h$ in $H$, let its inverse $h^{-1}$ in $G$ be in $H$. Then $H$ is a subgroup in $G$.*

***Proof***

Associativity for $G$ automatically implies associativity, under the same operation, for any subset of $G$. Since $H$ is nonempty, there is some $h$ in $H$ and the hypothesis tells us that $h^{-1}$ is in $H$. Then $\mathbf{e} = hh^{-1}$ is in $H$ by the closure assumption. Hence $H$ is a group under the operation of $G$; that is, $H$ is a subgroup in $G$.

The single element subset $\{\mathbf{e}\}$ is easily seen to be a subgroup in every group $G$. It is also clear that every group $G$ is a subgroup in itself. Therefore, a group with more than one element has at least two subgroups.

Frequently, the following result provides the easiest means of verifying that a given subset is a subgroup.

### *Theorem 1 Subgroup Conditions*

*In a group $G$, let $H$ be a nonempty subset closed under division; that is, let $hk^{-1}$ be in $H$ whenever $h$ and $k$ are in $H$. Then $H$ is a subgroup in $G$.*

***Proof***

Since $H$ is nonempty it has at least one element $a$. It is given that $hk^{-1}$ is in $H$ when $h$ and $k$ are in $H$. Letting each of $h$ and $k$ equal $a$, we see that the identity $\mathbf{e} = aa^{-1}$ of $G$ is in $H$.

For every $k$ in $H$ the hypothesis now tells us that $\mathbf{e}k^{-1} = k^{-1}$ is in $H$. It then implies that for all $h$ and $k$, we have $h(k^{-1})^{-1} = hk$ in $H$; that is, $H$ is closed under multiplication. Now it follows from Lemma 2 that $H$ is a subgroup in $G$.

The definitions and theorems that follow provide means for obtaining some subgroups in noncommutative groups. They also furnish information concerning the structure of such groups.

### *Definition 2 Center in a Group*

*The* **center** *$C$ in a group $G$ is the subset consisting of the $c$ in $G$ such that $cx = xc$ for all $x$ in $G$.*

### *Theorem 2 Center as a Subgroup*

*The center $C$ in a group $G$ is a subgroup in $G$.*

The proof of this theorem is left to the reader as Problem 19 of this section.

### *Definition 3 Centralizer of $a$ in $G$*

*Let $a$ be a fixed element of a group $G$. The* **centralizer** *of $a$ in $G$ is the subset $C_a$ of $G$ such that $g$ is in $C_a$ if and only if $ag = ga$.*

### *Theorem 3 Centralizers and the Center*

*In a group $G$, let $C$ be the center and $C_a$ be the centralizer of $a$. Then:*
*(a) $C$ is a subgroup in $C_a$ for every $a$ in $G$.*
*(b) $C_a$ is a subgroup in $G$ for every $a$ in $G$.*
*(c) An element $g$ is in $C$ if and only if $g$ is in $C_a$ for every $a$ in $G$.*
*(d) An element $g$ is in $C$ if and only if $C_g = G$.*

The proof of this result is left to the reader as Problem 21 of this section.

### *Example 1*

In the symmetric group $\mathbf{S}_3$, find the center $C$ and the centralizer $C_\alpha$ of $\alpha = (132)$.

### *Solution*

With the help of Table 2.1, one sees that the identity (1) is the only element of $\mathbf{S}_3$ which commutes with every element of $\mathbf{S}_3$; that is, $C = \{(1)\}$. Also, one sees that in $\mathbf{S}_3$ only (1), (123), and (132) commute with (132); that is, $C_\alpha = \{(1), (123), (132)\}$. We note that $C$ is a subgroup in $C_\alpha$ and $C_\alpha$ is a subgroup in $G$. Thus this example illustrates some relationships stated in Theorem 3.

## *Problems*

1. Make a multiplication table for the subset

$$A = \{(1), (123), (132)\}$$

of $\mathbf{S}_3$ and show that $A$ is a subgroup in $\mathbf{S}_3$.

2. Show that the subset $B = \{(1), (12)\}$ is a subgroup in $\mathbf{S}_3$.

3. Show that the subset

$$D = \{(1), (123), (132), (23)\}$$

of $\mathbf{S}_3$ is not closed under multiplication and hence is not a subgroup in $\mathbf{S}_3$.

4. Let $E = \{(1), (123), (132), (23), (13)\}$. Show that $E$ is not a subgroup in $\mathbf{S}_3$.

5. Show that the inverse of one of the permutations in

$$F = \{(1), (123), (23), (13), (12)\}$$

is not in $F$ and hence that $F$ is not a subgroup in $\mathbf{S}_3$.

6. Is there a subgroup $H$ in $\mathbf{S}_3$ such that (132) is in $H$ but (123) is not? Explain.

7. Let $\beta = (1234)$ and $H = \{(1), \beta, \beta^2, \beta^3\}$.
   (i) Make the multiplication table for $H$.
   (ii) Use the table to show that $H$ is a subgroup in $\mathbf{S}_4$.
   (iii) Is $\beta$ an element of a subgroup in $\mathbf{S}_4$ that has fewer than four elements? Explain.
   (iv) Show that $\{(1), \beta^2\}$ is a subgroup in $H$. Is it also a subgroup in $\mathbf{S}_4$?

8. Let $\alpha = (123)$. Find the subgroup $H$ of lowest order in $\mathbf{S}_3$ such that $\alpha$ is in $H$.

9. Let $H$ be a subgroup having order 2 in $\mathbf{S}_3$.
   (i) Explain why $H$ must be of the form $\{(1), \beta\}$ with $\beta^2 = (1)$.
   (ii) List all the subgroups with order 2 in $\mathbf{S}_3$.

10. Let $H$ be a subgroup having order 3 in $\mathbf{S}_3$.
    (i) Explain why $H$ must be of the form $\{(1), \alpha, \alpha^2\}$ with $\alpha^3 = (1)$. (See Example 2 of Section 2.3.)
    (ii) Find all subgroups of order 3 in $\mathbf{S}_3$.

11. Let $G = \{1, -1, \mathbf{i}, -\mathbf{i}\}$ be the group of four complex numbers described in Example 1 of Section 2.3.
    (i) Show that $\{1, -1\}$ is a subgroup in $G$.
    (ii) Show that $\{1, \mathbf{i}\}$ is not a subgroup in $G$.

12. Show that $H = \{(1), (12), (34), (12)(34)\}$ is a subgroup in $\mathbf{S}_4$ and that $\theta^2 = (1)$ for all $\theta$ in $H$. (This shows that the table for Problem 11 of Section 2.3 is the table of a group; in particular, it shows that the operation given by that table is associative.)

13. Let $G = \{\mathbf{e}, a, a^2, \ldots, a^{s-1}\}$ be a group. Is $G$ necessarily commutative? Explain.

14. Let $G = \{x: x = a^n, n \text{ an integer}\}$ be a group. Is $G$ necessarily abelian? Explain.

15. Let $H$ be a subgroup in $G$.
    (i) If $G$ is abelian, must $H$ be abelian? Explain.
    (ii) If $H$ is abelian, must $G$ be abelian? Explain.

16. Let $K$ be a subgroup in $H$ and $H$ a subgroup in $G$. Explain why $K$ must be a subgroup in $G$.

17. For each of the following conditions, show that the set of all permutations $\theta$ on $\{1, 2, 3, 4\}$, which satisfy the given condition, is a subgroup in $\mathbf{S_4}$:
    (a) $\theta(4) = 4$.
    (b) $\theta(1)$ and $\theta(2)$ are both in $\{1, 2\}$.
    (c) $\theta(x) \leq x$ for all $x$ in $\{1, 2, 3, 4\}$.

18. For each of the following conditions, explain why the set of all permutations $\theta$ on $\mathbf{X_n} = \{1, 2, \ldots, n\}$, which satisfy the given condition, is or is not a subgroup in $\mathbf{S_n}$.
    (a) $\theta(1) = 1$.
    (b) $\theta(1) = 2$.
    (c) $\theta(1)$ is in the set $\{1, 2\}$. (Here let $n > 2$.)
    (d) $\theta(x) \geq x$ for all $x$ in $\mathbf{X_n}$.

19. Let $C$ consist of the elements $c$ of a group $G$ such that $cx = xc$ for all $x$ in $G$; that is, let $C$ be the center in $G$. Show that $C$ is a subgroup in $G$ (and thus prove Theorem 2 above).

20. Find the center in $\mathbf{S_4}$.

21. In a group $G$, let $C$ be the center and $C_a$ be the centralizer of $a$. Prove Theorem 3 by showing the following:
    (a) $C$ is a subgroup in $C_a$ for all $a$ in $G$.
    (b) $C_a$ is a subgroup in $G$ for all $a$ in $G$.
    (c) An element $g$ is in $C$ if and only if $g$ is in $C_a$ for all $a$ in $G$.
    (d) An element $g$ is in $C$ if and only if $C_g = G$.

22. Find the centralizers of each of the following elements in $\mathbf{S_3}$.
    (a) (1); (b) (12); (c) (123).

23. Let $a$ be a fixed element of a group $G$ and let $H$ be a subgroup in $G$. Let $K$ be the subset of $G$ consisting of all elements of the form $a^{-1}ha$ with $h$ in $H$. Prove the following:
    (i) $K$ is a subgroup in $G$. (The subgroup $K$ is called the conjugate subgroup of $H$ by $a$. In Section 3.2, this definition is given formally in a different context.)
    (ii) $H$ is the conjugate subgroup $\{aka^{-1}: k \in K\}$ of $K$ by $a^{-1}$.

24. Let $G$ be a group and let $H$ be the subset of all the elements $h$ of $G$ such that $h^{-1} = h$.
    (i) Prove that $H$ is a subgroup in $G$ whenever $G$ is abelian.
    (ii) If $G$ is the group $\mathbf{S}_3$, is the subset $H$ a subgroup? Explain.

25. Let $G$ be a group, let $s$ be a positive integer, and let $H$ be the subset of all the elements of the form $g^s$ with $g$ in $G$.
    (i) Prove that $H$ is a subgroup in $G$ whenever $G$ is abelian.
    (ii) Is $H$ a subgroup when $G$ is $\mathbf{S}_3$ and $s$ is 2? Explain.
    (iii) Is $H$ a subgroup when $G$ is $\mathbf{S}_3$ and $s$ is 3? Explain.

26. Let $G$ be a group, let $s$ be a positive integer, and let $H$ be the subset of all elements $h$ of $G$ such that $h^s = \mathbf{e}$.
    (i) Prove that $H$ is a subgroup in $G$ whenever $G$ is abelian.
    (ii) Show by an example that $H$ need not be a subgroup when $G$ is not commutative.

27. Let $G$ be the group $\{1, -1, \mathbf{i}, -\mathbf{i}\}$ of four complex numbers.
    (i) Find all the subsets of $G$ that are closed under multiplication.
    (ii) Is each of these subsets a subgroup in $G$?

28. Find all the subsets of $\mathbf{S}_3$ that are closed under multiplication. Is each of these subsets a subgroup in $\mathbf{S}_3$?

29. Let $\gamma$ be an $s$-cycle $(a_1 a_2 \ldots a_s)$ and let $H$ consist of all the powers $\gamma^n$ with $n$ an integer. Prove that $H$ is a subgroup of order $s$ in any group $G$ that has $\gamma$ as an element.

30. Let $a$ be any element of a group $G$ and let $H$ consist of the elements $a^n$ for all integers $n$. Prove that $H$ is a subgroup in $G$.

31. Let $\alpha = (12345)$ and $\beta = (12)$ be elements of a subgroup $H$ in $\mathbf{S}_5$. Prove that $H = \mathbf{S}_5$ by showing the following:
    (i) $\alpha^{-1}\beta\alpha = (23)$ and hence (23) is in $H$.
    (ii) $\alpha^{-1}(23)\alpha = (34)$ and hence (34) is in $H$.
    (iii) (45) and (15) are in $H$.
    (iv) $(23)\beta(23) = (13)$ and hence (13) is in $H$.
    (v) (24), (35), (14), and (25) are in $H$.
    (vi) $H = \mathbf{S}_5$. (See Theorem 3 of Section 2.4.)

*32. Given that there is a 2-cycle and a 5-cycle in a subgroup $H$ in $\mathbf{S}_5$, prove that $H = \mathbf{S}_5$.

## 2.6 *Additive Notation, Groups of Numbers*

A group is a set with an operation that satisfies the group axioms. So far, our notation for the operation has been the same as for multiplication in our familiar number systems. For some abelian groups, it is more convenient to use additive notation. We therefore rewrite the definition of a commutative group given in Section 2.3 as follows:

### *Definition 1 Additive Group*

*An additively written group $G$ is a set $\hat{G}$ with an operation, called addition, that has the following properties:*

1 **Closure** *If $a$ and $b$ are elements (not necessarily distinct) of $\hat{G}$, then there is a unique sum $a + b$ in $\hat{G}$.*

2 **Identity** *There is an element $z$ in $\hat{G}$ such that, for all $a$ in $\hat{G}$,*

$$a + z = a = z + a.$$

3 **Inverses** *For each $a$ in $\hat{G}$ there is an element $h$ in $\hat{G}$ such that*

$$a + h = z = h + a.$$

4 **Associativity** *If $a$, $b$, and $c$ are elements (not necessarily distinct) in $\hat{G}$, then*

$$(a + b) + c = a + (b + c).$$

5 **Commutativity** *If $a$ and $b$ are in $\hat{G}$, then*

$$a + b = b + a.$$

We follow the general custom of restricting the additive notation for groups solely to abelian groups. Results that apply to all multiplicative groups also apply to all additive groups. On the other hand, results that apply to all additive groups need not apply to multiplicative groups unless they are commutative.

### *Example 1*

The translation of Theorem 1 of Section 2.5 (Subgroup Conditions) into additive notation is the following:

> *In an additive group $G$, let $H$ be a nonempty subset closed under subtraction; that is, let $h - k$ be in $H$ whenever $h$ and $k$ are in $H$. Then $H$ is a subgroup in $G$.*

The special case in which $G$ is the additive group of the integers is dealt with in Theorem 2 of Section 1.3.

The identity of an additive group is generally written as 0. The inverse of $a$ in an additive group is denoted by $-a$ and is usually called the ***negative*** of $a$. ***Subtraction*** of $a$ from $b$ is defined by $b - a = b + (-a)$.

In multiplicative notation, the product $a \cdot a \cdots a$ of $s$ equal factors $a$ is designated as $a^s$.

***Notation* Sum of Equal Terms**

*Additively, the sum $a + a + \cdots + a$ of s equal terms a is written as $s \cdot a$.*

We also define $0 \cdot a$ to be 0 and let $(-s) \cdot a = -(s \cdot a)$ for all positive integers $s$. Then we have

$$m \cdot a + n \cdot a = (m + n) \cdot a \quad \text{and} \quad m \cdot (n \cdot a) = (mn) \cdot a$$

for every element $a$ of an additive group and all integers $m$ and $n$. These formulas are the translations into additive notation of the rules for powers stated in Section 2.3. In the notation $m \cdot a$, the symbol $a$ is used for an element of an additive group and $m$ stands for an integer (that is usually not in $G$); the centered dot between $m$ and $a$ helps to remind us that $m \cdot a$ does not mean the product of two group elements.

The familiar number systems of elementary mathematics are a source of additive and multiplicative abelian groups. We assume some familiarity with these number systems; in particular, we assume it known that the following sets of numbers are groups under the given operation:

***Groups under Addition***

(1) The set **Z** of all the integers
(2) The set **Q** of all the rational numbers
(3) The set **R** of all the real numbers
(4) The set **C** of all the complex numbers

***Groups under Multiplication***

(a) The subset $\{1, -1\}$ of the integers
(b) The nonzero rational numbers
(c) The nonzero real numbers; also, the positive real numbers
(d) The nonzero complex numbers

The two results that follow are helpful in dealing with multiplicative properties of the complex numbers.

***Theorem 1* Polar Form of a Complex Number**

*For all real numbers a and b, the complex number $a + b\mathbf{i}$ is expressible as*

$$a + b\mathbf{i} = r(\cos t + \mathbf{i} \sin t),$$

*with r and t real and $r \geq 0$.*

*Proof*

Let $r = \sqrt{a^2 + b^2}$. Then

$$\left(\frac{a}{r}\right)^2 + \left(\frac{b}{r}\right)^2 = \frac{a^2 + b^2}{r^2} = \frac{r^2}{r^2} = 1.$$

Hence there is a real number $t$ such that $\cos t = a/r$ and $\sin t = b/r$. Then

$$a + b\mathbf{i} = r\left(\frac{a}{r} + \mathbf{i}\frac{b}{r}\right) = r(\cos t + \mathbf{i} \sin t).$$

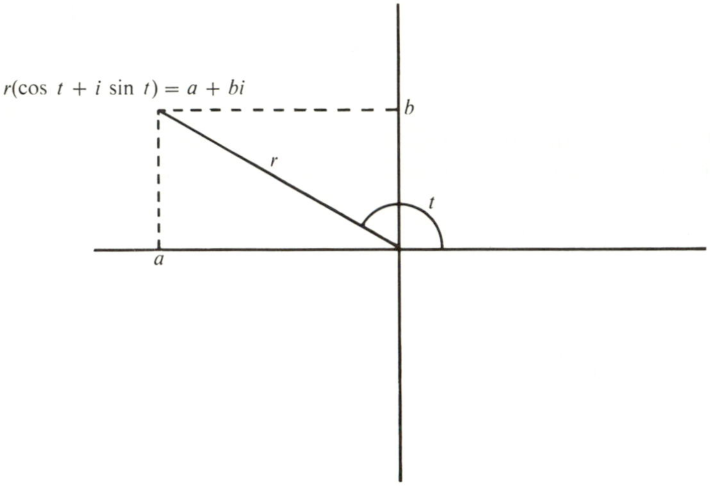


***Figure 2.1***

The nonnegative number $r = \sqrt{a^2 + b^2}$ is the ***absolute value*** (or ***modulus***) of the complex number $a + b\mathbf{i}$ and is denoted by $|a + b\mathbf{i}|$. A number $t$, such that $\cos t = a/r$ and $\sin t = b/r$, is called an ***argument*** (or ***amplitude***, or ***direction angle***) of $a + b\mathbf{i}$.

Since $2\pi$ is a period for the cosine and sine functions, we see that if $t$ is an argument for $a + b\mathbf{i}$ so is $t + 2n\pi$, for every integer $n$.

***Example 1***

Convert $4 + 4\mathbf{i}$ to polar form and $6[\cos(\pi/3) + \mathbf{i} \sin(\pi/3)]$ to the rectangular form $a + b\mathbf{i}$.

***Solution***

Using $a = 4$ and $b = 4$, we have $\sqrt{a^2 + b^2} = \sqrt{32} = 4\sqrt{2} = r$. An angle $t$ with $\cos t = a/r = 4/4\sqrt{2} = 1/\sqrt{2}$ and $\sin t = b/r = 1/\sqrt{2}$ is $t = \pi/4$. Hence $4 + 4\mathbf{i} = 4\sqrt{2}[\cos(\pi/4) + \mathbf{i} \sin(\pi/4)]$.

For the second part, we start with $r = 6$ and $t = \pi/3$. Then $\cos t = \cos(\pi/3) = 1/2$ and $\sin t = \sin(\pi/3) = \sqrt{3}/2$. Thus

$$6\left(\cos\frac{\pi}{3} + \mathbf{i}\sin\frac{\pi}{3}\right) = 6\left(\frac{1}{2} + \mathbf{i}\frac{\sqrt{3}}{2}\right) = 3 + 3\sqrt{3}\mathbf{i}.$$

***Theorem 2 Polar Multiplication***

*The absolute value of a product of two complex numbers is the product of their absolute values; that is, $|\alpha\beta| = |\alpha| \cdot |\beta|$. Also, the sum of arguments of complex numbers $\alpha$ and $\beta$ is an argument of $\alpha\beta$.*

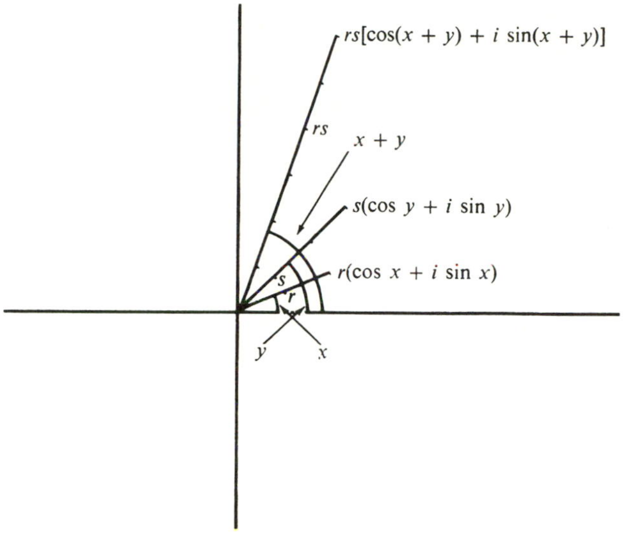


***Figure 2.2***

***Proof***

Using the addition formulas

$$\cos(x + y) = \cos x \cdot \cos y - \sin x \cdot \sin y,$$
$$\sin(x + y) = \sin x \cdot \cos y + \cos x \cdot \sin y,$$

we obtain the desired results as follows:

$$\begin{aligned}&[r(\cos x + \mathbf{i}\sin x)][s(\cos y + \mathbf{i}\sin y)] \\ &\quad = rs[(\cos x \cdot \cos y - \sin x \cdot \sin y) + \mathbf{i}(\sin x \cdot \cos y + \cos x \cdot \sin y)] \\ &\quad = rs[\cos(x + y) + \mathbf{i}\sin(x + y)].\end{aligned}$$

## Problems

Problems 5 and 6 below are cited in Section 2.7.

1. For each of the following sets, tell whether it is or is not a group under addition. If it is not a group, name a group axiom it does not satisfy.
   (a) All the integral multiples of 7, i.e., $\{\ldots, -7, 0, 7, 14, \ldots\}$.
   (b) All the polynomials $ax^2 + bx + c$ with $a, b$, and $c$ real numbers and $a \neq 0$.
   (c) All the polynomials $ax^3 + bx^2 + cx + d$ with $a$, $b$, $c$, and $d$ any real numbers.

2. Do as in Problem 1 for the following sets.
   (a) All the positive integers, i.e., $\mathbf{Z}^+ = \{1, 2, 3, \ldots\}$.
   (b) The nonnegative integers, i.e., $\{0, 1, 2, \ldots\}$.
   (c) All $2 \times 3$ real matrices, i.e., all arrays of real numbers $a, b, c, f, g$, and $h$ in 2 rows and 3 columns with addition defined by

$$\begin{pmatrix} a & b & c \\ f & g & h \end{pmatrix} + \begin{pmatrix} a' & b' & c' \\ f' & g' & h' \end{pmatrix} = \begin{pmatrix} a + a' & b + b' & c + c' \\ f + f' & g + g' & h + h' \end{pmatrix}.$$

3. For each of the following sets, tell whether it is or is not a group under multiplication. If it is not a group, name a group axiom it does not satisfy.
   (a) $\mathbf{Z} = \{\ldots, -2, -1, 0, 1, 2, 3, \ldots\}$.
   (b) The polynomials $ax + b$ with $a$ and $b$ real numbers.
   (c) All the positive rational numbers.

4. Do as in Problem 3 for the following sets.
   (a) All the rational numbers of the form $2^m 3^n$, with $m$ and $n$ integers.
   (b) All complex numbers $\alpha$ with $|\alpha| \leq 1$.
   (c) All $2 \times 2$ matrices of the form

$$\begin{pmatrix} \cos x & \sin x \\ -\sin x & \cos x \end{pmatrix}$$

   with $x$ a real number and multiplication defined by

$$\begin{pmatrix} a & b \\ c & d \end{pmatrix}\begin{pmatrix} f & g \\ h & k \end{pmatrix} = \begin{pmatrix} af + bh & ag + bk \\ cf + dh & cg + dk \end{pmatrix}.$$

5. Describe a subset $T$ of the additive group $\mathbf{Z}$ of the integers such that $T$ is closed under addition but $T$ is not a subgroup in $\mathbf{Z}$.

6. Describe a subset $S$, of the multiplicative group $V$ of the nonzero complex numbers, such that $S$ is closed under multiplication but $S$ is not a subgroup in $V$.

7. Restate Example 2 of Section 2.3 (without the solution) in the language of additive groups.
8. Restate Problem 30 of Section 2.5 (without the solution) in the language of additive groups.
9. Restate the multiplicative group result $(a^{-1})^{-1} = a$ in additive notation.
10. Explain why the multiplicative group formula $(ab)^{-1} = b^{-1}a^{-1}$ can be stated for additive groups as

$$-(a+b) = (-a) + (-b).$$

11. Find $2 \cdot M$, where $M$ is the $2 \times 3$ real matrix

$$\begin{pmatrix} 1 & 2 & 2 \\ 4 & -6 & 3 \end{pmatrix}.$$

[Note that, in an additive group, $2 \cdot M$ denotes $M + M$. Also see Problem 2(c).]

12. Find $(-2) \cdot M$ for the $M$ of the preceding problem.
13. Express each of the following complex numbers in the polar form $r(\cos t + \mathbf{i} \sin t)$:
(a) $-6 + 6\mathbf{i}$; (b) $9\mathbf{i}$; (c) $-5 = -5 + 0\mathbf{i}$.
14. Convert each of the following to polar form:
(a) $1 + \sqrt{3}\mathbf{i}$; (b) $\sqrt{3} - \mathbf{i}$; (c) 8.
15. Convert each of the following complex numbers from polar form to the rectangular form $a + b\mathbf{i}$:
(a) $8[\cos(\pi/6) + \mathbf{i} \sin(\pi/6)]$; (b) $\sqrt{2}[\cos(5\pi/4) + \mathbf{i} \sin(5\pi/4)]$.
16. Convert each of the following to $a + b\mathbf{i}$ form:
(a) $4[\cos(5\pi/6) + \mathbf{i} \sin(5\pi/6)]$; (b) $9[\cos(3\pi/2) + \mathbf{i} \sin(3\pi/2)]$;
(c) $6[\cos(-\pi/4) + \mathbf{i} \sin(-\pi/4)]$; (d) $7[\cos(6\pi) + \mathbf{i} \sin(6\pi)]$.
17. Express $(\sqrt{3} + \mathbf{i})^{10}(2 - 2\sqrt{3}\mathbf{i})^9$ in $a + b\mathbf{i}$ form by first converting $\sqrt{3} + \mathbf{i}$ and $2 - 2\sqrt{3}\mathbf{i}$ to polar form, then raising to powers and multiplying in polar form, and finally converting back.
18. Express $(5 + 5\mathbf{i})^{11}(-7\mathbf{i})^8$ in $a + b\mathbf{i}$ form.
19. Let $\omega = \cos(2\pi/3) + \mathbf{i} \sin(2\pi/3)$. Show the following:
(i) $\omega^3 = 1$.
(ii) $(\omega^3 - 1)/(\omega - 1) = 0$.
(iii) $\omega^2 + \omega + 1 = 0$.
(iv) $(x+1)^3 + (x+\omega)^3 + (x+\omega^2)^3 = 3(x^3 + 1)$.

20. Let $\omega$ be the complex cube root of unity $(-1 + \mathbf{i}\sqrt{3})/2$. Use the fact that $\omega^2 + \omega + 1 = 0$ to show that for all complex numbers $\alpha$, $\beta$, and $\gamma$ one has

$$\begin{aligned}(\alpha + \beta + \gamma)(\alpha + \omega\beta + \omega^2\gamma)(\alpha + \omega^2\beta + \omega\gamma) \\ &= (\alpha + \beta + \gamma)(\alpha^2 + \beta^2 + \gamma^2 - \beta\gamma - \alpha\gamma - \alpha\beta) \\ &= \alpha^3 + \beta^3 + \gamma^3 - 3\alpha\beta\gamma.\end{aligned}$$

21. Let $A$ consist of all nonzero complex numbers $r(\cos t + \mathbf{i} \sin t)$ with the argument $t$ in the set $\{0, 2\pi/3, 4\pi/3\}$. Prove that $A$ is a cyclic subgroup in the multiplicative group $V$ of the nonzero complex numbers.

22. Let $U$ be the set of all complex numbers $\alpha$ with $|\alpha| = 1$; that is, let

$$U = \{\cos t + \mathbf{i} \sin t : t \in \mathbf{R}\}.$$

Prove that $U$ is a subgroup in the multiplicative group $V$ of the nonzero complex numbers. (The group $U$ is called the ***circle group*** or the ***unit circle group***.)

23. Let $n$ be a positive integer. For $j = 0, 1, 2, \ldots, n - 1$, let

$$\alpha_j = \cos \frac{2\pi j}{n} + \mathbf{i} \sin \frac{2\pi j}{n}$$

and $U_n = \{\alpha_0, \alpha_1, \ldots, \alpha_{n-1}\}$. Show that:
(a) $(\alpha_j)^n = 1$ for all $j$.
(b) $U_n$ is a subgroup in the circle group $U$ of Problem 22.
(By (a), each number $U_n$ is an $n$th root of 1; one can see that $U_n$ consists of all $n$ $n$th roots of unity.)

24. Let $T$ consist of all the real numbers $t$ such that both $\cos t$ and $\sin t$ are in $\mathbf{Q}$ (that is, both are rational). Let $K$ consist of all complex numbers $\cos t + \mathbf{i} \sin t$ with $t$ in $T$. Prove that:
(a) $T$ is a subgroup in the additive group of the real numbers $\mathbf{R}$.
(b) $K$ is a subgroup in the circle group $U$ of Problem 22.
(c) If $\alpha$ and $\beta$ are in $K$, $|\alpha^2 - \beta^2|$ is in $\mathbf{Q}$.

*25. Let $G$ be a subgroup of order $n$ in the multiplicative group $V$ of the nonzero complex numbers. Prove that $G$ must be the group $U_n$ of Problem 23.

*26. Let $S$ be the set of all integers $s$ of the form $s = m^2 + n^2$ with $m$ and $n$ integers. Prove the following:
(i) $S$ is closed under multiplication.
(ii) If $s$ is in $S$, so are $2s$ and $5s$.

*A mathematician who is not also something*
*of a poet will never be a complete*
*mathematician.*

*Karl Weierstrass*

*A scientist worthy of the name, above all a mathematician, experiences in his work the same expression as an artist; his pleasure is as great and of the same nature.*

*Henri Poincaré*

---

**Leonhard Euler** (*1707–1783*) *Euler, the greatest Swiss mathematician and one of the greatest and most prolific mathematicians of all times, spent almost all of his life outside Switzerland. He was a student of Jean Bernoulli in Basel but in 1727 left for St. Petersburg to take a position (in medicine and physiology!) at the Imperial Academy there. In 1733, upon the departure of Daniel Bernoulli, the mathematician at the Academy, Euler became the chief mathematician and remained there until he accepted a comparable appointment at the Berlin Academy in 1741. In 1766 he returned to Russia, where he spent the rest of his life.*

*Euler's life represents a sharp contrast to those of Abel and Galois, both of whom died young and in rather dramatic and romantic circumstances. Euler had a large family, produced a prodigious number of books and papers in mathematics and died at the age of 76 while playing with one of his grandchildren. He was blind during the last 17 years of his life but this did not keep him from producing further mathematical results. His complete edited works run to approximately 70 folio volumes.*

*He did work in almost all the branches of mathematics and, in addition to creating new mathematical results, he also wrote textbooks, for example,* **Élémens d'Algèbre** (*1770*) *and* **Introductio in Analysin Infinitorum** (*1748*). *He is responsible for the wide use, if not the first use, of many modern mathematical notations:* $\pi$, $e$, $i$ (*all combined in his remarkable formula* $e^{\pi i} + 1 = 0$); *the abbreviations for the trigonometric functions;* $\sum$ *for summation; and* $f(x)$ *for function values. He was extremely facile at producing beautiful formulas. So many formulas and mathematical ideas are named for Euler, one can easily become confused: Euler's formula,*

$$e^{xi} = \cos x + i \sin x,$$

*for complex numbers with absolute value 1 (this appears on a Swiss postage stamp commemorating Euler); the Euler formula,*

$$V - E + F = 2,$$

*for polyhedra; the Euler constant,*

$$\gamma = \lim_{n\to\infty}\left(1 + \frac{1}{2} + \frac{1}{3} + \cdots + \frac{1}{n} - \log n\right);$$

*the Euler integrals (the so-called gamma and beta functions); the Euler $\phi$-function (see Sections 4.3 and 7.3); and so on.*

*Euler was the first to publish a proof of the little Fermat Theorem (Theorem 2 of Section 4.3 in this text) although Leibniz left a manuscript in which a demonstration appears. Euler also provided the generalization using his $\phi$-function.*

## 2.7 Cyclic Groups

One way to obtain subgroups $H$ in a group $G$ is to let $a$ be a fixed element of $G$ and let $H$ consist of all the elements.

$$\ldots, a^{-2}, a^{-1}, e, a, a^2, a^3, \ldots,$$

that is, of all $a^m$ with $m$ an integer. (If $G$ is written additively, the elements of $H$ are of the form $m \cdot a$, with $m$ an integer.)

A subset $H$ of this form is nonempty since $a$ is in $H$. Also $H$ is closed under division since the quotient $a^r(a^s)^{-1}$ of elements of $H$ is the element $a^{r-s}$ of $H$. Hence such an $H$ is a subgroup in $G$ by Theorem 1 of Section 2.5.

***Definition 1 Cyclic Group [a] Generated by a***

*Let $a$ be an element of a group $G$ and let $H$ be the subgroup in $G$ consisting of all $a^m$ with $m$ an integer. Then $H$ is called the* **cyclic group generated by** *$a$ and is designated as $[a]$.*

The subgroup $H = [a]$ may be $G$ itself; for example, if $G = \mathbf{S}_2 = \{(1), (12)\}$, we have $[(12)] = G$.

Let $K$ be a subgroup in $G$ and $a$ be an element of $K$. Then one can see that $[a]$ is a subgroup in $K$. In this sense, $[a]$ is the smallest subgroup in $G$ that contains $a$. (This concept will be generalized in Theorem 2 of Section 3.5.)

***Definition 2 Cyclic Group, Generator***

*If $G = [a]$, one says that $G$ is a* **cyclic group** *and that $a$ is a* **generator** *of $G$.*

A cyclic group may have more than one generator. For example, let $\beta = (1234)$ and $G = [\beta] = \{(1), \beta, \beta^2, \beta^3\}$. Clearly $\beta$ is a generator of this $G$; we see that $\beta^3$ is also a generator by noting that $[\beta^3]$ consists of

$$(1), \quad \beta^3, \quad (\beta^3)^2 = \beta^2, \quad \text{and} \quad (\beta^3)^3 = \beta.$$

The symmetric group $S_3$ is not cyclic, as one can show by finding all of its cyclic subgroups and noting that no one of them has order 6. (See Problem 9 of this section.)

***Theorem 1 Cyclic Groups Are Abelian***

*Every cyclic group $[a]$ is commutative.*

***Proof***

If $x$ and $y$ are elements of $[a]$, then $x = a^m$ and $y = a^n$ with $m$ and $n$ integers. Then

$$xy = a^m a^n = a^{m+n} = a^{n+m} = a^n a^m = yx,$$

since $m + n = n + m$ follows from the commutativity of addition of integers. Hence $[a]$ is abelian.

A cyclic group may be finite or infinite. For example, $[(1234)]$ is a cyclic group of order 4 and the additive group of the integers

$$[1] = [-1] = \{\ldots, -2, -1, 0, 1, 2, 3, \ldots\}$$

is an infinite cyclic group.

It is clear that $[a]$ is infinite when all the powers $a^m$ are different from each other. The following result shows that $[a]$ is finite if $a^u = a^v$ for one pair of integers $u$ and $v$ with $u \neq v$.

***Theorem 2 Repetition of Powers***

*Let $a$ be in a group $G$ and let $a^u = a^v$ for distinct integers $u$ and $v$. Then there is a smallest positive integer $s$ such that $a^s$ is the identity $\mathbf{e}$ of $G$ and*

$$[a] = \{\mathbf{e}, a, a^2, \ldots, a^{s-1}\}$$

*is a cyclic group of order $s$. Also, $s \mid (u - v)$.*

***Proof***

Let $a^u = a^v$ with $u \neq v$. There is no loss of generality in assuming that $u > v$. Then

$$a^{u-v} = a^u a^{-v} = a^v a^{-v} = a^0 = \mathbf{e}, \tag{1}$$

with $u - v$ positive. Having shown that there is at least one positive integer $m$ with $a^m = \mathbf{e}$, we may let $s$ be the smallest $m$ with this property.

Now let $n$ be any integer. Using the division algorithm, we express $n$ in the form $qs + r$ with $q$ and $r$ integers and $0 \leq r < s$. Then

$$a^n = a^{qs+r} = (a^s)^q a^r = \mathbf{e}^q a^r = a^r. \tag{2}$$

This shows that every $a^n$ in the cyclic group $[a]$ is one of the elements

$$\mathbf{e}, a, a^2, \ldots, a^{s-1}. \tag{3}$$

These elements are different from one another since

$$a^i = a^j \quad \text{with} \quad 0 \leq i < j < s$$

implies that $a^{j-i} = \mathbf{e}$ and $0 < j - i < s$, thus contradicting the minimal nature of $s$. Hence $[a]$ consists of the $s$ elements listed in (3).

Finally, we note that display (2) above shows that $a^n = a^0 = \mathbf{e}$ if and only if $s|n$. Then it follows from display (1) that $s|(u - v)$.

We recall that the order of a group $G$ is the number of elements in $G$ and now present the following terminology:

### *Definition 3 Order of an Element of a Group*

*The* **order** *of an element $a$ of a group $G$ is the order of the cyclic subgroup $[a]$ generated by $a$ in $G$. If $[a]$ is infinite, one says that $a$ has infinite order.*

### *Theorem 3 Test for the Order of an Element*

*Let $a$ be in a group $G$. If $s$ is a positive integer such that $a^s = \mathbf{e}$ and $a^m \neq \mathbf{e}$ for $1 \leq m < s$, then $s$ is the order of $a$.*

*Proof*

This follows immediately from part of Theorem 2.

We note that only the identity $\mathbf{e}$ has order 1. Also, if $G$ is a finite cyclic group, it can easily be shown that an element $g$ of $G$ is a generator of $G$ if and only if the order of $g$ equals the order of $G$. (The proof is left to the reader as Problem 26 of this section.)

***Example 1***

Let $G$ be the multiplicative group of the nonzero complex numbers and let $\alpha = (1 + \mathbf{i}\sqrt{3})/2$. Find $[\alpha]$ in the form $\{1, \alpha, \alpha^2, \ldots, \alpha^{s-1}\}$. Then give the order of each number in $[\alpha]$ and find a generator $\beta$ of $[\alpha]$ with $\beta \neq \alpha$.

***Solution***

Calculation shows that $\alpha^2 = (-1 + \mathbf{i}\sqrt{3})/2$, $\alpha^3 = -1$, $\alpha^4 = -\alpha$, $\alpha^5 = -\alpha^2$, and $\alpha^6 = (-1)^2 = 1$. Hence 6 is the smallest positive integer $s$ with $\alpha^s$ equal to the identity 1 of $G$; that is, $\alpha$ has order 6 and

$$[\alpha] = \{1, \alpha, \alpha^2, \alpha^3, \alpha^4, \alpha^5\}.$$

Similarly, one finds that

$$[\alpha^2] = \{1, \alpha^2, \alpha^4\} = [\alpha^4], \qquad [\alpha^3] = \{1, \alpha^3\}, \qquad [\alpha^5] = [\alpha].$$

Hence each of $\alpha^2$ and $\alpha^4$ has order 3, $\alpha^3$ has order 2, and $\alpha^5$ has order 6. Also, $\alpha^5$ is the only other generator of $[\alpha]$.

The following result shows that it is easier to prove that a subset of a group is a subgroup when the subset is finite rather than infinite.

***Theorem 4 Finite Subgroup Conditions***

*Let $H$ be a nonempty finite subset of a group $G$ and let $H$ be closed under the operation of $G$. Then $H$ is a subgroup in $G$.*

***Proof***

Since the single element subset $\{\mathbf{e}\}$ is a subgroup, we may assume that there is an $a$ in $H$ with $a \neq \mathbf{e}$. It then follows from closure of $H$ under multiplication that all of the elements

$$a, a^2, a^3, a^4, \ldots \tag{4}$$

are in $H$. But $H$ is finite; hence there must be repetitions in (4). This and Theorem 2 imply that $a$ has finite order $s$.

Since $a \neq \mathbf{e}$, we have $s > 1$. Then the identity $\mathbf{e} = a^s$ and the inverse $a^{-1} = a^{s-1}$ of $a$ are among the elements of (4) and hence are in $H$. It now follows from Lemma 2 of Section 2.5 that $H$ is a subgroup in $G$.

An infinite subset $S$ of a group $G$ may be closed under the operation of $G$ without being a subgroup in $G$. (In Problems 5 and 6 of Section 2.6 the reader was asked to give examples of such subsets.)

## Problems

1. State the order of each of the following:
   (i) (1); (ii) (12); (iii) (123); (iv) (1234); (v) (12345).

2. What is the order of an $s$-cycle $(a_1 a_2 \ldots a_s)$?

3. Put (123)(34) in standard form and find its order.

4. Put (12)(13)(14) in standard form and find its order.

5. Find the order of each of the following:
   (i) (12)(34); (ii) (12)(345); (iii) (12)(3456); (iv) (12)(34567).

6. Find the order of each of the following:
   (i) (123)(45); (ii) (123)(456); (iii) (123)(4567);
   (iv) (123)(45678); (v) (123)(456789).

7. For each of the following, give the order of a product of disjoint cycles of the given lengths.
   (i) 2 and 3; (ii) 2 and 4; (iii) 4 and 5;
   (iv) 4 and 6; (v) 4, 5, and 6; (vi) 2, 3, 4, 5, and 6.

8. State a generalization of the answers in Problem 7.

9. (i) List the elements of each of the cyclic subgroups $H$ of the symmetric group $\mathbf{S}_3$ and state the order of $H$.
   (ii) Explain why $\mathbf{S}_3$ is not cyclic.

10. Use the fact that an element of $\mathbf{S}_4$ has a standard form of the type (1), $(xy)$, $(xyz)$, $(xyzw)$, or $(xy)(zw)$ to give the set of positive integers that are orders of elements of $\mathbf{S}_4$.

11. (i) Find the orders of $\alpha = (-1 + \mathbf{i}\sqrt{3})/2$ and $\beta = (\sqrt{3} + \mathbf{i})/2$ in the multiplicative group $V$ of the nonzero complex numbers.
    (ii) Give a generator different from $\alpha$ for the cyclic group $[\alpha]$.
    (iii) Find all the generators of $[\beta]$.

12. (i) Find the order of $\gamma = \cos(3\pi/14) + \mathbf{i}\sin(3\pi/14)$ in the multiplicative group $V$ of the nonzero complex numbers.
    (ii) Find all the generators of $[\gamma]$.

13. Let $a$ be an element of order 12 in a group $G$.
    (i) Find the smallest positive integer $r$ such that $a^{9r} = \mathbf{e}$.
    (ii) What is the order of $a^9$?
    (iii) What is the order of $a^8$?
    (iv) What is the order of $a^7$?

14. Let $a$ be an element of order 7 in a group $G$. State the order of each of the following.
    (a) $a^2$; (b) $a^3$; (c) $a^4$; (d) $a^5$; (e) $a^6$; (f) $a^7$.

15. Let $G = [a]$ be a cyclic group of order 30. List the elements of a subgroup $H$ of order $q$ for each of the following values of $q$.
    (a) 2; (b) 3; (c) 5; (d) 6.
16. Let $a$ be an element of finite order $q$ in a group $G$. Show the following:
    (a) $a^{-1}$ also has order $q$.
    (b) For every integer $k$, $a^k$ and $a^{q-k}$ have the same order.
    (c) $[a^{-1}] = [a]$.
    (d) If $q$ is an even integer $2r$ and $r > 1$, $a^r$ is not a generator of $[a]$.
    (e) If $q > 2$, the number of generators of $[a]$ is even.
17. Let $G = [a]$ be a cyclic group of order 30.
    (a) List all the elements of order 2 in $G$.
    (b) List all the elements of order 3 in $G$.
    (c) List all the elements of order 10 in $G$.
18. (a) List the six integers that are orders of elements of the symmetric group $\mathbf{S}_5$.
    (b) List the sixteen integers that are orders of elements of a cyclic group of order 120.
19. Let $a^{15}$ equal the identity $\mathbf{e}$ of $G$. What are the possibilities for the order of $a$?
20. Let $a^m = \mathbf{e}$ in $G$, with $m$ an integer. What are the possibilities for the order of $a$?
21. Let $G$ have $m$ elements $g$ satisfying $g^2 = \mathbf{e}$.
    (i) How many elements of order 1 are there in $G$?
    (ii) How many elements of order 2 are there in $G$?
22. Let $g$ be in a group $G$.
    (i) Can $g$ have larger order than $G$ has? Explain.
    (ii) Can $g$ have infinite order while $G$ is finite?
23. List the four generators of $[(12345678)]$.
24. Let $n \in \mathbf{Z}^+$ and $\alpha = \cos(2\pi/n) + \mathbf{i}\sin(2\pi/n)$. Prove the following:
    (a) $\alpha$ has order $n$ in the multiplicative group of nonzero complex numbers.
    (b) The cyclic subgroup $[\alpha]$ consists of the $n$ complex $n$th roots of unity.
    (c) The generators of $[\alpha]$ are the complex numbers

    $$\cos(2k\pi/n) + \mathbf{i}\sin(2k\pi/n), \qquad \gcd(k, n) = 1, \quad 1 \le k \le n.$$

    (These generators are the primitive $n$th roots of unity. See Problem 23 of Section 2.6.)
25. Is a group of order 2 or 3 necessarily cyclic? Explain.
26. Let $G$ be a group of order $m$. Show that $G$ is cyclic if and only if it has an element of order $m$; that is, show both of the following:
    (i) If $G$ has an element $g$ of order $m$, $G$ is cyclic.
    (ii) If $G$ is cyclic, $G$ has an element $g$ of order $m$.

27. Let $a$ be an element of order 5 in a group. Define a function of two variables $t = F(r, s)$ by $a^r a^s = a^t$ and $0 \leq t < 5$. Tabulate $t$ for $r$ and $s$ ranging through $\{0, 1, 2, 3, 4\}$.

28. Do the previous problem for an $a$ of order 4 letting $r$, $s$, and $t$ range through $\{0, 1, 2, 3\}$.

29. Let $S$ be a nonempty subset of the integers and let $S$ be closed under subtraction. Show that $S$ must be a cyclic subgroup $[m]$ in the additive group $\mathbf{Z}$ of the integers for some integer $m$.

30. Let $H$ be a subgroup in the additive group $\mathbf{Z}$ of the integers. Show that $H$ is cyclic.

31. Let $a$ be an element of finite order in a group $G$. Let $S$ be the set of all integers $m$ such that $a^m = \mathbf{e}$. Show that $S$ contains at least one positive integer and is closed under subtraction and hence that $S$ consists of all the integral multiples of the order $q$ of $a$; that is, $S$ is the cyclic subgroup $[q]$ in the additive group of the integers.

32. (i) Explain why $[a]$ must always be a subgroup in the centralizer $C_a$.
 (ii) Use (i) and the fact that disjoint cycles commute to find the six permutations in the centralizer $C_\alpha$ of $\alpha = (345)$ in $\mathbf{S}_5$.

33. Let $S$ be a subset of the elements of a cyclic group $G = [a]$ and let $T$ consist of all the integers $m$ such that $a^m$ is in $S$. Show that $S$ is a subgroup in $G$ if and only if $T$ is a subgroup in the additive group $\mathbf{Z}$ of the integers.

34. Let $H$ be a subgroup in a cyclic group $G = [a]$. Show that $H$ must be cyclic.

35. Let $a$ be an element of order $q$ in a group $G$, let $s$ be an integer, and let $d = \gcd(q, s)$. Show that the order of $a^s$ is $q/d$.

36. Let $[a]$ be a cyclic group of order $q$. Show that $a^s$ is a generator of $[a]$ if and only if $\gcd(s, q) = 1$.

37. Let $G = [a]$ be a cyclic group of order $q$. For each of the following values of $q$, list the integers $m$ in $\{0, 1, \ldots, q - 1\}$ such that $a^m$ is a generator of $G$.
 (a) $q = 3$. (b) $q = 4$. (c) $q = 5$. (d) $q = 6$.
 (e) $q = 7$. (f) $q = 8$. (g) $q = 9$.

38. Let $G = [a]$ be of order $q$. For each of the following values of $q$, tell how many generators $G$ has.
 (a) $q = p$, a prime.
 (b) $q = p_1 p_2$ with $p_1$ and $p_2$ distinct positive primes.
 (c) $q = p^r$ with $p$ a positive prime and $r$ a positive integer.

39. Let $G = [a]$ be a cyclic group of order $de$, where $d$ and $e$ are positive integers. Prove that $a^k$ has order $d$ if and only if $k = fe$, with $f$ a positive integer relatively prime to $d$.

40. Let $d$ and $m$ be positive integers with $d \mid m$. Prove that there are the same number

of elements of order $d$ in a cyclic group of order $m$ as in a cyclic group of order $d$.

41. Let $\mathbf{Z}$ be the additive group of the integers. Show that each of 1 and $-1$ is a generator of $\mathbf{Z}$. Are there any other generators?
42. How many generators are there of an infinite cyclic group $[a]$? Explain.

## 2.8 *Even and Odd Permutations*

Among the symmetric groups $\mathbf{S_1}$, $\mathbf{S_2}$, $\mathbf{S_3}$, ... the first one $\mathbf{S_1} = \{(1)\}$ is exceptional in that it alone has odd order. Also $\mathbf{S_1}$ is the only $\mathbf{S_n}$ with no transpositions in it. Since 2-cycles are the basis for the material in this section, we restrict ourselves here to the $\mathbf{S_n}$ with $n > 1$.

Then Theorem 3 of Section 2.4 states that every permutation of $\mathbf{S_n}$ is either a transposition or a product of (not necessarily disjoint) transpositions. However, this factorization is not unique. For example,

$$(1234) = (14)(24)(34) = (12)(23)(31)(12)(14) = (12)(13)(14).$$

Although the factorization into 2-cycles is not unique, we can show that for a given permutation $\theta$ the number of transpositions in the representation has fixed parity; that is, it is either always even or always odd. This fact is needed in studying a subgroup $\mathbf{A_n}$ in $\mathbf{S_n}$ which has an essential role in the proof of the insolvability of the general fifth degree polynomial equation. (This is discussed further in Sections 2.13 and 5.12.)

In our proof of the invariance of parity property, Lemma 3 below, we use the convenient device of considering the effect of applying a permutation $\theta$ to the subscripts $i$ of the variables $x_i$ in a function $F(x_1, x_2, \ldots, x_n)$ of $n$ variables. Let $G(x_1, x_2, \ldots, x_n)$ be the result of replacing $x_1, \ldots, x_n$ in $F(x_1, \ldots, x_n)$ by

$$x_{\theta(1)}, \ldots, x_{\theta(n)},$$

respectively. Then we say that $\theta$ sends $F$ to $G$ or that $G$ is the result of applying $\theta$ on $F$.

For example, (1234) sends $(x_1 - x_2)(x_1 - x_4)$ to $(x_2 - x_3)(x_2 - x_1)$ and (123) sends $x_1^2 + x_2^2$ to $x_2^2 + x_3^2$.

We are particularly interested in the result of applying a permutation $\theta$ on the product $P_n$ of all the differences $x_j - x_i$ with $1 \le i < j \le n$. We note that the $P_n$ for $n = 2$, 3, and 4 are

$$P_2 = x_2 - x_1,$$

$$P_3 = (x_2 - x_1)(x_3 - x_1)(x_3 - x_2),$$

$$P_4 = (x_2 - x_1)(x_3 - x_1)(x_3 - x_2)(x_4 - x_1)(x_4 - x_2)(x_4 - x_3).$$

*Example 1*

Give the result of applying the transposition (13) on

$$(x_2 - x_1)(x_3 - x_1)(x_3 - x_2)\,(x_4 - x_1)(x_4 - x_2)(x_4 - x_3) = P_4.$$

*Solution*

$$(x_2 - x_3)(x_1 - x_3)(x_1 - x_2)\,(x_4 - x_3)(x_4 - x_2)(x_4 - x_1) = -P_4.$$

We generalize on this example in the following result.

**Lemma 1** ***A Transposition Changes the Sign of*** $P_n$

*Every 2-cycle* $(uv)$ *sends* $P_n$ *to* $-P_n$.

***Proof***

Since $(vu) = (uv)$, we assume that $u < v$. As an aid in finding the effect of $(uv)$ on $P_n$, we group some of the differences that are factors of $P_n$.

For $i < u$, we group $x_u - x_i$ and $x_v - x_i$ and note that $(uv)$ sends

$$A = (x_u - x_i)(x_v - x_i) \quad \text{to} \quad (x_v - x_i)(x_u - x_i) = A.$$

Similarly, for $v < i$, the transposition $(uv)$ sends

$$B = (x_i - x_u)(x_i - x_v) \quad \text{to} \quad (x_i - x_v)(x_i - x_u) = B.$$

For $u < i < v$, the transposition $(uv)$ sends

$$C = (x_i - x_u)(x_v - x_i) \quad \text{to} \quad [-(x_v - x_i)][-(x_i - x_u)] = C.$$

Also, $(uv)$ sends $(x_j - x_i)$ to itself if neither $i$ nor $j$ is $u$ or $v$.

The only factor of $P_n$ that we have not yet considered is $D = x_v - x_u$, which is sent by $(uv)$ to

$$x_u - x_v = -(x_v - x_u) = -D.$$

Together, all of these observations show that a transposition $(uv)$ sends $P_n$ to $-P_n$.

**Lemma 2** $\theta$ ***Sends*** $P_n$ ***to*** $\pm P_n$

*In* $\mathbf{S_n}$, *let* $\theta$ *be a product* $\gamma_1\gamma_2\cdots\gamma_r$ *of* $r$ *transpositions* $\gamma_i$. *Then* $\theta$ *sends* $P_n$ *to* $(-1)^r P_n$.

***Proof***

If $\alpha$ sends $F$ to $G$ and $\beta$ sends $G$ to $H$, it is clear that the composite $\alpha\beta$ sends $F$ to $H$. Since each $\gamma_i$ changes the sign of $P_n$, the composite $\theta$ of the $r$ transpositions $\gamma_i$ sends $P_n$ to $(-1)^r P_n$.

### *Lemma 3 Always Even or Always Odd*

*In* $\mathbf{S_n}$ *let* $\theta = \alpha_1\alpha_2 \cdots \alpha_r = \beta_1\beta_2 \cdots \beta_s$, *where each* $\alpha_i$ *and each* $\beta_j$ *is a transposition. Then* $r$ *and* $s$ *are both even or both odd.*

***Proof***

It follows from Lemma 2 that when we apply $\theta$ on $P_n$ we get both $(-1)^r P_n$ and $(-1)^s P_n$; thus we see that

$$(-1)^r P_n = (-1)^s P_n.$$

Hence $r$ and $s$ must have the same parity.

### *Definition 1 Even Permutation, Odd Permutation*

*A permutation that is expressible as a product of an even number of transpositions is an* **even permutation**. *A product of an odd number of transpositions is an* **odd permutation**.

Lemma 3 tells us that a given permutation is either even or odd, but not both. We also note that the identity $(1) = (12)(12)$ is even in all the $\mathbf{S_n}$ with $n > 1$.

### *Theorem 1 Permutation Parity*

*Let* $\theta = \gamma_1\gamma_2 \cdots \gamma_r$ *where* $\gamma_1, \ldots, \gamma_r$ *are cycles of lengths* $s_1, \ldots, s_r$, *respectively. Then* $\theta$ *is even or odd depending on whether*

$$s_1 + s_2 + \cdots + s_r - r \tag{A}$$

*is even or odd.*

The proof is left to the reader as Problem 7 of this section.

If the cycles $\gamma_i$ are disjoint, (A) is called the ***Cauchy Number*** of $\theta$.

## Problems

Problem 15 below is cited in Section 3.5 and Problem 9 is cited in Section 2.10.

1. In each of the following, write the result of applying $\theta$ on $F(x_1, \ldots, x_n)$.
   (a) $\theta = (123)$, $F(x_1, x_2, x_3) = x_1^2 + x_2^2 - x_3^2$.
   (b) $\theta = (12)$, $F(x_1, x_2, x_3) = x_1^2 + x_2^2 - x_3^2$.
   (c) $\theta = (123)$, $F(x_1, x_2, x_3, x_4) = x_1^2 + x_2^2 + x_4^2$.
2. In each of the following, write the result of applying $\theta$ on $F(x_1, \ldots, x_n)$.
   (a) $\theta = (124)$, $F(x_1, x_2, x_3, x_4) = x_1^2 + x_2^2 + x_4^2$.
   (b) $\theta = (23)$, $F(x_1, x_2, x_3) = (x_1 + x_2 - x_3)(x_2 + x_3 - x_1)(x_3 + x_1 - x_2)$.
   (c) $\theta = (12)(34)$, $F(x_1, x_2, x_3, x_4) = x_1^2 x_2^2 x_3 x_4$.
3. Which permutations $\theta$ of $\mathbf{S_4}$ leave $x_1^2 - x_2^2 + 2x_3 + x_4^2$ invariant, that is, send it to itself?
4. Which permutations $\theta$ of $\mathbf{S_5}$ leave $x_1 + x_2 + 3(x_3 + x_4) + 2x_5$ invariant?
5. Which of the following are even permutations? Which are odd?
   (a) (12345); (b) (123456); (c) (13)(12)(345); (d) (1234)(3456).
6. Show that a cycle of length $s$ is an even permutation if $s$ is odd and is an odd permutation if $s$ is even.
7. Let $\theta = \gamma_1\gamma_2 \cdots \gamma_r$, where $\gamma_1, \ldots, \gamma_r$ are cycles of lengths $s_1, \ldots, s_r$, respectively. Prove that $\theta$ is even or odd depending on whether $s_1 + s_2 + \cdots + s_r - r$ is even or odd (and thus prove Theorem 1 above).
8. State a necessary and sufficient condition on the lengths of the cycles for a product $\gamma_1\gamma_2 \cdots \gamma_r$ of cycles to be even and a condition for the product to be odd.
9. Show the following:
   (a) The product of two even permutations is even.
   (b) The product of two odd permutations is even.
   (c) The product of an even permutation and an odd permutation is odd.
   (d) The inverse of an odd permutation is odd.
10. Show that the set of even permutations in any group of permutations forms a subgroup.
11. Let $\alpha_1, \alpha_2, \ldots, \alpha_r$ be all the even permutations in $\mathbf{S_n}$ and let $\beta$ be an odd permutation in $\mathbf{S_n}$. Show the following:
    (i) The products $\beta\alpha_1, \beta\alpha_2, \ldots, \beta\alpha_r$ are distinct odd permutations.
    (ii) The products $\alpha_1\beta, \alpha_2\beta, \ldots, \alpha_r\beta$ are distinct odd permutations.
    (iii) If $n \geq 2$, there are at least as many odd permutations in $\mathbf{S_n}$ as there are even ones.

12. Let $\beta_1, \beta_2, \ldots, \beta_s$ be all the odd permutations in $\mathbf{S_n}$. Show the following:
    (i) $\beta_1\beta_1, \beta_1\beta_2, \ldots, \beta_1\beta_s$ are distinct even permutations.
    (ii) There are at least as many even permutations in $\mathbf{S_n}$ as odd ones.
    (iii) If $n \geq 2$, there are $(n!)/2$ even permutations and the same number of odd permutations in $\mathbf{S_n}$. [See Problem 11(iii).]
13. Give an example of a group consisting entirely of even permutations.
14. Let $H$ be a subgroup of $\mathbf{S_n}$. Show that either $H$ consists entirely of even permutations or $H$ has as many odd permutations as even ones.
15. Show that $\alpha^{-1}\beta^{-1}\alpha\beta$ is even for all permutations $\alpha$ and $\beta$ in $\mathbf{S_n}$.

---

***Baron Augustin-Louis Cauchy*** (*1789–1857*) *Cauchy was one of the greatest and most prolific of mathematicians, second only to Euler in mathematical output. Also, like Euler, he wrote not only research papers and books but also textbooks. His* **Cours d'Analyse de l'École Royale Polytechnique** *is a classic text which provided the model for calculus texts for many years. Unlike Euler, however, he was far more concerned with rigor, and hence our present-day treatment of limit and continuity in calculus courses is essentially that of Cauchy.*

*Although Cauchy's work was primarily in analysis, he also made significant contributions to algebra, the theory of numbers (he first proved the conjecture of Fermat, partially proved by Gauss, that every positive integer is the sum of at most three triangular numbers, four squares, five pentagonal numbers, and so on), and even geometry (he gave the first generally accepted proof of Euler's formula for polyhedra,* $V + F = E + 2$, *a formula actually discovered by Descartes).*

*Cauchy shared with Newton a tendency to write on religion as well as mathematics. He was very conservative in matters of religion and politics and was in voluntary exile for a number of years after the exile of the King, Charles X, to whom he was devoted. In 1830, unable to swear an oath to Charles' successor, Louis-Philippe, he was forced to leave France since he could not hold an academic post there. It was not until 1848, under Louis Napoleon, that he was awarded a professorship at the École Polytechnique.*

---

## 2.9 *Groups of Symmetries*

*There is no branch of mathematics,*
*however abstract, which may not someday*
*be applied to phenomena of the real world.*

*Nicolai Ivanovich Lobachevsky*

It would seem that this statement by Lobachevsky cannot be proved or disproved. However, some proposed counterexamples have not stood the test of time. For example, the famous physicist Sir James Jeans is quoted as having stated (in 1910): "We may as well cut out group theory. That is a subject that will never be of any use in physics."

But in 1964 the existence of the omega minus particle, predicted using group theory, was confirmed by laboratory test. Thus group theory became a central theme in another cycle of restoring order to previously chaotic data on elementary particles.

In this section, we examine a very small sample of the kind of group theory that has brought new insight into geometry and physics. (See the biographical note on Felix Klein after this section.)

Let us consider a square with vertices numbered 1, 2, 3, 4 in circular order, that is, with each of the pairs

$$\{1, 2\}, \quad \{2, 3\}, \quad \{3, 4\}, \quad \{4, 1\} \tag{1}$$

consisting of the numbers of the two vertices of a side of the square.

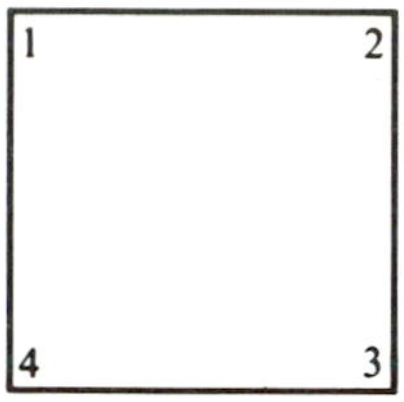


*Figure 2.3*

After a rigid motion of the square that ends with it covering the same area as in its starting position, let vertex $a_i$ be where vertex $i$ was originally. The rigid motion thus gives rise to a permutation

$$\theta: 1 \mapsto a_1, \quad 2 \mapsto a_2, \quad 3 \mapsto a_3, \quad 4 \mapsto a_4, \tag{2}$$

in which the pairs

$$\{a_1, a_2\}, \quad \{a_2, a_3\}, \quad \{a_3, a_4\}, \quad \{a_4, a_1\} \tag{3}$$

are the pairs of (1), although not necessarily in the same order.

***Definition 1 Symmetry of a Square***

*The permutation $\theta$ of (2) is a* **symmetry of a square** *if each pair of (3) is one of the pairs of (1).*

How many of the permutations of the symmetric group $\mathbf{S}_4$ are symmetries of a square? We note that, in (2), $a_1$ may be chosen to be any number in $\{1, 2, 3, 4\}$ and then for a given $a_1$ there are two ways of selecting $a_2$ and $a_4$ so that $\{a_1, a_2\}$ and $\{a_4, a_1\}$ are pairs of (1). After the choice of $a_1$, $a_2$, and $a_4$, one must have the remaining number as $a_3$. Hence there are $4 \cdot 2 = 8$ symmetries of a square.

In Figure 2.4 there is a square for each of these 8 symmetries with the new number $a_i$ for a vertex outside the square and the old number $i$ inside.

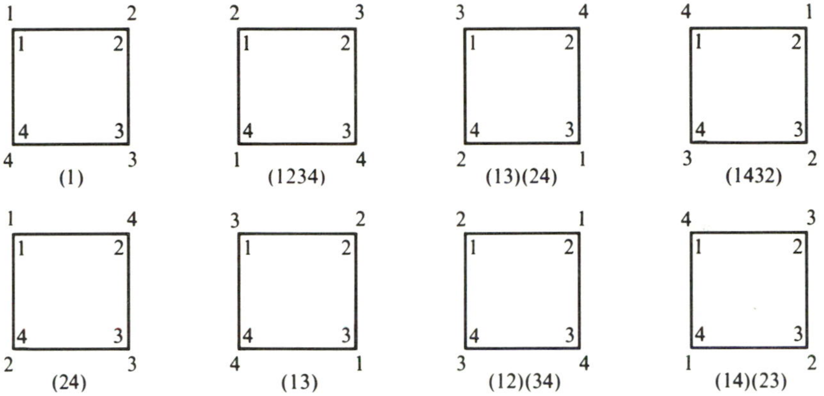


*Figure 2.4*

We now associate each of these symmetries with a rigid motion of the square. The permutations

$$\rho = (1234), \qquad \rho^2 = (13)(24), \qquad \rho^3 = (1432)$$

result from counterclockwise rotations of the square about its center through 90°, 180°, and 270°, respectively. One may associate the identity (1) with a rotation of 360° or any other motion that returns the square to its original position.

The symmetry $\phi = (24)$ results from a 180° flip of the square about the diagonal connecting vertices 1 and 3; the transposition (13) from a similar flip about the other diagonal.

The symmetry (12)(34) corresponds to a 180° rotation, in 3-dimensional space, of the square about the line through the midpoints of the sides with vertices $\{1, 2\}$ and $\{3, 4\}$, respectively. The permutation (14)(23) is associated with a similar rotation about the line through the midpoints of the other two sides.

A rigid motion of a square followed by another rigid motion is equivalent to some single rigid motion. This leads us to believe that the subset of $\mathbf{S}_4$, consisting of the eight symmetries of a square, is closed under multiplication. We now prove this, using the definition above.

Let $\alpha$ and $\beta$ be symmetries of a square, let $\gamma = \alpha\beta$, and let $\{a_1, a_2\}$ be any one of the pairs of (1). Let $\alpha(a_i) = b_i$ and $\beta(b_i) = c_i$ for $i = 1$ and 2. By the definition of a symmetry, $\{b_1, b_2\}$ is a pair of (1). Then it similarly follows that $\{c_1, c_2\}$ is a pair of (1). Since $\gamma(a_i) = c_i$, this means that the composite $\gamma$ is a symmetry of a square.

***Theorem 1 Subgroup of Symmetries***

*The 8 symmetries of a square form a subgroup in* $\mathbf{S}_4$.

***Proof***

These permutations form a nonempty finite subset closed under the operation of $\mathbf{S}_4$. Hence this subset is a subgroup by Theorem 4 of Section 2.7.

***Definition 2 Octic Group***

*The group of the 8 symmetries of a square is called the* **octic group**.

We next use a diagram to show some of the relationships among the subgroups in the octic group.

Subgroup Poset Diagram for the Octic Group

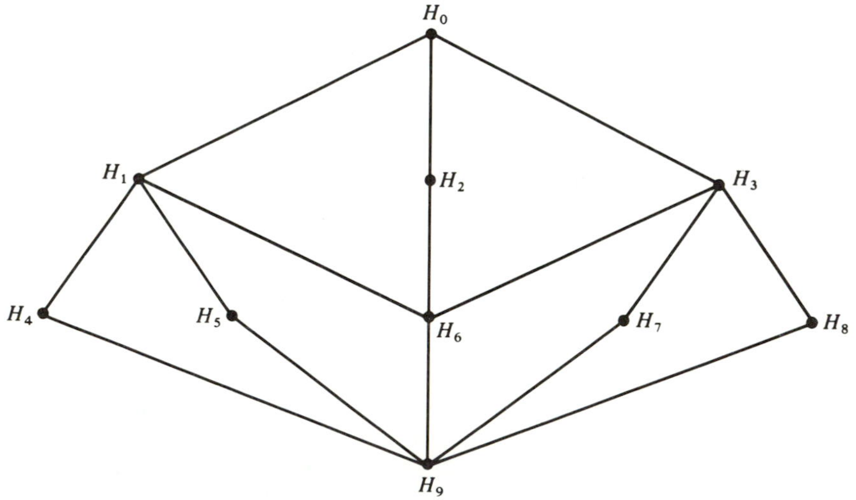


***Figure 2.5***

$H_0, H_1, \ldots, H_9$, shown in Figure 2.5, stand for the subgroups in the octic group. In Problems 7–10 below the reader is asked to identify the $H$'s with the subgroups in such a way that there is a rising line or broken line from $H_i$ to $H_j$ if and only if $H_i$ is a proper subgroup in $H_j$. This makes Figure 2.5 into a ***subgroup poset diagram*** for the octic group; the reason for the use of the word "poset" will be presented in Section 3.8.

## *Problems*

1. Let $\rho = (1234)$ and $\phi = (24)$. Show the following:
   (i) $\{(1), \rho, \rho^2, \rho^3, \phi, \rho\phi, \rho^2\phi, \rho^3\phi\}$ is the octic group.
   (ii) $\phi\rho = \rho^3\phi$.
2. Using the notation of Problem 1, list the even permutations in the octic group.
3. Using the notation of Problem 1, make a multiplication table for the octic group.
4. Give the order of each element of the octic group.
5. List the elements of the center of the octic group.
6. (i) List all the permutations $\alpha$ such that $\alpha = \beta^2$ with $\beta$ in the octic group.
   (ii) Show that the set $H$ of all the $\alpha$ of (i) is a subgroup in the octic group.
   (iii) List all the $\alpha$ such that $\alpha = \beta^3$ with $\beta$ in the octic group.
   (iv) Show that the set $K$ of all the $\alpha$ of (iii) is a subgroup in the octic group.
7. In the octic group, find the following:
   (i) A cyclic subgroup of order 4.
   (ii) Two noncyclic subgroups of order 4.
   (iii) Five subgroups of order 2.

*8. Prove that the octic group has no subgroup of order 3, 5, 6, or 7.

*9. Prove that the only subgroups in the octic group are $\{(1)\}$, the eight subgroups found in Problem 7, and the octic group itself.

10. Identify $H_0, H_1, \ldots, H_9$ in Figure 2.5 as the ten subgroups in the octic group in such a way that there is a rising line or broken line from $H_i$ to $H_j$ if and only if $H_i$ is a proper subgroup in $H_j$. (Do this in one of the several possible ways. List the elements of each subgroup in standard form.)
11. Let $\rho = (1234)$ and $\phi = (24)$. Let

$$H = \{(1), \rho^2, \rho\phi, \rho^3\phi\}$$

and let $K$ consist of the permutations $\phi^{-1}\theta\phi$ for all $\theta$ in $H$.

(i) Is $H$ a subgroup in the octic group? Explain.
(ii) List the elements of $K$.
(iii) Is $K$ a subgroup in the octic group? Explain.

12. List the elements of the centralizer of each of the following elements in the octic group:
(a) (24); (b) (13)(24).

13. Let $\rho = (12345)$ and $\phi = (25)(34)$. Show the following:
(i) $\phi\rho = \rho^{-1}\phi$.
(ii) The symmetries of a regular pentagon are

$$\{(1), \rho, \rho^2, \rho^3, \rho^4, \phi, \rho\phi, \rho^2\phi, \rho^3\phi, \rho^4\phi\}.$$

14. Show that every permutation of $\mathbf{S}_3$ corresponds to a symmetry of an equilateral triangle.

15. (i) Which permutations of $\mathbf{S}_4$ are associated with symmetries of a rectangle that is not a square?
(ii) Make the multiplication table for the set of permutations found in (i).

16. Describe a group of $2n$ symmetries of a regular polygon with $n$ sides.

---

***Felix Klein*** *(1849–1925) Klein was a student at Bonn and later taught at Erlangen, Munich, Leipzig, and Göttingen. It was upon his appointment to a professorship in 1872 at Erlangen that he outlined in his inaugural address what has come to be known as his Erlanger Programm. In this address he described different geometries as the studies of properties of figures which remain invariant under different groups of transformations. For example, plane euclidean geometry is the study of those properties of figures (length, angle size, area) which are unchanged under rotation and translation, such transformations in the plane forming a group. Similarly, affine geometry, projective geometry, etc., correspond to groups of transformations. This approach provided a very neat classification of various geometries in terms of the theory of groups.*

*Klein also provided a geometric interpretation to the problem of the solvability of the quintic (that is, fifth degree equation) by relating it to the group of symmetries of the regular icosahedron.*

*He had the reputation of being a great teacher and during his stay at Göttingen he attracted first-rate students, including many from the United States. He was a brilliant mathematician as well as teacher and provided a distinguished link in the chain of first-rank mathematicians at Göttingen in the nineteenth and early twentieth centuries: Gauss, Dirichlet, Riemann, Klein, and Hilbert.*

---

## 2.10 The Alternating Groups $\mathbf{A_n}$

For $n > 1$, let $\mathbf{A_n}$ denote the subset of all the even permutations in the symmetric group $\mathbf{S_n}$. It is easily seen that $\mathbf{A_n}$ is closed under composition. Since $\mathbf{A_n}$ is finite and nonempty, this suffices to make $\mathbf{A_n}$ a subgroup in $\mathbf{S_n}$.

### *Definition 1 Alternating Group*

*The group* $\mathbf{A_n}$ *of even permutations on* $\mathbf{X_n} = \{1, 2, \ldots, n\}$ *is the* **alternating group** *on* $\mathbf{X_n}$.

Table 2.2 is a multiplication table for $\mathbf{A_4}$. We note that $\mathbf{A_4}$ consists of the identity (1), three elements of the form $(ab)(cd)$, and eight 3-cycles $(abc)$. This checks with the fact that the order of $\mathbf{A_4}$ is $(4!)/2 = 24/2 = 12$. [See Problem 12(iii) of Section 2.8.]

It is left to the reader in Problem 23 below to identify $H_0, H_1, \ldots, H_9$, shown in Figure 2.6, as the subgroups in $\mathbf{A_4}$ so as to make the figure into a subgroup poset diagram for $\mathbf{A_4}$, that is, so that there is a rising line or broken line from $H_i$ to $H_j$ if and only if $H_i$ is a proper subgroup in $H_j$.

***Table 2.2 Multiplication Table for*** $\mathbf{A_4}$

In this table, the permutations of $\mathbf{A_4}$ are designated as $\alpha_1, \alpha_2, \ldots, \alpha_{12}$ and an entry $k$ inside the table represents $\alpha_k$.

| | $\alpha_1$ | $\alpha_2$ | $\alpha_3$ | $\alpha_4$ | $\alpha_5$ | $\alpha_6$ | $\alpha_7$ | $\alpha_8$ | $\alpha_9$ | $\alpha_{10}$ | $\alpha_{11}$ | $\alpha_{12}$ |
|---|---|---|---|---|---|---|---|---|---|---|---|---|
| $(1) = \alpha_1$ | 1 | 2 | 3 | 4 | 5 | 6 | 7 | 8 | 9 | 10 | 11 | 12 |
| $(12)(34) = \alpha_2$ | 2 | 1 | 4 | 3 | 8 | 7 | 6 | 5 | 11 | 12 | 9 | 10 |
| $(13)(24) = \alpha_3$ | 3 | 4 | 1 | 2 | 6 | 5 | 8 | 7 | 12 | 11 | 10 | 9 |
| $(14)(23) = \alpha_4$ | 4 | 3 | 2 | 1 | 7 | 8 | 5 | 6 | 10 | 9 | 12 | 11 |
| $(123) = \alpha_5$ | 5 | 6 | 7 | 8 | 9 | 10 | 11 | 12 | 1 | 2 | 3 | 4 |
| $(243) = \alpha_6$ | 6 | 5 | 8 | 7 | 12 | 11 | 10 | 9 | 3 | 4 | 1 | 2 |
| $(142) = \alpha_7$ | 7 | 8 | 5 | 6 | 10 | 9 | 12 | 11 | 4 | 3 | 2 | 1 |
| $(134) = \alpha_8$ | 8 | 7 | 6 | 5 | 11 | 12 | 9 | 10 | 2 | 1 | 4 | 3 |
| $(132) = \alpha_9$ | 9 | 10 | 11 | 12 | 1 | 2 | 3 | 4 | 5 | 6 | 7 | 8 |
| $(143) = \alpha_{10}$ | 10 | 9 | 12 | 11 | 4 | 3 | 2 | 1 | 7 | 8 | 5 | 6 |
| $(234) = \alpha_{11}$ | 11 | 12 | 9 | 10 | 2 | 1 | 4 | 3 | 8 | 7 | 6 | 5 |
| $(124) = \alpha_{12}$ | 12 | 11 | 10 | 9 | 3 | 4 | 1 | 2 | 6 | 5 | 8 | 7 |

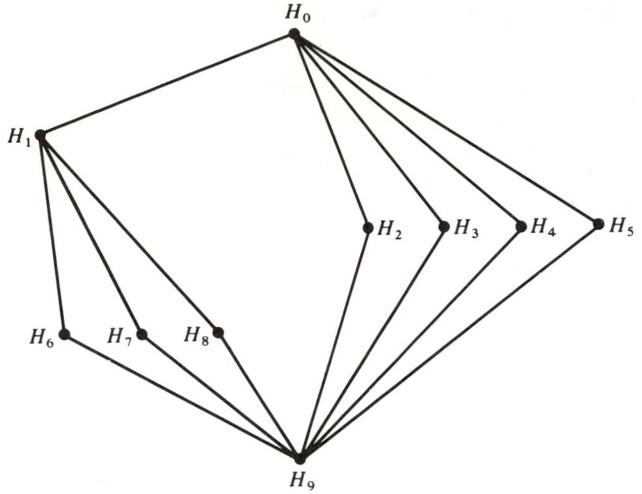


*Figure 2.6*

It is easily shown that an $s$-cycle $\gamma = (a_1 a_2 \ldots a_s)$ is an even permutation when $s$ is odd and that $\gamma$ is odd when $s$ is even. (See Problems 7 and 8 of Section 2.4.) Also, the product $\alpha\beta$ of two permutations is even when $\alpha$ and $\beta$ are both even or both odd, and the product is odd when one factor is an even permutation and the other is odd. (See Problem 9 of Section 2.8.) Thus a permutation with a standard form of the type

$$(abc)(defg)(hijk)$$

will always be even.

We next illustrate a technique for counting the permutations with a given type of standard form. Some easier counting problems of this nature occur in our investigation, in the next two sections, of the structure of $\mathbf{S_4}$ and $\mathbf{S_5}$.

***Example 1***

How many permutations $\alpha$ in $\mathbf{A_{15}}$ have a standard form of the type $(abc)(defg)(hijk)$?

***Solution***

We first note that a permutation with this type of standard form is even; hence $\alpha$ will be in $\mathbf{A_{15}}$ if $a, b, \ldots, k$ are 11 properly chosen elements of $\mathbf{X_{15}} = \{1, 2, 3, \ldots, 15\}$.

It is well known that one can choose a subset of $r$ elements from a set of $n$ elements in

$$\binom{n}{r} = \frac{n(n-1)(n-2)\cdots(n-r+1)}{1\cdot 2\cdot 3\cdots r}$$

ways; therefore the 4 elements of $\mathbf{X}_{15}$ *not* involved in $\alpha$ can be chosen in

$$\binom{15}{4} = \frac{15\cdot 14\cdot 13\cdot 12}{1\cdot 2\cdot 3\cdot 4} = 1365$$

ways. Then $a$ must be the smallest of the elements of $\mathbf{X}_{15}$ that are involved in $\alpha$. There are now 10 elements left from which to choose $b$ and after that 9 elements from which to choose $c$. Then $d$ must be the smallest of the remaining elements, $e$ can be any one of 7 elements, and so on. Thus the total number of permutations $\alpha$ with this type of standard form is

$$1365\cdot 10\cdot 9\cdot 7\cdot 6\cdot 5\cdot 3\cdot 2 = 154{,}791{,}000.$$

If we wanted to know how many of the permutations of $\mathbf{A}_{15}$ are products of three disjoint cycles, of which one has length 3 and the other two have length 4, we would also have to count the permutations with standard forms of the types $(abcd)\,(efg)(hijk)$ and $(abcd)(efgh)(ijk)$.

## *Problems*

1. Give the standard forms of all the elements of order two in $\mathbf{A}_4$.
2. List in standard form all the elements of order three in $\mathbf{A}_4$.
3. Are there any elements of order greater than three in $\mathbf{A}_4$?
4. List all the subgroups of order one or two in $\mathbf{A}_4$.
5. List all the subgroups of order three in $\mathbf{A}_4$.
6. Does $\mathbf{A}_4$ have a cyclic subgroup of order four or six? Explain.
7. Find the smallest subgroup in $\mathbf{A}_4$ that contains (12)(34) and (123).
8. Find the smallest subgroup in $\mathbf{A}_4$ that contains (14)(23) and (123).
9. Let $\alpha_1 = (1)$, $\alpha_5 = (123)$, and $\alpha_9 = (132)$.
   (i) For each $\theta$ of $\mathbf{A}_4$ find the set $\{\alpha_1\theta, \alpha_5\theta, \alpha_9\theta\}$.
   (ii) How many different subsets of $\mathbf{A}_4$ are there in the sets of (i)?
10. Let $\alpha_1 = (1)$, $\alpha_5 = (123)$, and $\alpha_9 = (132)$. Show that the sets

$$\{\alpha_1\theta, \alpha_5\theta, \alpha_9\theta\} \quad \text{and} \quad \{\theta\alpha_1, \theta\alpha_5, \theta\alpha_9\}$$

are different when $\theta = (124)$ and are the same when $\theta = (123)$.

11. Let $\alpha_1 = (1)$, $\alpha_2 = (12)(34)$, $\alpha_3 = (13)(24)$, and $\alpha_4 = (14)(23)$.
   (i) Find the subset $\{\alpha_1\theta, \alpha_2\theta, \alpha_3\theta, \alpha_4\theta\}$ for each $\theta$ in $\mathbf{A_4}$.
   (ii) How many different subsets are there in (i)?
   (iii) Show that

$$\{\alpha_1\theta, \alpha_2\theta, \alpha_3\theta, \alpha_4\theta\} = \{\theta\alpha_1, \theta\alpha_2, \theta\alpha_3, \theta\alpha_4\}$$

   for every $\theta$ in $\mathbf{A_4}$.

12. Which of the elements of $\mathbf{A_4}$ are squares of elements of $\mathbf{A_4}$? Which are cubes? Which are fifth powers?
13. How many elements of $\mathbf{A_5}$ are products of two disjoint 2-cycles? How many elements of $\mathbf{A_5}$ are 3-cycles? How many are 5-cycles?
14. Are there any other elements of $\mathbf{A_5}$ besides those counted in the parts of Problem 13?
15. Give the set of positive integers that are orders of elements of $\mathbf{A_5}$.
16. How many of the elements of $\mathbf{A_{10}}$ are products $(abcd)(efgh)$ of two disjoint 4-cycles?
17. Let $\alpha_2 = (12)(34)$. Complete the following table and note that $\theta^{-1}\alpha_2\theta$ is a product of two disjoint transpositions for every $\theta$ in $\mathbf{A_4}$:

| $\theta$ | $\alpha_1$ | $\alpha_2$ | $\alpha_3$ | $\alpha_4$ | $\alpha_5$ | $\alpha_6$ | $\cdots$ | $\alpha_{12}$ |
|---|---|---|---|---|---|---|---|---|
| $\theta^{-1}\alpha_2\theta$ | $\alpha_2$ | $\alpha_2$ | | | $\alpha_4$ | | | |

18. Let $\alpha_5 = (123)$. Show that $\theta^{-1}\alpha_5\theta$ is a 3-cycle for every $\theta$ in $\mathbf{A_4}$.
19. Find the centralizer of $\alpha_2 = (12)(34)$ in $\mathbf{A_4}$.
20. Find the centralizer of $\alpha_5 = (123)$ in $\mathbf{A_4}$.
21. (i) Is $(1234)(56)$ in $\mathbf{A_6}$?
   (ii) Is $(1234)(56)$ the square of an element of $\mathbf{S_6}$? Explain.
22. Let $a$, $b$, $c$, and $d$ be distinct positive integers. Use the fact that

$$(ab)(ac) = (abc) \text{ and } (ab)(cd) = (abc)(adc)$$

   to prove that every $\alpha$ in $\mathbf{A_n}$ is the product of 3-cycles when $n \geq 3$.

23. Identify $H_0, H_1, \ldots, H_9$, shown in Figure 2.6, as the subgroups in $\mathbf{A_4}$ so that there is a rising line or broken line from $H_i$ to $H_j$ if and only if $H_i$ is a proper subgroup in $H_j$. (Do this in one of the possible ways.)

## 2.11 *Cosets of a Subgroup*

Let us consider the orders of subgroups in the symmetric group $\mathbf{S}_4$. The cyclic subgroups [(1)], [(12)], [(123)], and [(1234)] have orders 1, 2, 3, and 4, respectively. The subset of all permutations $\theta$ in $\mathbf{S}_4$ such that $\theta(4) = 4$ is a subgroup of order 6. The octic group, the alternating group $\mathbf{A}_4$, and $\mathbf{S}_4$ itself are subgroups of order 8, 12, and 24, respectively. These orders

$$1, 2, 3, 4, 6, 8, 12, 24$$

of subgroups in $\mathbf{S}_4$ are all the positive integral divisors of the order 24 of $\mathbf{S}_4$.

In the alternating group $\mathbf{A}_4$, which has order 12, we have found subgroups with

$$1, 2, 3, 4, 12$$

as their orders. We shall prove that these are the only possibilities for the order of a subgroup in $\mathbf{A}_4$.

These examples, and all the others that we have encountered so far, might lead one to conjecture that the order of a subgroup $H$ in a finite group $G$ must be an integral divisor of the order of $G$ but that a positive integral divisor of the order of $G$ need not be the order of a subgroup in $G$. This conjecture will be proved to be correct with the help of the following concept:

***Definition 1*** *Coset for a of H in G*

*Let H be a subgroup in G and let a be a fixed element of G. The set of all products ah with h in H is called the* **left coset** *for a of H in G and is denoted by aH. The set Ha of all ha with h in H is the* **right coset** *for a of H in G.*

***Example 1***

Let $H$ be the subgroup $[(12)] = \{(1), (12)\}$ in $\mathbf{S}_3$. Find all the left cosets and all the right cosets of $H$ in $\mathbf{S}_3$.

***Solution***

$(123)H = \{(123)(1), (123)(12)\} = \{(123), (23)\}$ and $H(123) = \{(1)(123), (12)(123)\} = \{(123), (13)\}$. Similarly, one finds that

$$(1)H = (12)H = H = H(1) = H(12),$$

$$(23)H = (123)H = \{(23), (123)\}, \qquad (13)H = (132)H = \{(13), (132)\},$$
$$H(23) = H(132) = \{(23), (132)\}, \qquad H(13) = H(123) = \{(13), (123)\}.$$

We note that, in Example 1, $\theta H = H\theta$ for $\theta = (1)$ or $(12)$ but that $\theta H \neq H\theta$ for $\theta = (123), (132), (23)$, or $(13)$. We also see from $(13)H = (132)H$ that one can have $aH = bH$ with $a \neq b$.

### *Example 2*

Let $N$ be the subgroup $[(123)] = \{(1), (123), (132)\}$ in $\mathbf{S}_3$. Find all the left cosets and all the right cosets of $N$ in $\mathbf{S}_3$.

### *Solution*

$(23)N = \{(23)(1),\ (23)(123),\ (23)(132)\} = \{(23),\ (12),\ (13)\}$ and $N(23) = \{(1)(23),\ (123)(23), (132)(23)\} = \{(23), (13), (12)\}$. Similarly, one finds that

$$\theta N = N\theta = \{(1), (123), (132)\} = N \text{ for } \theta = (1), (123), \text{ and } (132);$$
$$\theta N = N\theta = \{(23), (13), (12)\} \text{ for } \theta = (23), (13), \text{ and } (12).$$

We see that $gH = Hg$ may be true for all $g$ in $G$, as in Example 2, or it may be true for some but not all elements of $G$, as in Example 1. One can easily show that $hH = Hh = H$ for each $h$ in $H$. [See Problems 17(iii) and 18 of this section.]

Since the identity $\mathbf{e}$ of a group $G$ is in every subgroup $H$, an element $a = a\mathbf{e} = \mathbf{e}a$ is always in both its left coset $aH$ and its right coset $Ha$.

The following lemmas are part of the proof of the important Theorem of Lagrange on the orders of subgroups of finite groups. (See Theorem 1 below. Also see the biographical note on Lagrange after this section.)

### *Lemma 1*

*Let $H$ be a finite subgroup in $G$. Then the number of elements in any left coset $aH$ (or any right coset $Ha$) is equal to the order of $H$.*

### *Proof*

Let $H = \{h_1, h_2, \ldots, h_s\}$. The cancellation theorem tells us that either $ah_i = ah_j$ or $h_i a = h_j a$ implies that $h_i = h_j$. This shows that there are exactly $s$ distinct elements in each of

$$aH = \{ah_1, \ldots, ah_s\} \quad \text{and} \quad Ha = \{h_1 a, \ldots, h_s a\}.$$

*Lemma 2*

*Let H be a subgroup in G and let a and b be elements of G. Then the left cosets aH and bH either have no elements in common or they are identical subsets of G. The same is true of the right cosets Ha and Hb.*

***Proof***

Let $c$ be in both $aH$ and $bH$; that is, let $c = ah = bh_1$ with $h$ and $h_1$ in $H$. Then $a = bh_1h^{-1}$.

Now any element $ah_2$ of $aH$ may be expressed as

$$ah_2 = (bh_1h^{-1})h_2 = b(h_1h^{-1}h_2) = bh_3,$$

where $h_3 = h_1h^{-1}h_2$ is a product of elements of $H$ and hence is in $H$ by group closure. Therefore, $ah_2 = bh_3$ is in $bH$; and we have shown that every element of $aH$ is in $bH$. Similarly, every element of $bH$ is in $aH$ and so $aH = bH$. This completes the proof for left cosets; that for right cosets is similar.

***Theorem 1 Lagrange's Theorem***

*Let G be a finite group with order r. Then the order of each subgroup H in G and the order of each element a of G is an integral divisor of r. Also,* $g^r = \mathbf{e}$ *for every g in G.*

***Proof***

Let $H$ be a subgroup in $G$. Also, let ord $G = r$ and ord $H = s$. Since $G$ is finite and each left coset of $H$ in $G$ has $s$ elements with no overlapping, there are a finite number, say $t$, of left cosets of $H$ in $G$. Every $a$ of $G$ is in its left coset $aH$.

Together the $t$ nonoverlapping left cosets, each with $s$ elements, have all $r$ elements of $G$; hence $r = st$ and (ord $H$)|(ord $G$).

Let $a$ be an element of order $u$ in $G$. Then the cyclic subgroup $[a]$ has order $u$ and hence $u|r$. Then $r = uv$, with $v$ an integer, and

$$a^r = a^{uv} = (a^u)^v = \mathbf{e}^v = \mathbf{e}.$$

Thus we see that $g^r = \mathbf{e}$ for all $g$ in $G$.

***Corollary Groups of Prime Order***

*Let the order of G be a prime p. Then each element of G, except the identity, has order p; G is cyclic; G has* $p - 1$ *generators; and G is abelian.*

***Definition 2 Index of a Subgroup***

*The number of right cosets of a subgroup $H$ in $G$ is called the* **index** *of $H$ in $G$.*

In the proof of Lagrange's Theorem we have shown that the index of a subgroup $H$ in a finite group $G$ is the quotient of the order of $G$ by the order of $H$, that is,

$$(\text{index of } H \text{ in } G) = \frac{\text{ord } G}{\text{ord } H}, \qquad \text{ord } G = (\text{ord } H)(\text{index of } H \text{ in } G).$$

Hence the index of $H$ in $G$ is an integral divisor of the order of $G$.

For example, the subgroup $[(12)] = \{(1), (12)\}$ has order 2 and index 3 in the symmetric group $\mathbf{S}_3$ of order 6.

Beginning with the next section, we shall show that an especially important role in group theory is played by the type of subgroup characterized in the following:

***Definition 3 Normal Subgroup***

*Let $N$ be a subgroup in $G$ such that $aN = Na$ for each $a$ in $G$. Then $N$ is a* **normal subgroup** *in $G$.*

Examples 1 and 2 show that in $\mathbf{S}_3$ the subgroup $[(123)]$ is normal but the subgroup $[(12)]$ is not.

If one wishes to show that a given subgroup $H$ is not normal in a group $G$, it is not enough just to produce elements $h$ in $H$ and $g$ in $G$ such that $hg \neq gh$. One must show that there are elements $h$ in $H$ and $g$ in $G$ such that the element $hg$ of the right coset $Hg$ is not in the left coset $gH$ or such that $gh$ is not in $Hg$. A very useful variation of this technique is included in the following result:

***Theorem 2 Equivalent Conditions for Normality***

*A subgroup $H$ is normal in a group $G$ if and only if $g^{-1}hg$ is in $H$ for every $h$ in $H$ and $g$ in $G$.*

The proof of Theorem 2 is left to the reader as Problem 20 of this section.

This result frequently provides the easiest way to show that a given subgroup $H$ is not normal in $G$, as we next illustrate.

***Example 3***

Using Theorem 2, one sees that $H = \{(1), (12)\}$ is not a normal subgroup in $\mathbf{S}_3$ since $(12)$ is in $H$, $(13)$ is in $\mathbf{S}_3$, and $(13)^{-1}(12)(13) = (23)$ is not in $H$.

A normal subgroup may be called an ***invariant subgroup***, or a ***self conjugate subgroup***, in some texts; this terminology will conform with the definition of "conjugate subgroup" in Section 3.2.

It is not necessary to distinguish between left and right cosets in an abelian group; this includes all additively written groups. Hence every subgroup in an abelian group is a normal subgroup. However, a subgroup $H$ in $G$ need not be normal when $H$ is abelian and $G$ is not. (See Example 1.)

In an additive group $G$, the coset for $a$ of $N$ in $G$ is denoted by $a + N$. For example, let $H$ be the cyclic subgroup $[3]$ in the additive group $\mathbf{Z}$ of the integers. Then we easily see that there are exactly 3 cosets of $H$ in $\mathbf{Z}$ and that they are

$$0 + H = 3 + H = -3 + H = \cdots = H = \{\ldots, -6, -3, 0, 3, 6, \ldots\},$$
$$1 + H = 4 + H = -2 + H = \cdots = \{\ldots, -5, -2, 1, 4, 7, \ldots\},$$
$$2 + H = 5 + H = -1 + H = \cdots = \{\ldots, -4, -1, 2, 5, 8, \ldots\}.$$

Also, if we let $N$ be the subgroup $[2]$, in $\mathbf{Z}$, then the two cosets of $N$ in $\mathbf{Z}$ are $0 + N = N$, which is the set of the even integers, and $1 + N$, which is the set of the odd integers.

## *Problems*

Problem 3 below is cited in Section 2.13.

1. Let $H = \{(1), (12)(34)\}$ and $K = \{(1), (12)(34), (13)(24), (14)(23)\}$.
   (i) Find two left cosets of $H$ in the alternating group $\mathbf{A}_4$ that are also right cosets of $H$ in $\mathbf{A}_4$.
   (ii) Show that $H$ is not a normal subgroup in $\mathbf{A}_4$.
   (iii) Show that $H$ is a normal subgroup in $K$.
2. Let $H = [(123)] = \{(1), (123), (132)\}$.
   (i) What is the index $t$ of $H$ in $\mathbf{A}_4$?
   (ii) Find the $t$ left cosets of $H$ in $\mathbf{A}_4$. (See the table in Section 2.10.)
   (iii) Find the $t$ right cosets of $H$ in $\mathbf{A}_4$.
   (iv) Is $H$ a normal subgroup in $\mathbf{A}_4$? Explain.
3. Let $N = \{(1), (12)(34), (13)(24), (14)(23)\}$.
   (i) Does $N$ contain every $\alpha$ of $\mathbf{A}_4$ with $\alpha^2 = (1)$?
   (ii) Is the cube $\beta^3$ of every $\beta$ of $\mathbf{A}_4$ in $N$?
   (iii) What is the index $t$ of $N$ in $\mathbf{A}_4$?
   (iv) Find the $t$ left cosets and the $t$ right cosets of $N$ in $\mathbf{A}_4$.
   (v) Is $N$ a normal subgroup in $\mathbf{A}_4$? Explain.
4. Find all the cosets of $N = [a^3]$ in a cyclic group $G = [a]$ of order 12.
5. Explain why a subgroup in the center of $G$ is always a normal subgroup in $G$.

6. Show that a subgroup $H = \{\mathbf{e}, h\}$ of order 2 is a normal subgroup in $G$ if and only if $H$ is contained in the center of $G$.
7. For $n > 1$, describe the left cosets and the right cosets of the alternating subgroup $\mathbf{A_n}$ in the symmetric group $\mathbf{S_n}$.
8. Prove that a subgroup of order $m$ in a group of order $2m$ is always a normal subgroup.
9. (i) Use Lagrange's Theorem to show that a group $G$ of order 4 is either cyclic or such that $g^2 = \mathbf{e}$ for all $g$ in $G$.
   (ii) Explain why every group of order 4 must be abelian.
10. Show that a group $G$ of order 27 is either cyclic or such that $g^9 = \mathbf{e}$ for all $g$ in $G$. Generalize on this and Problem 9(i).
11. Explain why the following are all the subgroups in the symmetric group $\mathbf{S}_3$: $\{(1)\}$, three cyclic subgroups of order 2, one cyclic subgroup of order 3, and $\mathbf{S}_3$ itself.
12. (a) Explain why the octic group has no subgroup of order 3, 5, 6, or 7.
    (b) Find all the normal subgroups in the octic group. [*Hint*: See Problem 9 of Section 2.9 and Problems 5, 6, and 8 of this section.]
13. Show the following:
    (i) A cyclic group of order 22 has one element of order 2 and ten elements of order 11.
    (ii) A noncyclic group of order 22 has 21 elements each of which has order 2 or 11.
    (iii) A group of order 110 must have a cyclic subgroup of order 2, 5, or 11.
14. Generalize on part (iii) of Problem 13.
15. (i) Give the set of orders of elements of the alternating group $\mathbf{A}_6$.
    (ii) What is the smallest positive integer $m$ such that $\alpha^m = (1)$ for all $\alpha$ in $\mathbf{A}_6$?
16. Do both parts of Problem 15 with $\mathbf{A}_6$ replaced by a cyclic group of order 360 [and (1) replaced by $\mathbf{e}$].
17. Let $H$ be a subgroup in $G$.
    (i) Show that $aH = bH$ if and only if $a^{-1}b$ is in $H$.
    (ii) If $aH = bH$, does $b^{-1}a$ have to be in $H$?
    (iii) Show that $hH = eH = H$ for each $h$ in $H$.
18. Do the analogue of Problem 17 for right cosets.
19. Let $H$ be a subgroup in $G$. Prove that $H$ is normal in $G$ if and only if for every $g$ in $G$ and $h$ in $H$ there exist $h_1$ and $h_2$ in $H$ such that

$$hg = gh_1 \quad \text{and} \quad gh = h_2 g.$$

20. Let $H$ be a subgroup in $G$. Prove that $H$ is normal in $G$ if and only if for every $g$ in $G$ and $h$ in $H$ the element $g^{-1}hg$ is in $H$. (This proves Theorem 2.)
21. Let $N$ be a normal subgroup in $G$. Let $a$ and $b$ be in $G$ and let $n_1$ and $n_2$ be in $N$. Prove that

$$an_1bn_2 = abn_3$$

with $n_3$ in $N$.

22. Let $n \geq 3$ and let every 3-cycle $(abc)$ of $\mathbf{S_n}$ be in a subgroup $H$ in $\mathbf{S_n}$. Prove the following:
    (i) Every even permutation of $\mathbf{S_n}$ is in $H$.
    (ii) $H$ is either $\mathbf{A_n}$ or $\mathbf{S_n}$.
23. Let $H$ be the subgroup $\{(1), (123), (132)\}$ in $\mathbf{A_4}$. Show that there are 9 distinct products $\alpha\alpha'$ in which $\alpha$ is in $H$ and $\alpha'$ is in the left coset $(12)(34)H$. (The table in Section 2.10 may be helpful.)
24. Let $N$ be the subgroup $\{(1), (12)(34), (13)(24), (14)(23)\}$ in $\mathbf{A_4}$. Show that there are only 4 distinct permutations among the 16 products $\alpha\alpha'$ in which $\alpha$ is in $N$ and $\alpha'$ is in the left coset $(123)N$.

*25. Let $s$ be a fixed positive integer. Let $H$ consist of all elements $a^s$ with $a$ in a group $G$. Given that $H$ is a subgroup in $G$, prove that $H$ is normal in $G$.

26. Let $K$ be a subgroup in $H$ and let $H$ be a subgroup in $G$. What are the possibilities for ord $H$ if ord $K = 6$ and ord $G = 288$?
27. In a nonabelian group $G$, let $a$ be an element that is not in the center $C$ and let $C_a = \{g : g \in G \text{ and } ga = ag\}$ be the centralizer of $a$ in $G$. Explain why $C$ must be a proper subgroup in $C_a$ and $C_a$ must be a proper subgroup in $G$.
28. Use Lagrange's Theorem to prove that the center $C$ cannot have prime index in a finite group $G$.
29. In a group $G$, let $H$ be a subgroup and $a$ an element of order 2. Explain why the left coset $aH$ consists of the inverses of the elements of the right coset $Ha$.
30. Let $H$ and $K$ be subgroups in $G$. Show the following:
    (a) The intersection $I = H \cap K$ is a subgroup in $G$.
    (b) If $H$ and $K$ are normal subgroups in $G$, so is $I$.

---

***Comte Joseph Louis de Lagrange*** *(1736–1813)* *Lagrange, though of French parentage and generally considered to be a French mathematician, was born in Turin, Italy, and spent his early years there. In 1766 he succeeded Euler as mathematician at the Berlin Academy. He was appointed to the post by Frederick the Great, who suggested in the invitation that the greatest of European geometers should be close to the greatest of kings!*

*Lagrange made important contributions to analysis* (**Théorie des Fonctions Analytiques**, *1797) and the calculus of variations. In these areas he introduced levels of rigor previously unknown in analysis. Much of his work parallels that of Euler but Euler's style was less concerned with rigor. Lagrange was the first to prove that any positive integer can be expressed as the sum of at most four squares. This is the first case of Waring's Problem, a problem to which Euler had made* *contributions. He also was the first to prove (in 1770) the so-called Wilson's Theorem, a result first announced by Waring in his* **Meditationes Algebraicae** *(1770) as being the work of Waring's student, John Wilson.*

*Also in 1770 he published a method for solving algebraic equations, which essentially reduced the problem of solving a given equation to that of solving a related equation called a resolvent. In the case of the quadratic, cubic, and quartic equations, the resolvent was of lower degree and this provided a systematic technique for solving polynomial equations of degree 2, 3, and 4. However, the Lagrange resolvent for the fifth degree polynomial has degree 6. This may have helped to dispel the misguided confidence of those who expected all polynomial equations to be solvable by radicals. However, it was still possible to search for a solution of the general quintic by other methods until Ruffini, Abel, and Galois settled this question completely.*

*The work of Lagrange on permutations of the roots of a polynomial equation provided a foundation for the group-theoretic approach of Galois.*

## 2.12 Quotient Groups

We show below that the collection of cosets of a normal subgroup $N$ in a group $G$ becomes a group when the product of such cosets is defined appropriately. This concept helps us to study the structure of groups. In Chapter 4 we shall use an extension of this process to create new algebraic structures from the integers and from polynomials.

First we present a result that is helpful in several places.

**Theorem 1 Subgroups with Index 2**

*Let $H$ be a subgroup of order $m$ in a group $G$ of order $2m$; that is, let $H$ have index 2 in the finite group $G$. Then $H$ is a normal subgroup in $G$.*

***Proof***

Let $K$ be the subset of the $m$ elements of $G$ that are not in $H$. If $a \in H$, both $aH$ and $Ha$ are $H$ and we have $aH = Ha$. If $a$ is in $G$ but not in $H$, it follows from Lemmas 1 and 2 of Section 2.11 that each of $aH$ and $Ha$ is a subset of $G$ with $m$ elements all different from every element of $H$; in this case $aH = K = Ha$. Hence $aH = Ha$ for all $a$ in $G$ and $H$ is normal in $G$.

Next we look at two examples. In the first one, we let $A$ denote the alternating subgroup in a symmetric group $\mathbf{S_n}$ with $n > 1$. Also, we let $B$ consist of the permutations of $\mathbf{S_n}$ not in $A$, namely, the odd permutations. Since $A$ has index 2 in $\mathbf{S_n}$, it follows from Theorem 1 and its proof that $A$ is a normal subgroup in $\mathbf{S_n}$ and that the two (left or right) cosets of $A$ in $\mathbf{S_n}$ are $A$ and $B$.

The product of two elements of $A$, or of two elements of $B$, is in $A$; while the product of an element of $A$ and an element of $B$ (in either order) is in $B$. (See Problems 10–12 of Section 2.8.) These statements may be represented by the following table:

| | $A$ | $B$ |
|---|---|---|
| $A$ | $A$ | $B$ |
| $B$ | $B$ | $A$ |

This clearly is the table for a group $\{A, B\}$ of order 2 with $A$ as the identity.

Next let $N$ be the normal subgroup

$$[3] = \{\ldots, -9, -6, -3, 0, 3, 6, 9, \ldots\}$$

in the additive group $\mathbf{Z}$ of the integers. We use the notation

$$P = 1 + N = \{\ldots, -8, -5, -2, 1, 4, 7, 10, \ldots\},$$

$$Q = 2 + N = \{\ldots, -7, -4, -1, 2, 5, 8, 11, \ldots\}$$

for the other cosets of $N$ in $\mathbf{Z}$. Then the table

| + | $N$ | $P$ | $Q$ |
|---|---|---|---|
| $N$ | $N$ | $P$ | $Q$ |
| $P$ | $P$ | $Q$ | $N$ |
| $Q$ | $Q$ | $N$ | $P$ |

for a group $\{N, P, Q\}$ of order 3, with $N$ as identity, may be interpreted as summarizing certain statements about the three cosets $N$, $P$, and $Q$. One of these statements is that the sum of an integer in $P$ and an integer in $Q$ is always an integer in $N$.

We now generalize on these examples. In the following definition and lemma, we use multiplicative notation for the operation in $G$ and leave it to the reader to make the appropriate changes for additive groups.

***Definition 1 Product of Left Cosets***

*Let $H$ be a subgroup in $G$. Then the product $aH \cdot bH$ of left cosets $aH$ and $bH$ is the subset of $G$ consisting of all products $uv$ with $u$ in $aH$ and $v$ in $bH$.*

***Lemma 1 Product of Cosets of a Normal Subgroup***

*Let $N$ be a normal subgroup in $G$. Then the product $aN \cdot bN$ of cosets $aN$ and $bN$ is the coset $(ab)N$.*

***Proof***

Since the identity **e** of $G$ is in $N$, the product $ab \cdot n = a\mathbf{e} \cdot bn$ is in the coset product $aN \cdot bN$ for every $n$ in $N$; thus every element of $(ab)N$ is in $aN \cdot bN$. For the converse, we let $an$ and $bn'$ be any elements of $aN$ and $bN$, respectively. Since $N$ is normal in $G$, $Nb = bN$ and hence $nb = bn_1$ for some $n_1$ in $N$. Then

$$an \cdot bn' = a(nb)n' = a(bn_1)n' = (ab)(n_1 n') = (ab)n_2,$$

where $n_2 = n_1 n'$ is in $N$. This shows that each element of the coset product $aN \cdot bN$ is in the coset $(ab)N$. Since we have already shown that every element of $(ab)N$ is in $aN \cdot bN$, we have the desired multiplication formula

$$aN \cdot bN = (ab)N$$

for the cosets of a normal subgroup $N$ in $G$.

It should be kept in mind that if $H$ is not normal in $G$, then the product $aH \cdot bH$ (of left cosets of $H$ in $G$) need not be a coset of $H$ in $G$. (For an example see Problem 1 of this section.)

If $N$ is a normal subgroup in $G$, it follows from Lemma 1 that the product $AB$, of cosets $A$ and $B$ of $N$, can be determined by performing one multiplication $ab$ of elements $a$ and $b$ of $A$ and $B$, respectively, and noting that $AB$ is the coset $(ab)N$ containing $ab$. The lemma shows that the coset product is well defined by the formula

$$aN \cdot bN = (ab)N.$$

***Theorem 2 Quotient Group $G/N$***

*Let $N$ be normal in $G$ and let $G/N$ denote the collection of all the cosets of $N$ in $G$. Then the operation of coset multiplication makes $G/N$ into a group. (This group is called the* **quotient group**, *or* **factor group**, *of $G$ by $N$.)*

***Proof***

Let $\mathbf{e}$ be the identity of $G$. Then

$$\mathbf{e}N \cdot bN = (\mathbf{e}b)N = bN = (b\mathbf{e})N = bN \cdot \mathbf{e}N$$

shows that $\mathbf{e}N = N$ is the identity of $G/N$. Also,

$$aN \cdot a^{-1}N = (aa^{-1})N = \mathbf{e}N = (a^{-1}a)N = a^{-1}N \cdot aN$$

tells us that each $aN$ in $G/N$ has an inverse $a^{-1}N$ in $G/N$.

For associativity in $G/N$ we note that $(aN \cdot bN)cN$ and $aN(bN \cdot cN)$ consist, respectively, of all products $(uv)w$ and all products $u(vw)$ with $u$ in $aN$, $v$ in $bN$, and $w$ in $cN$. Since $(uv)w = u(vw)$ by associativity in $G$, the triple products of cosets are equal and $G/N$ is a group.

Lagrange's Theorem and the concepts of this section are powerful tools for investigating the possibilities for groups whose orders are fairly small. As an example, we now illustrate how one could characterize all groups of order 8.

Let $G$ be such a group. By Lagrange's Theorem, an element of $G$ must have 1, 2, 4, or 8 as its order. A slightly stronger statement is that each of the seven elements of $G$ that are not the identity has order 2, 4, or 8.

If $G$ has an element with order 8, $G$ is cyclic and its structure is known to us. So we now restrict ourselves to groups of order 8 with no element of order 8.

If $a$ is an element of order 4 in a group, $a^3$ also has order 4 and $a^3 \neq a$. Hence the number of elements of order 4 in a group is even, when it is finite. This means that the number of elements of order 4 in a group of order 8 is 0, 2, 4, or 6.

In the example that follows, we characterize one of these groups of order 8 and leave the other cases to the reader as several problems below.

***Example 1***

Assume that there exists a group $G$ of order 8 with six elements having order 4 and construct its multiplication table.

***Solution***

Let $a$ be one of the elements with order 4 in $G$. Then

$$N = [a] = \{\mathbf{e}, a, a^2, a^3\}$$

is a subgroup with index 2 in $G$. It follows from Theorem 1 that $N$ is normal in $G$ and hence that the quotient group $G/N$ exists.

Since $\mathbf{e}$ has order 1 and $a^2$ has order 2, the other six elements of $G$ are those with order 4. Let $b$ be any one of the four elements of $G$ not in $N$. Then $b$ has order 4 and $b^2$ has order 2. But $a^2$ is the only element with order 2 in $G$; hence $b^2 = a^2$. Also, the elements of the coset $Nb = \{b, ab, a^2b, a^3b\}$ together with those of $N$ make up all of

G. That is,

$$G = \{\mathbf{e}, a, a^2, a^3, b, ab, a^2b, a^3b\}. \tag{1}$$

The key to showing that now there is only one possibility for the table of $G$ is the proof that only one of the elements displayed in (1) can equal $ba$.

Since $N$ is the identity coset in the quotient group $G/N = \{N, Nb\}$, we have $Nb \cdot N = Nb$. One of the elements in the coset product $Nb \cdot N$ is $ba$; hence

$$ba \in Nb = \{b, ab, a^2b, a^3b\}.$$

We next show that $ba = a^3b$ by demonstrating that $ba$ cannot equal $b$, $ab$, or $a^2b$ without introducing a contradiction.

The assumption that $ba = b$ leads to the contradiction $a = \mathbf{e}$; hence $ba \neq b$. Since $b^2 = a^2$, the equality $ba = ab$ would imply that

$$(ab)^2 = ab \cdot ab = ab \cdot ba = aa^2a = a^4 = \mathbf{e}.$$

But $(ab)^2 = \mathbf{e}$ contradicts the fact that $ab$ has order 4; hence $ba \neq ab$. Under the assumption $ba = a^2b$, we would have

$$ba^2 = (ba)a = (a^2b)a = a^2(ba) = a^2a^2b = b$$

and the equality $ba^2 = b$ would imply $a^2 = \mathbf{e}$, which is also a contradiction.

As the other cases lead to contradictions, we must have $ba = a^3b$. Now multiplication is completely determined in $G$. For example,

$$ab \cdot a^2 = a(ba)a = a(a^3b)a = a^4(ba) = a^3b.$$

Similarly, one obtains all the entries of Table 2.3.

***Table 2.3***

| | $\mathbf{e}$ | $a$ | $a^2$ | $a^3$ | $b$ | $ab$ | $a^2b$ | $a^3b$ |
|---|---|---|---|---|---|---|---|---|
| $\mathbf{e}$ | $\mathbf{e}$ | $a$ | $a^2$ | $a^3$ | $b$ | $ab$ | $a^2b$ | $a^3b$ |
| $a$ | $a$ | $a^2$ | $a^3$ | $\mathbf{e}$ | $ab$ | $a^2b$ | $a^3b$ | $b$ |
| $a^2$ | $a^2$ | $a^3$ | $\mathbf{e}$ | $a$ | $a^2b$ | $a^3b$ | $b$ | $ab$ |
| $a^3$ | $a^3$ | $\mathbf{e}$ | $a$ | $a^2$ | $a^3b$ | $b$ | $ab$ | $a^2b$ |
| $b$ | $b$ | $a^3b$ | $a^2b$ | $ab$ | $a^2$ | $a$ | $\mathbf{e}$ | $a^3$ |
| $ab$ | $ab$ | $b$ | $a^3b$ | $a^2b$ | $a^3$ | $a^2$ | $a$ | $\mathbf{e}$ |
| $a^2b$ | $a^2b$ | $ab$ | $b$ | $a^3b$ | $\mathbf{e}$ | $a^3$ | $a^2$ | $a$ |
| $a^3b$ | $a^3b$ | $a^2b$ | $ab$ | $b$ | $a$ | $\mathbf{e}$ | $a^3$ | $a^2$ |

Does Table 2.3 make the $G$ of display (1) into a group; that is, does there exist a group of order 8 with six elements having order 4? Let us see if the group axioms are satisfied by the operation of this table.

It is clear from the table that $G$ is closed under multiplication, $\mathbf{e}$ is the identity of $G$, each of $\mathbf{e}$ and $a^2$ is its own inverse, and the other six elements (the elements of order 4) pair off with their inverses as $\{a, a^3\}$, $\{b, a^2b\}$, and $\{ab, a^3b\}$.

To complete the proof that Table 2.3 is a group table, we need to show associativity; that is, $(xy)z = x(yz)$ for all $x$, $y$, and $z$ in $G$. There are 8 choices for each of $x$, $y$, and $z$ and hence a total of $8 \cdot 8 \cdot 8 = 512$ cases. Many of these cases are trivial but this task does not appeal to us. Instead, we present a group of 8 permutations in which there are 6 elements of order 4. (The technique for finding such a group is described in Section 3.4.)

Let $\alpha = (1234)(5876)$ and $\beta = (1537)(2648)$. It is relatively easy to verify the following:

(1) $\alpha$ and $\beta$ have order 4.

(2) $\beta$ is not equal to any permutation in $[\alpha] = \{\boldsymbol{\varepsilon}, \alpha, \alpha^2, \alpha^3\}$.

(3) $\beta\alpha = \alpha^3\beta$.

Then $G' = \{\boldsymbol{\varepsilon}, \alpha, \alpha^2, \alpha^3, \beta, \alpha\beta, \alpha^2\beta, \alpha^3\beta\}$ has 8 distinct elements and its multiplication table is Table 2.3 with the $\mathbf{e}$'s, $a$'s, and $b$'s replaced by $\boldsymbol{\varepsilon}$'s, $\alpha$'s, and $\beta$'s, respectively. But now associativity is no problem since we know that composition of permutations is associative.

## *Problems*

1. Let $H$ be the subgroup $\{(1), (12)(34)\}$ in the alternating group $\mathbf{A_4}$. Let $\alpha_3 = (13)(24)$ and $\alpha_5 = (123)$. Show that there are four distinct permutations in the product

$$\alpha_3 H \cdot \alpha_5 H$$

of left cosets and hence that this product is not a coset of $H$ in $\mathbf{A_4}$. (This should not be surprising since $H$ is not a normal subgroup in $\mathbf{A_4}$.)

2. Let $G = \{(1), (12)(34), (13)(24), (14)(23)\}$ and $H = \{(1), (12)(34)\}$. Use the definition of the product of left cosets and Table 2.2 of Section 2.10 to show that the product of any two left cosets of $H$ in $G$ is also a left coset of $H$ in $G$. (This should not be surprising since here $H$ is a normal subgroup.)

3. Let $N$ be the normal subgroup {(1), (12)(34), (13)(24), (14)(23)} in $\mathbf{A_4}$. Let $P$ and $Q$ be the cosets $\alpha_5 N$ and $\alpha_9 N$, respectively, where $\alpha_5 = (123)$ and $\alpha_9 = (132)$. Make the multiplication table for the quotient group $A_4/N = \{N, P, Q\}$.

4. Let $G$ be the octic group $\{\varepsilon, \rho, \rho^2, \rho^3, \phi, \rho\phi, \rho^2\phi, \rho^3\phi\}$, where $\rho = (1234)$ and $\phi = (24)$, and let $C = \{\varepsilon, \rho^2\}$ be the center of $G$.
   (i) Explain why $C$ is normal in $G$ and hence why the quotient group $G/C$ exists.
   (ii) Is $G/C$ cyclic? Explain.
   (iii) Is $G/C$ abelian? Explain.

5. For a fixed positive integer $m$, let $[m]$ be the cyclic group generated by $m$ in the additive group $\mathbf{Z}$ of the integers; that is, let $[m]$ consist of the integral multiples of $m$.
   (i) Explain why $[m]$ is a normal subgroup in $\mathbf{Z}$.
   (ii) What is the index of $[m]$ in $\mathbf{Z}$?
   (iii) Explain why the cosets of [4] in $\mathbf{Z}$ are

$$0 + [4],\ 1 + [4],\ 2 + [4], \text{ and } 3 + [4].$$

   (iv) Tabulate the operation of the quotient group $\mathbf{Z}/[4]$, using $\bar{0}$, $\bar{1}$, $\bar{2}$, and $\bar{3}$ to denote the cosets listed in (iii).

6. Do the analogue of Problem 5(iv) for $\mathbf{Z}/[5]$.

7. Let $G = \mathbf{S_3}$ and $H = [(123)]$. Show that $H$ is a normal subgroup in $G$ of index 2 and tabulate the operation of $G/H$.

8. Do Problem 7 with $G$ the octic group and $H = [(1234)]$.

9. Do Problem 7 with $G$ the multiplicative group of the nonzero real numbers and $H$ the subgroup consisting of all the positive real numbers.

10. Do Problem 7 with $G$ the additive group of the integers and $H$ the subgroup [2] of all the even integers.

11. In each of Problems 7, 8, and 9, show that $a^2$ is in $H$ for all $a$ in $G$.

12. Let $N$ be the normal subgroup {(1), (12)(34), (13)(24), (14)(23)} in $\mathbf{A_4}$. Show that all cubes of elements of $\mathbf{A_4}$ are in $N$.

13. Let $N$ be a normal subgroup of prime index $p$ in a group $G$. Explain why the quotient group $G/N$ is cyclic.

14. Let $N$ be a normal subgroup in $G$. Show that $G/N$ is abelian if and only if $a^{-1}b^{-1}ab$ is in $N$ for all $a$ and $b$ in $G$.

15. Let $\mathbf{R}$ be the additive group of the real numbers, let $N$ be its cyclic subgroup $[2\pi]$ (where $\pi$ is the number 3.14159...), and let $T$ be the quotient group $\mathbf{R}/N$. In $T$, let $\bar{a}$ denote the coset $a + N$. Explain why the following are true:

(a) $\overline{2m\pi} = \overline{0}$ for all integers $m$.
(b) For every $\bar{a}$ in $T$ there is a real number $b$ with $-\pi < b \leq \pi$ and $\bar{a} = \bar{b}$.
(c) The order of $\overline{\pi/12}$ in $T$ is 24.
(d) $\bar{1}$ has infinite order in $T$.

16. Let **R** be the additive group of the real numbers, **Z** be its cyclic subgroup $[1] = \{\ldots, -2, -1, 0, 1, 2, \ldots\}$, and $W$ be the quotient group **R**/**Z**.
(a) What is the order of the coset $(-2/5) + \mathbf{Z}$ in $W$?
(b) Use the fact that $\sqrt{3}$ is irrational to show that the coset $\sqrt{3} + Z$ does not have finite order in $W$.

17. Let $a$ be an element of order 4 in a group $G$ of order 8. Let $b$ be in $G$ but not in the subgroup $N = [a] = \{\mathbf{e}, a, a^2, a^3\}$. Show that $b^2$ is in $N$.

18. Let $a$, $b$, and $G$ be as in Problem 17. Show that $G$ is cyclic, with $b$ as a generator, if $b^2$ equals $a$ or $a^3$.

19. Let $a$, $b$, and $G$ be as in Problem 17. Show that $b^2 = a^2$ if $b$ has order 4.

*20. Let $G = \{\mathbf{e}, a, a^2, a^3, b, ab, a^2b, a^3b\}$ be a group of order 8 with $a$ having order 4. Tabulate the operation of $G$ under each of the following assumptions:
(a) $ba \neq ab$ and the only elements of order 4 in $G$ are $a$ and $a^3$.
(b) The elements of order 4 in $G$ are $a$, $a^3$, $ab$, and $a^3b$.

*21. Let $G = \{\mathbf{e}, u, v, w, vw, uw, uv, uvw\}$ be a group of order 8 with $g^2 = \mathbf{e}$ for all $g$ in $G$. Make a multiplication table for $G$.

***Nicolo of Brescia*** (***Tartaglia***) *(1499?–1557)* *Tartaglia (which means "the stammerer") was a teacher of mathematics in Brescia and Venice. In 1541 he discovered a general method for solving cubic equations. He made the mistake of delaying publication, however, and instead passed along the method to Hieronimo Cardano (1501–1576), a physician and professor of mathematics, who published the result (with due credit to Tartaglia) in his* **Ars Magna** *of 1545. The method was thereafter unfairly referred to as Cardano's method. Cardano's student, Ludovico Ferrari of Bologna (1522–1565), discovered the general method for solving quartics, that is, fourth degree equations, a method also first described in Cardano's* **Ars Magna.** *It was the success of these men in solving by radicals equations of degree higher than two that encouraged many to seek solutions of equations of degree higher than four. It was not until 1799 that Ruffini first proved this to be a futile search.*

***Paolo Ruffini*** *(1765–1822)* *Ruffini was an Italian physician who did mathematics on the side. He anticipated the idea of a group (which he called a permutation) in his book* **Teoria Generale delle Equazioni**, *published in Bologna in 1799. In this he claims to have proved the insolvability of the quintic, that is, fifth degree equation, by radicals, although his proof was not universally accepted. The full title of the book (translated) is: General Theory of Equations in Which the Algebraic Solution of General Equations of Degree Higher than Four Is Demonstrated To Be Impossible.*

*The most efficient modern proofs of insolvability of the general quintic are still very difficult. Also, mathematics of any period is written with assumptions that certain things are known to prospective readers. Hence there would be a large subjective element in any present day evaluation of Ruffini's work. However, there is little doubt that it was Abel's paper that settled the matter for the mathematical community.*

## 2.13 Solvable Groups

> *"There is no use trying," she said: "one can't believe impossible things." "I daresay you haven't had much practice," said the Queen. "When I was your age, I always did it for half-an-hour a day. Why, sometimes I've believed as many as six impossible things before breakfast."*
>
> *Charles Lutwidge Dodgson (Lewis Carroll)*

Techniques for solving second degree polynomial equations

$$ax^2 + bx + c = 0$$

go back at least to the times of the ancient Greeks. In the first half of the sixteenth century, Italian mathematicians found formulas expressing the roots of the general third and fourth degree polynomial equations,

$$ax^3 + bx^2 + cx + d = 0,$$

$$ax^4 + bx^3 + cx^2 + dx + e = 0,$$

in terms of the coefficients using addition, subtraction, multiplication, division, and extraction of roots. (Techniques for solving polynomial equations of degree 2, 3, and 4 are given in Section 5.7.)

Then for approximately 300 years, some of the greatest mathematicians of all time tried to find a formula of this type for the roots of the general fifth degree polynomial equation.

But no such formula is possible, as was proved by Niels Henrik Abel in a paper published in 1824. Paolo Ruffini gave a prior demonstration, but it was not generally accepted as a convincing proof. (See the biographical note on Ruffini at the end of Section 2.12.)

Modern treatments of solvability of polynomial equations are based on the work of Évariste Galois, in which group theory plays a major role. We shall go into this matter further in Section 5.12. A key concept is the following:

***Definition 1 Solvable Finite Group***

*A finite group G is* **solvable** *if there exists a chain*

$$G_0, G_1, G_2, \ldots, G_r$$

*of groups such that* $G_0 = G$, $G_{i+1}$ *is a normal subgroup with prime index in* $G_i$ *for* $0 \le i \le r - 1$, *and* $G_r = \{\mathbf{e}\}$ *has order* 1.

Note that, in Definition 1, the ratio (ord $G_i$)/(ord $G_{i+1}$) must be a prime number.

In Example 1 we shall show that the symmetric group $\mathbf{S_4}$ is solvable. Theorem 2 proves that the alternating group $\mathbf{A_5}$ is not solvable. It is left to the reader, in the problems for this section, to show that $\mathbf{A_4}$ is solvable and $\mathbf{S_5}$ is not.

Our proofs of nonsolvability are based on the counting of the number of squares, cubes, etc. in a given group and the following simple but powerful result:

***Theorem 1 Normal Subgroups of Index** m **Contain All** m**th Powers***

*Let N be a normal subgroup of index m in G. Then:*

*(a) In G/N,* $C^m = N$ *for all cosets C of N in G.*

*(b) For all g in G,* $g^m$ *is in N.*

*Proof*

The order of the quotient group $G/N$ is the number of cosets of $N$ in $G$, that is, the index of $N$ in $G$. It therefore follows from Lagrange's Theorem that $C^m$ is the identity coset $N$, for all cosets $C$.

Since $C^m$ consists of all products $c_1c_2\cdots c_m$ with each $c_i$ in $C$, we have as a special case that $c^m$ is in $N$ for each $c$ in every coset $C$ of $N$. But each $g$ of $G$ is in some coset $C$; hence $g^m$ is in $N$ for all $g$ in $G$.

### *Corollary Subgroups with Index 2*

*Let $H$ be a subgroup with index 2 in $G$. Then $g^2$ is in $H$ for all $g$ in $G$.*

*Proof*

$H$ is normal in $G$ by Theorem 1 of Section 2.12. Then it follows from Theorem 1 in the present section, with $m = 2$, that $H$ contains all squares of elements of $G$.

Theorem 1 helps one to find normal subgroups of a given index, when they exist (see Problem 1 below), and also can be used to prove that no such subgroup exists in certain groups. (See Lemmas 1, 2, and 3 below.)

Now we return to questions of solvability.

*Example 1*

Show that the symmetric group $\mathbf{S_4}$ is solvable.

*Solution*

Let $G_0 = \mathbf{S_4}$, $G_1 = \mathbf{A_4}$,

$$G_2 = \{(1), (12)(34), (13)(24), (14)(23)\},$$

$G_3 = \{(1), (12)(34)\}$, and $G_4 = \{(1)\}$. For $i = 0$, 2, and 3, $G_{i+1}$ has index 2 in $G_i$ and hence is normal in $G_i$ by Theorem 1 of Section 2.12. Also, $G_2$ is a normal subgroup with index 3 in $G_1$. (See Problem 3 of Section 2.11.) Since $G_{i+1}$ is a normal subgroup with prime index in $G_i$ for $0 \le i \le 3$, we have proved that $\mathbf{S_4}$ is solvable.

### *Definition 2 Simple Group*

*A group $G$ is* **simple** *if it has only itself and $\{\mathbf{e}\}$ as normal subgroups and $G \neq \{\mathbf{e}\}$.*

For example, it follows from Lagrange's Theorem that every group of prime order is simple.

***Theorem 2 Unsolvable Simple Groups***

*Let G be a finite simple group whose order is not prime. Then G is not solvable.*

***Proof***

This follows readily from Definitions 1 and 2.

It can be shown that the alternating groups $\mathbf{A_n}$ are simple for $n \geq 5$ and hence that these groups are not solvable. We content ourselves here with the outline of a proof, using a combinatorial technique, that $\mathbf{A_5}$ is not solvable. The proof that $\mathbf{A_n}$ is not solvable for $n \geq 5$ is asked for in Problem 17 of Section 3.5.

***Theorem 3 Unsolvable Alternating Group***

$\mathbf{A_5}$ *is not solvable.*

***Proof***

It follows from Definition 1 that a finite solvable group $G$ must have a normal subgroup of prime index in $G$. Since the order of $\mathbf{A_5}$ is $60 = 2^2 \cdot 3 \cdot 5$, a subgroup with prime index would have to have index 2, 3, or 5. It therefore suffices to show that $\mathbf{A_5}$ has no normal subgroup with index 2, 3, or 5. With the help of Theorem 1, we deal with these cases in the following lemmas.

***Lemma 1***

$\mathbf{A_5}$ *has no normal subgroup with index 2.*

***Proof***

For every $\alpha$ in $\mathbf{A_5}$, $\alpha^2$ must be in any normal subgroup with index 2 in $\mathbf{A_5}$ by Theorem 1. The twenty 3-cycles $(abc)$ in $\mathbf{A_5}$ are squares in $\mathbf{A_5}$ since $(abc) = (acb)^2$. The formula $(abcde) = (adbec)^2$ shows that the twenty-four 5-cycles in $\mathbf{A_5}$ are squares. With the identity, which is a square, this makes a total of at least 45 squares in $\mathbf{A_5}$. Since a subgroup with index 2 in $\mathbf{A_5}$ would have only 30 elements, there are, therefore, too many squares to fit into a subgroup of index 2. This proves the lemma.

We leave the proofs of the following two results to the reader as Problems 5 and 6 of this section.

*Lemma 2*

$\mathbf{A}_5$ *has no normal subgroup with index 3.*

*Lemma 3*

$\mathbf{A}_5$ *has no normal subgroup with index 5.*

We also leave the proof of the following result to the reader as Problem 11 below.

***Theorem 4 Unsolvable Symmetric Group***

$\mathbf{S}_5$ *is not solvable.*

Lagrange's Theorem states that the order of a subgroup in a finite group $G$ is an integral divisor of the order of $G$. Paolo Ruffini was the first to show that the converse is not true, that is, there exist finite groups $G$ for which some positive integral divisor of the order of $G$ is not the order of a subgroup in $G$. We illustrate this in the following example (also see Problem 3 below).

***Example 2***

Prove that the group $\mathbf{A}_5$, of order 60, has no subgroup $H$ of order 30.

***Solution***

A subgroup $H$ of order 30 would have index 2 and hence be normal in $\mathbf{A}_5$ by Theorem 1 of Section 2.12. But no normal subgroup with index 2 exists by Lemma 1 above. Hence there is no such $H$.

## Problems

1. (i) Which permutations of $\mathbf{A}_4$ are cubes of elements of $\mathbf{A}_4$ ?
   (ii) Is there a normal subgroup with index 3 (that is, order 4) in $\mathbf{A}_4$ besides $\{(1), (12)(34), (13)(24), (14)(23)\}$? Explain.
2. Prove that $\mathbf{A}_4$ is the only subgroup with index 2 in $\mathbf{S}_4$ by finding the subset of squares of permutations $\theta$ in $\mathbf{S}_4$.
3. Prove that there is no subgroup of order 6 in the group $\mathbf{A}_4$ of order 12.

4. Show that $\mathbf{S}_3$ is solvable by producing a chain of subgroups with the necessary properties.
5. Explain why the following are true:
   (i) Every permutation of $\mathbf{A}_5$ is a 1-cycle, 3-cycle, 5-cycle, or product of two disjoint 2-cycles.
   (ii) 40 of the 60 elements of $\mathbf{A}_5$ are cubes of elements of $\mathbf{A}_5$.
   (iii) There is no normal subgroup with index 3 in $\mathbf{A}_5$. (This is the proof of Lemma 2 left to the reader.)
6. Prove that there is no normal subgroup with index 5 in $\mathbf{A}_5$ (and thus prove Lemma 3).
7. Let $N$ be a normal subgroup with index 2 in $\mathbf{S}_5$. Prove that $N = \mathbf{A}_5$ by showing the following:
   (i) Every 3-cycle $(xyz)$ of $\mathbf{S}_5$ is in $N$.
   (ii) Every 5-cycle $(abcde)$ of $\mathbf{S}_5$ is in $N$.
   (iii) Every product of two disjoint transpositions of $\mathbf{S}_5$ is in $N$. [*Hint:* Use $(ab)(cd) = (abc)(adc)$.]
8. Prove that there is no normal subgroup with index 3 in $\mathbf{S}_5$.
9. Prove that there is no normal subgroup with index 5 in $\mathbf{S}_5$.
10. Explain why the only normal subgroup with prime index in $\mathbf{S}_5$ is $\mathbf{A}_5$. (*Hint:* Use Problems 7–9.)
11. Use Problem 10 and Theorem 3 to prove that $\mathbf{S}_5$ is not solvable (and thus prove Theorem 4).
12. Prove the following:
    (i) No one of the three subgroups with index 6 in $\mathbf{A}_4$ is a normal subgroup in $\mathbf{A}_4$.
    (ii) $\mathbf{A}_4$ has no normal subgroup with index 4.
    (iii) $\mathbf{A}_4$ has exactly three normal subgroups. (See Problems 1 and 3.)
13. Let $G = [a]$ be a cyclic group of order 25. Find a subgroup $G_1$ of index 5 in $G$, explain why $G_1$ is a normal subgroup, and show that $G$ is solvable.
14. Let $G = [a]$ be a cyclic group of order 1001. Find a subgroup $G_1$ of index 7 in $G$, a subgroup $G_2$ of index 11 in $G_1$, and a subgroup $G_3$ of index 13 in $G_2$. Then show that $G$ is solvable.
15. Let $G = [a]$ be a cyclic group of order $m$ and let

$$m = p_1 p_2 \cdots p_r,$$

    where the $p_i$ are positive primes. Show that $G$ is solvable.
16. Use Problem 15 and a factorization theorem of Section 1.7 to explain why every finite cyclic group is solvable.

*17. Prove that $\mathbf{A}_5$ is simple.

***W. Burnside*** *(1852–1927) Burnside, an English mathematician, conjectured early in the twentieth century that all groups of odd order are solvable. This remained an open question until 1963, when a paper "Solvability of Groups of Odd Order" appeared in the* **Pacific Journal of Mathematics.** *In this paper the authors, two Americans, John Thompson and Walter Feit, proved that all groups of odd order are solvable. Thus nonabelian finite simple groups have even order; the proof required over 250 pages. In 1965 they were awarded the Cole Prize of the American Mathematical Society for this achievement and in 1970 John Thompson received the Fields Medal at the International Congress of Mathematicians in Nice. (The Fields Medal is taken to be the equivalent in mathematics of a Nobel Prize. It is reported that there is no Nobel Prize in mathematics because Alfred Nobel felt a personal animosity toward the Swedish mathematician G. M. Mittag-Leffler, and since, if there had been a Nobel Prize in mathematics, Mittag-Leffler might have won it, Nobel decreed that there would be no Nobel Prize in mathematics.)*

*If a finite group G is not simple, it has a nontrivial normal subgroup N and much information concerning the structure of G is obtainable from the structures of N and G/N, which have lower orders than G. If N or G/N is not simple, the process can be continued. Hence simple groups are in a sense the fundamental building blocks for a theory of the structure of finite groups. The determination of all finite simple groups was generally considered to be practically impossible until Thompson and Feit proved that all such groups have even order. Then an enormous effort was begun by a number of first class algebraists; it reached a successful conclusion in 1980. M. Aschbacher, in his description of this effort in the* **Mathematical Intelligencer**, *3 (1981), 59–65, wrote that the "proof of the Classification Theorem is made up of thousands of pages in various mathematical journals with at least another thousand pages still left to appear in print."*

## *2.14 *The Sylow Theorems (without proofs)*

In this section, $G$ is always a finite group with order $m$. By Lagrange's Theorem, $(\text{ord } H) \mid m$ for every subgroup $H$ in $G$. However, there is not necessarily a subgroup with order $d$ for every positive integral divisor $d$ of $m$. (See Problem 3 of Section 2.13.) But there are partial converses of Lagrange's Theorem; we state here without proof a theorem of this nature and some related results, all due to the Norwegian mathematician Ludwig Sylow (1832–1918).

***Theorem 1 Sylow's Theorem, Part I***

*Let $p$ and $n$ be positive integers with $p$ a prime. Then $G$ has a subgroup of order $p^n$ if and only if $p^n \mid (\text{ord } G)$.*

The case $n = 1$ of this theorem is known as ***Cauchy's Theorem***. For each positive prime divisor $p$ of $m = \text{ord } G$, the case with the largest possible exponent $n$ is especially noteworthy.

***Definition 1 Sylow p-Subgroup***

*Let $p$ and $n$ be positive integers with $p$ a prime. Let the order of $G$ be an integral multiple of $p^n$ but not of $p^{n+1}$. Then a subgroup of order $p^n$ in $G$ is a* **Sylow p-subgroup** *in $G$.*

***Theorem 2 Sylow's Theorem, Part II***

*Let $H$ and $K$ be any two Sylow $p$-subgroups in $G$ for the same prime $p$. Then there is an element $a$ in $G$ such that*

$$K = \{a^{-1}ha : h \in H\}, \qquad H = \{aka^{-1} : k \in K\};$$

*that is, $H$ and $K$ are conjugate subgroups.* (See Problem 23 of Section 2.5.)

***Theorem 3 Sylow's Theorem, Part III***

*For each positive prime divisor $p$ of $m = \text{ord } G$, the number of Sylow $p$-subgroups in $G$ is an integral divisor $d$ of $m$ having the form $d = 1 + kp$, with $k$ a nonnegative integer.*

In Section 3.16, the structure of finite commutative groups is characterized completely in terms of the Sylow $p$-subgroups.

## *Problems*

1. Do the following for the alternating group $\mathbf{A}_4$:
   (a) Find a Sylow 2-subgroup and explain why it is the only Sylow 2-subgroup.
   (b) Show that the Sylow 3-subgroups $H = [(123)]$ and $K = [(124)]$ are conjugate subgroups and find two other Sylow 3-subgroups.

2. Describe a Sylow 2-subgroup in the symmetric group $\mathbf{S_4}$.
3. Show that $\mathbf{S_5}$ has six Sylow 5-subgroups.
4. Show that $\mathbf{S_5}$ has ten Sylow 3-subgroups.
5. Describe a Sylow 2-subgroup in $\mathbf{S_5}$.

*6. How many Sylow 2-subgroups are there in $\mathbf{S_5}$?

7. Let $p$ and $q$ be primes with $p > q > 0$. Use Theorem 3 to show that a group $G$ of order $pq$ has only one Sylow $p$-subgroup and that this subgroup is normal.
8. Show that the number of Sylow 5-subgroups in a group $G$ of order 55 must be 1 or 11.

## *Review Problems*

1. Let $a$, $b$, and $c$ be in a group and let $b \neq c$. Show that $ab^{-1} \neq ac^{-1}$.
2. Find the standard form of each of the following products.
   (a) (123)(145).
   (b) (1234)(1567).
   (c) (12345)(16789).
   (d) (14)(24)(34)(12)(24)(23).
3. Let $\alpha = (13)(15)(16)(21)(24)(26)$ and $\beta = (125)(326)(14)$. Find the standard forms of $\alpha$, $\alpha^{-1}$, $\beta$, and $\beta^{99}$ and give the order of each of these four permutations.
4. (a) Express (132)(154)(123)(145) in standard form.
   (b) Let $\alpha = (axb)$ and $\beta = (bcy)$. Show that $\alpha^{-1}\beta^{-1}\alpha\beta$ is a 3-cycle.
5. Let $G = [a]$, a cyclic group of order 80.
   (a) What is the order of $a^{36}$? Explain.
   (b) List the elements of a subgroup $H$ of order 5.
   (c) List the elements of a subgroup $N$ of index 10 in $G$.
   (d) List all the elements of order 10 in $G$.
6. If $a$ is in a group $G$ and $a^{52}$ is the identity, what are the possibilities for the order of $a$?
7. Let $G = [a]$, a cyclic group of order 6. Give all the subsets of $G$ that are closed under the operation of $G$. Is each of these subsets a subgroup in $G$? Explain.
8. (a) Give the set of orders of elements of $\mathbf{A_7}$.
   (b) Give the set of orders of elements of $\mathbf{S_7}$.
9. How many of the elements of $\mathbf{S_{10}}$ are products $(abcd)(efghi)$ of two disjoint cycles, one of length 4 and the other of length 5?
10. Let $H$ be a subgroup in $\mathbf{A_4}$ and let the order of $H$ be at least 7. Must $H$ be $\mathbf{A_4}$? Explain.

11. (a) Let $G$ be a group of order 64 and let $G$ contain an element $a$ such that $a^{32}$ is not the identity. Must $G$ be cyclic? Explain.
    (b) Show that a group of order 64 must have an element of order 2.

12. Let $G$ be a group of order 96 and let $a$ be an element of $G$ such that neither $a^{48}$ nor $a^{32}$ is the identity. Explain why $G$ must be cyclic.

13. Let $G$ be a finite group. Show the following:
    (a) $G$ has an even number of elements of order 3.
    (b) The number of elements of order 5 is a multiple of 4.
    (c) $G$ has an even number of elements $x$ such that $x^2$ is not the identity.

14. Let $H$ be a subgroup of order $s$ in a group $G$ of order $t$ and let $s < t$. Explain why $2s \leq t$.

15. Let $r$ and $f$ be elements of order 5 and 2, respectively, in a group $G$. Let $fr = r^4f$. Show that the smallest subgroup in $G$ containing $r$ and $f$ is

$$H = \{\mathbf{e}, r, r^2, r^3, r^4, f, rf, r^2f, r^3f, r^4f\}.$$

16. Show that the group $H$ of Problem 15 is solvable.

17. Let $H$ be a subgroup in the symmetric group $\mathbf{S_5}$ and let each of (123), (1234), and (12345) be in $H$. Show that $H = \mathbf{S_5}$.

18. (a) What is the smallest positive integer $s$ such that there is a nonabelian group of order $s$?
    (b) What is the smallest positive integer $t$ such that there is a noncyclic group of order $t$?

19. What is the smallest positive integer $m$ such that $\alpha^m = (1)$ for every permutation $\alpha$ of $\mathbf{A_7}$?

20. In the real numbers, the equation $x^2 = x$ has 0 and 1 as solutions. How many solutions are there of $x^2 = x$ in a group $G$? Explain.

21. (a) Let $a$ be an element of order 3 in a group $G$. Show that $a$ is a square in $G$, that is, there is an element $b$ in $G$ such that $b^2 = a$.
    (b) Show that every element of order 5 in $G$ is a square.

22. Generalize on parts (a) and (b) of Problem 21.

23. (a) Let $a$ be an element of order 2 in a group $G$. Show that there exist elements $b$ and $c$ in $G$ with $a = b^3 = c^5$.
    (b) Let $d$ have order 3 in $G$. Show that $d = f^5$ for some $f$ in $G$.
    (c) Let $g$ have order 5 in $G$. Show that $g = h^3$ for some $h$ in $G$.

24. Generalize on Problems 21–23.

25. Let $m$ be the order of a finite cyclic group $G$. Let $g$ be any element of $G$ and let $s$ be a positive integer such that gdc($m$, $s$) > 1. Explain why $g^s$ is not a generator of $G$.

26. Let $H$ be a subgroup in $G$. Prove the following:
    (a) If $aH = Hb$, then $aH = Ha = bH = Hb$.
    (b) $H$ is normal in $G$ if and only if for every $g$ in $G$ there is a $g'$ in $G$ such that $gH = Hg'$ (or such that $Hg = g'H$).
    (c) $H$ is normal in $G$ if and only if for every $h$ in $H$ and $g$ in $G$ there is an $h'$ in $G$ such that $hg = gh'$ (or such that $gh = h'g$).

## Supplementary and Challenging Problems

1. Let $\gamma$ be an $s$-cycle $(a_1a_2 \dots a_s)$ and $\theta$ any permutation in $\mathbf{S_n}$. Let $\theta(a_i) = b_i$ for $i = 1, 2, \dots, s$. Show that

$$\theta^{-1}\gamma\theta = (b_1b_2 \cdots b_s).$$

2. Let $\gamma$ and $\gamma'$ be $s$-cycles in $\mathbf{S_n}$. Show that there exists at least one permutation $\theta$ in $\mathbf{S_n}$ such that

$$\gamma' = \theta^{-1}\gamma\theta.$$

3. Let the permutation $\alpha$ of $\mathbf{S_n}$ be a product $\gamma_1\gamma_2 \cdots \gamma_t$ of disjoint cycles $\gamma_1, \gamma_2, \dots, \gamma_t$ of lengths $k_1, k_2, \dots, k_t$, respectively. Show that for every $\theta$ of $\mathbf{S_n}$, $\theta^{-1}\alpha\theta$ is also a product of $t$ disjoint cycles whose lengths are $k_1, k_2, \dots, k_t$.

4. Let $a$ and $b$ be two distinct elements of order 2 in a finite group $G$ of order $m$ and let $ab = ba$. Show that $4|m$.

5. Show that the multiplicative group $\mathbf{R}^{\pm}$ of the nonzero real numbers has only one subgroup $N$ with index 2. (You may use the fact that every positive real number has real square roots.)

6. Let $H$ be a subgroup in $G$. Let $K$ consist of all the elements $a$ of $G$ such that $aH = Ha$. Prove that $K$ is a subgroup in $G$ and that $H$ is a normal subgroup in $K$.

7. Let $a$ be the only element of order 2 in a group $G$. Prove that $a$ is in the center of $G$.

8. Let $G$ be a group of finite order $u$. Show the following:
   (i) The number of elements of $G$ with order greater than 2 is even.
   (ii) If $u$ is odd, every element of $G$ is a square of an element of $G$.
   (iii) If $u$ is even, there are at least two elements $g$ of $G$ with $g^2 = \mathbf{e}$ and hence at least one element of $G$ is not a square.

9. Let $a^2b^2 = b^2a^2$ for all $a$ and $b$ in a group $G$. Let $H$ consist of the elements of odd order in $G$. Prove that $H$ is a commutative subgroup in $G$.

10. Let $r$ be a positive integer and let $a^rb^r = b^ra^r$ for all $a$ and $b$ in a group $G$. Let $H$ be the subset of $G$ consisting of the elements with order relatively prime to $r$. Prove that $H$ is a commutative subgroup in $G$.

11. Give the set of positive integers that are orders of subgroups in $\mathbf{S_5}$.

12. Prove that no 3-cycle $(abc)$ is the cube of a permutation in $\mathbf{S_n}$.

13. Does there exist a nonabelian group $G$ whose center $C$ has index 9 in $G$?

14. Prove that a group $G$ that has a finite number of subgroups must be finite.

15. In a finite group $G$, let $T$ be a subset with more than half of the elements of $G$. Prove that every element of $G$ is a product $tt'$, with $t$ and $t'$ in $T$.

16. For every positive integer $n$, let $m_n$ be the smallest positive integer $m$ such that $\theta^m = (1)$ for all $\theta$ in $\mathbf{S_n}$. Prove that:
   (i) $m_n = m_{n-1}$ if $n$ is an integral multiple of two distinct primes.
   (ii) $m_n = pm_{n-1}$ if $n$ is a power of a prime $p$.

17. Let ord $G = n$ and let $g_1, g_2, \ldots, g_n$ be $n$ not necessarily distinct elements of $G$. Show that there exist integers $i$ and $j$ such that $1 \le i \le j \le n$ and $g_i g_{i+1} \cdots g_j = \mathbf{e}$, the identity of $G$.

18. Let $T = \{t_1, t_2, \ldots, t_m\}$ be a subset with $m$ elements in an abelian group $G$. Let $t^{-1}$ not be in $T$ whenever $t$ is in $T$. Show that the product $t_i t_j$ is in $T$ for at most $m(m-1)/2$ of the $m^2$ choices of $i$ and $j$ with $1 \le i \le m$ and $1 \le j \le m$.

19. Give an example of a noncommutative group $G$ in which $(ab)^3 = a^3b^3$ for all $a$ and $b$ in $G$.

20. Prove that every subgroup in $\mathbf{S}_4$ is solvable.

# 3

## *Sets and Mappings*

*To see what is general in what is particular*
*and what is permanent in what is*
*transitory is the aim of scientific thought.*

*Alfred North Whitehead*

The *function* concept is an indispensable tool in all current branches of mathematics. Elementary calculus courses deal with real valued functions of a real variable, that is, functions from a set of real numbers to a set of real numbers. In Chapter 2 we used the one-to-one functions from a finite set $\mathbf{X_n} = \{1, 2, \ldots, n\}$ to itself to form the permutation group $\mathbf{S_n}$.

Here we generalize on these examples and define a function from any set $X$ to any set $Y$. We usually substitute the synonym *mapping* for the word *function* to lessen the possibility of confusing the general concept with the previously familiar special cases.

Mappings will be used to clarify the meaning of statements such as: "Groups $G_1$ and $G_2$ have essentially the same structure." Mappings will also provide new insight into the important concepts of normal subgroups and quotient groups and will help us to develop, in Chapter 4, the appropriate analogues of these concepts in the theory of rings.

### *3.1 Mappings*

#### *Definition 1 Mapping, Image*

*A* **mapping** *$\theta$ from a set $X$ to a set $Y$ is a rule or procedure that assigns to each element $x$ of $X$ a unique element $y$ of $Y$. The element $y$ of $Y$ assigned to a given $x$ of $X$ is called the* **image** *of $x$ under $\theta$ and is denoted by $\theta(x)$.*

#### *Definition 2 Domain, Codomain, Image Set*

*Let $\theta$ be a mapping from $X$ to $Y$. Then $X$ is the* **domain** *and $Y$ is the* **codomain** *of $\theta$. The* **image set** *of $\theta$ is the subset of $Y$ consisting of the images $y$ of all the elements $x$ of $X$.*

Many elementary calculus texts use *range* for *codomain*, many others use *range* for *image set*. Because of this ambiguity we shall avoid the word *range* in what follows.

### *Notation Mapping Arrows*

*One indicates that $\theta$ is a mapping from $X$ to $Y$ by writing*

$$\theta: X \to Y.$$

*An alternative way of indicating that $y = \theta(x)$ is to write*

$$\theta: x \mapsto y$$

*or to write*

$$x \overset{\theta}{\mapsto} y.$$

*Either of these may be read as "$\theta$ sends $x$ to $y$" or as "$x$ goes to $y$ under $\theta$."*

### *Example 1*

Let $X = \{1, 2, 3, 4\}$ and $Y = \{a, b, c\}$. Let $\theta: X \to Y$ be such that

$$\theta: 1 \mapsto a,\ 2 \mapsto c,\ 3 \mapsto a,\ 4 \mapsto a.$$

A pictorial representation of the mapping $\theta$ is given in Figure 3.1.

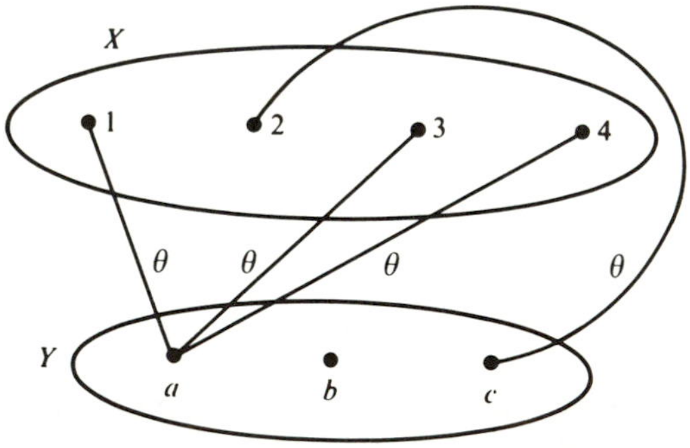


***Figure 3.1***

As is required by the definition of a function, each $x$ of $X$ has a unique image. The definition does not require that every element of $Y$ be the image of some $x$ in $X$; we see here that no element of $X$ has $b$ as its image. Also, an element of $Y$ may be the image of more than one element of $X$. For example, the element $a$ of $Y$ is the image of three elements of $X$ under this mapping.

The image set of $\theta$ is the proper subset $\{a, c\}$ of the codomain $\{a, b, c\}$ in this example.

***Example 2***

Let $G = [a]$ and $G' = [b]$ be cyclic groups of orders 6 and 3, respectively. Let $\theta$ be the mapping from $G$ to $G'$ with

$$\mathbf{e} \mapsto \mathbf{e}', \quad a \mapsto b, \quad a^2 \mapsto b^2, \quad a^3 \mapsto \mathbf{e}', \quad a^4 \mapsto b, \quad a^5 \mapsto b^2.$$

Here the image set of $\theta$ is the same set as the codomain $G' = \{\mathbf{e}', b, b^2\}$. This mapping is illustrated in Figure 3.2.

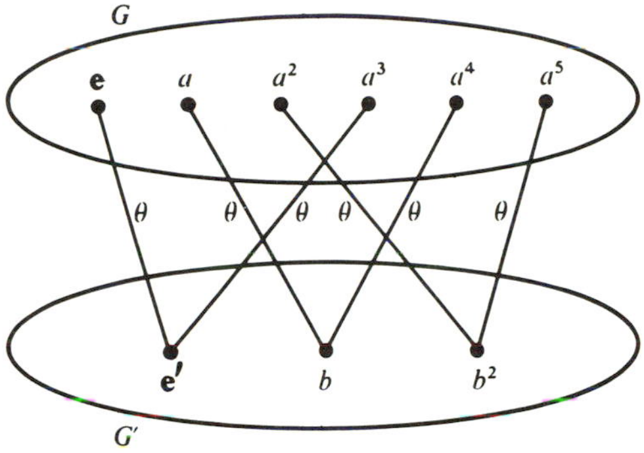


***Figure 3.2***

***Example 3***

Let $X$ be a proper subset of $Y$ and let $\theta$ be the mapping from $X$ to $Y$ with $\theta(x) = x$ for all $x$ in $X$. The image set is $X$, a proper subset of the codomain $Y$. A special case of this type of mapping is depicted in Figure 3.3.

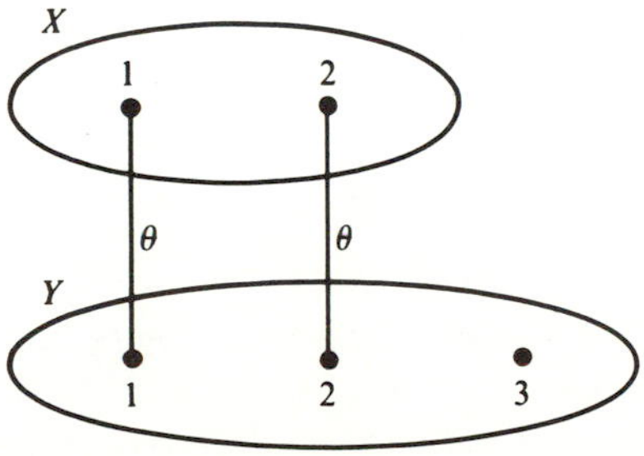


***Figure 3.3***

*Example 4*

Let $\theta$ be the mapping from the set **R** of the real numbers to itself with $\theta(r) = 2^r$. The image set of $\theta$ consists of the positive numbers and is a proper subset of the codomain **R**. Figure 3.4 is a partial representation of this mapping.

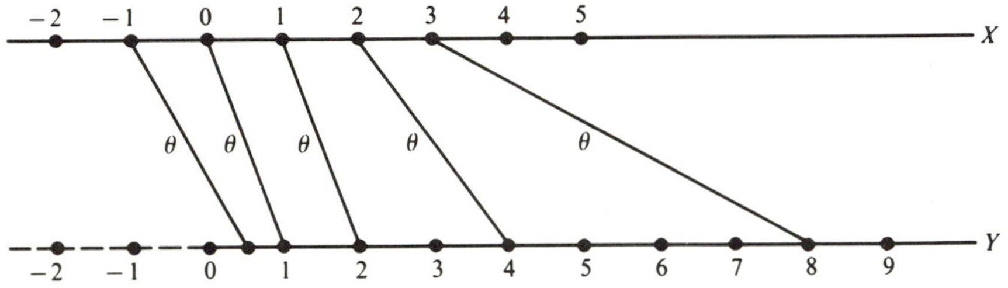


*Figure 3.4*

*Example 5*

Let $\theta$ be the mapping from the real numbers **R** to the positive real numbers $\mathbf{R}^+$ given by $\theta(r) = 2^r$. Here we have modified the mapping of Example 4 so that the codomain $\mathbf{R}^+$ is the same as the image set. Figure 3.4, excluding the dashed part of the line, depicts this $\theta$.

*Example 6*

Let $a$ be a fixed element of a group $G$ and let $\theta$ be the mapping from $G$ to itself with $\theta(x) = xa$ for all $x$ in $G$. If $b$ is in $G$, $\theta(ba^{-1}) = ba^{-1}a = b$; this shows that every element of the codomain $G$ is in the image set of this $\theta$.

Let $\theta$ be a fixed mapping from $X$ to $Y$. By definition of a mapping, a given element $x$ of $X$ always determines one and only one image $\theta(x)$. However, if $\theta(x)$ is given, one may not be able to determine $x$ uniquely. In Example 1, if $\theta(x)$ is known to be $a$, then all one can deduce is that $x$ is 1, 3, or 4.

The following terminology is applied to mappings $\theta$ such that each $\theta(x)$ in the image set determines $x$ uniquely.

***Definition 3 Injection***

*Let $\theta$ be a mapping such that distinct elements $x_1$ and $x_2$ of the domain always have distinct images $\theta(x_1)$ and $\theta(x_2)$; that is, let $x_1 \neq x_2$ imply $\theta(x_1) \neq \theta(x_2)$. Then $\theta$ is said to be an* **injection**, *or an* **injective mapping.**

An injection is also called a ***one-to-one mapping.*** The mapping of Example 1 is not injective since $1 \neq 3$ but $\theta(1) = \theta(3)$. (See Figure 3.1.) In Example 2, $\mathbf{e} \neq a^3$ and $\theta(\mathbf{e}) = \theta(a^3)$; hence, this $\theta$ is also not an injection. (See Figure 3.2.) The mappings of Examples 3, 4, 5, and 6 are injections (one-to-one mappings) since $\theta(x_1) = \theta(x_2)$ implies $x_1 = x_2$ for these mappings. (See Figures 3.3 and 3.4.)

***Definition 4 Surjection***

*If the image set of a mapping $\theta$ from $X$ to $Y$ is the same set as the codomain $Y$, the mapping is said to be* **surjective**, *or a* **surjection**.

In other words, a mapping is surjective when every element of the codomain is in the image set. A surjection from $X$ to $Y$ is also called a mapping from $X$ ***onto*** $Y$.

The mapping of Example 1 is not a surjection (mapping onto the codomain) since the element $b$ of the codomain is not in the image set. (See Figure 3.1.) The mappings of Examples 3 and 4 are also not surjective. The mappings of Examples 2, 5, and 6 are surjections. (See Figure 3.2 and Figure 3.4 without the dashed part.)

***Definition 5 Bijection, Permutation***

*A mapping that is both injective and surjective is said to be* **bijective**, *or a* **bijection**. *A bijection from $X$ to itself is a* **permutation** *on $X$.*

Of the examples above, only the mappings in Examples 5 and 6 are bijections (one-to-one mappings onto the codomain). The mapping in Example 6 is a permutation.

One may note that Figure 3.1 represents a mapping that is neither surjective nor injective, Figure 3.2 a surjection that is not injective, Figure 3.3 an injection that is not surjective, and Figure 3.4 (without the dashed part) a bijection.

***Definition 6 Pre-image, Complete Inverse Image***

*Let $\theta$ be a mapping from $X$ to $Y$. If $m'$ is an element of the image set of $\theta$, an element $m$ with $\theta(m) = m'$ is called a* **pre-image** *or an* **antecedent** *of $m'$. For every $y$ in $Y$, $\theta^{-1}(y)$ denotes the set of all $x$ in $X$ such that $\theta(x) = y$; the subset $\theta^{-1}(y)$ of $X$ is called the* **complete inverse image** *of $y$.*

We note that the set $\theta^{-1}(y)$ of all antecedents of $y$ is empty when $y$ is not in the image set of $\theta$ and hence $\theta$ is surjective if and only if $\theta^{-1}(y)$ is nonempty for all $y$ in the codomain $Y$. Also, $\theta$ is injective if and only if $\theta^{-1}(y)$ never has more than one element. Hence $\theta$ is bijective if and only if $\theta^{-1}(y)$ has exactly one element for every $y$ in $Y$. One may think of $\theta^{-1}$ as a mapping from $Y$ to the class of subsets of $X$.

If $\alpha$ is a bijection from $X$ to $Y$, there is a mapping $\beta$ from $Y$ to $X$ with $\beta(y)$ the unique pre-image of $y$; this $\beta$ is called the ***inverse mapping*** of $\alpha$. One can easily show that the inverse mapping $\beta$ of a bijection $\alpha$ is also a bijection. For this reason, one may use the two-headed arrow notation

$$x \overset{\alpha}{\leftrightarrow} y$$

to indicate that $x$ is sent by the bijection $\alpha$ to $y$.

The inverse mapping of a bijection $\alpha$ usually is designated as $\alpha^{-1}$; hence in what follows the context will indicate whether $\alpha^{-1}(y)$ means the image $x$ of $y$ under the inverse mapping $\alpha^{-1}$ or the complete inverse image $\{x\}$.

A finite sequence $a_1, a_2, \ldots, a_n$ of elements from a set $A$ characterizes the mapping $\alpha$ from $\{1, 2, \ldots, n\}$ to $A$ with $\alpha(i) = a_i$ for $1 \le i \le n$. An infinite sequence $b_1, b_2, b_3, \ldots$ of elements from a set $B$ is a way of looking at the mapping $\beta\colon \mathbf{Z}^+ \to B$ with $\beta(i) = b_i$ for each $i$ in $\mathbf{Z}^+$.

A numeral with $n$ digits can define a mapping $\alpha\colon X \to D$, where $X$ is any set $\{x_1, x_2, \ldots x_n\}$ with $n$ elements and $D$ is the set $\{0, 1, \ldots, 9\}$ of digits. For example, with $n = 6$ the numeral 010147 gives us the mapping with the table

| $x$ | $x_1$ | $x_2$ | $x_3$ | $x_4$ | $x_5$ | $x_6$ |
|---|---|---|---|---|---|---|
| $\alpha(x)$ | 0 | 1 | 0 | 1 | 4 | 7 |

When the elements of $X$ and $Y$ are denoted by Greek letters $\alpha$, $\beta$, etc., we shall generally use Latin letters $f$, $g$, etc. for mappings from $X$ to $Y$.

Some synonyms for the word mapping are *map, function, correspondence,* and *transformation.* We repeat that an injection may be referred to as a *one-to-one mapping* and that a surjection from $X$ to $Y$ may be called a mapping from $X$ *onto* $Y$, or an *onto mapping.* A bijection from $X$ to $Y$ is also called a *one-to-one correspondence* between $X$ and $Y$, or a *one-to-one* mapping from $X$ *onto* $Y$.

## *Problems*

Problem 12 below is cited in Section 3.14.

1. Describe a bijection $\theta$ from the set $\mathbf{Z}^+ = \{1, 2, 3, \ldots\}$ to its proper subset $Y = \{2, 3, 4, \ldots\}$.
2. Describe a bijection $\theta$ from the set $\mathbf{Z} = \{\ldots, -2, -1, 0, 1, 2, \ldots\}$ to its proper subset $E = \{\ldots, -4, -2, 0, 2, 4, \ldots\}$.

3. (i) Tabulate one of the injections (one-to-one mappings) $\theta$ from $X = \{1, 2, 3\}$ to $Y = \{1, 2, 3, 4\}$.
   (ii) Is $\theta$ surjective (a mapping onto $Y$)? Explain.
4. (i) Describe an injection $\theta$ from a set $X = \{x_1, x_2, \ldots, x_n\}$ with $n$ elements to the set **Z** of the integers.
   (ii) Is $\theta$ surjective?
5. Let $\theta$ be the mapping from the set **R** of the real numbers to the set $N$ of the nonnegative real numbers given by $\theta(x) = |x|$.
   (i) Is $\theta$ injective? Explain.
   (ii) Is $\theta$ surjective? Explain.
   (iii) List the elements of the complete inverse image $\theta^{-1}(5)$.
6. Let $G = [a]$ be a cyclic group of order 12.
   (i) Tabulate the mapping $\theta$ from $G$ to $G$ with $\theta(x) = x^3$.
   (ii) List the elements of the image set of $\theta$.
   (iii) Is $\theta$ injective? Explain.
   (iv) Is $\theta$ surjective? Explain.
   (v) List the elements of the complete inverse images $\theta^{-1}(a^9)$ and $\theta^{-1}(a^{10})$.
7. Let $X = \{x_1, x_2, x_3\}$ be a set with 3 elements and let $\theta$ be a mapping from $X$ to itself.
   (i) Can $\theta$ be injective without being surjective?
   (ii) Can $\theta$ be surjective without being injective?
8. (i) Describe an injection $\alpha$, from the set **Z** of the integers to itself, such that $\alpha$ is not a surjection.
   (ii) Describe a surjection $\beta$ from **Z** to **Z** that is not an injection.
9. How many mappings are there from a set $X$ with 2 elements to a set $Y$ with 3 elements?
10. How many mappings are there from a set $X$ with $m$ elements to a set $Y$ with $n$ elements?
11. Let $\theta$ be a mapping from a group $G$ to a set $Y$. An element $p$ of $G$, such that $\theta(xp) = \theta(x)$ for all $x$ in $G$, is called a ***right period*** of $\theta$. Prove that the set of all right periods of $\theta$ is a subgroup in $G$.
12. Let $a$ be a fixed element of a group $G$ and let $\theta$ be the mapping from $G$ to itself with $\theta(x) = xa$ for all $x$ in $G$. In Example 6 above we showed that $\theta$ is a surjection. Explain why $\theta$ is also an injection and hence is a bijection.
13. Let $\rho = (1234)$ and $\phi = (24)$. Let $G$ be the octic group

$$\{\varepsilon, \rho, \rho^2, \rho^3, \phi, \rho\phi, \rho^2\phi, \rho^3\phi\};$$

$N$ be its center $\{\varepsilon, \rho^2\}$; and $N$, $B = N\rho$, $C = N\phi$, and $D = N\rho\phi$ be the cosets of the normal subgroup $N$. Let $f$ be the mapping from $G$ to the quotient group $G/N$ such that the image of each element $\alpha$ of $G$ is its coset $N\alpha$.

(i) Does $f(\alpha\beta) = f(\alpha)f(\beta)$ for all $\alpha$ and $\beta$ in $G$? Explain.
(ii) Tabulate the mapping $f$.
(iii) Is $f(\varepsilon)$ the identity of $G/N$? Explain.
(iv) Does $f(\alpha^{-1}) = [f(\alpha)]^{-1}$ for all $\alpha$ in $G$? Explain.
(v) Which elements of $G$ are in the complete inverse image $f^{-1}(N)$?

14. Let $N$ be a normal subgroup in $G$ and let $\theta$ be the mapping from $G$ to $G/N$ with $\theta(a) = aN$.
(i) Does $\theta(ab) = \theta(a)\theta(b)$ for all $a$ and $b$ in $G$? Explain.
(ii) Does $\theta$ send the identity of $G$ to the identity of $G/N$? Explain.
(iii) Does $\theta(a^{-1}) = [\theta(a)]^{-1}$ for all $a$ in $G$? Explain.
(iv) Which elements of $G$ are in the complete inverse image $\theta^{-1}(N)$?
(v) Is $\theta$ surjective? Explain.
(vi) What must be true of $N$ for $\theta$ to be injective?

15. Let $\alpha$ and $\beta$ be bijections from $A$ to $B$ and from $B$ to $C$, respectively. Describe a bijection $\gamma$ from $A$ to $C$.

*16. Let an injection $\alpha: A \to B$ have $C$ as its image set. Use $\alpha$ to characterize a bijection $\beta: A \to C$, that is, a one-to-one mapping from $A$ onto $C$.

## 3.2 Group Isomorphisms

*Mathematicians do not study objects, but relations among objects; they are indifferent to the replacement of objects by others as long as relations do not change. Matter is not important, only form interests them.*

*Henri Poincaré*

There are many groups of order 3, for example, the groups

$$G_1 = \{(1), (123), (132)\}, \qquad G_2 = \{1, (-1 + \mathbf{i}\sqrt{3})/2, (-1 - \mathbf{i}\sqrt{3})/2\},$$

with composition of permutations the operation for $G_1$ and multiplication of complex numbers the operation for $G_2$. However, the differences between these groups are superficial in that one can obtain the table for the operation of $G_2$ by replacing (1) with 1, (123) with $(-1 + \mathbf{i}\sqrt{3})/2$, and (132) with $(-1 - \mathbf{i}\sqrt{3})/2$ in the row headings, column headings, and entries of the table for $G_1$. Actually, the construction of the table for any group $G = \{\mathbf{e}, b, c\}$ of order 3, given in Example 2 of Section 2.3, shows that all groups of order 3 have essentially the same structure.

On the other hand, two groups of order 4 may differ in more than the names for their elements and the names for their operations; for example, one can be cyclic and the other noncyclic.

We shall be able to make these statements more precise after introducing the terminology which follows.

***Definition 1 Group Homomorphism***

*A mapping $\theta$ from a group $G$ to a group $G'$ with*

$$\theta(ab) = \theta(a)\theta(b) \text{ for all } a \text{ and } b \text{ in } G, \tag{P}$$

*is a* **group homomorphism** *from $G$ to $G'$.*

The property of $\theta$ given in formula (P) is called ***preservation of the operations*** of $G$ and $G'$; it states that the image of a product is the product of the images. As we see in Figure 3.5, the multiplication $ab$ is performed in $G$ while $\theta(a)\theta(b)$ denotes a product in $G'$.

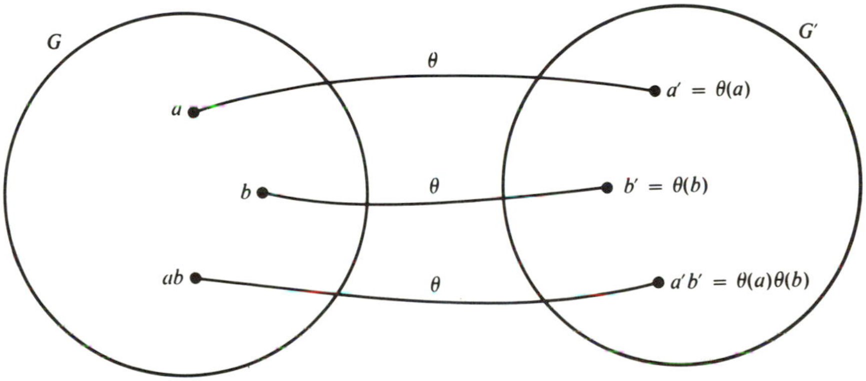


***Figure 3.5***

***Definition 2 Group Isomorphism***

*A bijective homomorphism is called an* **isomorphism**; *that is, a group isomorphism from $G$ to $G'$ is a bijection $\theta$ such that $\theta(ab) = \theta(a)\theta(b)$ for all $a$ and $b$ in $G$.*

We study isomorphisms in this section and take up the more general concept of homomorphisms in the next section.

One says that $G$ is ***isomorphic*** to $G'$ if there exists an isomorphism $\theta$ from $G$ to $G'$. The inverse mapping $\theta^{-1}$ of such a $\theta$ is easily seen to be an isomorphism from $G'$ to $G$. Hence the groups in an isomorphism play interchangeable roles and we may replace *$G$ is isomorphic to $G'$* with *$G$ and $G'$ are isomorphic.*

Preservation of the operations implies that whenever elements $a$, $b$, and $c$ of a group $G$ are such that $ab = c$, their images $a'$, $b'$, and $c'$ under an isomorphism $\theta$ from $G$ to $G'$ must satisfy $a'b' = c'$ in $G'$. If the operation of $G$ is specified by a table, it follows that replacement of the row and column headings and the entries of this table by their images under the isomorphism $\theta$ gives one the table for the operation of $G'$. This means that $G$ and $G'$ have essentially the same structure; the only possibilities for differences are in the names for their elements and for their operations.

***Example 1***

Let $G = \{\mathbf{e}, b, c, d\}$ and $G' = \{\mathbf{e}' \ u, v, w\}$ be any two noncylic groups of order 4. Show that $G$ and $G'$ are isomorphic.

***Solution***

Since $G$ and $G'$ are noncyclic, it follows from Lagrange's Theorem that each $x$ in $G$ or $G'$ has order 1 or 2 and hence that $x^2$ is the identity. Then one shows easily that the tables for the operations of $G$ and $G'$ are:

| | **e** | *b* | *c* | *d* |
|---|---|---|---|---|
| **e** | **e** | *b* | *c* | *d* |
| *b* | *b* | **e** | *d* | *c* |
| *c* | *c* | *d* | **e** | *b* |
| *d* | *d* | *c* | *b* | **e** |

| | **e**′ | *u* | *v* | *w* |
|---|---|---|---|---|
| **e**′ | **e**′ | *u* | *v* | *w* |
| *u* | *u* | **e**′ | *w* | *v* |
| *v* | *v* | *w* | **e**′ | *u* |
| *w* | *w* | *v* | *u* | **e**′ |

We see that the table on the left goes over into the one on the right by applying the bijection $\theta$ with

$$\mathbf{e} \mapsto \mathbf{e}', \quad b \mapsto u, \quad c \mapsto v, \quad d \mapsto w.$$

Hence this bijection satisfies $\theta(x)\theta(y) = \theta(xy)$ for all $x$ and $y$ in $G$ and is an isomorphism.

This example shows that there is essentially only one noncyclic group of order 4; such a group is called a ***Klein 4-group*** or a ***4-group*** and is often denoted by $V$, the initial letter of "viergruppe," which is "4-group" in German. (See the biographical note on Felix Klein after Section 2.9.)

The following result will help us to set up isomorphisms, when they exist, and to prove that certain pairs of groups are not isomorphic.

***Theorem 1 Properties of Isomorphisms***

*Let $\theta$ be a group isomorphism from G to G′. Then:*
(a) $\mathbf{e} \overset{\theta}{\mapsto} \mathbf{e}'$, *where* **e** *and* **e**′ *are the identities of G and G′.*
(b) *If* $a \overset{\theta}{\mapsto} a'$, *then* $a^n \overset{\theta}{\mapsto} (a')^n$ *for all integers n. In particular,* $a^{-1} \overset{\theta}{\mapsto} (a')^{-1}$.
(c) *If* $a \overset{\theta}{\mapsto} a'$, *then a and a′ have equal orders.*

***Proof***

Using preservation of the operations and properties of the identities, we have

$$\theta(\mathbf{e})\theta(\mathbf{e}) = \theta(\mathbf{ee}) = \theta(\mathbf{e}) = \mathbf{e}'\theta(\mathbf{e}).$$

Then the equality $\theta(\mathbf{e})\theta(\mathbf{e}) = \mathbf{e}'\theta(\mathbf{e})$ and right cancellation give us $\theta(\mathbf{e}) = \mathbf{e}'$; this proves (a).

For (b) we start with $aa^{-1} = \mathbf{e}$. Preservation of the operations and the result in (a) give us

$$\theta(a)\theta(a^{-1}) = \theta(aa^{-1}) = \theta(\mathbf{e}) = \mathbf{e}'.$$

The equality $\theta(a)\theta(a^{-1}) = \mathbf{e}'$ shows that $\theta(a^{-1}) = [\theta(a)]^{-1}$, that is, $a^{-1} \mapsto (a')^{-1}$. It is left to the reader to complete a proof of (b) by mathematical induction.

For (c), we let $a \mapsto a'$ and denote the orders of $a$ and $a'$ by $q$ and $q'$, respectively. From (a) we have $\mathbf{e} \mapsto \mathbf{e}'$, while $\mathbf{e} = a^q \mapsto (a')^q$ by (b). Hence $(a')^q = \mathbf{e}'$. Since $q$ is a positive integer and $q'$ is the least positive integer $m$ with $(a')^m = \mathbf{e}'$, we have $q \geq q'$. Similarly, $q' \geq q$. Together, these inequalities imply that $q = q'$.

In the proof of Theorem 1, we used the fact that a group isomorphism $\theta$ preserves products to show that it also preserves the identity, (integral) powers, and the inverse and order of all elements.

If groups $G$ and $G'$ are isomorphic and either one has finite order, the existence of a bijection from $G$ to $G'$ implies that the two groups have the same order. The following example shows that the converse is not true; that is, groups of the same order need not be isomorphic.

***Example 2***

Show that a cyclic group of order $q$ is not isomorphic to a noncyclic group of order $q$ and hence that there exist nonisomorphic groups of order 4.

*Solution*

Let $G$ be a cyclic group $[a]$ of order $q$. Then the generator $a$ has order $q$. On the other hand, a noncyclic group $G'$ of order $q$ has no element of order $q$. Hence there is no element in $G'$ to serve as the image of $a$ under an isomorphism [see Theorem 1(c)] and therefore $G$ and $G'$ are not isomorphic.

In particular, the group [(1234)] is not isomorphic to the noncyclic group {(1), (12), (34), (12)(34)}.

*Example 3*

Show that two cyclic groups of the same order are isomorphic.

*Solution*

Let $G = [a] = \{\mathbf{e}, a, a^2, \ldots, a^{q-1}\}$ and $G' = [b] = \{\mathbf{e}', b, b^2 \ldots, b^{q-1}\}$ be cyclic groups of order $q$. If $\theta$ is to be an isomorphism from $G$ to $G'$, $\theta(a)$ must have order $q$, and hence be a generator of $G'$, by Theorem 1(c). This motivates us to let $\theta(a) = b$. Theorem 1(b) then tells us that we must let

$$\theta(a^j) = b^j \text{ for } j = 0, 1, \ldots, q - 1. \tag{1}$$

Clearly the mapping $\theta$ from $G$ to $G'$ given by (1) is a bijection. Also, this $\theta$ preserves the operations since

$$\theta(a^j a^k) = \theta(a^{j+k}) = b^{j+k} = b^j b^k = \theta(a^j)\theta(a^k)$$

and hence $\theta(xy) = \theta(x)\theta(y)$ for all $x$ and $y$ in $G$. Therefore, $G$ and $G'$ are isomorphic.

The groups $G$ and $G'$ of an isomorphism need not be distinct. For example, if $G = \{\mathbf{e}, b, c\}$ is a group of order 3, then $G = [b] = [c]$ and it follows from Example 3 that the mapping $\theta$ with

$$\mathbf{e} \mapsto \mathbf{e}, \quad b \mapsto c, \quad c \mapsto b$$

is an isomorphism from $G$ onto itself.

***Definition 3 Group Automorphism***

*A group isomorphism from G to itself is a* **group automorphism** *of G.*

If $G$ is a group, the mapping $\theta$ with $\theta(g) = g$ for all $g$ in $G$ is always an automorphism of $G$; this is called the ***trivial***, or ***identity***, automorphism of $G$. A nontrivial automorphism $\theta$ of $G$ is one with $\theta(a) \neq a$ for at least one $a$ in $G$.

The following result provides both a source of automorphisms and a tool for studying normal subgroups.

***Theorem 2 Conjugation***

*Let a be a fixed element of a group G and let $\theta$ be the mapping from G to itself with $\theta(x) = a^{-1}xa$ for all x in G. Then $\theta$ is an automorphism of G.*

***Proof***

We see that $\theta$ is surjective by noting that for every $b$ in $G$ one has $\theta(aba^{-1}) = a^{-1}(aba^{-1})a = b$. Also, $\theta$ is injective since $a^{-1}xa = a^{-1}ya$ implies that $x = y$. Hence $\theta$ is bijective. Finally,

$$\theta(x)\theta(y) = (a^{-1}xa)(a^{-1}ya) = a^{-1}(xy)a = \theta(xy)$$

shows that $\theta$ preserves the operations. This completes the proof.

***Definition 4 Inner Automorphism, Conjugate***

*An automorphism having the form of the $\theta$ in Theorem 2 is called an* **inner automorphism**; *the specific inner automorphism $x \mapsto a^{-1}xa$ is* **conjugation** *by a. The image $\theta(b) = a^{-1}ba$ is the* **conjugate** *of b by a. If b is in a group G,*

$$\{g^{-1}bg : g \in G\}$$

*is the set of* **conjugates** *of b in G.*

Let $H$ be a subgroup in $G$ and $\theta$ be the inner automorphism of conjugation by $a$. Then one can show [see Problem 23(i) of Section 2.5] that

$$K = \{a^{-1}ha : h \in H\} = \{\theta(h) : h \in H\} \tag{2}$$

is also a subgroup in $G$.

***Definition 5 Conjugate Subgroup***

*If H is a subgroup in G and a is an element of G, the subgroup K of (2) is the* **conjugate subgroup** *of H by a.*

We note that if $K$ is the conjugate subgroup of $H$ by $a$, then $H$ is the conjugate subgroup of $K$ by $a^{-1}$ [see Problem 23(ii) of Section 2.5]; that is, $H$ and $K$ are conjugate subgroups of each other.

***Theorem 3 Tests for a Subgroup To Be Normal***

*A subgroup H is normal in G if and only if any one of the following is true:*

*(a) The conjugate $g^{-1}hg$ is in H for every h in H and g in G.*

(*b*) *The conjugate subgroup* $\{g^{-1}hg : h \in H\}$, *of* $H$ *by* $g$, *is contained in* $H$ *for every* $g$ *in* $G$.
(*c*) *The conjugate subgroup of* $H$ *by* $g$ *is* $H$ *for every* $g$ *in* $G$.

***Proof***

Part (a) is Problem 20 of Section 2.11 and (b) is a rewording of (a). We leave (c) to the reader as Problem 29 of this section.

Because of the result in Theorem 3(c), a normal subgroup in $G$ is frequently called a ***self conjugate subgroup*** or an ***invariant subgroup***.

## *Problems*

1. Let $G = [a]$ and $G' = [b]$ be cyclic groups of order 5.
   (i) Complete the following table so that each $\theta_i$ is an isomorphism from $G$ to $G'$:

| $x$ | **e** | $a$ | $a^2$ | $a^3$ | $a^4$ |
|---|---|---|---|---|---|
| $\theta_1(x)$ | | $b$ | | | |
| $\theta_2(x)$ | | $b^2$ | | | |
| $\theta_3(x)$ | | $b^3$ | | | |
| $\theta_4(x)$ | | $b^4$ | | | |

   [Note that $\theta(a^2) = \theta(a)\theta(a) = [\theta(a)]^2$, $\theta(a^3) = [\theta(a)]^3$, etc. for an isomorphism $\theta$.]
   (ii) Explain why there are exactly 4 isomorphisms from $G$ to $G'$.
2. Let $G = [a]$ and $G' = [b]$ be cyclic groups of order 6.
   (i) Complete the following table so that $\alpha$ and $\beta$ are isomorphisms from $G$ to $G'$:

| $x$ | **e** | $a$ | $a^2$ | $a^3$ | $a^4$ | $a^5$ |
|---|---|---|---|---|---|---|
| $\alpha(x)$ | | $b$ | | | | |
| $\beta(x)$ | | $b^5$ | | | | |

   (ii) Explain why there are just 2 isomorphisms from $G$ to $G'$.
3. Let $G = \{\mathbf{e}, b, c, d\}$ and $G' = \{\mathbf{e}', u, v, w\}$ be noncyclic groups of order 4. The isomorphism from $G$ to $G'$ with

$$\mathbf{e} \mapsto \mathbf{e}', \quad b \mapsto u, \quad c \mapsto v, \quad d \mapsto w$$

is given in Example 1. Tabulate five other isomorphisms from $G$ to $G'$.

4. In an isomorphism between groups of permutations, is the image of an odd permutation necessarily odd? [*Hint:* Consider the groups $G = \{(1), (12), (34), (12)(34)\}$ and $G' = \{(1), (12)(34), (13)(24), (14)(23)\}$.]

5. Explain why the following are true.
   (a) The only automorphism of a group of order 2 is the trivial automorphism.
   (b) A group of order 3 has exactly 2 automorphisms.

6. (a) Write the formula which expresses preservation of the operations in a mapping $\theta$ from a multiplicative group $G$ to an additive group $G'$.
   (b) Is the mapping $\theta$ with $\theta(x) = 10^x$ an isomorphism from the additive group $\mathbf{R}$ of the real numbers to the multiplicative group $\mathbf{R}^+$ of the positive numbers? Explain.

7. Let $G$ be isomorphic to $G'$ and let $G$ be abelian. Does $G'$ have to be abelian? Explain.

8. Let $\theta$ be an isomorphism from $G$ to $G'$. Let $C$ and $C'$ be the centers of $G$ and $G'$, respectively. Are $C$ and $C'$ isomorphic? Explain.

9. Complete the following table so that $f$, $g$, and $h$ are automorphisms of the symmetric group $\mathbf{S}_3$:

| $\alpha$ | (1) | (123) | (132) | (23) | (13) | (12) |
|---|---|---|---|---|---|---|
| (a) $f(\alpha)$ | | (123) | | (23) | | |
| (b) $g(\alpha)$ | | (123) | | (13) | | |
| (c) $h(\alpha)$ | | (123) | | (12) | | |

[Use preservation of the operations. For example, if $\rho = (123)$ and $\phi = (23)$, then under the mapping $g$,

$$(13) = (123)(23) = \rho\phi \mapsto g(\rho\phi) = g(\rho)g(\phi) = (123)(13) = (12).]$$

10. Complete the following table so that $f$, $g$, and $h$ are automorphisms of the symmetric group $\mathbf{S}_3$:

| $\alpha$ | (1) | (123) | (132) | (23) | (13) | (12) |
|---|---|---|---|---|---|---|
| (a) $f(\alpha)$ | | (132) | | (23) | | |
| (b) $g(\alpha)$ | | (132) | | (13) | | |
| (c) $h(\alpha)$ | | (132) | | (12) | | |

11. Explain why every automorphism of $\mathbf{S}_3$ must be one of the 6 automorphisms of Problems 9 and 10.
12. Explain why $\mathbf{S}_3$ is not isomorphic to [(123456)].
13. For the automorphism $g$ of Problem 9, find a $\beta$ in $\mathbf{S}_3$ such that $g(\theta) = \beta^{-1}\theta\beta$ for all $\theta$ in $\mathbf{S}_3$.
14. For the automorphism $h$ of Problem 9, find a $\gamma$ in $\mathbf{S}_3$ such that $h(\theta) = \gamma^{-1}\theta\gamma$ for all $\theta$ in $\mathbf{S}_3$.
15. Let $G = [a]$ and $G' = [b]$ be cyclic groups of the same order. Show that among the isomorphisms $\theta$ from $G$ to $G'$ there is exactly one with $\theta(a) = c$ if and only if $c$ is a generator of $G'$.
16. Let $G$ and $G'$ be cyclic groups of order 8. How many isomorphisms are there from $G$ to $G'$?
17. Explain why any two groups of the same prime order must be isomorphic.
18. How many automorphisms are there of a group of prime order $p$?
19. Describe two automorphisms of the additive group of the integers.
20. Let $G$ and $G'$ be infinite cyclic groups. How many isomorphisms are there from $G$ to $G'$? Explain.
21. Let $g$ be an element of a group $G$. Show the following:
    (i) Conjugates $a^{-1}ga$ and $b^{-1}gb$ of $g$ are equal if and only if $ba^{-1}$ is in the centralizer $C_g$. (The definition of centralizer is in Section 2.5.)
    (ii) $a^{-1}ga = b^{-1}gb$ if and only if $a$ and $b$ are in the same right coset of the centralizer $C_g$ in $G$.
    (iii) If $G$ is finite, the number $n_g$ of conjugates of $g$ in $G$ equals the index of the centralizer $C_g$ in $G$; that is, $n_g = (\text{ord } G)/(\text{ord } C_g)$. Also, $n_g = 1$ if and only if $g$ is in the center in $G$.
22. Show that the set of conjugates of (12) in $\mathbf{S}_4$ is

    $$\{(12), (13), (14), (23), (24), (34)\},$$

    and hence that the centralizer of (12) in $\mathbf{S}_4$ has index 6 (and order 4). (See Problem 21.)
23. Let $\alpha$ and $\beta$ be the inner automorphisms of a group $G$ with $\alpha(x) = a^{-1}xa$ and $\beta(x) = b^{-1}xb$.
    (i) Show that $\alpha = \beta$ (that is, $\alpha(x) = \beta(x)$ for all $x$ in $G$) if and only if $ba^{-1}$ is in the center $C$ in $G$.
    (ii) Show that $\alpha = \beta$ if and only if $Ca = Cb$.
    (iii) Let $C$ have index $q$ in $G$. Explain why there are exactly $q$ inner automorphisms of $G$.
24. Explain why the only inner automorphism of an abelian group is the trivial automorphism $\theta$ with $\theta(x) = x$ for all $x$ in $G$.
25. How many inner automorphisms are there of the octic group?

26. How many inner automorphisms are there of the alternating group $\mathbf{A}_4$?

27. Let $M$ be the multiplicative group of $2 \times 2$ matrices of the form

$$m(x) = \begin{pmatrix} \cos x & \sin x \\ -\sin x & \cos x \end{pmatrix}$$

with $x$ a real number. Let $U$ be the multiplicative group of complex numbers of the form

$$\cos x + \mathbf{i} \sin x,$$

with $x$ real. [See Problems 4(c) and 22 of Section 2.6. $U$ is called the circle group.] Prove that $M$ and $U$ are isomorphic.

28. Let $U$ be as in Problem 27. Let $T$ be the quotient group $\mathbf{R}/[2\pi]$ of the additive group $\mathbf{R}$ of the real numbers by its cyclic subgroup $[2\pi]$. Show that $U$ and $T$ are isomorphic.

29. Let $N$ be a normal subgroup in $G$. Show every element $m$ of $N$ is a conjugate $g^{-1}ng$ of an element $n$ of $N$ by an element $g$ of $G$. Then show that the previous statement and Theorem 3(b) imply Theorem 3(c).

30. Show that the subgroup $H = \{(1), (12)\}$ is not normal in the symmetric group $\mathbf{S}_3$ by producing a conjugate subgroup $K$ of $H$ that is different from $H$.

31. Let $H$ be a subgroup in $G$ and $a$ be an element of $G$.
    (a) Show that $Ha = aK$, where $K$ is the conjugate subgroup of $H$ by $a$.
    (b) Show that $aH = La$ for some subgroup $L$ in $G$.

32. Let each of $G$ and $G'$ be a group of order 8 with six elements of order 4. Prove that $G$ and $G'$ are isomorphic. (Such a group is called a ***quaternion group***. See Example 1 of Section 2.12.)

33. Let $G$ be a group of order 8 with two elements of order 4 and five elements of order 2. Prove that $G$ is isomorphic to the octic group.

*34. Let $G$ be a noncyclic group of order 6. Prove that:
    (i) $G$ has at least one element of order 3.
    (ii) $G$ has three elements of order 2.
    (iii) $G$ is isomorphic to the symmetric group $\mathbf{S}_3$.

*35. Prove that the octic group has 8 automorphisms, of which 4 are inner automorphisms.

*36. Prove that the alternating group $\mathbf{A}_4$ has 24 automorphisms, of which 12 are inner automorphisms.

37. Let $G$ be a finite abelian group of order $n$ and let $\theta$ be the mapping from $G$ to itself with $\theta(g) = g^m$ for all $g$ in $G$, where $m$ is an integer such that $\gcd(m, n) = 1$. Prove the following:
    (i) There exists a fixed integer $r$ such that $g^{rm} = g$ for all $g$ in $G$.
    (ii) $\theta$ is injective and surjective; that is, $\theta$ is a one-to-one mapping from $G$ onto itself.
    (iii) $\theta$ is an automorphism of $G$.

38. Let $\theta$ be a bijection from a set $S$ to a group $G$. Prove that $S$ becomes a group isomorphic to $G$ when multiplication is defined by

$$ab = \theta^{-1}[\theta(a)\theta(b)]$$

for all $a$ and $b$ in $S$.

## 3.3 Group Homomorphisms

We recall from the previous section that a group homomorphism $\theta$ from $G$ to $G'$ is a mapping from $G$ to $G'$ with $\theta(ab) = \theta(a)\theta(b)$ for all $a$ and $b$ in $G$ and note that the multiplication $ab$ is performed in $G$ while $\theta(a)\theta(b)$ is performed in $G'$.

The preservation of operations property of a homomorphism from $G$ to $G'$ helps in deducing properties of one of the groups from properties of the other. The work here with group homomorphisms will be extended and applied to rings in Chapters 4 and 5.

In this section, we shall see that group homomorphisms are closely related to normal subgroups and quotient groups.

### *Definition 1 Kernel of a Homomorphism*

*The* **kernel** *of a group homomorphism $\theta$ from $G$ to $G'$ is the set of all $k$ in $G$ such that $\theta(k) = \mathbf{e}'$, where $\mathbf{e}'$ is the identity of $G'$.*

In other words, the kernel of a group homomorphism $\theta$ from $G$ to $G'$ is the complete inverse image $\theta^{-1}(\mathbf{e}')$.

### *Definition 2 Natural Map*

*Let $N$ be a normal subgroup in $G$. Then the* **natural map** *from $G$ to the quotient group $G/N$ is the mapping $\theta$ with $a \mapsto aN$, that is, the mapping that sends each element $a$ of $G$ to the coset for $a$ of $N$ in $G$.*

We are now ready for our first result.

### *Theorem 1 The Natural Map is a Surjective Homomorphism*

*Let $N$ be a normal subgroup in $G$. Then the natural map $\theta$ from $G$ to $G/N$ is a surjective homomorphism with $N$ as its kernel.*

*Proof*

Using the formula $aN \cdot bN = (ab)N$ of Section 2.12 for multiplication of cosets of a normal subgroup we see that

$$\theta(ab) = (ab)N = aN \cdot bN = \theta(a)\theta(b);$$

that is, the natural map $\theta$ is a homomorphism. It is surjective since a given coset $gN$ in $G/N$ is the image of $g$ under $\theta$.

The identity of $G/N$ is the coset $N$. Also, $\theta(g) = N$ if and only if $g$ is in $N$. These two statements tell us that the kernel of $\theta$ is N.

*Example 1*

The subgroup $N = [(123)] = \{(1), (123), (132)\}$ has index 2 in the symmetric group $\mathbf{S}_3$ and hence it follows from Theorem 1 of Section 2.12 that $N$ is a normal subgroup in $\mathbf{S}_3$. Let $\phi = (23)$. Then the natural map $f$ from $\mathbf{S}_3$ to $\mathbf{S}_3/N$ has the table

| $\alpha$ | (1) | (123) | (132) | (23) | (13) | (12) |
|---|---|---|---|---|---|---|
| $f(\alpha)$ | $N$ | $N$ | $N$ | $\phi N$ | $\phi N$ | $\phi N$ |

*Example 2*

Let $g$ be the mapping from $\mathbf{S}_3$ to $\mathbf{S}_4$ given by the table

| $\alpha$ | (1) | (123) | (132) | (23) | (13) | (12) |
|---|---|---|---|---|---|---|
| $g(\alpha)$ | (1) | (1) | (1) | (23) | (23) | (23) |

Explain why $g$ is a group homomorphism.

*Solution*

The image set of $g$ is the subgroup $\{(1), (23)\}$ of order 2 in the codomain $\mathbf{S}_4$. Each of the tables in Examples 1 and 2 may be considered to be the table for a mapping from $\mathbf{S}_3$ to a group $\{\mathbf{e}, b\}$ of order 2, with $\mathbf{e}$ and $b$ given different names in the two examples. Since the natural map $f$ is a homomorphism, this implies that $g$ is also a homomorphism.

We could also verify preservation of the operations by noting that $g(\alpha) = (1)$ when $\alpha$ is an even permutation and $g(\alpha) = (23)$ when $\alpha$ is odd.

The homomorphism $g$ of Example 2 is not surjective; for example, the element $\rho = (1234)$ of the codomain $\mathbf{S}_4$ is not in the image set $\{\varepsilon, \phi\}$, where $\phi = (23)$. Hence the complete inverse image $g^{-1}(\rho)$ is the empty set. We note that, for the elements of the image set, the complete inverse images are the cosets

$$g^{-1}(\varepsilon) = \{(1), (123), (132)\}, \qquad g^{-1}(\phi) = \{(23), (13), (12)\}$$

of the kernel $K = [(123)]$ of $g$.

The parts of the following result generalize on this observation and other properties of the homomorphisms of Examples 1 and 2. (One might compare this result with Theorem 1 of Section 3.2.)

***Theorem 2 Properties of Group Homomorphisms***

*Let $\theta$ be a group homomorphism from $G$ to $G'$. Then:*

(a) $\theta(\mathbf{e}) = \mathbf{e}'$; *that is, the identity of $G$ is sent by $\theta$ to the identity of $G'$. (This also implies that $\mathbf{e}$ is in the kernel $K$ and $\mathbf{e}'$ is in the image set $M$ of $\theta$.)*

(b) $\theta(g^n) = [\theta(g)]^n$ *for every integer $n$ and each $g$ in $G$.*

(c) *The image set $M$ of $\theta$ is a subgroup in $G'$.*

(d) *The kernel $K$ of $\theta$ is a normal subgroup in $G$.*

(e) *If $\theta(a) = a'$, the complete inverse image $\theta^{-1}(a')$ is the coset $aK$. Also, the order of $a'$ is an integral divisor of the order of $a$.*

(f) *$G/K$ is isomorphic to the image group $M$.*

(g) *If $\theta$ is injective, $G$ is isomorphic to $M$.*

***Proof***

The proofs of (a) and (b) are similar to the analogous results for isomorphisms in Theorem 1 of Section 3.2.

For (c), let $a'$ and $b'$ be in the image set $M$. Then there are $a$ and $b$ in $G$ such that $\theta(a) = a'$ and $\theta(b) = b'$. Now

$$\theta(ab^{-1}) = \theta(a)\theta(b^{-1}) = \theta(a)[\theta(b)]^{-1} = a'(b')^{-1},$$

using preservation of the operations and (b). This means that $a'(b')^{-1}$ is in $M$. Hence $M$ is closed under division. Also, $M$ is nonempty since $\mathbf{e}'$ is in $M$ by (a). The last two statements and Theorem 1 of Section 2.5 tell us that $M$ is a subgroup in $G'$.

For (d), let $j$ and $k$ be in the kernel $K$. Then

$$\theta(jk^{-1}) = \theta(j)\theta(k^{-1}) = \theta(j)[\theta(k)]^{-1} = \mathbf{e}'(\mathbf{e}')^{-1} = \mathbf{e}'$$

shows that $jk^{-1}$ is in $K$; that is, $K$ is closed under division. Also $K$ is nonempty by (a). Hence $K$ is a subgroup in $G$.

Next we see that for every $g$ in $G$ and $k$ in the kernel $K$,

$$\begin{aligned}\theta(g^{-1}kg) &= \theta(g^{-1})\theta(k)\theta(g) = [\theta(g)]^{-1}\mathbf{e}'\theta(g)\\ &= [\theta(g)]^{-1}\theta(g) = \mathbf{e}'.\end{aligned}$$

This means that $g^{-1}kg$ is an element $k_1$ of $K$. Then $kg = gk_1$, which shows that the right coset $Kg$ is contained in the left coset $gK$. Similarly, one sees that $gK$ is contained in $Kg$. Hence $Kg = gK$ for all $g$ in $G$; that is, $K$ is a normal subgroup in $G$.

For (e) we first let $b$ be in $aK$. Then $b = ak$ with $k$ in $K$ and

$$\theta(b) = \theta(ak) = \theta(a)\theta(k) = \theta(a)\mathbf{e}' = \theta(a) = a';$$

that is, $b$ is in $\theta^{-1}(a')$. Conversely, let $\theta(b) = a'$. Then

$$\theta(a^{-1}b) = \theta(a^{-1})\theta(b) = [\theta(a)]^{-1}\theta(b) = (a')^{-1}a' = \mathbf{e}',$$

hence $a^{-1}b$ is an element $k$ of $K$ and $b = ak$ is in $aK$.

Now let $\theta(a) = a'$ and let the orders of $a$ and $a'$ be $q$ and $q'$, respectively. Using (a) and (b) we have

$$\mathbf{e}' = \theta(\mathbf{e}) = \theta(a^q) = (a')^q.$$

The equality $\mathbf{e}' = (a')^q$ implies that $q' \mid q$, as desired.

For (f), we note that (e) implies that $\theta(a) = \theta(b)$ if and only if $aK = bK$. This means that the mapping $\alpha$ from $G/K$ to $M$ with $\alpha(gK) = \theta(g)$ is well defined and is injective. Also, for any $c'$ in $M$ there is a $c$ in $G$ with $\theta(c) = c'$. Then $\alpha(cK) = \theta(c) = c'$; that is, $\alpha$ is surjective. Finally,

$$\alpha(aK \cdot bK) = \alpha(abK) = \theta(ab) = \theta(a)\theta(b) = \alpha(aK)\alpha(bK)$$

shows that $\alpha$ preserves the operations. Together, these statements tell us that $\alpha$ is an isomorphism.

For (g) we let $\beta$ be the mapping from $G$ to $M$ with $\beta(g) = \theta(g)$ for all $g$ in $G$; that is, $\beta$ is the modification of $\theta$ in which one ignores the elements of the codomain $G'$ that are not in the image set $M$. This changes the injection $\theta$ into the bijection $\beta$. Also, $\beta$ preserves the operations since it has the same effect as the homomorphism $\theta$. Hence $\beta$ is an isomorphism from $G$ to $M$.

### Theorem 3 Normal Subgroups ↔ Kernels

*A subset $S$ of a group $G$ is a normal subgroup in $G$ if and only if $S$ is the kernel of a homomorphism from $G$ to some group $G'$.*

***Proof***

Theorem 1 tells us that every normal subgroup $N$ in $G$ is the kernel of a homomorphism, the natural map, from $G$ to the quotient group $G/N$. Conversely, Theorem 2(d) states that the kernel of a homomorphism from $G$ to $G'$ is a normal subgroup in $G$.

We next present some notation and preliminary results concerning the quotient group $\mathbf{Z}/[m]$ of the additive group $\mathbf{Z}$ of the integers by a cyclic subgroup $[m]$; this will facilitate our applications of the concept of natural maps to $\mathbf{Z}$. The material given here will be very helpful in Chapter 4.

Let $m$ be a fixed positive integer. Since $\mathbf{Z}$ is abelian, the cyclic subgroup

$$[m] = \{\ldots, -3m, -2m, -m, 0, m, 2m, 3m, \ldots\}$$

is normal in $\mathbf{Z}$ and the quotient group $\mathbf{Z}/[m]$ exists. This quotient group consists of cosets

$$a + [m] = \{\ldots, a - 2m, a - m, a, a + m, a + 2m, \ldots\}.$$

We introduce the notation $\bar{a}$ for the coset $a + [m]$.

### Lemma 1

$\bar{a} = \bar{b}$ in $\mathbf{Z}/[m]$ *if only if* $m \mid (b - a)$ in $\mathbf{Z}$.

#### Proof

Let $\bar{a} = \bar{b}$. Then $b$ is in the coset $\bar{a} = a + [m]$ and hence $b = a + sm$, with $s$ an integer. Transposing, we have $b - a = sm$, which means that $m \mid (b - a)$.

Conversely, let $m \mid (b - a)$. Then $b - a = sm$, with $s$ an integer, and $b = a + sm$ is in the coset $\bar{a}$. Now the cosets $\bar{a}$ and $\bar{b}$ have $b$ as a common element and hence $\bar{a} = \bar{b}$.

### Lemma 2

*In* $\mathbf{Z}/[m]$, *the cosets* $\bar{0}, \bar{1}, \bar{2}, \ldots, \overline{m-1}$ *are distinct.*

#### Proof

Let $\bar{j} = \bar{k}$ with $0 \leq j \leq m - 1$ and $0 \leq k \leq m - 1$. There is no loss of generality is assuming that $j \leq k$. Then $0 \leq k - j < m$. But Lemma 1 tells us that $k - j$ is a multiple of $m$; hence $k - j = 0$ and $k = j$. Therefore, the cosets $\bar{0}, \bar{1} \ldots, \overline{m-1}$ are distinct.

### Lemma 3

*Let* $a$, $m$, $q$, *and* $r$ *be integers with* $a = qm + r$. *Then* $\bar{a} = \bar{r}$; *that is,* $\overline{qm + r} = \bar{r}$, *in* $\mathbf{Z}/[m]$.

#### Proof

Since $a = qm + r$, we have $a - r = qm$ and $m \mid (a - r)$. Then $\bar{a} = \bar{r}$ by Lemma 1.

### Lemma 4

$\mathbf{Z}/[m] = \{\bar{0}, \bar{1}, \bar{2}, \ldots, \overline{m-1}\}$.

*Proof*

Lemma 2 tells us that the $m$ cosets

$$\overline{0}, \overline{1}, \overline{2}, \ldots, \overline{m-1} \tag{1}$$

are distinct. By the division algorithm, any integer $a$ is expressible as $a = qm + r$ with $q$ and $r$ integers and $0 \le r \le m - 1$. Using Lemma 3, we then have $\overline{a} = \overline{r}$; that is, every coset $\overline{a}$ is equal to one of those in display (1).

The natural map $\theta$ from $\mathbf{Z}$ to the quotient group $\mathbf{Z}/[m]$ has $\theta(n) = \overline{n}$ for all integers $n$. Since the homomorphism $\theta$ preserves addition,

$$\overline{a + b} = \theta(a + b) = \theta(a) + \theta(b) = \overline{a} + \overline{b}.$$

Hence $\overline{2} = \overline{1 + 1} = \overline{1} + \overline{1} = 2 \cdot \overline{1}$. Similarly, $\overline{3} = 3 \cdot \overline{1}$, $\overline{4} = 4 \cdot \overline{1}$, etc. Also, $m \cdot \overline{1} = \overline{m} = \overline{0}$, by Lemma 1 (or Lemma 3). Thus we see that the quotient group $\mathbf{Z}/[m]$ is a cyclic group $[\overline{1}]$ of order $m$.

We next give some further terminology for special kinds of homomorphisms.

***Definition 3 Homomorphic***

*If there exists a surjective homomorphism from $G$ to $G'$, one says that $G$ is* **homomorphic** *to $G'$.*

In Problem 25 below, the reader is asked to show that $G$ can be homomorphic to $G'$ without $G'$ being homomorphic to $G$.

***Definition 4 Endomorphism, Monomorphism, Epimorphism***

*A homomorphism from $G$ to itself is an* **endomorphism**. *A homomorphism $\theta$ from $G$ to $G'$ may be called a* **monomorphism** *if $\theta$ is injective and an* **epimorphism** *if $\theta$ is surjective.*

We see that a bijective endomorphism is an automorphism.

## Problems

1. Let $G = [a]$ be a cyclic group of order 9 and let $G' = [(123)]$.
   (i) Complete the following table so as to make $\theta$ a homomorphism from $G$ to $G'$: [*Hint*: Use $\theta(a^2) = \theta(a)\theta(a)$, $\theta(a^3) = \theta(a^2)\theta(a)$, etc.]

| $g$ | $\mathbf{e}$ | $a$ | $a^2$ | $a^3$ | $a^4$ | $a^5$ | $a^6$ | $a^7$ | $a^8$ |
|---|---|---|---|---|---|---|---|---|---|
| $\theta(g)$ | (1) | (123) | | | | | | | |

(ii) What is the kernel of $\theta$?
(iii) Give the complete inverse image of (123) and of (132).

2. Show that the $f$ of the following table is not a homomorphism from $\mathbf{S}_3$ to [(123)] by finding $\beta_1$ and $\beta_2$ in $\mathbf{S}_3$ such that

$$f(\beta_1\beta_2) \neq f(\beta_1)f(\beta_2).$$

| $\beta$ | (1) | (123) | (132) | (23) | (13) | (12) |
|---|---|---|---|---|---|---|
| $f(\beta)$ | (1) | (123) | (132) | (132) | (123) | (1) |

3. Let $G = [a]$ and $G' = [b]$ be cyclic groups of orders 3 and 2, respectively. Prove that there is no homomorphism $\theta$ from $G$ to $G'$ with $\theta(a) = b$.
4. Let $G$ be the octic group and let $G' = \{\mathbf{e}, b, c, d\}$ be a noncyclic group of order 4 (Klein 4-group) with $b^2 = c^2 = d^2 = \mathbf{e}$. Also, let $\rho = (1234)$ and $\phi = (24)$.
   (i) Complete the following table for a homomorphism $f$ from $G$ to $G'$.

| $\alpha$ | $\varepsilon$ | $\rho$ | $\rho^2$ | $\rho^3$ | $\phi$ | $\rho\phi$ | $\rho^2\phi$ | $\rho^3\phi$ |
|---|---|---|---|---|---|---|---|---|
| $f(\alpha)$ | | $b$ | $\mathbf{e}$ | | $c$ | | | |

   (ii) Is $f$ injective? Is $f$ surjective? Explain each answer.
   (iii) Give the complete inverse image for each element of the codomain.
5. Let $N$ be the cyclic subgroup [3] in the additive group $\mathbf{Z}$ of the integers. Let the three cosets in the quotient group $\mathbf{Z}/[N]$ be $\bar{0} = 0 + [3]$, $\bar{1} = 1 + [3]$, $\bar{2} = 2 + [3]$.
   (i) Make the addition table for $\mathbf{Z}/N = \{\bar{0}, \bar{1}, \bar{2}\}$.
   (ii) Complete the following partial table for the natural map $\theta$ from $\mathbf{Z}$ to $\mathbf{Z}/[N]$:

| $x$ | $-5$ | $-4$ | $-3$ | $-2$ | $-1$ | 0 | 1 | 2 | 3 | 4 | 5 |
|---|---|---|---|---|---|---|---|---|---|---|---|
| $\theta(x)$ | | | | $\bar{1}$ | $\bar{2}$ | $\bar{0}$ | $\bar{1}$ | $\bar{2}$ | $\bar{0}$ | | |

6. Let $G$ be the quotient group $\mathbf{Z}/[6] = \{\bar{0}, \bar{1}, \bar{2}, \bar{3}, \bar{4}, \bar{5}\}$ and let $G' = \{\mathbf{e}, b\}$ be a group of order 2.

(i) Tabulate a homomorphism $\theta$ from $G$ to $G'$ with $\theta(\bar{5}) = b$.
(ii) What is the kernel of $\theta$?
(iii) What set is the complete inverse image $\theta^{-1}(b)$?

7. Let $\mathbf{R}^{\pm}$ and $\mathbf{R}^{+}$ denote the multiplicative groups of the nonzero real numbers and the positive real numbers, respectively. Let $\theta$ be the mapping from $\mathbf{R}^{\pm}$ to $\mathbf{R}^{+}$ with $\theta(x) = |x|$.
   (i) Explain why $\theta$ is a group homomorphism.
   (ii) What is the kernel of $\theta$?
   (iii) List the elements of the complete inverse image $\theta^{-1}(6)$.

8. Let $\mathbf{R}^{\pm}$ and $V$ be the multiplicative groups of the nonzero real numbers and the nonzero complex numbers, respectively. Let $\theta$ be the mapping from $V$ to $\mathbf{R}^{\pm}$ with $\theta(z) = |z|$; that is, let $\theta$ send $r(\cos x + \mathbf{i} \sin x)$ to $r$.
   (i) Explain why $\theta$ is a group homomorphism.
   (ii) What is the image set of $\theta$?
   (iii) What is the kernel of $\theta$?
   (iv) Explain why $\theta^{-1}(2)$ consists of all complex numbers $a + b\mathbf{i}$ with $a^2 + b^2 = 4$.
   (v) What is $\theta^{-1}(-2)$?

9. Let $G$ and $G'$ be groups with $\mathbf{e}'$ the identity of $G'$. Let $\theta$ be the mapping from $G$ to $G'$ such that $\theta(g) = \mathbf{e}'$ for all $g$ in $G$.
   (i) Show that $\theta$ is a homomorphism.
   (ii) What is the kernel of $\theta$?

10. Let $H$ be a subgroup in $G$. Describe an injective homomorphism from $H$ to $G$.

11. Let $N$ be a normal subgroup in $G$. Let $q$ be the order of an element $a$ of $G$ and let $q'$ be the order of the coset $aN$ in the quotient group $G/N$. Explain why $q'|q$.

12. Let $K$ and $M$ be the kernel and image set, respectively, of a group homomorphism $\theta$ from $G$ to $G'$. Let the orders of $G$, $G'$, $K$, and $M$ be $r$, $s$, $t$, and $u$, respectively. Explain why the following are true:
   (i) $u|s$. [See Theorem 2(c).]
   (ii) $r = tu$. [See Theorem 2(f).]
   (iii) $r|(st)$.

13. Let $\theta$ be a group homomorphism from $G$ to $G'$. Let $a$ have order $q$ in $G$ and let the image set of $\theta$ be a group $M$ of order $m$. Let $\gcd(q. m) = 1$. Use Theorem 2(e) above and Lagrange's Theorem to prove that $a$ is in the kernel of $\theta$.

14. Let $f$ be a homomorphism from the alternating group $\mathbf{A}_4$ to $[(123)]$. Use Problem 13 to show that $f(\alpha) = (1)$ for each $\alpha$ in the subset $\{(12)(34), (13)(24), (14)(23)\}$ of $\mathbf{A}_4$.

15. Let $s$ be a fixed integer and let $\theta$ be the mapping from a group $G$ to itself with $\theta(g) = g^s$ for all $g$ in $G$.
    (i) Show that $\theta$ is a homomorphism when $G$ is abelian.
    (ii) Show by an example that $\theta$ need not be a homomorphism when $G$ is not commutative.

16. Let $\theta$ be the mapping from a group $G$ to itself with $\theta(g) = g^{-1}$ for all $g$ in $G$. Prove that $\theta$ is an isomorphism if and only if $G$ is abelian.

17. Let $\mathbf{Z}$ be the additive group of the integers and let $a$ be a fixed element of a group $G$. Let $\theta$ be the mapping from $\mathbf{Z}$ to $G$ with $\theta(n) = a^n$ for all integers $n$.
    (i) Show that $\theta$ is a homomorphism.
    (ii) What is the kernel of $\theta$ when $a$ has order $q$?
    (iii) What is the kernel of $\theta$ when $a$ has infinite order?

18. Let $\theta$ be a homomorphism from a cyclic group $G = [a]$ to a group $G'$. Let $\theta(a) = a'$.
    (i) Explain why the image set $M$ is the cyclic group $[a']$.
    (ii) Explain why the kernel $K$ is cyclic. (See Problem 34 of Section 2.7.)
    (iii) Explain why the quotient group $G/K$ is cyclic. (*Hint*: Use the natural map.)

19. Let $G$ be a cyclic group $[a]$. Let $b'$ be any element of a group $G'$.
    (i) Show that there is at most one homomorphism from $G$ to $G'$ with $\theta(a) = b'$.
    (ii) Show that there is a homomorphism $\theta$ from $G$ to $G'$ with $\theta(a) = b'$ if and only if the order of $b'$ is an integral divisor of the order of $a$.
    (iii) State a condition on the orders of $a$ and $b'$ for the homomorphism of (ii) to be injective.

20. Let each of $G$ and $G'$ have order 4 with $G$ cyclic and $G'$ noncyclic. Explain why there are exactly 4 homomorphisms from $G$ to $G'$.

21. Let $G = [a]$ be a cyclic group of order $q$. Explain why there are exactly $q$ endomorphisms of $G$, that is, homomorphisms from $G$ to itself.

22. Let $G$ be a Klein 4-group (noncyclic group of order 4). Show that there are exactly 16 endomorphisms of $G$, that is, homomorphisms from $G$ to itself.

23. Let $G$ consist of all $2 \times 2$ matrices $\alpha = \begin{pmatrix} a & b \\ c & d \end{pmatrix}$ with $a$, $b$, $c$, and $d$ complex numbers and with $ad - bc \neq 0$. Let multiplication in $G$ be defined by

$$\begin{pmatrix} a & b \\ c & d \end{pmatrix}\begin{pmatrix} a' & b' \\ c' & d' \end{pmatrix} = \begin{pmatrix} aa' + bc' & ab' + bd' \\ ca' + dc' & cb' + dd' \end{pmatrix}.$$

Let $V$ be the multiplicative group consisting of all the complex numbers except 0 and let $f$ be the mapping from $G$ to $V$ given by $f(\alpha) = ad - bc$.

(a) Show that $G$ is an infinite nonabelian group.
(b) Show that $f$ is a homomorphism from $G$ to $V$.
(c) Let $H$ be the subset of all $\alpha$ in $G$ with $f(\alpha)$ in $\{1, -1\}$. Show that $H$ is a subgroup in $G$ and describe a subgroup $K$ of index 2 in $H$.

24. Let $\theta$ be a homomorphism from $\mathbf{S_5}$ to $\mathbf{S_2}$. Explain why the following are true:
 (i) The image set of $\theta$ has either 1 or 2 elements.
 (ii) The kernel $K$ of $\theta$ has index 1 or 2 in $\mathbf{S_5}$. [See (i) of this problem and (ii) of Problem 12.]
 (iii) $K$ is either $\mathbf{S_5}$ or $\mathbf{A_5}$. (See Problem 7 of Section 2.13.)
 (iv) There are only two homomorphisms from $\mathbf{S_5}$ to $\mathbf{S_2}$.

• 25. Give an example of groups $G$ and $G'$ such that $G$ is homomorphic to $G'$ but $G'$ is not homomorphic to $G$.

## 3.4 *Cayley's Theorem*

Here we prove Cayley's Theorem, which states that every group of order $n$ is isomorphic to a subgroup in the symmetric group $\mathbf{S_n}$. (See the biographical note on Cayley after this section.)

First we reconsider a problem that helps to motivate the proof of Cayley's Theorem. We use the notation

$$G = \{q_1, q_2, q_3, q_4, q_5, q_6, q_7, q_8\}$$

for the elements of the group of order 8, with six elements having order 4, dealt with in Example 1 of Section 2.12. With the new notation, Table 2.3 for that example becomes Table 3.1.

***Table 3.1***

In this table, an entry $k$ inside the table stands for $q_k$.

| | $q_1$ | $q_2$ | $q_3$ | $q_4$ | $q_5$ | $q_6$ | $q_7$ | $q_8$ |
|---|---|---|---|---|---|---|---|---|
| $q_1$ | 1 | 2 | 3 | 4 | 5 | 6 | 7 | 8 |
| $q_2$ | 2 | 3 | 4 | 1 | 6 | 7 | 8 | 5 |
| $q_3$ | 3 | 4 | 1 | 2 | 7 | 8 | 5 | 6 |
| $q_4$ | 4 | 1 | 2 | 3 | 8 | 5 | 6 | 7 |
| $q_5$ | 5 | 8 | 7 | 6 | 3 | 2 | 1 | 4 |
| $q_6$ | 6 | 5 | 8 | 7 | 4 | 3 | 2 | 1 |
| $q_7$ | 7 | 6 | 5 | 8 | 1 | 4 | 3 | 2 |
| $q_8$ | 8 | 7 | 6 | 5 | 2 | 1 | 4 | 3 |

How can one verify that $G$ is a group under the operation given by Table 3.1? It is easy to see that $G$ is closed under the operation, has an identity, and has inverses for each of its elements.

Associativity is the remaining axiom to be checked. We recall that associativity has been proved for composition of permutations in $\mathbf{S_n}$. Hence we solve our problem by setting up a bijection $\theta$, which preserves the operations, from $G$ to a subgroup $G'$ in the symmetric group $\mathbf{S_8}$.

To each element $q_j$ of $G$ we associate the permutation $\beta_j$ on $\mathbf{X_8} = \{1, 2, \ldots, 8\}$ such that $\beta_j$ sends $x$ to $y$ when $q_x q_j = q_y$. This means that $\theta(q_j) =$

$$\beta_j = \begin{pmatrix} 1 & 2 & 3 & \cdots & 8 \\ b_1 & b_2 & b_3 & \cdots & b_8 \end{pmatrix},$$

where the second row $b_1, \ldots, b_8$ is taken from the entries (from top to bottom) in the column headed by $q_j$ in Table 3.1. For example,

$$q_2 \overset{\theta}{\mapsto} \beta_2 = \begin{pmatrix} 1 & 2 & 3 & 4 & 5 & 6 & 7 & 8 \\ 2 & 3 & 4 & 1 & 8 & 5 & 6 & 7 \end{pmatrix} = (1234)(5876).$$

One can also see that the images of $q_1$, $q_3$, $q_4$, etc. are $\beta_1 = (1)$, $\beta_3 = (13)(24)(57)(68)$, $\beta_4 = (1432)(5678)$, $\beta_5 = (1537)(2648)$, $\beta_6 = (1638)(2745)$, $\beta_7 = (1735)(2846)$, and $\beta_8 = (1836)(2547)$.

We now use the technique illustrated above to prove the following result.

***Theorem 1*** ***Cayley's Theorem***

*Every group $G$ of order $n$ is isomorphic to a subgroup in the symmetric group $\mathbf{S_n}$.*

***Proof***

Let $G = \{g_1, g_2, \ldots, g_n\}$. For $1 \le j \le n$, let $\beta_j$ be the permutation on $\mathbf{X_n} = \{1, 2, \ldots, n\}$ such that

$$\beta_j(x) = y \text{ when } g_x g_j = g_y.$$

Let $\alpha$ be the mapping from $G$ to $\mathbf{S_n}$ with $\alpha(g_j) = \beta_j$ and let $M$ be the image set of $\alpha$. We next show that $\alpha$ is injective.

Let $\beta_a$ and $\beta_b$ be in the image set $M$ of $\alpha$. Let $\beta_a(1) = c$ and $\beta_b(1) = d$; that is, let $g_1 g_a = g_c$ and $g_1 g_b = g_d$. If $\beta_a = \beta_b$, then 1 has the same image under $\beta_a$ and under $\beta_b$; that is, $c = d$. It follows that

$$g_c = g_d, \qquad g_1 g_a = g_1 g_b, \qquad \text{and} \qquad g_a = g_b.$$

The fact that $\beta_a = \beta_b$ implies $g_a = g_b$ shows that $\alpha$ is injective.

To prove that $\beta$ preserves the operations, we assume that $g_a g_b = g_c$ and show, as follows, that this implies $\beta_a \beta_b = \beta_c$. Let $x$ be any integer in $\mathbf{X_n} = \{1, 2, \ldots, n\}$. Also let

$$g_x g_a = g_y \quad \text{and} \quad g_y g_b = g_z. \tag{1}$$

Then

$$g_x g_c = g_x g_a g_b = g_y g_b = g_z. \tag{2}$$

The equations in (1) and (2) and the definition of the $\beta$'s tell us that

$$x \overset{\beta_a}{\mapsto} y \overset{\beta_b}{\mapsto} z \quad \text{and} \quad x \overset{\beta_c}{\mapsto} z.$$

Since $x$ is any integer in $\mathbf{X_n}$, this means that $\beta_a \beta_b = \beta_c$. Hence $\alpha$ is a homomorphism. Finally, it follows from parts (c) and (g) of Theorem 2 of Section 3.3 that the image set $M$ of $\alpha$ is a subgroup in $\mathbf{S_n}$ and that $G$ is isomorphic to $M$.

The reader may have observed in the above example and proof that each element of the group $G$ is associated with the permutation characterized by the corresponding column of the group table. Thus, given a group table, one can easily furnish the isomorphic permutation group.

In Problem 8 below, the reader is asked to identify the $H_i$ in Figure 3.6 so as to make it the subgroup poset diagram for the group whose operation is given by Table 3.1. This group, or any group isomorphic to it, is called a ***quaternion group***. (Another description is given in Theorem 7 of Section 4.7.)

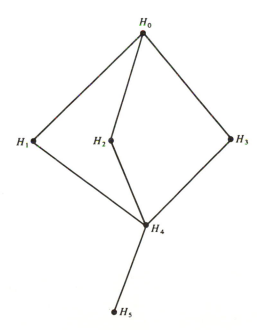

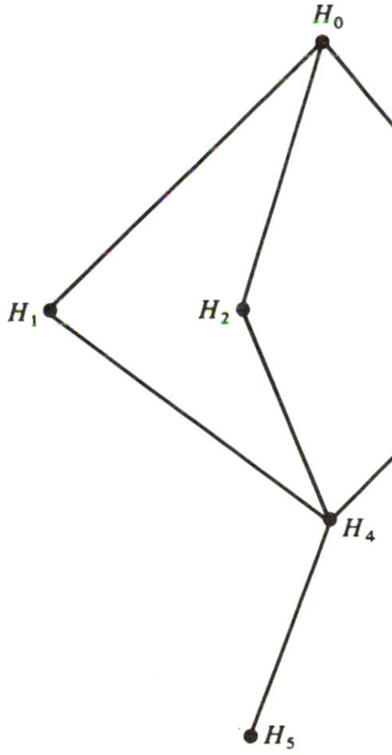


***Figure 3.6***

### Problems

1. Use the techique in Cayley's Theorem to obtain an isomorphism $f$ from $\mathbf{S_3}$ to a subgroup in $\mathbf{S_6}$. Number the elements of $\mathbf{S_3}$ as $\theta_1 = (1)$, $\theta_2 = (123)$, $\theta_3 = \theta_2^2$, $\theta_4 = (23)$, $\theta_5 = \theta_2 \theta_4$, and $\theta_6 = \theta_3 \theta_4$.
2. Use the technique in Cayley's Theorem to obtain an isomorphism $\alpha$ from a Klein 4-group $G = \{g_1, g_2, g_3, g_4\}$, with $g_1$ the identity, to a subgroup in $\mathbf{S_4}$.
3. List the orders $q$ for which a cyclic group of order $q$ is isomorphic to a subgroup in $\mathbf{S_5}$.
4. Give some examples of noncyclic groups $G$ such that $G$ is isomorphic to a subgroup in $\mathbf{S_n}$ with ord $G > n$.
5. Let $G = \{q_1, q_2, \ldots, q_8\}$ be the quaternion group of Table 3.1. Find the following:
   (i) All the elements of order 2.
   (ii) All the elements of order 4.
   (iii) A subgroup of order 2.
   (iv) Three subgroups of order 4.
6. Show that the quaternion group $G$ has only one subgroup of order 2.
7. Show that the quaternion group $G$ has only three subgroups of order 4.
8. Identify $H_0, H_1, \ldots, H_5$, in Figure 3.6, as the six subgroups in the quaternion group so that there is a rising line or broken line from $H_i$ to $H_j$ if and only if $H_i$ is a proper subgroup in $H_j$.
9. Find the center of the quaternion group.
10. Explain why every subgroup is normal in the quaternion group $G$.

---

***Arthur Cayley*** *(1821–1895)* *Cayley, like Fermat, was a lawyer as well as a mathematician. He was educated at Trinity College, Cambridge, a college which, incidentally, has produced an extraordinary number of mathematicians and physicists, including Sir Isaac Newton, Isaac Barrow, Roger Cotes, Augustus de Morgan, James Clerk-Maxwell, Sir J. J. Thomson, Lord Rayleigh, Lord Rutherford, Sir Arthur Eddington, Sir James Jeans, and G. H. Hardy.*

*Cayley is generally credited with the first discussion of abstract groups, although earlier references appear in the work of Cauchy and formal definitions were given only later by Kronecker, Weber, and Frobenius. Prior to Cayley's (and probably Cauchy's) work, groups were thought of as permutation groups and were considered only in connection with specific applications.*

*Today Cayley is largely remembered for his contributions to the theory of matrices, notably his definition of matrix multiplication, and for his work on invariants. Much of his work in algebra was done with J. J. Sylvester, another English mathematician.*

## 3.5 Unions, Intersections, Partitions

A set whose elements are themselves sets will be called a *collection of sets.*

In this section, we extend the terminology and notations given on the inside front cover for unions and intersections of finite collections of sets to include infinite collections.

These concepts are applied to groups here and to rings and fields in later chapters.

### *Definition 1 Union of the Sets of a Collection*

*Let $\Gamma$ be a collection of sets. Then the set consisting of those objects that are in at least one of the sets $A$ of $\Gamma$ is the* **union** *of the sets of $\Gamma$ and is denoted by*

$$\bigcup_{A \in \Gamma} A$$

### *Definition 2 Intersection of the Sets of a Collection*

*The set consisting of those objects that are in all of the sets of $\Gamma$ is the* **intersection** *of the sets of $\Gamma$ and is denoted by*

$$\bigcap_{A \in \Gamma} A$$

For example, it follows from Theorem 3 of Section 2.5 that

$$C = \bigcap_{a \in G} C_a,$$

where $G$ is any group, $C$ is the center in $G$, and $C_a$ denotes the centralizer of $a$ in $G$.

### *Definition 3 Smallest Set of a Collection*

*If the intersection $D$ of the sets of $\Gamma$ is one of the sets of $\Gamma$, one calls $D$ the* **smallest set** *of $\Gamma$.*

*Example 1*

Let $G = \{\varepsilon, \rho, \rho^2, \rho^3, \phi, \rho\phi, \rho^2\phi, \rho^3\phi\}$ be the octic group of symmetries of the square. We recall that $\varepsilon = (1)$, $\rho = (1234)$, and $\phi = (24)$. Let

$$H = \{\varepsilon, \rho, \rho^2, \rho^3\}, \qquad K = \{\varepsilon, \phi, \rho^2, \rho^2\phi\},$$
$$L = \{\varepsilon, \rho\phi, \rho^2, \rho^3\phi\}.$$

Each of $H$, $K$, and $L$ is a finite subset of $G$ that is closed under multiplication; hence each of these subsets is a subgroup in $G$. Then we easily see that the union and intersection of the sets of the collection $\{H, K, L\}$ are

$$H \cup K \cup L = G \quad \text{and} \quad H \cap K \cap L = \{\varepsilon, \rho^2\} = [\rho^2],$$

respectively. We also note that

$$[\rho] \cup [\phi] \cup [\rho\phi] = \{\varepsilon, \rho, \rho^2, \rho^3, \phi, \rho\phi\}$$

and that this union of subgroups is not a subgroup.

Example 1 shows that the union of a collection of subgroups in a group $G$ may or may not be a subgroup in $G$. We can be more definite about the intersection of subgroups.

**Theorem 1 The Intersection of Subgroups Is a Subgroup**

*Let $\Gamma$ be a nonempty collection of subgroups in a group $G$ and let $D$ be the intersection of the subgroups of $\Gamma$. Then $D$ is also a subgroup in $G$.*

The proof of Theorem 1 is left to the reader as Problem 6(a) of this section.

**Theorem 2 Subgroup Generated by a Subset**

*Let $T$ be a subset of a group $G$, let $\Gamma$ be the collection of all the subgroups $H$ in $G$ such that $T \subseteq H$, and let $D$ be the intersection of the subgroups of $\Gamma$. Then $D$ is a subgroup in $G$ and hence is the smallest subgroup of $\Gamma$.*

***Proof***

Since $T \subseteq G$, the collection $\Gamma$ is nonempty and it follows from Theorem 1 that $D$ is a subgroup in $G$. Also, $T \subseteq D$. Then $D$ is the smallest subgroup of $\Gamma$ by definition of the smallest set of a collection.

The intersection $D$ of all the subgroups $H$ in $G$ that contain a subset $T$ is called the ***subgroup generated by*** the subset $T$. Clearly, a cyclic subgroup $[a]$ is the subgroup generated by the single element subset $\{a\}$.

***Example 2***

Show that $\mathbf{A}_4$ is the subgroup generated by the subset $T = \{(123), (143)\}$ in the group $\mathbf{S}_4$.

***Solution***

Let $\Gamma$ consist of all subgroups $H$ in $\mathbf{S}_4$ with $T \subseteq H$. Since $(123)(143) = (12)(34)$, $(143)(123) = (14)(23)$, and $(12)(34) \cdot (14)(23) = (13)(24)$, every $H$ in $\Gamma$ must have the groups $[(123)]$ and

$$\{(1), (12)(34), (14)(23), (13)(24)\}$$

of orders 3 and 4, respectively, as subgroups. Then it follows from Lagrange's Theorem that the order of every $H$ in $\Gamma$ is a multiple of 3 and 4, and hence of 12. Since 12 and 24 are the only positive integers that are both multiples of 12 and divisors of the order 24 of $\mathbf{S}_4$, such an $H$ must be $\mathbf{S}_4$ or have index 2 in $\mathbf{S}_4$. But $\mathbf{A}_4$ is the only subgroup with index 2 in $\mathbf{S}_4$. (See Problem 2 of Section 2.13.) Hence $\mathbf{A}_4$ and $\mathbf{S}_4$ are the only possibilities for subgroups $H$ of $\Gamma$. Now we note that $T \subseteq \mathbf{A}_4$ and $T \subseteq \mathbf{S}_4$. It then follows that $\Gamma = \{\mathbf{A}_4, \mathbf{S}_4\}$ and that the subgroup generated by $T$ is the smallest subgroup $\mathbf{A}_4$ of $\Gamma$.

One can easily show that $ab = ba$ in a group $G$ if and only if $a^{-1}b^{-1}ab$ is the identity of $G$. For this reason, elements of the form $a^{-1}b^{-1}ab$ play an important role in noncommutative groups such as the symmetric groups $\mathbf{S}_n$ with $n > 2$.

***Definition 4 Commutator, Commutator Subgroup***

*If $c = a^{-1}b^{-1}ab$ with $a$ and $b$ in a group $G$, $c$ is a* **commutator** *of $G$. The* **commutator subgroup** *in $G$ is the subgroup generated by the subset of all the commutators of $G$; that is, the commutator subgroup is the smallest subgroup containing all the commutators.*

***Example 3***

Show that the commutator subgroup in an abelian group is the trivial subgroup $\{\mathbf{e}\}$.

***Solution***

$a^{-1}b^{-1}ab = a^{-1}ab^{-1}b = \mathbf{e}$ for all $a$ and $b$ in a commutative group.

*Example 4*

Show that every 3-cycle $\gamma = (abc)$ is a commutator of the alternating group $\mathbf{A_n}$ (or the symmetric group $\mathbf{S_n}$) for $n \geq 5$.

*Solution*

For $n \geq 5$, there are numbers $d$ and $e$ in $\mathbf{X_n} = \{1, 2, \ldots, n\}$ that are different from $a$, $b$, and $c$. Then $\alpha = (acd)$ and $\beta = (abe)$ are in $\mathbf{A_n}$ and one easily verifies that $\gamma$ is the commutator $\alpha^{-1}\beta^{-1}\alpha\beta$.

It can be shown that every 3-cycle of $\mathbf{S_4}$ is a commutator of $\mathbf{S_4}$ but that no 3-cycle is a commutator of $\mathbf{A_4}$. (See Example 6 and Problem 12 of this section.)

*Example 5*

Explain why every commutator in $\mathbf{S_n}$ is an even permutation and hence the commutator subgroup in $\mathbf{S_n}$ is a subgroup in $\mathbf{A_n}$.

*Solution*

First we note that the inverse of an even permutation is even and the inverse of an odd permutation is odd. Then a consideration of cases in which each of $\alpha$ and $\beta$ is either even or odd shows that a commutator $\alpha^{-1}\beta^{-1}\alpha\beta$ is always even. (See Problem 15 of Section 2.8.) Now $\mathbf{A_n}$ is a subgroup, in $\mathbf{S_n}$, that contains the commutators and hence $\mathbf{A_n}$ contains the smallest subgroup with this property.

*Example 6*

Show that $\mathbf{A_4}$ is the commutator subgroup in $\mathbf{S_4}$.

*Solution*

We easily verify that

$$(13)^{-1}(124)^{-1}(13)(124) = (123) \quad \text{and} \quad (13)^{-1}(142)^{-1}(13)(142) = (143);$$

that is, (123) and (143) are commutators in $\mathbf{S_4}$. Then it follows from Example 2 that every permutation of $\mathbf{A_4}$ is in the commutator subgroup in $\mathbf{S_4}$. Conversely, Example 5 tells us that every element of the commutator subgroup is in $\mathbf{A_4}$. Together, the last two statements give us the desired result.

For examples of multiplicative groups $G$ of 3 by 3 matrices with polynomial entries such that the commutator subgroup in $G$ contains noncommutators, see "Products of commutators are not always commutators: an example" by Phyllis Joan Cassidy in the *American Mathematical Monthly*, 86

(1979), 772. (We will deal with matrices in Section 4.7 and polynomials in Chapter 5.)

We next present some additional terminology.

**Definition 5 Disjoint Sets**

*Sets $A$ and $B$ are* **disjoint** *if they have no elements in common, i.e., if $A \cap B = \varnothing$.*

**Definition 6 Partition of a Set**

*Let $X$ be a nonempty set and let $\Gamma$ be a collection of subsets of $X$. Then $\Gamma$ is a* **partition** *of $X$ if the three following conditions are met:*
*(a) The empty set $\varnothing$ is not a member of $\Gamma$.*
*(b) The union of the sets of $\Gamma$ is $X$.*
*(c) Any two distinct sets $A$ and $B$ of $\Gamma$ are disjoint; that is, if $A$ and $B$ are in $\Gamma$, then either $A \cap B = \varnothing$ or $A = B$.*

**Example 7**

Let $H$ be a subgroup in $G$ and let $\Gamma$ consist of all the left cosets (or of all the right cosets) of $H$ in $G$. Explain why $\Gamma$ is a partition of $G$.

**Solution**

This follows from Lemma 2 of Section 2.11.

**Example 8**

Explain why the collection $\Gamma = \{H, K, L\}$, given in Example 1, is not a partition of the octic group $G$.

**Solution**

Since $H$ and $K$ are distinct members of $\Gamma$ with the permutations $\varepsilon$ and $\rho^2$ in common, $\Gamma$ is not a partition of the octic group. (The conditions that $\varnothing$ not be in $\Gamma$ and that $H \cup K \cup L = G$ are met.)

**Example 9**

For every cyclic subgroup $H$ in a group $G$, let $g(H)$ denote the set of generators of $H$. Then the collection $\Gamma$ of all these subsets $g(H)$ of $G$ is a partition of $G$.

The proof of the statement in Example 9 is straightforward and is left to the reader. (See Problems 19, 20, and 21 of this section.)

*Example 10*

For every element $a$ of a group $G$, let $M(a)$ denote the set of all conjugates $g^{-1}ag$ of $a$ by elements $g$ of $G$. Then the collection $\Gamma$ of all these $M(a)$ is a partition of $G$.

The special case of Example 10, in which $G$ is the alternating group $\mathbf{A}_4$, is dealt with in Problem 33 below.

*Example 11*

Let $\alpha: X \to Y$ and let $\Gamma$ be the collection of complete inverse images $\alpha^{-1}(y)$ for all $y$ in the image set of $\alpha$. Then $\Gamma$ is a partition of $X$.

In Section 3.7, we shall reformulate partitions of a set $X$, using the concept of "equivalence relations" in $X$.

## Problems

1. Let $G = \{\mathbf{e}, b, c, d\}$ be a Klein 4-group, that is, a noncyclic group of order 4. Show that $G$ is the union of its three subgroups that have order 2.
2. Let $\rho = (1234)$ and $\phi = (24)$. Show that the octic group
$$G = \{(1), \rho, \rho^2, \rho^3, \phi, \rho\phi, \rho^2\phi, \rho^3\phi\}$$
is the union of five cyclic subgroups.
3. Find the subgroup generated by the subset $\{\rho^2, \phi\}$ in the octic group of Problem 2.
4. For each of the following choices of $G$ and $T$ find the subgroup in $G$ generated by the subset $T$.
(a) $G = \mathbf{S}_3$ and $T = \{(23), (13)\}$.
(b) $G = \mathbf{S}_5$ and $T = \{(12), (345)\}$.
(c) $G = \mathbf{S}_6$ and $T = \{(12), (34), (56)\}$.
(d) $G = \mathbf{S}_5$ and $T = \{(25)(34), (12345)\}$.
5. (a) Let $a$ be in $G$. Explain why $[a]$ is the subgroup generated by $\{a\}$.
(b) Let $T$ be a subset of a subgroup $H$ in $G$. Explain why $T$ generates the same subgroup in $H$ as it does in $G$.
6. Let $\Gamma$ be a nonempty collection of subgroups in $G$. Let $D$ and $U$ be the intersection and union, respectively, of the sets of $\Gamma$.
(a) Prove that $D$ is a subgroup in $G$ (and thus prove Theorem 1).
(b) Let $H$ be the subgroup in $G$ generated by $U$. Prove that every group $K$ of $\Gamma$ is a subgroup in $H$.

7. Let $G$ have subgroups $H$ and $K$ of orders $r$ and $s$, respectively. Let $d = \gcd(r, s)$ and $m = \operatorname{lcm}[r, s]$.
   (a) Explain why the order of $H \cap K$ must be a positive integral divisor of $d$.
   (b) Let $L$ be the subgroup in $G$ generated by $H \cup K$. Explain why $m|(\text{ord } L)$.

8. Prove that a quotient group $G/N$ is abelian if and only if the commutator subgroup in $G$ is a subgroup in $N$.

9. Let $H$ be a subgroup in $G$ such that every commutator of $G$ is an element of $H$. Show the following:
   (i) $H$ is normal in $G$. [*Hint*: Use $ha = a(a^{-1}hah^{-1})h$ to show that $Ha \subseteq aH$ for every $a$ in $G$.]
   (ii) The quotient group $G/H$ is abelian. (See Problem 8 above or Problem 14 of Section 2.12.)
   (iii) Let $\theta$ be the natural map from $G$ to $G/H$. Then $\theta(a)\theta(b) = \theta(b)\theta(a)$ for all $a$ and $b$ in $G$.
   (iv) Let $a_1, a_2, \ldots, a_s$ be any elements of $G$ and let $m$ be a product of the $2s$ elements

$$a_1, a_2, \ldots, a_s, a_1^{-1}, a_2^{-1}, \ldots, a_s^{-1}$$

   in any order. Then $\theta(m)$ is the identity of $G/H$; that is, $m$ is in the kernel $H$ of the natural map $\theta$.

10. Let $C$ be the commutator subgroup in $G$. Explain why the following are true:
    (a) $C$ is normal in $G$.
    (b) $C$ is the smallest normal subgroup $N$ in $G$ such that $G/N$ is abelian.

11. Find the commutator subgroup in the octic group.

12. Show that the commutator subgroup in $\mathbf{A_4}$ has order 4.

13. Let $n \geq 3$ and let $T$ consist of all the 3-cycles $(abc)$ of $\mathbf{S_n}$. Prove that the subgroup generated by $T$ in $\mathbf{S_n}$ is $\mathbf{A_n}$.

14. Show that $\mathbf{A_n}$ is the commutator subgroup in $\mathbf{S_n}$ for $n = 2$ and 3.

15. Prove that $\mathbf{A_n}$ is the commutator subgroup in $\mathbf{S_n}$ for $n \geq 2$. (Use Problems 13 and 14 and Examples 4 and 6 above.)

16. Prove that $\mathbf{A_n}$ is its own commutator subgroup for $n \geq 5$.

17. Use Problems 8 and 16 to prove that $\mathbf{A_n}$ is not solvable for $n \geq 5$.

18. Let $G = \{a_1, a_2, \ldots, a_{2t+1}\}$ be a finite group of odd order. Let $m$ be a product of the $2t + 1$ elements of $G$ in any order. Prove that $m$ is in the commutator subgroup of $G$. [*Hint*: First explain why no $a_j$ has order 2 and then use Problem 9(iv).]

19. Let $S$ be the subset of all elements of order 7 in a group $G$. For every $s$ in S let $A(s) = \{s, s^2, \ldots, s^6\}$.

(i) Show that the collection $\Gamma$ of all the sets $A(s)$ is a partition of $S$. (Do not use the result in Example 9.)
(ii) If $S$ has a finite number $m$ of elements, explain why $6|m$.

20. Do the analogue of Problem 19(ii) in which 7 is replaced by a general prime $p$.

21. For every cyclic subgroup $A$ in $G$ let $g(A)$ denote the set of generators in $A$. Prove that the collection of all these subsets $g(A)$ is a partition of $G$ (and thus prove the statement in Example 9.)

22. Let $\Gamma$ consist of the centralizers $C_a$ for all the elements $a$ in a group $G$. Explain why the intersection of the sets of $\Gamma$ is the center $C$ in $G$.

23. Explain why a collection $\Gamma$ of subgroups in $G$ is not a partition of $G$, unless $\Gamma = \{G\}$.

24. Show that a cyclic group $G = [a]$ is the union of a collection $\Gamma$ of subgroups in $G$ if and only if $G$ itself is a member of $\Gamma$.

25. Let $K$ be a subgroup in $H$ and $H$ a subgroup in $G$. Show that every left coset of $H$ in $G$ is the union of a collection of left cosets of $K$ in $G$.

26. State the analogue for right cosets of Problem 25.

27. Let $H$ be a subgroup in $G$ and let $U$ be the union of the left cosets of $H$ in $G$ that are also right cosets. Prove that $U$ is a subgroup in $G$.

28. Let $H$, $G$, and $U$ be as in Problem 27. Prove that $H$ is a normal subgroup in $U$.

29. Let $H$ and $K$ be subgroups of $G$. Let neither one of $H$ and $K$ be a subset of the other; that is, let there be an $h$ that is in $H$ but is not in $K$ and a $k$ that is in $K$ but is not in $H$. Explain why the following are true.
(i) $hk$ is not in $H$ and is not in $K$.
(ii) $H \cup K$ is not closed under the operation of $G$ and therefore is not a subgroup in $G$.

30. Show that the union $U = H \cup K$ of subgroups $H$ and $K$ in $G$ is a subgroup in $G$ if and only if one of $H$ and $K$ is a subgroup in the other.

31. Let $G_1, G_2, G_3, \ldots$ be an infinite sequence of groups such that $G_i$ is a subgroup in $G_{i+1}$ for all $i$. Let $G$ be the union of all the groups of this sequence. Define multiplication in $G$ so that $G$ becomes a group with each $G_i$ as a subgroup.

32. Let $S$ be the set of standard forms of permutations in all the symmetric groups $\mathbf{S_n}$. Motivated by Problem 31, define multiplication in $S$ so as to make it into an infinite noncommutative group.

33. In Example 10, let $G$ be the alternating group $\mathbf{A_4}$. Give the partition $\{B, C, D, E\}$ of $\mathbf{A_4}$ in which $C$ contains four conjugate 3-cycles, $D$ contains four other conjugate 3-cycles, and $E$ contains three other conjugates of each other.

34. Let there be a finite number $N(q)$ of elements having order $q$ in a group $G$. Let $\phi(q)$ be the number of generators of a cyclic group of order $q$. Use Problem 21 above to prove that $\phi(q)|N(q)$.

35. Let $T$ be the subset $\{(123), (12)(34)\}$ of $\mathbf{S_4}$. Show that the subgroup in $\mathbf{S_4}$ generated by $T$ is $\mathbf{A_4}$.

36. Let $U$ be the subset $\{(12), (12345)\}$ of $\mathbf{S_5}$. Use Problem 31 of Section 2.5 to show that the subgroup in $\mathbf{S_5}$ generated by $U$ is $\mathbf{S_5}$ itself.

37. Give the partition $\Gamma$ of Example 7 in the special case in which $G = \mathbf{A_4}$ and $H = \{(1), (12)(34), (13)(24), (14)(23)\}$.

*38. Let $H$ and $K$ be two different subgroups each of index 2 in a group $G$. Prove that

$$D = H \cap K$$

is a normal subgroup of index 4 in $G$ and that $G/D$ is not cyclic.

*39. Let $H$ and $K$ be subgroups in $G$ and let $H$ have index $m$ in $G$. Prove that the index of $H \cap K$ in $K$ must be in the set $\{1, 2, \ldots, m\}$.

## 3.6 Cartesian Products, Direct Products

We assume that the reader is familiar with ordered pairs from coordinate geometry. Here we use this concept as an aid in constructing new groups from known groups.

If $S$ and $T$ are sets, the set of all ordered pairs $(s, t)$ with $s$ in $S$ and $t$ in $T$ is called the *cartesian product* of $S$ by $T$ and is denoted by $S \times T$ (read as "$S$ cross $T$"). The name is in honor of René Descartes, one of the founders of analytic geometry. (See the biographical note after this section.)

We note that ordered pairs $(a, b)$ differ from pairs $\{a, b\}$ in the two following ways:

(1) A pair $\{a, b\}$ is a set with two distinct elements and hence the notation $\{a, b\}$ should only be used when $a \neq b$. However, the ordered pair $(a, a)$ is in $S \times T$ when $a$ is in both $S$ and $T$.

(2) If $a \neq b$, $(a, b) \neq (b, a)$ but $\{a, b\} = \{b, a\}$.

In $S \times T$, $(a, b) = (c, d)$ if and only if both $a = c$ and $b = d$. One can also show that $S \times T \neq T \times S$ unless $S = T$. (See Problems 1 and 2 of this section.)

**Theorem 1 Cartesian Product of Groups**

*Let $G_1$ and $G_2$ be groups and let $G$ be the cartesian product $G_1 \times G_2$ of their sets. Then $G$ is a group under the operation given by*

$$(a_1, a_2)(b_1, b_2) = (a_1 b_1, a_2 b_2).$$

The proof is left to the reader as Problem 7 of this section.

***Definition 1 Direct Product of Groups***

*If $G_1$ and $G_2$ are groups, the group $G_1 \times G_2$ of Theorem 1 is called the* **direct product** *of $G_1$ and $G_2$.*

Now we extend the concepts of cartesian product and direct product as follows: If $S_1, S_2, \ldots, S_n$ are sets, the set of all ordered $n$-tuples

$$(x_1, x_2 \ldots, x_n), \quad x_j \text{ in } S_j$$

is called the *cartesian product* of $S_1, S_2, \ldots, S_n$ and is designated as

$$S_1 \times S_2 \times \cdots \times S_n.$$

An ordered $n$-tuple $(x_1, \ldots, x_n)$ differs from a set $\{y_1, \ldots, y_n\}$ with $n$ elements in the same ways that an ordered pair differs from a pair. Specifically, the same element may be used more than once among the coordinates $x_j$ of $(x_1, \ldots, x_n)$; also,

$$(a_1, \ldots, a_n) = (b_1, \ldots, b_n)$$

if and only if $a_1 = b_1, a_2 = b_2, \ldots, a_n = b_n$.

If $G_1, G_2, \ldots, G_n$ are groups, then the cartesian product

$$G = G_1 \times G_2 \times \cdots \times G_n$$

of their sets is a group under the operation given by

$$(a_1, a_2, \ldots, a_n)(b_1, b_2, \ldots, b_n) = (a_1 b_1, a_2 b_2, \ldots, a_n b_n);$$

this group is called the ***direct product*** of $G_1, \ldots, G_n$.

The mapping $\alpha_i$ from a cartesian product

$$P = S_1 \times S_2 \times \cdots \times S_n$$

to $S_i$ with $\alpha_i(s_1, s_2, \ldots, s_n) = s_i$ is called the ***projection*** of $P$ on $S_i$.

Many mathematicians characterize a mapping $\theta$ from $S$ to $T$ using the subset $U$ of $S \times T$ consisting of all the ordered pairs $(s, t)$ with $\theta(s) = t$. Such a subset $U$ of $S \times T$ has the property that for every $s$ in $S$ there is exactly one ordered pair $(x, y)$ in $U$ with $x = s$. If one is given a subset $U$ of $S \times T$ with this property, one can define the associated mapping from $S$ to $T$ as the $\theta$ with $\theta(a) = b$ whenever $(a, b)$ is in $U$.

## Problems

1. Let $S = \{1, 2\}$ and $T = \{1, 3, 4\}$.
   (i) List the six ordered pairs of $S \times T$.
   (ii) List the six ordered pairs of $T \times S$.
   (iii) Does $S \times T = T \times S$ for these sets $S$ and $T$?
2. Explain why $S \times T = T \times S$ if and only if $S = T$.
3. How many elements are there in $S \times T$ when $S$ has $m$ elements and $T$ has $n$ elements?
4. Describe a bijection from $(S \times T) \times U$ to $S \times (T \times U)$.
5. Let $G_1 = [a] = \{e, a, a^2\}$ and $G_2 = [b] = \{e', b\}$ be cyclic groups of order 3 and 2, respectively.
   (i) Make the multiplication table for the direct product $G = G_1 \times G_2$.
   (ii) Is $G$ cyclic? Explain.
   (iii) Is $G$ isomorphic to the symmetric group $\mathbf{S}_3$? Explain.
   (iv) Tabulate a surjective homomorphism from $G$ onto $G_1$.
6. Show that every noncyclic group of order 4 (that is, every Klein 4-group) is isomorphic to a direct product $G_1 \times G_2$ with $G_1$ and $G_2$ each of order 2. (See Example 1 of Section 3.2.)
7. Give the details of showing that the operation

$$(a_1, a_2)(b_1, b_2) = (a_1 b_1, a_2 b_2)$$

   makes a cartesian product of groups into a group. (This is the proof of Theorem 1.)
8. (i) Explain why the direct product of abelian groups is also abelian.
   (ii) Is the octic group (of symmetries of a square) isomorphic to a direct product of two groups, having 2 and 4 as their orders? Explain.
9. Is the projection of a direct product $G_1 \times G_2$ on $G_1$ a homomorphism? Explain.
10. Let $G = G_1 \times G_2 \times G_3$, where each of $G_1$, $G_2$, and $G_3$ has order 2. Find the smallest positive integer $m$ such that $g^m$ is the identity of $G$ for all $g$ in $G$.
11. Let $\mathbf{R}^{\pm}$, $\mathbf{R}^{+}$, and $M$ denote the multiplicative groups of the nonzero real numbers, the positive real numbers, and the invertibles $\{1, -1\}$ of $\mathbf{Z}$, respectively. Explain why $\mathbf{R}^{\pm}$ is isomorphic to the direct product $\mathbf{R}^{+} \times M$.
12. Let $\mathbf{R}^{\pm}$ and $M$ be as in Problem 11 and let $\mathbf{R}$ denote the additive group of real numbers. Explain why $\mathbf{R}^{\pm}$ is isomorphic to the direct product $\mathbf{R} \times M$.
13. Let $\mathbf{C}$ and $\mathbf{R}$ denote the additive groups of the complex and of the real numbers, respectively. Explain why $\mathbf{C}$ is isomorphic to $\mathbf{R} \times \mathbf{R}$.

14. Let $V$, $\mathbf{R}^+$, and $U$ denote the multiplicative groups of the nonzero complex numbers, the positive real numbers, and the complex numbers of absolute value 1, respectively. Explain why $V$ is isomorphic to $\mathbf{R}^+ \times U$.

15. Let $H$ and $K$ be groups isomorphic to $H'$ and $K'$, respectively. Explain why $H \times K$ is isomorphic to $H' \times K'$.

16. Let $V$ and $\mathbf{R}^+$ be as in Problem 14. Let $T$ denote the quotient group $\mathbf{R}/[2\pi]$ of the additive group $\mathbf{R}$ of the real numbers by its cyclic subgroup $[2\pi]$. Explain why $V$ is isomorphic to $\mathbf{R}^+ \times T$.

17. Let $M = \{1, -1\}$ be a group of order 2 with 1 as identity and let $G = \mathbf{S_n} \times M$. Let $H$ be the subset of $G$ consisting of all $(\alpha, m)$ such that $m = 1$ when $\alpha$ is even and $m = -1$ when $\alpha$ is odd. Show that $H$ is a subgroup in $G$ and that $H$ is isomorphic to $\mathbf{S_n}$.

*18. Let $s$ and $t$ be positive integers and let $G = [a]$ be a cyclic group of order $st$. Show that $G$ is isomorphic to $[a^s] \times [a^t]$ if and only if $s$ and $t$ are relatively prime.

---

***René Descartes*** *(1596–1650)* *Descartes, the French philosopher, has a place in the history of mathematics largely due to his work* **La Géométrie**, *which was published as part of his masterpiece* **Discours de la Méthode** *(1637) and which contained the elements of analytic geometry. To say that Descartes invented analytic geometry is, as with most such statements, an oversimplification. One can indeed trace the ideas of analytic geometry back as far as Apollonius (c. 250* B.C.*) and the application of algebra to geometrical problems was certainly in the works of Viète (1540–1603) and Oughtred (1574–1660). Fermat wrote a treatise on analytic geometry at roughly the same time as* **La Géométrie**. *However, it was not published until later and Descartes, therefore, receives the credit for the discovery.* **La Géométrie** *is not a systematic treatment of analytic geometry but instead a series of problems in which the techniques of analytic geometry are used.*

*In the third book of* **La Géométrie** *he points out that if a cubic equation with rational coefficients has a rational root, then its roots can be constructed by straightedge and compass. He derived the cubic equation on which the angle trisection problem is based (see Section 6.3); however, he does not go on to consider the nonconstructability with straightedge and compass, but instead he proceeds to perform the trisection with the aid of a parabola and circle.*

*In addition to philosophy and mathematics, physics claimed much of his time during his most productive years, which he spent in Holland, 1629 to 1649. In 1649 he accepted an invitation of Queen Christina of Sweden to come to the Swedish court. He did not long withstand the rigors of the Stockholm winter and the perverse conviction of the Queen that 5 o'clock in the morning was the proper time for the study of philosophy and that was the time for Descartes to give her lessons. He died one year after going to Sweden.*

*Seventeen years after his death his bones were returned to Paris for entombment in the Panthéon. Because his political and philosophical views were potentially embarrassing to the government, there was no public oration. The mathematician Carl G. J. Jacobi commented: "It is often more convenient to possess the ashes of great men than to possess the men themselves during their lifetime."*

---

## 3.7 Relations

In the paragraph preceding the problems for Sections 3.6, we gave the characterization of a mapping from $S$ to $T$ as a special kind of subset of the cartesian product $S \times T$. Here we study general subsets of the cartesian product of a set with itself.

### Definition 1 Relation

*A relation on a set $S$ is a subset of the cartesian product $S \times S$.*

### Notation

*If $R$ stands for a relation on $S$, $aRb$ denotes that the ordered pair $(a, b)$ is in the subset $R$ of $S \times S$ and $a\bar{R}b$ means that $(a, b)$ is not in $R$. The letter $R$ may be replaced by other letters or by an appropriate symbol such as $<$, $>$, $\leq$, $\geq$, $\subset$, $\subseteq$, $\supset$, or $\supseteq$.*

One may read $aRb$ as "$a$ is related to $b$ (under $R$)" and $a\bar{R}b$ as "$a$ is not related to $b$ (under $R$)."

### Example 1

Let $S$ be the set of integers $\{1, 2, 3\}$. Then the $<$ relation on $S$ is the subset $\{(1, 2), (1, 3), (2, 3)\}$ of $S \times S$ and the $\leq$ relation on $S$ is

$$\{(1, 1), (1, 2), (1, 3), (2, 2), (2, 3), (3, 3)\}.$$

If $P$ stands for $\leq$, then $2P2$ tells us that 2 is less than or equal to itself and $3\bar{P}2$ tells us that 3 is not less than or equal to 2.

A mapping $\theta$ from a set $S$ to itself may be thought of as a relation $F$ on $S$ with the special property that for every $a$ in $S$ there is exactly one $b$ in $S$ with $aFb$. This unique $b$ is of course $\theta(a)$.

### *Definition 2 Reflexive, Transitive, Symmetric, Antisymmetric*

*Let $R$ be a relation on a set $S$. Then*
(i) *$R$ is* **reflexive** *if $aRa$ for all $a$ in $S$;*
(ii) *$R$ is* **transitive** *if $aRb$ and $bRc$ together imply $aRc$;*
(iii) *$R$ is* **symmetric** *if $aRb$ implies $bRa$;*
(iv) *$R$ is* **antisymmetric** *if $aRb$ and $bRa$ together imply $a = b$.*

#### *Example 2*

Let $A$, $B$, $C$, $D$, and $E$ be the relations on the integers **Z** in which $xAy$ means that $x < y$, $xBy$ means that $x \leq y$, $xCy$ means that $x - y$ is an even integer, $xDy$ means that $x|y$, and $xEy$ means that $x = y$. All five of these relations are transitive. All but $A$ are reflexive. (For $D$, this implies that we have defined 0 to be an integral divisor of itself.) $C$ and $E$ are symmetric while $A$, $B$, and $E$ are antisymmetric. ($A$ is vacuously antisymmetric, since $x < y$ and $y < x$ are contradictory.) Thus $E$ is both symmetric and antisymmetric. Also we note that $D$ is neither symmetric nor antisymmetric; $D$ is not symmetric since $3D6$ but $6\bar{D}3$, while $D$ is not antisymmetric since $(-3)D3$ and $3D(-3)$ but 3 and $-3$ are not equal.

The relations in Example 2 show that "antisymmetric" has a different meaning from "not symmetric."

### *Definition 3 Equivalence Relation*

*A relation that is reflexive, symmetric, and transitive is an* **equivalence relation.**

#### *Example 3*

Let $H$ be a subgroup in $G$ and let $R$ be the relation on $G$ for which $aRb$ means that $a$ and $b$ are in the same left coset of $H$ in $G$; that is, $aH = bH$. This relation is easily seen to be reflexive, symmetric, and transitive; hence it is an equivalence relation. The special case in which $G$ is the additive group **Z** of the integers and $H$ is a cyclic subgroup $[m]$ is called the relation of "congruence modulo $m$." This topic will be discussed in some detail in Section 4.3.

***Theorem 1*** ***Partitions and Equivalence Relations***

*Let $\Gamma$ be a partition of a set $S$ and let $E$ be the relation on $S$ for which $aEb$ means that $a$ and $b$ are in the same member set $C$ of the partition $\Gamma$. Then $E$ is an equivalence relation.*

The proof is left to the reader as Problem 7 of this section.

***Theorem 2*** ***Equivalence Classes and Partitions***

*Let $E$ be an equivalence relation on a nonempty set $S$. For every $a$ in $S$, let $C(a)$ be the subset of all elements $x$ of $S$ satisfying $aEx$. Then the collection $\Gamma$ of all these subsets $C(a)$ is a partition of $S$. [The $C(a)$ are called the* **equivalence classes** *of the equivalence relation $E$.]*

We leave the proof to the reader as Problem 8 of this section.

The transitive relation $L$ on the integers $\mathbf{Z}$ in which $xLy$ means $x < y$ has the property that if $a$ and $b$ are in $\mathbf{Z}$, then exactly one of the following holds:

$$\text{(i)}\quad a = b; \qquad \text{(ii)}\quad aLb; \qquad \text{(iii)}\quad bLa. \tag{T}$$

We next generalize on this relation.

***Definition 4*** ***Trichotomy***

*A relation $L$ on a set $S$ has the* **trichotomy** *property if whenever $a$ and $b$ are in $S$, then one and only one of the statements (i), (ii), (iii) of display (T) above is true.*

***Definition 5*** ***Linear Ordering***

*If a relation $R$ on a set $S$ is transitive and has the trichotomy property, $R$ is a* **linear ordering** *and $(S, R)$ is a* **linearly ordered set.**

If $S$ is any set of real numbers (for example, $\mathbf{Z}^+$, $\mathbf{Z}$, $\mathbf{Q}^+$, $\mathbf{Q}$, $\mathbf{R}^+$, or $\mathbf{R}$ itself), then clearly $<$ is a linear ordering on $S$ and $(S, <)$ is a linearly ordered set; also $(S, >)$ is a linearly ordered set.

*Notation*

*If $<$ is a linear ordering on a set $S$ and $x$ and $y$ are in $S$, $x \leq y$ means that either $x < y$ or $x = y$.*

***Definition 6 Inverse of a Relation***

*Let $A$ be a relation on a set $S$. Then the relation $B$ on $S$, such that $xBy$ if and only if $yAx$, is called the* **inverse relation** *of $A$ and is denoted by $A^{-1}$.*

## *Problems*

1. Let $H$ be a subgroup in $G$. Let $L$ be the relation on $G$ for which $aLb$ means that $a$ is in the left coset for $b$ of $H$ in $G$. Without using Theorem 1, show that $L$ is reflexive, symmetric, and transitive and hence is an equivalence relation.
2. Let $E$ be the relation on a group $G$ for which $aEb$ means that $a$ and $b$ have the same order. Show that $E$ is an equivalence relation.
3. Let $E$ be the relation on a group $G$ for which $aEb$ means that either $b = a$ or $b = a^{-1}$. Show that $E$ is an equivalence relation.
4. Let $C$ be the relation on a group $G$ such that $aCb$ means $a = g^{-1}bg$ for some $g$ in $G$; that is, $aCb$ means that $a$ is a conjugate of $b$. Show that $C$ is an equivalence relation.
5. Let $E$ be the relation on a group $G$ for which $aEb$ means that $[a] = [b]$. Show that $E$ is an equivalence relation.
6. Let $L$ be the relation on a group $G$ such that $aLb$ means that $ab = ba$. Is $L$ always an equivalence relation? Explain.
7. Prove Theorem 1 of this section.
8. Prove Theorem 2 of this section.
9. Explain why the relation $L$ on the integers $\mathbf{Z}$ for which $aLb$ means $a \leq b$ is not an equivalence relation.
10. Explain why the $L$ of Problem 9 is not a linear order relation.
11. How many distinct relations are there in a set $S$ with two elements?
12. How many distinct relations are there in a set $S$ with three elements?

13. Let $M$ be the relation on the integers $\mathbf{Z}$ for which $aMb$ means $b|a$. Is $M$ reflexive? Is it symmetric? Is it transitive? Is it a linear order relation?

14. Let $P$ consist of all the subsets of a set $S$ and let $I$ be the relation on $P$ for which $\alpha I \beta$ means $\alpha \subset \beta$. Is $I$ reflexive? Symmetric? Transitive? A linear order relation?

15. Let $P$ be the relation on $\mathbf{Z}$ such that $aPb$ means that $\gcd(a, b) = 1$. Is $P$ reflexive? Symmetric? Transitive? A linear order relation?

16. Let $A$ be the relation on $\mathbf{Z}$ such that $mAn$ means that both $m|n$ and $n|m$. Is $A$ an equivalence relation?

17. Let $L$ be the relation on the rational numbers $\mathbf{Q}$ for which $aLb$ means that $a > b$. Is $L$ a linear order relation?

18. Let $R$ be a linear order relation on a set $S$ and let $R^{-1}$ be the inverse relation of $R$. ($R^{-1}$ is such that $xR^{-1}y$ if and only if $yRx$.) Is $R^{-1}$ also a linear order relation? Explain.

19. Show that there are exactly 6 linear order relations on a set $S = \{a, b, c\}$ with 3 elements.

20. How many linear order relations are there on a set with $n$ elements?

21. Let $\alpha$ be a mapping from $S$ to itself, let $R$ be the relation on $S$ for which $xRy$ means $y = \alpha(x)$, and let $R^{-1}$ be the inverse relation of $R$. Show that there exists a mapping $\beta$ from $S$ to itself such that $uR^{-1}v$ means $v = \beta(u)$ if and only if $\alpha$ is bijective. (In other words, if either $\beta$ or the inverse function $\alpha^{-1}$ exists, the other exists and $\beta = \alpha^{-1}$.)

22. Let $R$ be a relation on a set $S$. Prove that $R$ is its own inverse if and only if $R$ is symmetric.

23. Let $L$ be a relation on a set $S$ such that $L$ is symmetric and transitive but is not reflexive.
    (i) Prove that there is an element $a$ in $S$ such that $a\bar{L}s$ for all $s$ in $S$.
    (ii) Give an example to show that there exists such a relation $L$.

24. Let $\Gamma$ be a collection of sets and let $\rho$ be the relation on $\Gamma$ such that $A\rho B$ means that there exists a bijection from $A$ to $B$. Prove that $\rho$ is an equivalence relation.

25. Let $\Gamma$ be the collection of all subgroups in a group $G$. Also let $\rho$ be the relation on $\Gamma$ such that $H\rho K$ if and only if $K$ is the conjugate subgroup of $H$ by some $g$ in $G$. Show that $\rho$ is an equivalence relation.

*26. Let $P$ be the set of all nonnegative real numbers and let $R$ be the relation on $P$ for which $aRb$ means that

$$a - \sqrt{a} \leq b - \tfrac{1}{4} \leq a + \sqrt{a}.$$

Show that $R$ is reflexive and symmetric but is not transitive.

## 3.8 Partially Ordered Sets

Here we generalize on the relation $\leq$ on the set of real numbers (or on any of its subsets).

### *Definition 1 Partial Ordering, Poset*

*Let $R$ be a reflexive, transitive, and antisymmetric relation on a set $S$; then $R$ is a* **partial ordering** *on $S$ and $(S, R)$ is a* **partially ordered set** *or (for short)* **poset**.

For example, $(S, \leq)$ or $(S, \geq)$ is a poset if $S$ is any subset of the real numbers.

### *Notation*

*If $(S, \leq)$ is a poset and $x$ and $y$ are in $S$, $x < y$ denotes that $x \leq y$ and $x \neq y$.*

### *Definition 2 Poset Diagram*

*Let $(S, \leq)$ be a poset, with $S$ finite. Let $D$ be a diagram having a vertex for every element of $S$ and having a rising line or broken line from the vertex for an element $s$ to that for an element $s'$ if and only if $s < s'$; then $D$ is a* **poset diagram** *for $(S, \leq)$. (Some of the vertices may be isolated, that is, have no lines emanating from them. Also it is customary to use as few segments as possible while satisfying this definition.)*

### *Example 1*

Let $G$ be a group and $\Gamma$ be the collection of all the subgroups in $G$. Then $(\Gamma, \subseteq)$ is a poset; there are poset diagrams for such posets in Sections 2.9, 2.10, and 3.4. Similarly, $(\Gamma, \supseteq)$ is a poset. Also, if $H \lhd K$ denotes that $H$ is a normal subgroup in $K$, $(\Gamma, \lhd)$ is a poset.

### *Definition 3. Minimal Element, Maximal Element*

*Let $(X, \beta)$ be a poset. An $m$ in $X$ is a* **minimal element** *if $m\beta x$ for all $x$ in $X$. An $M$ in $X$ is a* **maximal element** *if $x\beta M$ for all $x$ in $X$.*

We will see in Problems 9 and 10 below that when a minimal element exists it is unique and that the same is true for a maximal element. The poset $(\mathbf{Z}^+, \leq)$ has 1 as its minimal element and has no maximal element. The poset $(\mathbf{Z}, \leq)$ has neither a minimal nor a maximal element.

*Example 2*

Let $S = \{1, 2, 3, 4, 6, 12\}$ and $D$ be the relation on $S$ for which $xDy$ means that $x$ is an integral divisor of $y$. Then $(S, D)$ is a poset with 1 as the minimal element and 12 as the maximal element.

### Problems

1. Let $D$ be the relation on $S = \{2, 3, 4, 6, 12\}$ for which $xDy$ means that $x|y$. Which ordered pairs of $S \times S$ are in $D$?
2. Do Problem 1 with $\{2, 3, 4, 6, 12\}$ replaced by $\{1, 2, 3, 4, 6\}$.
3. Does the poset $(S, D)$ of Problem 1 have a minimal element? A maximal element?
4. Does the poset $(S, D)$ of Problem 2 have a minimal element? A maximal element?
5. Let $m$ be a minimal element for a poset $(S, \leq)$. What is the role of $m$ for the poset $(S, \geq)$?
6. Do Problem 5 with $m$ replaced by a maximal element $M$ for $(S, \leq)$.
7. In a poset $(X, \leq)$, let $x \leq y$ and $y \leq x$. Explain why one must have $x = y$.
8. In a poset $(X, \leq)$, is it possible to have $x < y$ and $y < x$ simultaneously? [*Hint*: See the meaning of the notation $x < y$ in $(X, \leq)$.]
9. Prove that there is at most one minimal element in a poset $(X, \leq)$.
10. Do Problem 9 with "minimal" replaced by "maximal."
11. (i) Construct a poset diagram for the poset $(S, D)$ of Problem 1.
    (ii) Why does the diagram of Part (i) have fewer segments than there are ordered pairs in $D$?
12. Give an example of a poset $(S, \leq)$ in which $S$ is infinite and there is both a minimal element and a maximal element.
13. Let $M$ be the relation on the set $\mathbf{Z}^+$ of positive integers for which $aMb$ means $b|a$, that is, $a$ is an integral multiple of $b$. Is $M$ reflexive? Is it symmetric? Is it antisymmetric? Is it transitive? Is it a linear ordering? Is it a partial ordering?

14. Let $P$ consist of all the subsets of a set $S$ and let $I$ be the relation on $P$ for which $\alpha I \beta$ means $\alpha \subset \beta$, that is, $\alpha$ is a proper subset of $\beta$. Is $I$ reflexive? Symmetric? Antisymmetric? Transitive? A linear ordering? A partial ordering?

## 3.9 Power Sets

*The essence of mathematics is its freedom.*

*Georg Cantor*

*No one can expel us from the paradise which Cantor created for us.*

*David Hilbert*

The main topic of this section has applications to fields such as computer science and logic. We will also see that it furnishes important examples for the other topics of this chapter.

### *Definition 1 Power Set, Universe, Complement*

*The collection of all the subsets of a set $S$ is called the* **power set** *for $S$ and is denoted by $P(S)$. $S$ is called the* **universe** *for $P(S)$. If $A$ is an element of $P(S)$, that is, a subset of $S$, then the subset consisting of the elements of $S$ not in $A$ is the* **complement** *of $A$ (in $S$) and is denoted by $\bar{A}$ (or by $S \backslash A$ when one wishes to show the role of the universe explicitly).*

### *Example 1*

Let $S$ be a set $\{s_1, s_2, s_3\}$ with three elements. Then

$$P(S) = \{\varnothing, \{s_1\}, \{s_2\}, \{s_3\}, \{s_1, s_2\}, \{s_1, s_3\}, \{s_2, s_3\}, S\}.$$

***Theorem 1 Power Set Poset***

*For any set S, $(P(S), \subseteq)$ is a poset with $\emptyset$ as the minimal element and S as the maximal element.*

***Proof***

Since $A \subseteq A$ for all sets $A$, the relation $\subseteq$ on $P(S)$ is reflexive. If $A \subseteq B$ and $B \subseteq C$, one has $A \subseteq C$; hence $\subseteq$ is transitive. Also, $A \subseteq B$ and $B \subseteq A$ together imply that $A = B$; hence $\subseteq$ is antisymmetric. Thus $(P(S), \subseteq)$ is a poset. Since $\emptyset \subseteq A \subseteq S$ for all $A$ in $P(S)$, $\emptyset$ is the minimal element and $S$ is the maximal element for this poset.

We next introduce a few concepts that are somewhat related to power sets.

***Notation $b^S$, Base b Numeral***

*Let S be a set and b be an integer greater than 1. Then* $\mathbf{b^S}$ *denotes the set of all mappings from S into $\{0, 1, 2, \ldots, b-1\}$. If S is a finite set $\{s_1, s_2, \ldots, s_n\}$, the* **base b numeral** *for a mapping $\alpha$ in $b^S$ is the n-digit numeral $d_1 d_2 \ldots d_n$ with $d_i = \alpha(s_i)$ for $1 \le i \le n$.*

The special case of this notation with $b = 2$ is used in the following:

***Definition 2 Characteristic Function, Binary Numeral***

*Let A be in the power set P for S; that is, A is a subset of S. Then the* **characteristic function** *for A is the mapping $\alpha$ in $2^S$ with $\alpha(s) = 1$ when s is in A and $\alpha(s) = 0$ when s is not in A. If S is finite, the base 2 numeral for $\alpha$ is also called the* **binary numeral** *for $\alpha$ or for A.*

***Example 2***

Let $A$ be the subset $\{s_2, s_4, s_5\}$ of $S = \{s_1, s_2, s_3, s_4, s_5\}$. Then the characteristic function for $A$ is the mapping $\alpha$ from $S$ into $\{0, 1\}$ with $\alpha\{s_2\} = \alpha(s_4) = \alpha(s_5) = 1$ and $\alpha(s_1) = \alpha(s_3) = 0$. Hence 01011 is the binary numeral for $A$ or for $\alpha$.

***Notation*** $\boldsymbol{P_n}$

*We let $P_n$ denote the set of all n-digit binary numerals.*

For example, $P_1 = \{0, 1\}$ and $P_2 = \{00, 01, 10, 11\}$.

***Example 3***

Let $S = \{s_1, s_2, \ldots, s_n\}$. For any subset $A$ of $S$, that is, any element $A$ of the power set $P(S)$, let $f(A)$ be the binary numeral for $A$. Then $f$ is a bijection from $P(S)$ onto $P_n$. If we fix $n$ as 3, the table for this bijection is

| $A$ | $\varnothing$ | $\{s_3\}$ | $\{s_2\}$ | $\{s_2, s_3\}$ | $\{s_1\}$ | $\{s_1, s_3\}$ | $\{s_1, s_2\}$ | $S$ |
|---|---|---|---|---|---|---|---|---|
| $f(A)$ | 000 | 001 | 010 | 011 | 100 | 101 | 110 | 111 |

and the poset diagram for $(P(S), \subseteq)$ is given in Figure 3.7, with the number for $A$ as the label of the vertex $A$.

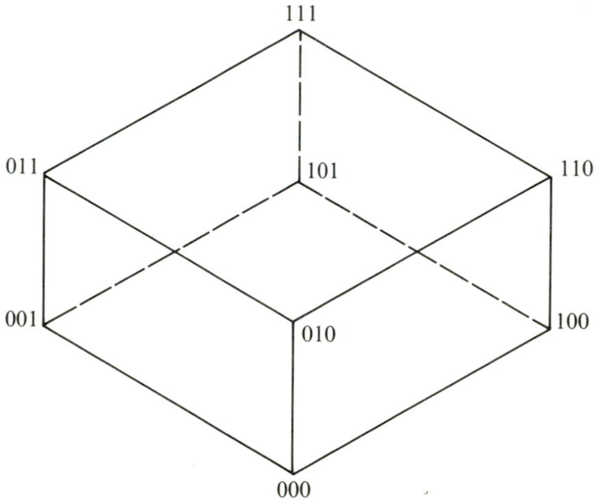


***Figure 3.7***

***Problems***

1. Let $S = \{s_1, s_2, s_3, s_4, s_5\}$ and $P$ be its power set.
   (a) What is the binary numeral for $A = \{s_1, s_4\}$? Also, give the complement $\bar{A}$ and its binary numeral.

(b) Give the element $B$ of $P$ with 01110 as its binary numeral and also its complement $\bar{B}$.
(c) Give the numeral for the minimal element of $(P, \subseteq)$.
(d) Tell how to obtain the numeral $c_1c_2c_3c_4c_5$ for $A \cup B$ from the numerals $a_1a_2a_3a_4a_5$ and $b_1b_2b_3b_4b_5$ for $A$ and $B$.
(e) Does $A \cup \bar{A} = S$ for all $A$ in $P$?

2. Let $P$ be the power set for $S = \{s_1, s_2, \ldots s_n\}$.
(a) Describe the binary numeral for the maximal element of $(P, \subseteq)$.
(b) Tell how to obtain the numeral $c_1c_2 \ldots c_n$ for $\bar{A}$ from the numeral $a_1a_2 \ldots a_n$ for $A$.
(c) Tell how to obtain the numeral $d_1d_2 \ldots d_n$ for $A \cap B$ from the numerals $a_1a_2 \ldots a_n$ and $b_1b_2 \ldots b_n$ for $A$ and $B$.
(d) Does $A \cap \bar{A} = \varnothing$ for all $A$ in $P$?

3. Give the answer to Problem 2(d) when $S$ is an infinite set.

4. Give the answer to Problem 1(e) when $S$ is an infinite set.

5. In a power set $P(S)$, let $B = \bar{A}$. Does $\bar{B} = A$?

6. Let $A$ and $B$ be in $P(S)$, $A \cup B = S$, and $A \cap B = \varnothing$. Is each of $A$ and $B$ the complement of the other in $S$?

7. Is $(P(S), \supseteq)$ a poset for every set $S$? If so, what is its minimal element?

8. What is the maximal element for $(P(S), \supseteq)$?

9. Let $S = \{s_1, s_2, \ldots, s_n\}$. How many mappings are there in $3^S$?

10. Do Problem 9 with $3^S$ replaced by $b^S$.

11. Let $S$ be a set. Is every mapping $\alpha$ in $2^S$ the characteristic function for some subset $A$ of $S$?

12. If $S$ is a set with $n$ elements, how many elements are there in $P(S)$?

13. Let $\leq$ be the relation on the set $P_n$ of $n$-digit binary numerals for which $a_1a_2 \ldots a_n \leq b_1b_2 \ldots b_n$ means that $a_i \leq b_i$ for $i = 1, 2, \ldots, n$.
(a) Explain why $(P_n, \leq)$ is a poset.
(b) What is the minimal element of this poset?
(c) What is the maximal element?

14. Let $T$ be a subset of $P_n$ and $\leq$ be as in Problem 13.
(a) Explain why $(T, \leq)$ is a poset.
(b) Does it always have a minimal element? Explain.
(c) Does it always have a maximal element? Explain.

15. Let $S$ be a set and $T$ consist of some of the subsets of $S$ in $P(S)$. Explain why $(T, \subseteq)$ is a poset.

16. Give an example in which the $(T, \subseteq)$ of Problem 15 has neither a minimal nor a maximal element.

## *3.10 Operations

Some material concerning operations has been presented in Chapter 2; we now give a more formal and more general treatment.

### *Definition 1 Operation on a Set*

*Let $S^n$ be the cartesian product $S \times S \times \cdots \times S$ of n copies of a set S. An* **n-ary operation** *on S is a mapping F from $S^n$ to S. In particular, a* **unary operation** *on S is a mapping from S to itself and a* **binary operation** *on S is a mapping from $S \times S$ to S.*

With this definition, $S$ must be closed under any operation on $S$.

### *Notation*

*The image of an ordered n-tuple $(s_1, s_2, \ldots, s_n)$ in $S^n$ under an n-ary operation F on S may be denoted by $F(s_1, s_2, \ldots, s_n)$. The image $B(x, y)$ under a binary operation may also be written as $x + y$ when B represents addition, as $x \cdot y$ or $xy$ when B is multiplication, as $x - y$ when B is subtraction, and as $x/y$ or $x \div y$ when B is division. The symbols $\circ$ and $*$ are used frequently to represent abstract binary operations.*

It is clear from the above definition and notation that an $n$-ary operation on the set $\mathbf{R}$ of real numbers is the same as a real valued function of $n$ real variables.

### *Example 1*

Let $c$ be a fixed element of a group $G$. Then the mappings $\alpha$ and $\beta$ of $G$ into $G$ with $\alpha(x) = x^{-1}$ and $\beta(x) = cx$ are unary operations on $G$.

### *Example 2*

Let $S$ be the subset $\{1, 2, 5, 7, 10, 14, 35, 70\}$ of the integers and let $L(x, y)$ mean $\mathrm{lcm}[x, y]$. One can easily check that $L(x, y)$ is in $S$ whenever $x$ and $y$ are in $S$; hence $L$ is a binary operation on $S$.

***Example 3***

Let $A(x_1, x_2, \ldots, x_n) = (x_1 + x_2 + \cdots + x_n)/n$ whenever the $x_i$ are in **Q**. Clearly, this averaging function $A$ is an $n$-ary operation on **Q**.

***Definition 2 Associativity, Commutativity, Identity, Inverse***

*Let $\theta$ be a binary operation on $S$. Then $\theta$ is* **associative** *if $\theta(\theta(x, y), z) = \theta(x, \theta(y, z))$ for all $x$, $y$, $z$ in $S$. $\theta$ is* **commutative** *if $\theta(x, y) = \theta(y, x)$ for all $x$, $y$ in $S$. An element $e$ of $S$ is an* **identity** *under the operation $\theta$ if $\theta(x, e) = x = \theta(e, x)$ for all $x$ in $S$. If $e$ is an identity under $\theta$ and $s$ is in $S$, an* **inverse** *of $s$ with respect to $\theta$ and $e$ is an element $s^{-1}$ of $S$ such that $\theta(s, s^{-1}) = e = \theta(s^{-1}, s)$.*

Using the symbol $\circ$ for an abstract binary operation on a set $S$, we can rewrite these definitions as follows:

**Associativity** *If $(x \circ y) \circ z = x \circ (y \circ z)$ for all $x$, $y$, $z$ in $S$, $\circ$ is associative.*
**Commutativity** *If $x \circ y = y \circ x$ for all $x$, $y$ in $S$, $\circ$ is commutative.*
**Identity** *If $x \circ e = x = e \circ x$ for all $x$ in $S$, $e$ is an identity under $\circ$.*
**Inverse** *If $e$ is an identity under $\circ$ and $s \circ t = e = t \circ s$, then $t$ is an inverse of $s$ with respect to $\circ$.*

One can readily see that the binary operation $L$ of Example 2 is associative and commutative. Also the element 1 is an identity for $L$ since $\text{lcm}[x, 1] = x = \text{lcm}[1, x]$ for all $x$ in $S$. With respect to this operation, only the element 1 has an inverse. For example, the integer 2 in $S$ does not have an inverse since $\text{lcm}[2, x]$ is not the identity 1 for any $x$ in $S$.

***Theorem 1 Uniqueness of the Identity***

*There is at most one identity under a binary operation $\circ$ on a set $S$.*

The proof is the same as that for Lemma 1 of Section 2.3.

Having this result, we can now write "the identity" instead of "an identity."

***Example 4***

Let $P$ be the power set for a universe $S$. Then the operation of taking the complement with respect to $S$ is a unary operation on $S$. Also the operation $\cup$ of forming the union of two sets and the operation $\cap$ of forming the intersection of two sets are associative and commutative binary operations on $P$. In the special case in which $S = \{s_1\}$ is a single element set, $P = \{\emptyset, S)\}$. For this case (and using the binary numerals 0 and 1 for $\emptyset$ and $S$, respectively) the unary operation of taking complements and the binary operations of forming unions and forming intersections are given by the following tables:

| $A$ | 0 | 1 |
|---|---|---|
| $\bar{A}$ | 1 | 0 |

| $\cup$ | 0 | 1 |
|---|---|---|
| 0 | 0 | 1 |
| 1 | 1 | 1 |

| $\cap$ | 0 | 1 |
|---|---|---|
| 0 | 0 | 0 |
| 1 | 0 | 1 |

***Definition 3*** ***Min and Max Operations***

*Let $(S, <)$ be a linearly ordered set. For $x$ and $y$ in $S$, let* **min**$(x, y)$ *be $x$ if $x \le y$ and be $y$ if $y < x$. Also, let* **max**$(x, y)$ *be $x$ if $y \le x$ and be $y$ if $x < y$.*

***Example 5***

Let $S$ be the set $\{2, 3, 4, 5\}$ of integers. Let $\circ$ and $*$ be the operations on $S$ with $x \circ y = \min(x, y)$ and $x * y = \max(x, y)$. Then each of $\circ$ and $*$ is commutative and associative. The largest element 5 is the identity for the operation $\circ$ (that is, for the operation min). The smallest element 2 is the identity for $*$ (that is, for max).

## Problems

1. Let $S$ have 3 elements.
   (a) How many unary operations are there on $S$?
   (b) How many binary operations are there on $S$?
   (c) How many of the binary operations are commutative?
   (d) How many $n$-ary operations are there on $S$?
2. Do Problem 1 with $S$ a set having $m$ elements.
3. Let $*$ be the operation on $\mathbf{Z}^+$ with $x * y = \max(x, y)$. Is there an identity under $*$? Does each element of $\mathbf{Z}^+$ have an inverse with respect to this operation?
4. Do Problem 3 with $\mathbf{Z}^+$ replaced by the set $\mathbf{Z}$ of all the integers.
5. Do Problem 3 with $*$ replaced by the operation $\circ$ with $x \circ y = \min(x, y)$.

6. Answer the questions of Problem 3 for the operation $\circ$ on $\mathbf{Z}$ with $x \circ y = \min(x, y)$.
7. Let $P$ be the power set for $S = \{1, 2\}$.
   (a) On $P$, tabulate the unary operation $A \mapsto \bar{A}$ of complementation.
   (b) Tabulate the binary operation $\cup$ on $P$.
   (c) What is the identity for $\cup$?
8. Do as in Problem 7(b) and (c) with $\cup$ replaced by $\cap$.
9. Let $P$ be the power set for $S = \{s_1, s_2, s_3\}$.
   (a) In terms of the numerals of

$$P_3 = \{000, 001, 010, 100, 011, 101, 110, 111\}$$

   tabulate the binary operation $\cap$ on $P$.
   (b) What is the identity under the operation of (a)?
10. Do Problem 9 with $\cap$ replaced by $\cup$.
11. For each of the following, tell whether the statement is true for all subsets in the power set $P(S)$ for a fixed set $S$:
   (a) *Idempotent law for intersections:*

$$A \cap A = A.$$

   (b) *Idempotent law for unions:*

$$A \cup A = A.$$

   (c) *Associativity for intersections:*

$$A \cap (B \cap C) = (A \cap B) \cap C.$$

   (d) *Associativity for unions:*

$$A \cup (B \cup C) = (A \cup B) \cup C.$$

   (e) *Distributivity of intersections over unions:*

$$A \cap (B \cup C) = (A \cap B) \cup (A \cap C).$$

   (f) *Distributivity of unions over intersections:*

$$A \cup (B \cap C) = (A \cup B) \cap (A \cup C).$$

12. Let $\mathbf{Z}^+ = \{1, 2, 3, \ldots\}$ and let $\alpha$ and $\beta$ be the operations in $\mathbf{Z}^+$ with

$$(m, n) \overset{\alpha}{\mapsto} \gcd(m, n), \qquad (m, n) \overset{\beta}{\mapsto} \operatorname{lcm}[m, n].$$

For each of the following, tell whether or not the statement is true for all positive integers:

(a) *Idempotent law for* $\alpha$:

$$\gcd(a, a) = a.$$

(b) *Idempotent law for* $\beta$:

$$\operatorname{lcm}[a, a] = a.$$

(c) *Associativity for* $\alpha$:

$$(a, (b, c)) = ((a, b), c).$$

(d) *Associativity for* $\beta$:

$$[a, [b, c]] = [[a, b], c].$$

(e) *Distributivity of* $\alpha$ *over* $\beta$:

$$(a, [b, c]) = [(a, b), (a, c)].$$

(f) *Distributivity of* $\beta$ *over* $\alpha$:

$$[a, (b, c)] = ([a, b], [a, c]).$$

13. Let $P$ be a power set $P(S)$. For each $A$ in $P$ (that is, subset $A$ of $S$), let $\bar{A}$ be the complement of $A$ in $S$. For each of the following, tell whether the statement is true or false:
(a) The complement of $\bar{A}$ is $A$ for all $A$ in $P$.
(b) $\overline{A \cup B} = \bar{A} \cap \bar{B}$ for all $A$ and $B$ in $P$.
(c) $\overline{A \cap B} = \bar{A} \cup \bar{B}$ for all $A$ and $B$ in $P$.

14. Let $P$ be a power set $P(S)$. For $A$ in $P$, let $\bar{A}$ be the complement of $A$ in $S$ and $f$ be the mapping from $P$ to $P$ with $f(A) = \bar{A}$.
(a) Is $f$ a bijection? Explain.
(b) What is the inverse function of $f$?

15. (a) What is the identity for the operation $\cup$ in Problem 11?
(b) What is the identity for the operation $\cap$ in Problem 11?

16. (a) Is there an identity for the operation $\alpha$ in Problem 12?
    (b) Is there an identity for the operation $\beta$ in Problem 12?
17. Let $P$ be the power set for $S = \{s_1, s_2, \ldots, s_n\}$. In $P$, let $A \cup B = C$ and let $a_1a_2 \ldots a_n$, $b_1b_2 \ldots b_n$, and $c_1c_2 \ldots c_n$ be the binary numerals for $A$, $B$, and $C$. Tell how to find $c_i$ directly from $a_i$ and $b_i$. [*Hint*: Use one of the operations min and max of Example 5 above.]
18. Do Problem 17 with $\cup$ replaced by $\cap$.
19. Let $M$ be the set of nonzero real numbers and let $C = M \times M$. Let $\theta$ be the operation in $M$ (mapping from $C$ to $M$) such that the image of $(a, b)$ under $\theta$ is the real number product of $a$ and $|b|$.
    (i) Explain why $\theta$ is associative but not commutative.
    (ii) Show that $M$ has two "right identities" under $\theta$, that is, two elements $j$ such that, for all $a$ in $M$,

$$(a, j) \overset{\theta}{\mapsto} a$$

    (iii) Show that $M$ has no left identity, that is, no element $k$ such that, for all $b$ in $M$,

$$(k, b) \overset{\theta}{\mapsto} b.$$

    (iv) Let $j$ be either one of the right identities. Show that every element $b$ of $M$ has a left inverse (with respect to $j$) under $\theta$; that is, show the existence of a $b^{-1}$ with

$$(b^{-1}, b) \overset{\theta}{\mapsto} j.$$

*20. Let $S$ be a set and let $\theta$ be an operation in $S$, that is, a mapping from $S \times S$ to $S$. Let $\theta$ be associative and let $e$ be a left identity under $\theta$. For every $a$ in $S$ let there be a left inverse $a^{-1}$ under $\theta$ (with respect to $e$). Prove that $e$ is also a right identity and that a left inverse $a^{-1}$ is also a right inverse. Note that this proves that $S$ is a group under the operation $\theta$.

## *3.11 Algebraic Structures

An ***algebraic structure*** is a set $S$, called the ***carrier*** of the structure, with one or more operations on $S$. The type of structure depends on the axioms satisfied by these operations. With the formal definition that an $n$-ary operation on $S$ is a mapping from $S^n$ to $S$, the set $S$ is automatically closed under any operation on $S$. Binary operations are most common and associativity is the most useful axiom; therefore we begin with a type of structure that involves a single binary operation and has associativity as the sole axiom.

### *Definition 1 Semigroup*

*A* **semigroup** *is an ordered pair* $(S, \circ)$, *with* $S$ *a set and with* $\circ$ *an associative binary operation on* $S$.

We note that every group is a semigroup since associativity is one of the axioms for the binary operation of a group. A semigroup may or may not be commutative; thus a commutative semigroup is an ordered pair $(S, \circ)$ such that $\circ$ is a commutative and associative binary operation on $S$.

#### *Example 1*

Let $E$ be the set of even integers. Since the product $x \cdot y$ of even integers is always an even integer, $\cdot$ is a binary operation on $E$. As this operation is associative, $(E, \cdot)$ is a semigroup; it is not a group. Since $\cdot$ is also commutative, $(E, \cdot)$ is a commutative semigroup.

#### *Example 2*

$(\mathbf{Z}, -)$ is not a semigroup since the binary operation of subtraction is not associative, as we see in the example

$$(10 - 6) - 3 = 4 - 3 = 1, \qquad 10 - (6 - 3) = 10 - 3 = 7.$$

$(\mathbf{Z}, \div)$ is not a semigroup since $\mathbf{Z}$ is not closed under division and hence $\div$ is not an operation on $\mathbf{Z}$. Let $\mathbf{Q}^+$ be the set of positive rational numbers; then $\div$ is an operation on $\mathbf{Q}^+$ but $(\mathbf{Q}^+, \div)$ is not a semigroup since $\div$ is not associative.

### *Definition 2 Subsemigroup*

*If* $(T, *)$ *and* $(S, \circ)$ *are semigroups with* $T$ *a subset of* $S$ *and with* $x * y = x \circ y$ *for all* $x$ *and* $y$ *in* $T$, *then* $(T, *)$ *is a* **subsemigroup** *in* $(S, \circ)$.

If $(T, *)$ is a subsemigroup of $(S, \circ)$, we will usually denote the operation of the subsemigroup by the same symbol as for the semigroup since $x * y = x \circ y$ for all $x$ and $y$ in the subset $T$ of $S$.

***Definition 3 Powers***

*Let* $(S, \circ)$ *be a semigroup and* $a \in S$. *Then* $a^2$ *denotes* $a \circ a$, $a^3$ *denotes* $a \circ a^2$, *and so on. Also* $\langle a \rangle$ *denotes the set* $\{a^n : n \in \mathbf{Z}^+\}$ *of positive integral powers of* $a$.

***Theorem 1 Principal Subsemigroup Generated by an Element***

*Let* $(S, \circ)$ *be a semigroup and* $a \in S$. *Then* $(\langle a \rangle, \circ)$ *is a subsemigroup in* $(S, \circ)$.

The result follows readily from the definitions; $\langle a \rangle$ is called the ***principal subsemigroup generated by a.***

***Example 3***

Let $H = \{0, 1, 2, \ldots, 99\}$ and $\circ$ be multiplication modulo 100, that is, $x \circ y$ is the number formed by the two rightmost digits of the ordinary product $xy$ of integers. Then $(H, \circ)$ is a semigroup and the principal subsemigroup $\langle 6 \rangle$ is $\{6, 36, 16, 96, 76, 56\}$. Note that the powers $6, 6^2, 6^3, \ldots$ form the sequence 6, 36, 16, 96, 76, 56, 36, 16, ... in which the first term, 6, does not reappear and the block of the next five terms repeats endlessly.

***Definition 4 Monoid***

*A* **monoid** *is a semigroup* $(S, \circ)$ *having an identity under* $\circ$.

Thus a monoid is an ordered triple $[S, \circ, e]$ in which $\circ$ is an associative binary operation on $S$ and $e$ is an identity for the operation $\circ$. The two axioms for this type of algebraic structure are as follows:

$M_1$ **Associativity** $(x \circ y) \circ z = x \circ (y \circ z)$ for all $x, y, z$ in $S$.

$M_2$ **Identity** $x \circ e = x = e \circ x$ for all $x$ in $S$.

Since these axioms are included among the group axioms, every group is a monoid. In fact a group is a monoid $[G, \circ, e]$ such that every element $g$ of $G$ has an inverse (with respect to $\circ$).

*Example 4*

$[\mathbf{Z}^+, \cdot, 1]$ is a monoid but is not a group since, for example, the positive integer 2 does not have a multiplicative inverse in $\mathbf{Z}^+$. Let $N = \{0, 1, 2, \ldots\}$; then $[N, +, 0]$ is a monoid and is not a group since, for example, 1 does not have an inverse under $+$ in $N$.

## Problems

1. (a) Is $(\mathbf{Z}^+, +)$ a semigroup? Is it a monoid? Is it a group?
   (b) Do Part (a) with $(\mathbf{Z}^+, +)$ replaced by $(\mathbf{Z}^+, \cdot)$.
2. Let $\mathbf{Q}^+$ be the set of positive rational numbers.
   (a) Is $(\mathbf{Q}^+, +)$ a semigroup? Is it a monoid? Is it a group?
   (b) Do Part (a) with $(\mathbf{Q}^+, +)$ replaced by $(\mathbf{Q}^+, \cdot)$.
3. Let $\mathbf{R}$ be the real numbers and $-$ be the binary operation of subtraction. Is $(\mathbf{R}, -)$ a semigroup? Is it a monoid? Is it a group?
4. Let $\mathbf{R}^+$ be the positive real numbers. Is $(\mathbf{R}^+, \div)$ a semigroup? Is it a monoid? Is it a group?
5. Let $S$ be a general set and $P$ be its power set.
   (a) Is $[P, \cup, \varnothing]$ a commutative monoid?
   (b) Is $[P, \cup, \varnothing]$ a group? Explain.
6. Do Problem 5 with $[P, \cup, \varnothing]$ replaced by $[P, \cap, S]$.
7. (a) Give an example of a semigroup that is not a monoid.
   (b) Does there exist a monoid that is not a semigroup? Explain.
8. (a) Give an example of a monoid that is not a group.
   (b) Does there exist a group that is not a monoid? Explain.
9. Let $(H, \circ)$ be the semigroup of Example 3. List the elements of the principal subsemigroup $\langle 2 \rangle$.
10. (a) Do Problem 9 with $\langle 2 \rangle$ replaced by $\langle 8 \rangle$.
    (b) Do Problem 9 with $\langle 2 \rangle$ replaced by $\langle 5 \rangle$.

## *3.12 Boolean Algebra

The structures studied in this section were introduced by the British mathematician George Boole (1815–1864) in his effort to provide an algebraic basis for logic. These structures have applications in situations featuring dichotomy. In logic the dichotomy is that of "true" or "false." In electrical circuits it is that of switches being in the "on" or "off" position and of the current "flowing" or "not flowing." In set theory it is that of an element being "in" or "out" of a set.

We start by considering an algebraic structure $(S, \circ, *)$ consisting of a carrier set $S$ and binary operations $\circ$ and $*$ on $S$.

### *Definition 1 Distributivity*

*If* $x \circ (y * z) = (x \circ y) * (x \circ z)$ *for all* $x, y, z$ *in* $S$, *then* $\circ$ *is* **distributive** *over* $*$.

#### *Example 1*

For the structure $(\mathbf{Z}, +, \cdot)$, we know that multiplication is distributive over addition, that is, $x(y + z) = xy + xz$ for all $x, y, z$ in $\mathbf{Z}$. However, addition is not distributive over multiplication, since, for example, $1 + (2 \cdot 3)$ does not equal $(1 + 2) \cdot (1 + 3)$.

#### *Example 2*

Let $P$ be the power set for a general universe $S$. We saw in Problems 5 and 6 of Section 3.11 that $[P, \cup, \varnothing]$ and $[P, \cap, S]$ are commutative monoids. The unary operation $\bar{}$ of forming the complement is related to the identities of these monoids by the properties that

$$A \cup \bar{A} = S \quad \text{and} \quad A \cap \bar{A} = \varnothing \quad \text{for all } A \text{ in } P.$$

In the structure $(P, \cup, \cap)$, each of the binary operations $\cup$ and $\cap$ is distributive over the other; that is,

$$A \cup (B \cap C) = (A \cup B) \cap (A \cup C),$$

$$A \cap (B \cup C) = (A \cap B) \cup (A \cap C),$$

for all $A, B, C$ in $P$.

### *Definition 2 Boolean Algebra, Join, Meet, Complement*

*A* **boolean algebra** *is an ordered 6-tuple* $[X, \vee, \wedge, ', O, I]$ *in which* $X$ *is a set,* $\vee$ *and* $\wedge$ *are binary operations on* $X$, *each of which is distributive over the other,* $O$ *and* $I$ *are distinct elements of* $X$ *such that* $[X, \vee, O]$ *and* $[X, \wedge, I]$ *are commutative monoids, and* $'$ *is a unary operation on* $X$ *with* $x \vee x' = I$ *and* $x \wedge x' = O$ *for all* $x$ *in* $X$. *One calls* $x \vee y$ *the* **join** *of* $x$ *and* $y$, $x \wedge y$ *the* **meet** *of* $x$ *and* $y$, *and* $x'$ *the* **complement** (*or* **negation**) *of* $x$.

The statements in Example 2 imply that $[P(S), \cup, \cap, ^{-}, \varnothing, S]$ is a boolean algebra for every nonempty set $S$.

***Definition 3 Power Set Boolean Algebras***

*We call* $[P(S), \cup, \cap, ^{-}, \varnothing, S]$ *the* **P(S) boolean algebra.**

These structures furnish examples of boolean algebras with infinite carriers and with carriers having $2^n$ elements, where $n$ is any positive integer. In particular, when $S$ has only one element, the carrier consists just of the identities $\varnothing$ and $S$ under $\cup$ and $\cap$, respectively. Hence, we call this the ***doubleton boolean algebra*** and usually write its carrier as $\{O, I\}$ or $\{0, 1\}$; it is fundamental in applications. See Example 4 of Section 3.10 for tables of its operations.

***Theorem 1 Complements of O and I***

*In a boolean algebra,* $O' = I$ *and* $I' = O$.

***Proof***

Since $O$ is the identity under $\vee$, $O' = O \vee O'$. Also $O \vee O' = I$ by one of the axioms on complementation. Hence $O' = I$. It is left to the reader as Problem 3 below to give the similar proof that $I' = O$.

***Theorem 2 Duality***

*Any true equation in a boolean algebra* $B$ *remains true when each element* $x$ *is replaced by its complement* $x'$, *each* $\vee$ *is replaced by* $\wedge$, *and each* $\wedge$ *is replaced by* $\vee$. *(The result of these replacements is called the* **dual** *of the original equation.)*

The proof is outlined here but several details are left to the reader as problems below. We observe that the dual of any axiom (in the definition of a boolean algebra) is also an axiom. For example, commutativity of $\vee$, i.e., $x \vee y = y \vee x$ for all $x$ and $y$ in the carrier $X$, has as its dual $x' \wedge y' = y' \wedge x'$. But $x'$ and $y'$ can be any elements of $X$ since complementation is a surjective mapping from the carrier onto itself. [See Problem 22(ii) below.] Thus the dual of the axiom that $\vee$ is commutative is essentially the axiom that $\wedge$ is commutative. This self-dual nature of the axioms for a boolean algebra then implies that the proof of any result becomes the proof of its dual when each step is replaced by its dual and each justification (that is, reference to an axiom or previous result) is also replaced by its dual.

There are many examples of duality in the problems for this section; see especially Problems 25 and 26, in which each of DeMorgan's Laws is the dual of the other.

### *Definition 4 Boolean Subalgebra*

*Let $B = [X, \vee, \wedge, ', O, I]$ be a boolean algebra. Let $Y$ be a subset of $X$; $O$ and $I$ be in $Y$; and $Y$ be closed under $\vee$, $\wedge$, and $'$. Then $[Y, \vee, \wedge, ', O, I]$ is a* **boolean subalgebra** *in $B$.*

### *Definition 5 Boolean Homomorphisms and Isomorphisms*

*Let $A = (X, \vee, \wedge, ', O, I)$ and $B = (Y, \vee, \wedge, ', O, I)$ be boolean algebras. A* **boolean homomorphism** *from $A$ to $B$ is a mapping $f$ from $X$ to $Y$ such that $f(x_1 \vee x_2) = f(x_1) \vee f(x_2)$ and $f(x_1 \wedge x_2) = f(x_1) \wedge f(x_2)$ for all $x_1$ and $x_2$ in $X$, $f(x') = [f(x)]'$ for all $x$ in $X$, $f(O) = O$, and $f(I) = I$. A* **boolean isomorphism** *is a bijective boolean homomorphism.*

### *Definition 6 Boolean Cartesian Product Operations*

*Let $B_i$ be a boolean algebra $(Y_i, \vee, \wedge, ', O, I)$ for $i = 1, 2, \ldots, n$. Then the operations $\vee$, $\wedge$, and $'$ are defined on the cartesian product $Y = Y_1 \times Y_2 \times \cdots \times Y_n$ as follows:*

$$(a_1, a_2, \ldots, a_n) \vee (b_1, b_2, \ldots, b_n) = (a_1 \vee b_1, a_2 \vee b_2, \ldots, a_n \vee b_n)$$

$$(a_1, a_2, \ldots, a_n) \wedge (b_1, b_2, \ldots, b_n) = (a_1 \wedge b_1, a_2 \wedge b_2, \ldots, a_n \wedge b_n)$$

$$(a_1, a_2, \ldots, a_n)' = (a_1', a_2', \ldots, a_n').$$

### *Theorem 3 Boolean Direct Product*

*Under the operations of Definition 6, the cartesian product of the carriers of n boolean agebras is also a boolean algebra. (We call it the* **boolean direct product**.)

The proof is straightforward and the details are left to the reader.

## Problems

In Problems 1, 2, 3, and 11 through 26 below, $B$ denotes an arbitrary boolean algebra and $X$ is its carrier.

1. In $B$, show the following:
   (i) $O \wedge (O \vee O') = O$.
   (ii) $(O \wedge O) \vee (O \wedge O') = O \wedge O$.
   (iii) $O \wedge O = O$.
2. In the work for Problem 1, replace $O$, $\wedge$, and $\vee$ wherever each such symbol appears with $I$, $\vee$, and $\wedge$, respectively, and thus prove that $I \vee I = I$.
3. Prove that $I' = O$ and thus do the part of Theorem 1 left to the reader. [*Hint*: Dualize the proof that $O' = I$ in Theorem 1.]
4. In a boolean algebra, let the carrier be $\{O, I\}$. Tabulate the unary operation of complementation.
5. Tabulate the binary operation $\vee$ for the boolean algebra of Problem 4.
6. Tabulate the operation $\wedge$ for the boolean algebra of Problem 4.
7. Let $\vee$, $\wedge$, and $'$ be the operations on $X = \{1, 3, 5, 15\}$ with $x \vee y = \text{lcm}[x, y]$, $x \wedge y = \text{gcd}(x, y)$, and $x' = 15/x$.
   (a) Tabulate the operations $\vee$, $\wedge$, and $'$ on $Y$.
   (b) Is $\vee$ distributive over $\wedge$?
   (c) Is $\wedge$ distributive over $\vee$?
   (d) Is $[X, \vee, \wedge, ', 1, 15]$ a boolean algebra?
8. Let $Y = \{1, 3, 5, 7, 15, 21, 35, 105\}$, $y' = 105/y$ and $\vee$ and $\wedge$ be as in Problem 7. Is $[Y, \vee, \wedge, ', 1, 105]$ a boolean algebra?
9. Let $\vee$ and $\wedge$ be the operations on $Y = \{1, 2, 4, 8\}$ with $x \vee y = \text{lcm}[x, y]$ and $x \wedge y = \text{gcd}(x, y)$.
   (i) Find the identities for the algebraic structures $(Y, \vee)$ and $(Y, \wedge)$.
   (ii) Is it possible to define a unary operation $'$ on $Y$ such that $y \vee y'$ is the identity under $\wedge$ and $y \wedge y'$ is the identity under $\vee$ for each $y$ in $Y$?
10. Can $\mathbf{Z}^+ = \{1, 2, 3, \ldots\}$ be the carrier of a boolean algebra having the binary operations given by $x \vee y = \text{lcm}[x, y]$ and $x \wedge y = \text{gcd}(x, y)$?
11. In a general boolean algebra, prove the following:
    (i) $x \vee (x' \wedge x) = x$ for all $x$ in $X$ (the carrier of $B$).
    (ii) $(x \vee x') \wedge (x \vee x) = x \vee x$ for all $x$ in $X$.
    (iii) $x = x \vee x$ for all $x$ in $X$.
12. Prove the following for all $x$ in $X$:
    (i) $x = x \wedge x$;
    (ii) $x \neq x'$ (using the fact that $O \neq I$);
    (iii) $O \wedge x = O$;
    (iv) $I \vee x = I$.

13. Prove that $(x \vee y) \wedge x = x$ for all $x$ and $y$ in $X$.
14. Prove that $(x \wedge y) \vee x = x$ for all $x$ and $y$ in $X$.
15. Let the relation $\leq$ on the carrier $X$ of $B$ be defined so that $x \leq y$ if and only if $x \vee y = y$. Prove that $\leq$ is a partial ordering of $X$.
16. Let the relation $\leq$ on the carrier $X$ of $B$ be such that $x \leq y$ if and only if $x \wedge y = x$. Prove that $\leq$ is a partial ordering of $X$.
17. What are the minimal and maximal elements for the poset of Problem 15?
18. What are the minimal and maximal elements for the poset of Problem 16?
19. What is the meaning of the relation $\leq$ of Problem 15 in the special case of the boolean algebra $[P(S), \cup, \cap, \bar{\ }, \varnothing, S]$ of Example 2?
20. Do Problem 19 with the $\leq$ of Problem 15 replaced by the $\leq$ of Problem 16.
21. In $B$, let $a \wedge c = O = b \wedge c$ and $a \vee c = I = b \vee c$. Explain why:
    (i) $a = a \wedge (b \vee c) = (a \wedge b) \vee (a \wedge c) = a \wedge b$.
    (ii) $b = b \wedge a$.
    (iii) $a = b$.
22. In $B$, prove the following:
    (i) $(x')' = x$ for all $x$ in the carrier $X$.
    (ii) Complementation is a bijection from $X$ onto $X$.
23. In $B$, explain why the following is true for all $x$ and $y$ in $X$:

$$(x \vee y) \wedge (x' \wedge y') = [x \wedge (x' \wedge y')] \vee [y \wedge (x' \wedge y')] = O \vee O = O.$$

24. In $B$, prove that $(x \vee y) \vee (x' \wedge y') = I$ for all $x$ and $y$ in $X$.
25. In $B$, prove that $(x \vee y)' = x' \wedge y'$ for all $x$ and $y$ in $X$.
26. In $B$, prove that $(x \wedge y)' = x' \vee y'$ for all $x$ and $y$ in $X$.
    (The formulas of this and the preceding problem are known as ***DeMorgan's Laws***.)
27. Let $L$ and $M$ be boolean algebras each with a carrier having two elements. Is there a boolean isomorphism from $L$ onto $M$? Explain.
28. Let $L$ be the boolean algebra of Problems 4–6 and $L^n$ be the direct product of $n$ copies of $L$. Let $S = \{s_1, s_2, \ldots, s_n\}$. Is there a boolean isomorphism from $L^n$ onto the $P(S)$ boolean algebra? Explain.
29. Let $S = \{s_1, s_2, s_3\}$. Using binary numbers for $P(S)$ complete the following table for a boolean isomorphism $f$ from the $P(S)$ boolean algebra onto the boolean algebra of Problem 8:

| $x$ | 000 | 001 | 010 | 100 | 011 | 101 | 110 | 111 |
|---|---|---|---|---|---|---|---|---|
| $f(x)$ | | 3 | 5 | 7 | | | | |

30. Do Problem 29 with "isomorphism" replaced by "homomorphism" and the table replaced with the following:

| $x$ | 000 | 001 | 010 | 100 | 011 | 101 | 110 | 111 |
|---|---|---|---|---|---|---|---|---|
| $f(x)$ | | 1 | 3 | 5 | | | | |

## *3.13 Boolean Functions and Their Applications

In this section, $D$ stands for the doubleton boolean algebra whose carrier is $\{0, 1\}$.

### *Definition 1 Boolean Function*

*An* **n-ary boolean function** *is an n-ary operation on the doubleton boolean algebra D that can be expressed in terms of the operations* $\vee$, $\wedge$, *and* $'$.

### *Example 1*

$\alpha(x, y, z) = x \wedge (y \vee z)$, $\beta(x, y, z) = (x \wedge y) \vee (x \wedge z)$, and $\gamma(x, y, z) = x \wedge [(x' \vee y) \vee z]$ are ternary boolean functions.

We will see below that a boolean function may represent such diverse things as an electrical circuit or an insurance policy. Thus the ability to prove that two boolean functions are equal may help one to design a simpler (and thus more reliable and less expensive) circuit that is equivalent to a complicated circuit or may help one to rewrite a legal document in more understandable form.

When a boolean function is applied to an electrical circuit, each of the variables $x$, $y$, and so on stands for one or more switches; and the possible values 0 and 1 for a variable denote the open or closed positions, respectively, for its switches. Also $\theta \vee \phi$ indicates that the $\theta$ and $\phi$ parts of the circuit are in parallel, $\theta \wedge \phi$ indicates that these parts are in series, and $\theta'$ denotes a part of the circuit that is open (in other words, a part through which current cannot flow) when the $\theta$ part is closed and is closed when $\theta$ is open. (Here $\theta$ or $\phi$ may represent a single switch or a portion of the circuit.)

The circuits depicted in Figures 3.8a, 3.8b, and 3.8c are associated with the boolean functions $\alpha$, $\beta$, and $\gamma$ of Example 1. If one of the switches labeled with $x$ in Figure 3.8b is open, the other one labeled with $x$ must also be open.

$\alpha(x, y, z) = x \wedge (y \vee z)$

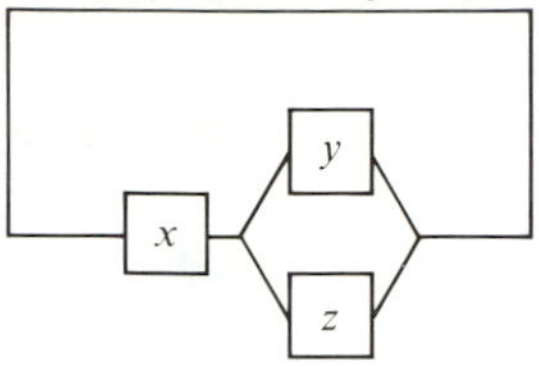


*Figure 3.8a*

$\beta(x, y, z) = (x \wedge y) \vee (x \wedge z)$

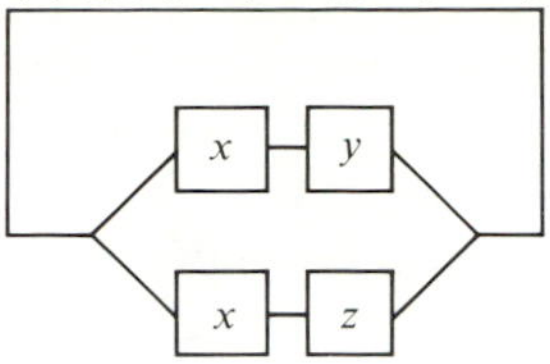


*Figure 3.8b*

$\gamma(x, y, z) = x \wedge (x' \vee y \vee z)$

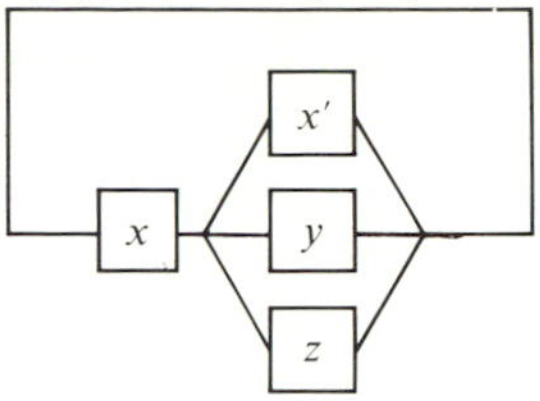


*Figure 3.8c*

In Figure 3.8c, the $x'$ switch must be open when the $x$ switch is closed and be closed when the $x$ switch is open. Looking at any one of the figures, one can easily see that current will flow only when the $x$ switch is closed (or, in Figure 3.8b, the $x$ switches are closed) and at least one of the $y$ and $z$ switches is closed. That is, current will flow only when the positions of the switches are given by $(x, y, z) = (1, 0, 1)$, $(1, 1, 0)$, or $(1, 1, 1)$. Clearly, the three circuits are equivalent and this tells us that $\alpha$, $\beta$, and $\gamma$ are equal to each other as boolean functions. The circuit of Figure 3.8a seems to be the simplest of these equivalent circuits.

When a circuit is complicated, it may be useful to analyze it by tabulating its boolean function. The tables for the operations $\vee$, $\wedge$, and $'$ on $D$ are:

| $\vee$ | 0 | 1 |
|---|---|---|
| 0 | 0 | 1 |
| 1 | 1 | 1 |

| $\wedge$ | 0 | 1 |
|---|---|---|
| 0 | 0 | 0 |
| 1 | 0 | 1 |

| $x$ | 0 | 1 |
|---|---|---|
| $x'$ | 1 | 0 |

Using these tables one can tabulate the boolean function $\gamma$ of Example 1 (and Figure 3.8c) as shown in Table 3.2.

***Table 3.2***

| $(x, y, z)$ | $x'$ | $x' \vee y$ | $(x' \vee y) \vee z$ | $\gamma = x \wedge [(x' \vee y) \vee z]$ |
|---|---|---|---|---|
| (0, 0, 0) | 1 | 1 | 1 | 0 |
| (0, 0, 1) | 1 | 1 | 1 | 0 |
| (0, 1, 0) | 1 | 1 | 1 | 0 |
| (0, 1, 1) | 1 | 1 | 1 | 0 |
| (1, 0, 0) | 0 | 0 | 0 | 0 |
| (1, 0, 1) | 0 | 0 | 1 | 1 |
| (1, 1, 0) | 0 | 1 | 1 | 1 |
| (1, 1, 1) | 0 | 1 | 1 | 1 |

Similarly one could tabulate the boolean functions $\alpha$ and $\beta$ of Example 1 and Figures 3.8a and 3.8b. This would again show us that these functions are all equal; that is, $\alpha(x, y, z) = \beta(x, y, z) = \gamma(x, y, z)$ for each of the eight possible choices of the ordered triple $(x, y, z)$.

Boolean functions may also be helpful in formal logic or in the analysis of compound statements in important documents. We now introduce meanings for $\vee$, $\wedge$, and $'$ in such applications.

A statement that is either true or false (but not both) will be called a proposition or an assertion. One manufactures compound propositions from simpler ones as follows:

If $p$ and $q$ are propositions, let $p'$ denote the assertion that $p$ is false, let $p \vee q$ denote the assertion that at least one of $p$ and $q$ is true, and let $p \wedge q$ denote the assertion that both $p$ and $q$ are true.

Note that $\vee$ has the meaning of "and/or"; it is called the "logical or" and also is called the "inclusive or." One may read $\wedge$ as "and." Also, $p'$ is called the negation of $p$. Here $O$ denotes a false proposition such as "$3 + 4 = 5$" and $I$ denotes some fixed true proposition.

### *Example 2*

Let $p$ stand for the assertion that $3 + 4 = 5$ and let $q$ represent the assertion that $3^2 + 4^2 = 5^2$. Then $p \vee q$ is the true assertion that at least one of $p$ and $q$ is true and $p \wedge q$ is the false assertion that $p$ and $q$ are both true. Also $p'$ and $q'$ are the assertions that $p$ and $q$, respectively, are false. One can translate $p'$ as "$3 + 4$ is not equal to 5." Since $p$ is false, $p'$ is true; since $q$ is true, $q'$ is false.

### *Example 3*

Valid checks for the account of a certain organization must be signed by the president and also by either the secretary or the treasurer of the organization. Thus a check is valid only if $\alpha(p, s, t) = 1$ where $\alpha(p, s, t) = p \wedge (s \vee t)$ and $p$, $s$, or $t$ has the value 1 when the check is signed by the president, secretary, or treasurer, respectively, and has the value 0 if the respective signature is not present.

### *Example 4*

For a fixed employee of some university, let $g$ and $n$ denote the employee's age and years of service at the institution. The employee is eligible for retirement benefits if and only if at least one of the following conditions is satisfied:

(i) $g + n \geq 75$;
or (ii) $g \geq 65$ and $n \geq 5$.

Letting $x$ be the assertion that $g + n \geq 75$, $y$ be the assertion that $g \geq 65$, $z$ be the assertion that $n \geq 5$, and $\alpha(x, y, z) = x \vee (y \wedge z)$, one sees that eligibility occurs if and only if $\alpha(x, y, z) = 1$. We note that when $(g, n) = (62, 11)$ one has $(x, y, z) = (0, 0, 1)$ and $\alpha(0, 0, 1) = 0 \vee (0 \wedge 1) = 0 \vee 0 = 0$; hence the employee is not eligible. However, when $(g, n) = (66, 6)$, one has $(x, y, z) = (0, 1, 1)$ and $\alpha(0, 1, 1) = 0 \vee (1 \wedge 1) = 0 \vee 1 = 1$; in this case the employee is eligible. Also the employee is eligible when $(g, n) = (50, 25)$ since then $(x, y, z) = (1, 0, 1)$ and $\alpha(1, 0, 1) = 1 \vee (0 \wedge 1) = 1 \vee 0 = 1$.

Next we introduce a symbol which is useful in applying boolean functions to formal logic and to the statements and proofs of theorems.

### *Definition 2 Implication*

*If $P$ and $Q$ are assertions (or are boolean functions), $P \Rightarrow Q$ is an alternate symbolism for the assertion (or boolean function) $P' \vee Q$. One calls $\Rightarrow$ the* **implication sign**. *An assertion $P \Rightarrow Q$ is called* **an implication** *with $P$ as the* **hypothesis** *or* **premise** *and $Q$ as the* **conclusion**; *$P \Rightarrow Q$ is read in any of the following ways:*

(i) *If $P$, then $Q$.*
(ii) *$P$ only if $Q$.*
(iii) *$P$ implies $Q$.*
(iv) *$P$ is a sufficient condition for $Q$.*
(v) *$Q$ is a necessary condition for $P$.*

### *Example 5*

Let $a$, $b$, and $c$ be elements of a group. Let $P$, $Q$, and $R$ be the assertions that $ab = ac$, that $ba = ca$, and that $b = c$, respectively. Then the Cancellation Theorem of Section 2.3 can be expressed in the form $(P \vee Q) \Rightarrow R$. By definition of implication, this has the same meaning as $(P \vee Q)' \vee R$, which in turn can be rewritten as $(P' \wedge Q') \vee R$ by using one of DeMorgan's Laws. (See Problem 25 of Section 3.12.)

### *Definition 3 Equivalence*

*For assertions (or boolean functions) $P$ and $Q$, $P \Leftrightarrow Q$ denotes the assertion (or boolean function) $[P \Rightarrow Q] \wedge [Q \Rightarrow P]$. The* **equivalence** *$P \Leftrightarrow Q$ is read in any of the following ways:*

(i) *$P$ is equivalent to $Q$.*
(ii) *$P$ if and only if $Q$.*
(iii) *$P$ is a necessary and sufficient condition for $Q$.*

*Example 6*

For elements $a$ and $b$ of a group, let $P$ be the assertion that $(ab)^{-1} = a^{-1}b^{-1}$ and $Q$ be the assertion that $ab = ba$. Problem 20 of Section 2.3 asks the reader to "show that $(ab)^{-1} = a^{-1}b^{-1}$ if and only if $ab = ba$." This can be rewritten as "show that $P \Leftrightarrow Q$."

**Definition 4 Converse, Contrapositive**

*Let $P$ and $Q$ be assertions. The* **converse** *of the implication $P \Rightarrow Q$ is the implication $Q \Rightarrow P$. The* **contrapositive** *of $P \Rightarrow Q$ is $Q' \Rightarrow P'$.*

Let $P \Rightarrow Q$ be a true implication (that is, a theorem). Then its converse may be true, as in Example 6 above, or its converse may be false, as in Example 7 below. Theorem 1 below shows that $P \Rightarrow Q$ is true if and only if its contrapositive $Q' \Rightarrow P'$ is true.

*Example 7*

For a finite group $G$ of order $r$, let $P$ be the assertion that "$d$ is the order of a subgroup in $G$" and $Q$ be the assertion that "$d$ is a positive integral divisor of $r$." Then Lagrange's Theorem in Section 2.11 tells us that $P \Rightarrow Q$, that is, "if $d$ is the order of a subgroup in $G$, then $d$ is a positive integral divisor of $r$." The converse $Q \Rightarrow P$ of $P \Rightarrow Q$ is the false statement that "if $d$ is a positive integral divisor of $r$, then $d$ is the order of a subgroup in $G$;" see Problem 3 of Section 2.13 for a counterexample demonstrating that this particular converse is false. The contrapositive $Q' \Rightarrow P'$ of $P \Rightarrow Q$ is the true statement that "if $d$ is not a positive integral divisor of $r$, then $d$ is not the order of a subgroup in $G$." Thus it follows from the contrapositive that a subset with 5 elements is not a subgroup in the octic group.

**Definition 5 Tautology, Contingency, Contradiction**

*An $n$-ary boolean function $\alpha$ is called a* **tautology** *if $\alpha(x_1, x_2, \ldots, x_n) = 1$ for each of the $2^n$ choices of $(x_1, x_2, \ldots, x_n)$. Also, $\alpha$ is a* **contradiction** *if $\alpha(x_1, x_2, \ldots, x_n) = 0$ for all choices of the $x_i$, and $\alpha$ is a* **contingency** *if it has 1 as a value for some choice of the $x_i$ and has 0 as a value for another choice of the $x_i$.*

We note that a given boolean function is either a tautology or a contingency or a contradiction but cannot be two of these.

***Example 8***

Let $\alpha(x, y) = (x \Rightarrow y) \Leftrightarrow (x' \vee y)$, $\beta(x, y) = x \wedge y$, and $\gamma(x) = x \wedge x'$. Using the definitions of implication and equivalence, one sees that $\alpha$ is a tautology. Since $\beta(1, 1) = 1$ and $\beta(1, 0) = 0$, $\beta$ is a contingency. Since $\gamma(x) = 0$ for $x = 1$ and for $x = 0$, $\gamma$ is a contradiction. Also, $(x \wedge x') \wedge y$ is a contradiction.

***Theorem 1 Contrapositive***

(i) $(x \Rightarrow y)$ *and* $(y' \Rightarrow x')$ *are equal boolean functions.*
(ii) $[(x \Rightarrow y) \Leftrightarrow (y' \Rightarrow x')]$ *is a tautology.*

The proof is left to the reader as Problems 5 and 10 of this section.

Theorem 1 enables us to prove a theorem $P \Rightarrow Q$ and then to apply its contrapositive $Q' \Rightarrow P'$ when it is convenient to do so; see Example 7 above for a specific illustration.

## *Problems*

1. Suppose that the functioning of a circuit with three switches is described by the boolean function $x \vee (y \wedge y')$. Is there an equivalent circuit with just one switch? Explain.
2. Let the functioning of a circuit with four switches be characterized by $(x \wedge y) \vee (x \wedge z)$. Is there an equivalent circuit with three switches? Explain.
3. Let $\alpha(x, y, z) = x \wedge (y \vee z)$ and $\beta(x, y, z) = (x \wedge y) \vee z$. Are $\alpha$ and $\beta$ equal boolean functions? Explain.
4. Let $\gamma(x, y, z) = z \vee (y \wedge x)$ and $\beta(x, y, z) = (x \wedge y) \vee z$. Are $\gamma$ and $\beta$ equal boolean functions? Explain.
5. Let $\alpha(x, y) = (x \Rightarrow y)$ and $\beta(x, y) = (y' \Rightarrow x')$. Show that $\alpha$ and $\beta$ are equal boolean functions by tabulating each for the four possible ordered pairs $(x, y)$. [This is the proof of Theorem 1(i).]
6. Let $\alpha(x, y) = (x \Rightarrow y)$ and $\gamma(x, y) = (y \Rightarrow x)$.
   (i) Find an ordered pair $(x_1, y_1)$ such that $\alpha(x_1, y_1) = 1$ and $\gamma(x_1, y_1) = 0$.
   (ii) Find an ordered pair $(x_2, y_2)$ such that $\alpha(x_2, y_2) = 1 = \gamma(x_2, y_2)$.
   (iii) Is $[(x \Rightarrow y) \Rightarrow (y \Rightarrow x)]$ a tautology, a contingency, or a contradiction?
7. Let $\alpha(x, y) = x \Rightarrow y$ and $\beta(x, y) = x' \vee y$.
   (i) Evaluate $\beta(0, 0)$.
   (ii) Use the answer to Part (i) and the definition of implication to evaluate $\alpha(0, 0)$.

(iii) Use the answer to Part (ii) to tell whether or not the following assertion is true: "If $n^2$ is negative for some integer $n$, then $\sqrt{2}$ is a rational number."

8. For each of the following implications, tell whether it is true or false:
(a) If $3 = -3$, then $3^2 = -3^2$.
(b) If $3 = -3$, then $3^2 = (-3)^2$.
(c) If $3 = 1 + 2$, then $3^2 = -3^2$.
(d) If $3 = 1 + 2$, then $3^2 = (-3)^2$.

9. Let $\alpha(x, y) = (x \Leftrightarrow y)$ and $\beta(x, y) = [(x \wedge y) \vee (x' \wedge y')]$. Are $\alpha$ and $\beta$ equal boolean functions? Explain.

10. Prove that boolean functions $\alpha$ and $\beta$ are equal if and only if $\alpha \Leftrightarrow \beta$ is a tautology.

11. Let $\alpha(0, 0) = 0 = \alpha(1, 1)$ and $\alpha(1, 0) = 1 = \alpha(0, 1)$. Give a formula for $\alpha(x, y)$ which uses only symbols chosen from the set $\{x, y, ', \Leftrightarrow\}$.

12. Let $\beta(0, 0) = \beta(0, 1) = \beta(1, 0) = 1$ and $\beta(1, 1) = 0$. Give a formula for $\beta(x, y)$ which uses only symbols chosen from the set $\{x, y, ', \Rightarrow\}$.

13. Let $\alpha(p, s, t) = p \wedge (s \vee t)$. Evaluate $\alpha(0, 1, 1)$ and use this to tell whether in Example 3 a check is valid when signed by the secretary and treasurer but not by the president.

14. Using the $\alpha$ of Problem 13, evaluate $\alpha(1, 0, 0)$ and then translate that result into a statement concerning the validity of a check in Example 3.

15. In Example 4, would an employee with 20 years of service be eligible for retirement benefits at age 50? Explain.

16. In Example 4, what is the lowest age at which an employee would be eligible for retirement with 4 years of service?

17. Let $a$, $b$, $p$ be integers with $p$ being prime. Assign assertions involving $a$, $b$, $p$ as meanings of $R$, $S$, and $T$ so that Lemma 1 of Section 1.7 is expressible as $R \Rightarrow (S \vee T)$.

18. Assign assertions as meanings for $P$, $Q_1$, $Q_2$, ..., $Q_n$ so that Lemma 2 of Section 1.7 is expressible as $P \Rightarrow (Q_1 \vee Q_2 \vee \cdots \vee Q_n)$.

19. For each of the following boolean functions, tell whether it is a tautology, a contingency, or a contradiction and explain your answer:
(a) $(x \vee y) \Rightarrow (x \wedge y)$.
(b) $[(x \Rightarrow y) \wedge (z \Rightarrow w)] \Rightarrow [(x \wedge z) \Rightarrow (y \wedge w)]$.
(c) $(x \vee y) \wedge (x \vee y') \wedge x'$.
(d) $[(x \Rightarrow y) \wedge (y \Rightarrow z)] \Rightarrow (x \Rightarrow z)$.
(e) $[(x \wedge y) \Rightarrow (z \wedge w)] \Rightarrow (x \Rightarrow z)$.

20. Do as in Problem 11 for each of the following:
(a) $(x \wedge y) \Rightarrow (x \vee y)$.

(b) $(x \vee y) \wedge (x \vee y') \wedge (x' \vee y) \wedge (x' \vee y')$.
(c) $[(x \wedge y) \Rightarrow z] \Rightarrow (x \Rightarrow z)$.
(d) $[x \Rightarrow (y \wedge z)] \Rightarrow (x \Rightarrow y)$.
(e) $[(x \wedge z) \Rightarrow (y \wedge z)] \Rightarrow (x \Rightarrow y)$.

## 3.14 Composition of Mappings, Groups of Bijections

Here we generalize on composition of permutations by defining the composite of a mapping from $S$ to $T$ and a mapping from $T$ to $U$, where $S$, $T$, and $U$ are any (finite or infinite) sets.

We then generalize on the symmetric groups $\mathbf{S}_n$ by showing that the set of bijections from any (finite or infinite) set $X$ to itself becomes a group $\mathbf{B}(X)$ under composition of mappings. Also, it is noted that the automorphisms of a group $G$ form a subgroup in $\mathbf{B}(G)$.

Let $S$, $T$, and $U$ be any sets, not necessarily distinct. The ***composite*** of a mapping $\alpha$ from $S$ to $T$ and a mapping $\beta$ from $T$ to $U$ is the mapping $\gamma = \alpha\beta$ with

$$s \overset{\gamma}{\mapsto} u \quad \text{whenever } s \overset{\alpha}{\mapsto} t \overset{\beta}{\mapsto} u.$$

The composite $\beta\alpha$ in the other order does not have meaning unless $S = U$ and need not equal $\alpha\beta$ even when it has meaning.

### *Example 1*

Let $\alpha$ and $\beta$ be group homomorphisms from $G_1$ to $G_2$ and from $G_2$ to $G_3$, respectively. Then the composite $\gamma = \alpha\beta$ is a homomorphism from $G_1$ to $G_3$.

The proof is left to the reader as Problem 2 of this section.

In Theorem 1, we shall use the special case of the following lemmas in which all the sets are the same.

### *Lemma 1 Surjections, Injections, Bijections*

*Let $\alpha$ and $\beta$ be mappings from $S$ to $T$ and from $T$ to $U$, respectively. Then:*
*(a) If $\alpha$ and $\beta$ are surjections, so is $\alpha\beta$.*
*(b) If $\alpha$ and $\beta$ are injections, so is $\alpha\beta$.*
*(c) If $\alpha$ and $\beta$ are bijections, so is $\alpha\beta$.*

***Proof***

For (a), let $u$ be any element of $U$. If $\beta$ and $\alpha$ are surjections, there is an element $t$ in $T$ with $\beta(t) = u$ and an $s$ in $S$ with $\alpha(s) = t$. Now the composite $\alpha\beta$ sends $s$ to $u$; that is, every $u$ of $U$ is in the image set of $\alpha\beta$.

For (b), let $s_1$ and $s_2$ be elements of $S$ with $s_1 \neq s_2$. Let $\alpha(s_i) = t_i$ and $\beta(t_i) = u_i$ for $i = 1$ and 2. If $\alpha$ and $\beta$ are injections, it follows that $t_1 \neq t_2$ and then that $u_1 \neq u_2$. That is, distinct elements $s_1$ and $s_2$ of $S$ have distinct images $u_1$ and $u_2$ under $\alpha\beta$.

Putting the results of (a) and (b) together, we have (c).

### *Lemma 2 Composition Is Associative*

*Let $\alpha$, $\beta$, and $\gamma$ be mappings from $S$ to $T$, $T$ to $U$, and $U$ to $V$, respectively. Then $(\alpha\beta)\gamma = \alpha(\beta\gamma)$.*

***Proof***

Let $s$ be any element of $S$ and let

$$s \overset{\alpha}{\mapsto} t \overset{\beta}{\mapsto} u \overset{\gamma}{\mapsto} v.$$

Then

$$s \overset{\alpha\beta}{\longmapsto} u, \qquad t \overset{\beta\gamma}{\longmapsto} v, \qquad s \overset{(\alpha\beta)\gamma}{\longmapsto} v, \qquad s \overset{\alpha(\beta\gamma)}{\longmapsto} v.$$

That is, every $s$ of $S$ has the same image under both $(\alpha\beta)\gamma$ and $\alpha(\beta\gamma)$.

### *Theorem 1 Group* $\mathbf{B}(X)$ *of Bijections of* $X$

*Let $X$ be any (finite or infinite) set and let $\mathbf{B}(X)$ consist of all the bijections from $X$ to itself. Then $\mathbf{B}(X)$ is a group under composition of mappings.*

***Proof***

Lemma 1(c) implies that $\mathbf{B}(X)$ is closed under composition and Lemma 2 tells us that this operation is associative. The identity of $\mathbf{B}(X)$ is the bijection $\boldsymbol{\varepsilon}$ with $\boldsymbol{\varepsilon}(x) = x$ for all $x$ in $X$.

In Section 3.1, we noted that a bijection $\alpha$ from $S$ to $S$ has an inverse mapping $\alpha^{-1}$ that is also a bijection from $S$ to $S$. Also, $\alpha\alpha^{-1} = \varepsilon = \alpha^{-1}\alpha$. Hence composition of mappings in $\mathbf{B}(X)$ satisfies the group axioms.

**Theorem 2** ***Subgroup of Automorphisms*** $\mathbf{A}(G)$

*Let* $\mathbf{A}(G)$ *consist of all the automorphisms of a group* $G$. *Then* $\mathbf{A}(G)$ *is a subgroup in the group of bijections* $\mathbf{B}(G)$.

The proof is left to the reader as Problem 3 of this section.

The following result states that the structure of $\mathbf{B}(X)$ is determined by the number of elements in the set $X$.

**Theorem 3** ***Condition for*** $\mathbf{B}(X)$ ***and*** $\mathbf{B}(Y)$ ***To Be Isomorphic***

*Let there exist a bijection* $\beta$ *from* $X$ *to* $Y$. *Then for every bijection* $\theta$ *from* $X$ *to itself, the mapping* $\beta^{-1}\theta\beta$ *is a bijection from* $Y$ *to itself. Also, the mapping* $f$ *with* $f(\theta) = \beta^{-1}\theta\beta$ *is a group isomorphism from* $\mathbf{B}(X)$ *to* $\mathbf{B}(Y)$.

The proof is left to the reader as Problem 7 of this section.

**Theorem 4** ***Generalized Cayley Theorem***

*Let* $G$ *be a (finite or infinite) group. Then the group* $\mathbf{B}(G)$ *of bijections of* $G$ *has a subgroup* $M$ *that is isomorphic to* $G$.

***Proof***

For a fixed element $a$ of $G$, let $\beta_a$ be the mapping from $G$ to $G$ with $\beta_a(x) = xa$ for all $x$ in $G$. We noted in Section 3.1 that $\beta_a$ is a bijection from $G$ to itself. (See Example 6 and Problem 12 of Section 3.1.)

Hence the mapping $\theta$ with $\theta(a) = \beta_a$ is a mapping from $G$ to $\mathbf{B}(G)$. We next show that $\theta$ is an injective homomorphism.

For any $x$ in $G$,

$$x \overset{\beta_a}{\mapsto} xa \overset{\beta_b}{\mapsto} (xa)b = x(ab).$$

This shows that the $\beta_a\beta_b = \beta_{ab}$. Then

$$\theta(ab) = \beta_{ab} = \beta_a\beta_b = \theta(a)\theta(b).$$

Thus $\theta$ is a homomorphism.

Now let $\theta(a) = \theta(b)$; that is, let $\beta_a = \beta_b$. Then for all $x$ in $G$, we have $\beta_a(x) = \beta_b(x)$, or $xa = xb$. Clearly this implies $a = b$. Hence $\theta$ is injective. Now it follows from parts (g) and (c) of Theorem 2 of Section 3.3 that $G$ is isomorphic to the image set $M$ of $\theta$ and that $M$ is a subgroup in $\mathbf{B}(G)$.

If $G$ is a group of finite order $n$, Theorem 3 tells us that $\mathbf{B}(G)$ is isomorphic to the symmetric group $\mathbf{S_n}$. Then Theorem 4 implies that $G$ is isomorphic to a subgroup in $\mathbf{B}(G)$, and hence to a subgroup in $\mathbf{S_n}$. This shows that Theorems 3 and 4 contain Cayley's Theorem as a special case.

## *Problems*

1. Let $G_1, G_2, G_3$ be cyclic groups $[a]$, $[b]$, $[c]$ of orders 12, 6, and 2, respectively. Let $\alpha$ be the mapping from $G_1$ to $G_2$ with $\alpha(a^s) = b^s$ for $s = 0, 1, \ldots, 11$ and let $\beta$ be the mapping from $G_2$ to $G_3$ with $\beta(b^t) = c^t$ for $t = 0, 1, \ldots, 5$. Tabulate the composite mapping $\gamma = \alpha\beta$ from $G_1$ to $G_3$ and tell which of $\alpha$, $\beta$, and $\gamma$ are homomorphisms.
2. Let $G_1$, $G_2$, $G_3$ be any three groups. Let $\alpha$ and $\beta$ be homomorphisms from $G_1$ to $G_2$ and from $G_2$ to $G_3$, respectively. Prove that the composite $\alpha\beta$ is a homomorphism from $G_1$ to $G_3$ (and thus give the proof left to the reader in Example 1).
3. Prove Theorem 2; that is, prove that the set $\mathbf{A}(G)$ of automorphisms of a group $G$ is a subgroup in the group $\mathbf{B}(G)$ of bijections from $G$ to $G$.
4. Prove that the set of inner automorphisms of a group $G$ is a subgroup in $\mathbf{A}(G)$ and hence is also a subgroup in $\mathbf{B}(G)$.
5. Prove that $\mathbf{B}(X)$ is nonabelian when $X$ has at least three elements.
6. Give an example of an infinite noncommutative group.
7. Prove Theorem 3; that is, prove that $\mathbf{B}(X)$ and $\mathbf{B}(Y)$ are isomorphic if there is a bijection $\beta$ from $X$ to $Y$.
8. Let $S$ be the set consisting of all the real numbers except 0 and 1. Let $\theta_1, \theta_2, \theta_3, \theta_4, \theta_5$, and $\theta_6$ be the mappings from $S$ to $S$ such that

$$\theta_1(x) = x, \quad \theta_2(x) = 1/x, \quad \theta_3(x) = 1 - x,$$
$$\theta_4(x) = 1/(1 - x), \qquad \theta_5(x) = (x - 1)/x,$$
$$\theta_6(x) = x/(x - 1)$$

for all $x$ in $S$. Let $T$ be the set of these six mappings. Prove the following:
(i) Each of the $\theta_i$, $1 \le i \le 6$, is a bijection from $S$ to $S$.
(ii) The set $T$ is closed under the operation of composition of mappings.
(iii) $T$ is isomorphic to the symmetric group $S_3$.

9. Let $C$ be the center in a group $G$.
   (i) Explain why $C$ is normal in $G$ and hence $G/C$ exists.
   (ii) Prove that $G/C$ is isomorphic to the group of inner automorphisms of $G$.
10. For each element $g$ of a group $G$, let $\sigma_g$ be the inner automorphism of $G$ given by $\sigma_g(x) = g^{-1}xg$ for all $x$ in $G$. Let $F$ be the mapping from $G$ to $\mathbf{A}(G)$ given by $F(g) = \sigma_g$ for all $g$ in $G$. Show the following:
   (i) $F$ is a homomorphism.
   (ii) The kernel of $F$ is the center $C$ in $G$.

*11. Let $\mathbf{B}(\mathbf{Z})$ be the group of bijections of the set $\mathbf{Z}$ of integers. Let $\beta$ be the bijection in $\mathbf{B}(\mathbf{Z})$ with $\beta(n) = n + 1$ for all $n$ in $\mathbf{Z}$. Is the cyclic subgroup $[\beta]$ isomorphic to the additive group of $\mathbf{Z}$? Explain.

12. Let $X$ and $Y$ be nonempty sets. Let $\alpha$ and $\beta$ be mappings from $X$ to $Y$ and from $Y$ to $X$, respectively, such that $\alpha\beta$ is a bijection from $X$ to itself. Prove that $\alpha$ is an injection and $\beta$ is a surjection.
13. Let $\alpha$ be an injection from $X$ to $Y$ and let $u$ be a known element of $X$. Show that there is at least one mapping $\beta$ from $Y$ to $X$ such that $\alpha\beta$ is the identity mapping in $\mathbf{B}(X)$ and that $\beta$ must be a surjection.

*14. Let $\beta$ be a surjection from $Y$ to $X$. Show that there is at least one (injective) mapping $\alpha$ from $X$ to $Y$ such that $\alpha\beta$ is the identity of $\mathbf{B}(X)$. (This problem is starred because its solution requires something called the "Axiom of Choice.")

## *3.15 Conjugacy Classes and the Class Equation

Here we use a method of partitioning groups and a counting technique to study the structure of finite noncommutative groups.

If $a$ is an element of a group $G$, the set $\{g^{-1}ag : g \in G\}$, of all conjugates of $a$ by elements of $G$, is called the ***conjugacy class*** of $a$ in $G$ and we denote this set by $\mathrm{CC}(a)$. Since $a$ is its own conjugate by the identity $\mathbf{e}$, it is clear that one always has $a$ in $\mathrm{CC}(a)$. It can also be shown that for any elements $a$ and $b$ in $G$, $\mathrm{CC}(a)$ and $\mathrm{CC}(b)$ are either identical or are disjoint. These facts imply the following:

***Theorem 1 Partition by Conjugacy***

*In each finite group $G$, there exist elements $g_1, g_2, \ldots, g_r$ such that the conjugacy classes $\mathrm{CC}(g_i)$ for $i = 1, 2, \ldots, r$ form a partition of $G$.*

We have outlined a proof above and also refer the reader to Problems 4 and 8 of Section 3.7 for another approach.

For any finite set $S$, let $|S|$ denote the number of elements in $S$. In particular, $|G| = \text{ord } G$ for any finite group $G$.

**Theorem 2 Number of Conjugates**

*Let $C_a$ and CC($a$) be the centralizer and conjugacy class, respectively, of $a$ in a finite group $G$.*
*(a) $|\text{CC}(a)| = |G|/|C_a|$, that is, the number of conjugates of $a$ in $G$ is the index of the centralizer of $a$ in $G$.*
*(b) $|\text{CC}(a)| = 1$, or $\text{CC}(a) = \{a\}$, if and only if $a$ is in the center $C$ in $G$.*

The proof is left to the reader. [See Problem 21(iii) of Section 3.2.]

**Theorem 3 The Class Equation**

*Let $g_1, \ldots, g_r$ be elements of a finite group $G$ whose conjugacy classes $\text{CC}(g_i)$ form a partition of $G$. Let $n_i$ be the index of the centralizer of $g_i$ in $G$. Then*

$$|G| = n_1 + n_2 + \cdots + n_r. \tag{1}$$

***Proof***

Since each element of $G$ is in one and only one of the $\text{CC}(g_i)$,

$$|G| = |\text{CC}(g_1)| + |\text{CC}(g_2)| + \cdots + |\text{CC}(g_r)|.$$

Using Theorem 2(a), we substitute $n_i$ for $|\text{CC}(g_i)|$ and thus obtain (1).

***Example 1***

Let $G$ be the octic group $\{\varepsilon, \rho, \rho^2, \rho^3, \phi, \rho\phi, \rho^2\phi, \rho^3\phi\}$ with $\rho = (1234)$ and $\phi = (24)$. Using the middle table on the inside back cover, one can show that the conjugacy classes in this $G$ are:

$$\text{CC}(\varepsilon) = \{\varepsilon\}, \quad \text{CC}(\rho) = \{\rho, \rho^3\} = \text{CC}(\rho^3), \quad \text{CC}(\rho^2) = \{\rho^2\},$$
$$\text{CC}(\phi) = \{\phi, \rho^2\phi\} = \text{CC}(\rho^2\phi), \text{CC}(\rho\phi) = \{\rho\phi, \rho^3\phi\} = \text{CC}(\rho^3\phi).$$

Thus we have both of the following:

$$G = \mathrm{CC}(\varepsilon) \cup \mathrm{CC}(\rho) \cup \mathrm{CC}(\rho^2) \cup \mathrm{CC}(\phi) \cup \mathrm{CC}(\rho\phi),$$

$$8 = |G| = |\mathrm{CC}(\varepsilon)| + |\mathrm{CC}(\rho)| + |\mathrm{CC}(\rho^2)| + |\mathrm{CC}(\phi)| + |\mathrm{CC}(\rho\phi)| = 1 + 2 + 1 + 2 + 2.$$

We illustrate the use of the class equation in the proof of the following result.

***Theorem 4 Groups with Prime Power Order***

*Let* ord $G = p^m$, *with* $p$ *and* $m$ *positive integers and* $p$ *prime. Let* $C$ *be the center in* $C$. *Then* $p\,|\,(\mathrm{ord}\ C)$.

***Proof***

The only conjugate of an element $c$ in the center $C$ is $c$ itself; in other words, the only conjugacy class that contains $c$ is $\mathrm{CC}(c) = \{c\}$. Therefore, if $g_1, \ldots, g_r$ are elements of $G$ such that the $\mathrm{CC}(g_i)$ partition $G$, each element of $C$ must be one of these $g_i$. Let ord $C = s$; then we may assume that $g_1, \ldots, g_s$ are elements of $C$ and that $g_{s+1}, g_{s+2}, \ldots, g_r$ are those $g_i$ that are not in $C$. Then it follows from Theorem 2(b) that in the class equation

$$|G| = n_1 + n_2 + \cdots + n_s + n_{s+1} + \cdots + n_r$$

we have $n_1 = n_2 = \cdots = n_s = 1$ and $n_i > 1$ for $i = s + 1, \ldots, r$.

Using the fact that $n_i = 1$ for $1 \le i \le s$ and the hypothesis $|G| = p^m$, we then have

$$p^m = s + n_{s+1} + n_{s+2} + \cdots + n_r. \tag{2}$$

Each $n_i$ is the index of a subgroup in $G$ and hence each $n_i$ is one of the positive integral divisors $1, p, p^2, \ldots, p^m$ of $p^m =$ ord $G$. For $i > s$, we have $n_i > 1$ and thus these $n_i$ are of the form $p^k$, with $k$ a positive integer. From (2) we have

$$\mathrm{ord}\ C = s = p^m - n_{s+1} - n_{s+2} - \cdots - n_r.$$

Since $p\,|\,n_i$ for $i > s$, this implies that $p\,|\,(\mathrm{ord}\ C)$, as desired.

Theorem 4 implies that in a group of order $8 = 2^3$, the order of the center is 2, 4, or 8. Actually, only 2 and 8 are possible orders for the center in a group of order 8. Example 1 of this section shows that the center in the octic group has order 2 and a cyclic (or other abelian) group of order 8 has a center of order 8. In Problem 3 below, the reader is asked to prove that no group of order $p^3$, with $p$ prime, has a center of order $p^2$.

### Problems

1. Let the alternating group $\mathbf{A}_4$ have $\alpha_1, \ldots, \alpha_{12}$ as its elements, with the $\alpha_i$ as in the row headings for the table on the inside rear cover. In $\mathbf{A}_4$, show the following:
   (a) $CC(\alpha_1) = \{\alpha_1\}$.
   (b) $CC(\alpha_2) = \{\alpha_2, \alpha_3, \alpha_4\}$.
   (c) $CC(\alpha_5) = \{\alpha_5, \alpha_6, \alpha_7, \alpha_8\}$.
   (d) $CC(\alpha_9) = \{\alpha_9, \alpha_{10}, \alpha_{11}, \alpha_{12}\}$.
2. Show that in the symmetric group $\mathbf{S}_4$, the conjugacy class of (123) consists of all eight 3-cycles.
3. Let ord $G = p^3$, with $p$ prime. Use Theorem 4 of this section and Problem 27 of Section 2.11 to prove that the center $C$ in $G$ must have order $p$ or $p^3$.
4. Let $G$ be a nonabelian group of order $p^3$, with $p$ prime. Prove that the center $C$ in $G$ must have order $p$.

## *3.16 The Fundamental Theorem on Abelian Groups (without proof)

Let $G$ be a finite group and let the standard factorization of ord $G$ be

$$\text{ord } G = m = (p_1)^{r_1}(p_2)^{r_2} \cdots (p_s)^{r_s}. \tag{1}$$

For each prime $p_i$ in the factorization (1), there exists a Sylow $p_i$-subgroup $G_i$ of order $(p_i)^{r_i}$. (See Section 2.14.)

We now state without proof the following result:

#### Fundamental Theorem on Finite Abelian Groups

(a) *Let $G$ be a finite commutative group with ord $G$ as in* (1). *For each $p_i$ in* (1), *let $G_i$ be a Sylow $p_i$-subgroup in $G$. Then $G$ is isomorphic to the direct product $G_1 \times G_2 \times \cdots \times G_s$.*
(b) *Every finite commutative group is isomorphic to the direct product of a finite number of its cyclic subgroups.*

For example, an abelian group $G$ of order 360 is isomorphic to the direct product $G_1 \times G_2 \times G_3$ of subgroups $G_1$, $G_2$, and $G_3$ with ord $G_1 = 2^3$, ord $G_2 = 3^2$, and ord $G = 5$. Also, the subgroup $G_2$ is either cyclic or isomorphic to the direct product of two cyclic groups of order 3. There are three possibilities for $G_1$. One is that $G_1$ is cyclic, a second is that it is isomorphic to a direct product of two cyclic groups whose orders are 2 and 4, the third is that it is isomorphic to a direct product of three cyclic groups of order 2.

## *Review Problems*

1. Let $G = \{\mathbf{e}, a, a^2, a^3\}$ be a cyclic group of order 4 and let $G'$ be the multiplicative group $\{1, -1, \mathbf{i}, -\mathbf{i}\}$ of 4 complex numbers.
   (i) Tabulate two isomorphisms from $G$ to $G'$.
   (ii) Explain why there are no other isomorphisms from $G$ to $G'$.
2. What is the smallest positive integer $m$ such that there exist nonisomorphic groups of order $m$?
3. Let $\rho = (1234)$, $\phi = (24)$, and $G$ be the octic group

$$G = \{(1), \rho, \rho^2, \rho^3, \phi, \rho\phi, \rho^2\phi, \rho^3\phi\}.$$

   Let $f$ and $g$ be the mappings from $G$ to itself with $f(\theta) = \theta^2$ and $g(\theta) = \theta^3$ for all $\theta$ in $G$.
   (a) Is $f$ injective? Is $f$ surjective? Is $f$ a homomorphism? Give reasons for the answers.
   (b) Is $g$ injective? Is $g$ surjective? Is $g$ a homomorphism? Give reasons for the answers.
4. Let $G$ be the octic group (with its elements as in Problem 3) and let $G' = \{\mathbf{e}, b\}$ be a group of order 2.
   (i) Tabulate a homomorphism $f$ from $G$ to $G'$ with $f(\rho) = b$ and $f(\rho\phi) = \mathbf{e}$.
   (ii) What is the kernel of $f$?
   (iii) What is the complete inverse image $f^{-1}(b)$?
   (iv) What is the image set $M$ of $f$?
5. Let $f$ be the mapping from the octic group $G$ to the symmetric group $\mathbf{S}_4$ with $f(\theta) = (12) \cdot \theta \cdot (12)$ for all $\theta$ in $G$. Show that $f$ is an injective homomorphism.
6. Let $\mathbf{A}_4 = \{\alpha_1, \alpha_2, \ldots, \alpha_{12}\}$, with the $\alpha_i$ as in Table 2.2 of Section 2.10 and let $G = \{\mathbf{e}, b, b^2\}$ be a cyclic group of order 3. Given that $f$ is a homomorphism from $\mathbf{A}_4$ to $G$ with $f(\alpha_2) = \mathbf{e}$ and $f(\alpha_5) = b$, find the following complete inverse images:
   (i) $f^{-1}(\mathbf{e})$; (ii) $f^{-1}(b)$; (iii) $f^{-1}(b^2)$.
7. Let $G$ have subgroups $H$ and $K$, each of order 43. Explain why either $H = K$ or $H \cap K = \{\mathbf{e}\}$.
8. Explain why the union of all the subgroups in a group $G$ is a subgroup in $G$.
9. Let $G = \{\mathbf{e}, b, c\}$ be a cyclic group of order 3. Tabulate the products $(b, c)(x, y)$ for all the ordered pairs $(x, y)$ in the direct product $G \times G$.

10. Let $L$ be the relation on $\mathbf{Z}$ for which $aLb$ means that $a^2 = b^2$. Is $L$ symmetric? Reflexive? Transitive? An equivalence relation? A linear order relation? Explain.

11. How many antisymmetric relations are there on a set $\{a, b\}$ with 2 elements? Of these, how many are also reflexive?

12. Do Problem 11 for a set $\{a, b, c\}$ with 3 elements.

13. Let $D$ be the relation on $\mathbf{Z}^+$ for which $aDb$ means $a|b$. Is $D$ reflexive? Symmetric? Antisymmetric? Transitive? A partial ordering?

14. Do Problem 13 with $\mathbf{Z}^+$ replaced by the nonzero integers.

15. Let $S = \{1, 2, 3, 4\}$.
    (a) How many subsets of $S$ are there in the power set $P(S)$?
    (b) How many mappings are there from $P(S)$ to $S$?
    (c) How many mappings are there from $S$ to $P(S)$? Of these, how many are injective?

16. Do Problem 15 with $\{1, 2, 3, 4\}$ replaced by $\{1, 2, 3, 4, 5\}$.

17. Let $S = \{1, 2, 3, 4\}$.
    (a) How many unary operations are there on $S$?
    (b) How many binary operations are there on $S$?
    (c) How many of the binary operations are commutative?

18. How many $n$-ary operations are there on $S = \{1, 2, 3, 4\}$?

19. Let $T = \{0, 1, 2, \ldots, 999\}$ and $*$ be multiplication modulo 1000.
    (a) Is the semigroup $(T, *)$ a monoid? Is it a group? Explain.
    (b) How many elements are there in the principal subsemigroup $\langle 2 \rangle$?
    (c) Is $\langle 2 \rangle$ a monoid? Is it a group? Explain.

20. Do the analogues of (b) and (c) in Problem 19 with $\langle 2 \rangle$ replaced by $\langle 4 \rangle$.

21. Describe a boolean algebra whose carrier has 32 elements.

22. Describe a boolean algebra whose carrier has 64 elements.

23. Show that $x \wedge (y \vee x') = x \wedge y$ for all $x$ and $y$ in the carrier of a boolean algebra.

24. Does $x \vee (y \wedge x') = x \vee y$ for all $x$ and $y$ in the carrier of a boolean algebra?

25. For each of the following boolean functions, tell whether it is a tautology, a contingency, or a contradiction and explain your answer.
    (a) $\alpha(x, y) = (x \wedge y) \Rightarrow x$.
    (b) $\beta(x, y, z) = [(x \wedge y) \Rightarrow z] \Leftrightarrow [(x \Rightarrow z) \vee (y \Rightarrow z)]$.

26. Do as in Problem 25 for the following boolean functions:
    (a) $\alpha(x, y) = (x \wedge y) \vee (x \wedge y') \vee (x' \wedge y) \vee (x' \wedge y')$.
    (b) $\beta(x, y) = [(x \Rightarrow y) \wedge (x \Rightarrow y')] \Rightarrow x$.

27. Let $Y$ be a subset of $X$ and let $H$ be a subgroup in the group $\mathbf{B}(X)$ of all bijections from $X$ to itself. Let $K$ be the set of all $\theta$ in $H$ such that $\theta(y) = y$ for every $y$ in $Y$. Prove that $K$ is a subgroup in $H$.

## *Supplementary and Challenging Problems*

1. Let $\beta$ be a fixed odd permutation in $\mathbf{S_n}$.
   (i) Show that $\beta^{-1}\alpha\beta$ is in $\mathbf{A_n}$ for every $\alpha$ in $\mathbf{A_n}$.
   (ii) Show that the mapping $f$ with $f(\alpha) = \beta^{-1}\alpha\beta$ is an automorphism of $\mathbf{A_n}$.
   (iii) Explain why $f$ is not an inner automorphism.
2. Show the following:
   (i) A homomorphism $f$ from $\mathbf{A_4}$ to some group $G$ is completely determined once one knows the images of (12)(34) and (123) under $f$.
   (ii) $\mathbf{A_4}$ has exactly 24 automorphisms of which 12 are inner automorphisms.
3. (i) How many automorphisms are there of $\mathbf{S_4}$?
   (ii) How many inner automorphisms are there of $\mathbf{S_4}$?
4. A mapping $\theta$ from a group $G$ to a group $G'$ with

$$\theta(xy) = \theta(y)\theta(x) \text{ for all } x \text{ and } y \text{ in } G$$

   is a group *antihomomorphism*. Tabulate all the antihomomorphisms from the symmetric group $\mathbf{S_3}$ to a group $\{\mathbf{e}, b\}$ of order 2.
5. An *antiautomorphism* of a group $G$ is a bijective antihomomorphism from $G$ to itself. (See Problem 4.) Tabulate all the antiautomorphisms of the symmetric group $\mathbf{S_3}$.
6. Let $\mathbf{A}(G)$ and $A'(G)$ consist of all the automorphisms and all the antiautomorphisms, respectively, of a group $G$. (See Problem 5.) Let $A''(G) = \mathbf{A}(G) \cup A'(G)$. Show the following:
   (i) $A'(G)$ is not the empty set.
   (ii) If either $\mathbf{A}(G)$ or $A'(G)$ is finite, there are the same number of mappings in each of these sets.
   (iii) $A''(G)$ is a group, with $\mathbf{A}(G)$ as a normal subgroup, under composition of mappings.
7. Let $M$ and $N$ be normal subgroups in $G$ such that $M \cap N = \{\mathbf{e}\}$. Prove that $ab = ba$ for all $a \in M$ and $b \in N$.
8. Let $H$ and $K$ be subgroups in $G$ and let $D = H \cap K$. Let $h_1k_1 = h_2k_2$ with $h_1$ and $h_2$ in $H$ and $k_1$ and $k_2$ in $K$. Show that $h_1$ and $h_2$ are in the same left coset of $D$ in $H$.
9. Let $\theta$ be a surjective group homomorphism from $G$ to $G'$. Let $H'$ be a subgroup in $G'$ and let $H$ be the subset of $G$ consisting of all the elements $h$ of $G$ such that $\theta(h)$ is in $H'$. ($H$ is the union of the complete inverse images of the elements of $H'$.) Show that:
   (a) $H$ is a subgroup in $G$.
   (b) $H$ is normal in $G$ if $H'$ is normal in $G'$.
10. Let $a$ be an element different from the identity $\mathbf{e}$ of a noncyclic group $G$ of order 9. Explain why the following are true:
    (i) The order of $a$ is 3.
    (ii) The centralizer $C_a$ has index 1 or 3 in $G$.

(iii) Under the assumption that $C_a$ is not $G$, $a$ would have 3 conjugates (that is, there would be 3 distinct elements of the form $g^{-1}ag$ with $g$ in $G$).
(iv) $C_a = G$. (*Hint*: A set with 8 elements cannot be partitioned into subsets each having 3 elements.)
(v) Every group of order 9 is abelian.

11. Let $G$ be a group of order $p^2$, where $p$ is a prime.
(i) Prove that $G$ is abelian.
(ii) Prove that either $G$ is cyclic or $G$ is isomorphic to the direct product of two groups of order $p$.

12. (i) Prove that every group of order 15 is cyclic.
(ii) Explain why any two groups of order 15 are isomorphic.

13. Let $G$ be a group of order 10. Prove the following:
(i) Either $G$ is cyclic or $G$ is isomorphic to the group of symmetries of a regular pentagon (that is, to the group generated by {(12345), (25)(34)}).
(ii) $G$ has exactly one subgroup of order 5.
(iii) $G$ has 4 elements of order 5.
(iv) The number of elements of order 2 in $G$ is either 1 or 5.

14. Generalize on Problem 13 to groups of order $2p$, where $p$ is an odd prime.

15. Let $G$ be a group of order 8 and let $f_j$ designate the number of elements of order $j$ in $G$ (for $j = 1, 2, 4, 8$). Show the following:
(i) If $G$ is abelian, $(f_1, f_2, f_4, f_8)$ is (1, 1, 2, 4), (1, 3, 4, 0), or (1, 7, 0, 0).
(ii) If $G$ is not abelian, $(f_1, f_2, f_4)$ is (1, 1, 6) or (1, 5, 2).

16. What is the maximum number of groups of order 8 that one can have without two of them being isomorphic?

17. Let $T$ be a set $\{\gamma_1, \gamma_2, \ldots, \gamma_s\}$ of $s$ disjoint transpositions in $\mathbf{S_n}$. Let $H$ be the subgroup in $\mathbf{S_n}$ generated by $T$.
(i) Show that $H$ consists of the permutations of the form

$$\gamma_1^{e_1}\gamma_2^{e_2} \cdots \gamma_s^{e_s},$$

where $e_j$ is 0 or 1 for $j = 1, 2, \ldots, s$.
(ii) What is the order of $H$?

18. In Problem 17, let $\gamma_j$ be the transposition $(a_j b_j)$ with $a_j = 2j - 1$ and $b_j = 2j$ for $j = 1, 2, \ldots, s$. Explain why each left coset of the subgroup $H$ (of Problem 17) in $\mathbf{S_{2s}}$ has exactly one permutation

$$\alpha = \begin{pmatrix} 1 & 2 & \cdots & 2s \\ c_1 & c_2 & \cdots & c_{2s} \end{pmatrix}$$

such that in the listing $c_1, c_2, \ldots, c_{2s}$ one has $2j - 1$ to the left of $2j$ for $j = 1, 2, \ldots, s$.

19. Let $\Gamma$ be the collection of all the right cosets $Hg$ of a subgroup $H$ in $G$. For every $a$ in $G$, let $\theta(a)$ be the mapping from $\Gamma$ to $\Gamma$ with $Hg \mapsto H(ga)$. Let $M$ consist of the $\theta(a)$ for all $a$ in $G$. Prove the following:
    (i) $M$ is a subgroup in the group $\mathbf{B}(\Gamma)$ of bijections from $\Gamma$ to $\Gamma$.
    (ii) The mapping $f$ with $a \mapsto \theta(a)$ is a homomorphism from $G$ to $M$.
    (iii) Let $K$ be the kernel of $f$. Then $K$ is a subgroup in $H$, $K$ is a normal subgroup in $G$, and $K$ is the union of all the subgroups in $H$ which are normal in $G$.
    (iv) Let $H$ and $K$ have indices $t$ and $u$, respectively, in $G$. Then $t \mid u$ and $u \mid (t!)$.

20. Let $H$ be a subgroup of order 7 in a group $G$ of order 21. Show that $H$ is normal in $G$.

21. Prove that a group $G$ of order $pq$ is solvable if $p$ and $q$ are distinct primes in $\mathbf{Z}$.

22. Let $\theta$ be a mapping from an additive group $G$ to the real numbers $\mathbf{R}$. Show that there exist mappings $\alpha$ and $\beta$ from $G$ to $\mathbf{R}$ such that all of the following hold for every $g$ in $G$:
    (a) $\theta(g) = \alpha(g) + \beta(g)$.
    (b) $\alpha(-g) = \alpha(g)$.
    (c) $\beta(-g) = -\beta(g)$.

23. Let $T$ be a set of $mn + 1$ positive integers, where $m$ and $n$ are positive integers. Let $D$ be the relation on $T$ for which $aDb$ if and only if $a \mid b$. Also let $E$ be the relation on $T$ such that $aEb$ if and only if both $a\bar{D}b$ and $b\bar{D}a$. Prove that in $T$ there exists either a subset $\{a_1, a_2, \ldots, a_{m+1}\}$ with $a_i D a_{i+1}$ for $1 \le i \le m$ or a subset $\{b_1, b_2, \ldots, b_{n+1}\}$ with $b_j E b_k$ for $j \ne k$.

24. Do the analogue of Problem 23 for a collection $\Gamma$ of $mn + 1$ subsets of a set $S$ and the relation $\Delta$ in $\Gamma$ such that $\alpha \Delta \beta$ if and only if $\alpha \subset \beta$.

25. Let a boolean algebra have a finite carrier $X$. Prove that the number of elements in $X$ is a positive integral power of 2.

# 4

## Rings and Fields

*In most sciences one generation tears down what another has built, and what one has established another undoes. In mathematics alone each generation builds a new story to the old structure.*

*Hermann Hankel*

Group theory gives us information concerning addition (and its related operation subtraction) in the integers **Z**, the rational numbers **Q**, the real numbers **R**, and the complex numbers **C** since additively each of the above number systems is a group. One can also apply group theory to multiplication and division in **Q**, **R**, and **C** (but not in **Z**) by noting that the nonzero elements of any one of these three systems form a multiplicative group.

However, group theory is not sufficient to deal with all the algebraic properties of the familiar number systems and their generalizations. For example, group theory alone does not enable us to derive consequences of the distributive law since this axiom involves both addition and multiplication. Therefore, we now begin a general study of algebraic structures—among which are **Z**, **Q**, **R**, and **C**—that satisfy a set of axioms involving addition, multiplication, and (implicitly) subtraction.

### 4.1 Rings

#### Definition 1 Ring

*A ring $R$ is a set $\hat{R}$ with operations of addition and multiplication satisfying the three following axioms:*
($A$) *$R$ is an abelian group under addition.*
($M$) *$R$ is closed and associative under multiplication.*

(*D*) *Multiplication is distributive over addition, that is, for all a, b, and c in* $\hat{R}$ *one has*

$$a(b + c) = ab + ac$$

*and*

$$(b + c)a = ba + ca.$$

As was true in the notation for a group $G$ and its set $\hat{G}$, we shall usually write $R$ for $\hat{R}$ and leave it to the context to indicate whether the ring or its set is involved.

By definition all rings are commutative under addition; therefore, the phrase ***commutative ring*** is used to refer to a ring commutative under multiplication. The number systems, **Z**, **Q**, **R**, and **C** are examples of infinite commutative rings. In the next section we shall construct an infinite number of finite commutative rings. An infinite number of other (finite and infinite) commutative rings will be characterized in Chapter 5.

A finite noncommutative ring is given in Example 1 of this section. Many finite and infinite noncommutative rings are characterized in Section 4.7 below.

Additively, a ring $R$ is a group; hence $R$ always has an additive identity, called the ***zero element*** and written as 0. Also, every element $a$ of $R$ has an additive inverse, called the ***negative*** of $a$ and denoted by $-a$.

The ring axioms are such that a ring $R$ may or may not have a multiplicative identity. If there is a multiplicative identity in $R$, it is unique (see the proof of Theorem 2 of Section 2.3) and is called the ***unity*** of $R$. The unity is represented by 1 (and of course has the property $a \cdot 1 = a = 1 \cdot a$ for all $a$ in $R$).

The number systems **Z**, **Q**, **R**, and **C** are examples of rings with unity; an example of a ring that does not have a unity is the ring of even integers.

### *Definition 2 Invertible Element of a Ring with Unity*

*If $U$ is a ring with unity, an element $v$ of $U$, for which there is a $v^{-1}$ in $U$ with $vv^{-1} = 1 = v^{-1}v$, is called an* **invertible**, *or a* **unit**, *of $U$. The multiplicative inverse $v^{-1}$ is also called the* **reciprocal** *of $v$.*

Although the word "unit" is used widely, we prefer "invertible" since it seems more descriptive and less likely to be confused with "unity." Also, "invertible" is customary in linear algebra and matrix theory courses.

The invertibles of the ring **Z** of the integers are 1 and $-1$ since these are the only integers with reciprocals in **Z**. In **Q**, **R**, or **C** every nonzero number is an invertible.

**Theorem 1** ***Multiplicative Group of Invertibles***

*Let $U$ be a ring with unity and let $V$ consist of all the invertibles of $U$. Then $V$ is a multiplicative group.*

The proof of Theorem 1 is left to the reader as Problem 7 of this section.

Distributivity of multiplication over addition enables us to obtain the following multiplicative property of the additive identity 0 of a ring.

**Theorem 2** ***Multiplication by* 0**

*For all $r$ in a ring $R$, $r \cdot 0 = 0 = 0 \cdot r$.*

***Proof***

Since 0 is the additive identity, $0 + 0 = 0$ and so $r(0 + 0) = r \cdot 0$. Using distributivity, this becomes

$$r \cdot 0 + r \cdot 0 = r \cdot 0. \tag{1}$$

Additively, $R$ is a group and we may apply Theorem 2 of Section 2.3 (after translating that result into additive notation) to equation (1) and so find that $r \cdot 0 = 0$. We leave it to the reader (in Problem 10 of this section) to prove that $0 \cdot r = 0$.

The following result is a generalization of one of the "rules of signs" of elementary algebra.

**Theorem 3** ***A Rule of Signs***

*If $r$ and $s$ are elements of a ring $R$,*

$$r(-s) = (-r)s = -(rs).$$

***Proof***

By definition of the negative of an element, $s + (-s) = 0$ and so

$$r[s + (-s)] = r \cdot 0.$$

Using distributivity and Theorem 2, this becomes

$$rs + r(-s) = 0.$$

Hence each of $rs$ and $r(-s)$ is the negative of the other. Similarly, one shows that $(-r)s = -(rs)$.

### *Corollary*

*If $r$ and $s$ are in $R$, then $(-r)(-s) = rs$.*

The proof of this corollary is left to the reader as Problem 11 of this section.

We next define the analogue of a subgroup in a group.

### *Definition 3 Subring*

*A subset $S$ of a ring $R$ such that $S$ is a ring under the operations of $R$ is a* **subring** *in $R$.*

We note that a subring $S$ in a ring $R$ additively is a normal subgroup of the additive group of $R$. We consider the integers $\mathbf{Z}$ to be a subring in the rational numbers $\mathbf{Q}$, $\mathbf{Q}$ to be a subring in the real numbers $\mathbf{R}$, and $\mathbf{R}$ to be a subring in the complex numbers $\mathbf{C}$.

### *Theorem 4 Subring Conditions*

*A nonempty subset $S$ of a ring $R$ is a subring in $R$ if and only if $S$ is closed under both subtraction and multiplication, that is, if and only if $S$ contains $a - b$ and $ab$ whenever it contains $a$ and $b$.*

The proof of Theorem 4 is left to the reader as Problem 12 of this section.

Using distributivity and mathematical induction, one can establish the following result:

***Theorem 5 General Distributivity***

*In a ring R,* $(a_1 + a_2 + \cdots + a_m)(b_1 + b_2 + \cdots + b_n) = a_1b_1 + a_1b_2 + \cdots + a_1b_n + a_2b_1 + a_2b_2 + \cdots + a_2b_n + \cdots + a_mb_1 + a_mb_2 + \cdots + a_mb_n$.

***Corollary***

*If a and b are in a ring R and m and n are in* $\mathbf{Z}$, $(m \cdot a)(n \cdot b) = (mn) \cdot (ab)$.

The proof is left to the reader.

**Example 1*

Show that $R = \{0, b, c, d\}$ is a noncommutative ring under the operations given by the following tables:

| + | 0 | b | c | d |
|---|---|---|---|---|
| 0 | 0 | b | c | d |
| b | b | 0 | d | c |
| c | c | d | 0 | b |
| d | d | c | b | 0 |

| | 0 | b | c | d |
|---|---|---|---|---|
| 0 | 0 | 0 | 0 | 0 |
| b | 0 | 0 | b | b |
| c | 0 | 0 | c | c |
| d | 0 | 0 | d | d |

***Solution***

Additively, $R$ is a noncyclic group of order 4, that is, a Klein 4-group. (See Problem 11 of Section 2.3 and Problem 12 of Section 2.5.) It is also clear from the multiplication table that $R$ is closed under multiplication.

Now we consider associativity of multiplication. We let $u = (xy)z$ and $v = x(yz)$, where $x$, $y$, and $z$ are any elements of $R$. If at least one of $y$ and $z$ is in $\{0, b\}$, we see that $u = 0 = v$. In all other cases (when $y$ and $z$ are in $\{c, d\}$), each of $u$ and $v$ is equal to $x$. Hence $(xy)z = x(yz)$ for all $x$, $y$, and $z$ in $R$; that is, multiplication in $R$ is associative.

Next we turn to distributivity. Let $s = x(y + z)$ and $t = xy + xz$. If $y = z$, then the addition table shows that $y + z = y + y = 0$ and $t = xy + xz = xy + xy = 0$; in such a case $s = x \cdot 0 = 0$ and so $s = t$. If $y$ or $z$ is 0, it also follows that $s = t$. In the remaining cases, $y$ and $z$ are distinct elements of $S = \{b, c, d\}$ and for such $y$ and $z$ the addition table shows that $y + z$ is the third element of $S$; all these cases also lead to $s = t$.

Showing the other half of distributivity, $(y + z)x = yx + zx$, is entirely similar. Hence $R$ is a ring. Since $bc = b$ and $cb = c$, $R$ is not commutative.

A ring in which the elements are mappings is described in the following example.

*Example 2*

Let $\mathbf{R}$ be the set of all real numbers and $S$ the subset of all $x$ in $\mathbf{R}$ satisfying $0 < x < 1$. Let $M$ be the set of all mappings from $S$ to $\mathbf{R}$. For $\alpha$ and $\beta$ in $M$, we define the sum $\alpha + \beta$ to be the mapping $\sigma$ from $S$ to $\mathbf{R}$ such that

$$\sigma(x) = \alpha(x) + \beta(x)$$

and the product $\alpha\beta$ to be the mapping $\pi$ from $S$ to $\mathbf{R}$ with

$$\pi(x) = \alpha(x) \cdot \beta(x).$$

It can be shown that these operations make $M$ into a ring. (See Problem 31 below.)

## Problems

Problem 31 below is cited in Section 5.4.

1. Let $R = \{0, b\}$ be a ring with two elements.
   (i) Make the table for the additive group of $R$.
   (ii) Make a multiplication table for $R$ under the assumption that $b^2 = b$.
   (iii) Assume that $b^2 = 0$ and make a multiplication table for $R$.
   (iv) Does $R$ have a unity in (ii) or (iii)?
2. Let $U = \{0, 1, c\}$ be a ring with three elements (and, of course, with 1 as the unity).
   (i) Explain why $1 + 1 = c$. (See Example 2 of Section 2.3.)
   (ii) Explain why $3 \cdot 1 = 1 + 1 + 1 = 0$ in $U$.
   (iii) Explain why $c^2 = (1 + 1)(1 + 1) = 1$ in $U$.
   (iv) Make addition and multiplication tables for $U$.
3. Let $A = \{u, v, w\}$ have the following addition and multiplication tables:

| $+$ | $u$ | $v$ | $w$ |
|---|---|---|---|
| $u$ | $u$ | $v$ | $w$ |
| $v$ | $v$ | $w$ | $u$ |
| $w$ | $w$ | $u$ | $v$ |

| $\cdot$ | $u$ | $v$ | $w$ |
|---|---|---|---|
| $u$ | $u$ | $u$ | $u$ |
| $v$ | $u$ | $u$ | $u$ |
| $w$ | $u$ | $u$ | $u$ |

Is $A$ a ring under these operations? Explain.

4. Let $G$ be an additive group. Describe an operation of multiplication in $G$ that converts $G$ into a ring. (See Problem 3.)
5. Let $U$ be a ring with unity. Show the following:
   (i) If $0 = 1$, then $U$ consists of just one element.
   (ii) If 0 is an invertible (that is, unit), $U = \{0\}$.
   (iii) If $U = \{0\}$, then 0 is the unity and is an invertible.
6. Prove that in a ring $U$ with unity the reciprocal (that is, multiplicative inverse) of an invertible is unique. (Since $U$ need not be a group under multiplication, one should not use cancellation without proof that it is valid under the present hypothesis.)
7. Prove Theorem 1; that is, prove that the set $V$, of invertibles of a ring $U$ with unity, is a multiplicative group by showing the following:
   (i) The unity 1 of $U$ is an invertible.
   (ii) If $v$ and $w$ are in $V$, so is $vw^{-1}$.
8. Make a multiplication table for the group $V = \{1, -1\}$ of invertibles (units) of the integers $\mathbf{Z}$.
9. Let $v$ be an invertible in a ring $U$. Prove that either $uv = 0$ or $vu = 0$ in $U$ implies $u = 0$.
10. Prove that $0 \cdot r = 0$ for all $r$ in a ring $R$. (This is the half of Theorem 2 that was left to the reader.)
11. Prove the corollary to Theorem 3; that is, prove that $(-r)(-s) = rs$ for all $r$ and $s$ in a ring $R$.
12. Prove Theorem 4; that is, prove that a nonempty subset $S$ of a ring $R$ is a subring in $R$ if and only if $a - b$ and $ab$ are in $S$ whenever $a$ and $b$ are in $S$.
13. Which of the following subsets of the ring $\mathbf{Z}$ of the integers is a subring in $\mathbf{Z}$?
   (a) $\{0, 1, -1\}$.
   (b) $\{3, 6, 9, \ldots\}$, i.e., the positive integral multiples of 3.
   (c) $\{0, 5, -5, 10, -10, \ldots\}$, i.e., the integral multiples of 5.
   (d) $\{\ldots, -4, -2, 0, 2, 4, \ldots\}$, i.e., the even integers.
14. Which of the following subsets is a subring in the ring $\mathbf{C}$ of the complex numbers $a + b\mathbf{i}$?
   (a) The subset of all $a + b\mathbf{i}$ with $a$ and $b$ rational.

(b) The subset of all $a + b\mathbf{i}$ with $a^2 + b^2 \leq 1$.
(c) The subset $S$ of all the rational numbers of the form $m/2^n$ with $m$ an integer and $n$ a positive integer or zero. ($S$ consists of the rational numbers $0, \pm 1, \pm 2, \ldots; \pm 1/2, \pm 3/2, \pm 5/2, \ldots; \pm 1/4, \pm 3/4, \pm 5/4, \ldots;$ and so on.)

15. Let $m$ be a fixed integer and let $I$ consist of all the integral multiples of $m$. Prove the following:
    (i) For every $i$ in $I$ and $z$ in $\mathbf{Z}$, $iz$ is in $I$.
    (ii) $I$ is nonempty and closed under subtraction.
    (iii) $I$ is a subring in $\mathbf{Z}$ (and hence the additive group of $I$ is a normal subgroup in the additive group of $\mathbf{Z}$).
    (iv) Let $a$ and $b$ be any elements of $\mathbf{Z}$. Let $i$ and $i'$ be elements of $I$. Then the product $(a + i)(b + i')$ is in the coset $(ab) + I$ of the normal subgroup $I$ in the additive group $\mathbf{Z}$.
    *(v) The additive quotient group $\mathbf{Z}/I$ can be made into a ring by using

$$(a + I)(b + I) = (ab) + I$$

    as the formula for multiplication of cosets.

16. Let $I$ be the subring of all the integral multiples of 3 in the ring $\mathbf{Z}$ of the integers. Then the cosets of the additive quotient group $\mathbf{Z}/I$ are:

$$\overline{0} = 0 + I = \{\ldots, -6, -3, 0, 3, 6, \ldots\},$$
$$\overline{1} = 1 + I = \{\ldots, -5, -2, 1, 4, 7, \ldots\},$$
$$\overline{2} = 2 + I = \{\ldots, -4, -1, 2, 5, 8, \ldots\}.$$

    (i) Explain why $\overline{4} = \overline{1}$, i.e., $4 + I = \overline{1}$.
    (ii) Make the addition table for $\mathbf{Z}/I = \{\overline{0}, \overline{1}, \overline{2}\}$.
    (iii) Use the formula of Problem 15(v) to make a multiplication table for $\{\overline{0}, \overline{1}, \overline{2}\}$ and thus convert it into a ring.

17. Among the properties of a subring $S$ in a ring $R$ is that $S$ is nonempty and closed under subtraction. Use this to explain why every subring in the ring $\mathbf{Z}$ of the integers must be one of the subrings $I$ of Problem 15.

18. Let $a$ be a fixed element of a commutative ring $R$ and let $I$ consist of the products $sa$ for all $s$ in $R$.
    (i) Show that $ri$ is in $I$ for all $r$ in $R$ and $i$ in $I$.
    (ii) Use Problem 12 to prove that $I$ is a subring in $R$.

19. Let $S$ consist of all the real numbers of the form $a + b\sqrt{2}$ with $a$ and $b$ rational numbers.
    (i) Show that $S$ is a subring in the ring $\mathbf{R}$ of real numbers and hence is a subring in the complex numbers $\mathbf{C}$.
    (ii) Would $S$ be a subring in $\mathbf{R}$ if $a$ and $b$ were restricted to integer values?

20. Let $T$ consist of all the real numbers, $a + b\sqrt{3}$ with $a$ and $b$ integers. Show that $T$ is a subring in the ring **R** of real numbers.

21. Let $G$ consist of all the complex numbers $m + n\mathbf{i}$ with $m$ and $n$ integers. Show that $G$ is a subring in the ring **C** of complex numbers. (This subring is called the ring of ***gaussian integers*** after Carl Friedrich Gauss.)

22. Find all the invertibles of the ring of gaussian integers. (See Problem 21 above and Example 1 of Section 2.3.)

23. Let $U = \{0, 1, c, d\}$ be a ring (with four elements) and let $c$ and $d$ be invertibles in $U$. Make the multiplication table for $U$.

24. Let $U = \{0, 1, c, d\}$ be a ring and let $1 + 1 = c$ in $U$. Make the addition and multiplication tables for $U$.

25. Prove that every ring $R$ with 5 elements is commutative.

26. Prove that every ring with 7 elements is commutative.

27. Let two sequences $q_0, q_1, q_2, \ldots$ and $p_0, p_1, p_2, \ldots$ of nonnegative integers be defined by $q_0 = 1$, $p_0 = 0$, $q_{n+1} = q_n + 2p_n$, and $p_{n+1} = q_n + p_n$ for $n = 1, 2, 3, \ldots$. Prove the following for $n = 0, 1, 2, \ldots$:
    (i) $(1 + \sqrt{2})^n = q_n + p_n\sqrt{2}$.
    (ii) $(q_n + p_n\sqrt{2})(q_n - p_n\sqrt{2}) = (-1)^n$.
    (iii) $p_{n+2} = 2p_{n+1} + p_n$ and $q_{n+2} = 2q_{n+1} + q_n$.

*28. Let $S$ be the ring of Problem 19(ii) and let $q_n$ and $p_n$ be as in Problem 27.
    (i) Show that every invertible of $S$ is of the form $\pm(q_n \pm p_n\sqrt{2})$ for some nonnegative integer $n$.
    (ii) Let $V^+$ be the set of positive real numbers that are invertibles of $S$. Show that $V^+$ is a cyclic multiplicative group.
    (iii) List the two elements, each of which is a generator of $V^+$.

*29. Given that $a$ and $b$ are integers, show that $a + b\sqrt{3}$ is an invertible of the ring $T$ of Problem 20 if and only if $a^2 - 3b^2 = \pm 1$.

30. Let $R$ be the cartesian product $S \times T$ of rings $S$ and $T$. Define addition and multiplication in $R$ by

$$(s, t) + (s', t') = (s + s', t + t'), \; (s, t)(s', t') = (ss', tt').$$

    (a) Show that $R$ is a ring (called the ***direct product*** of $S$ and $T$).
    (b) Define addition and multiplication for the direct product $S_1 \times S_2 \times \cdots \times S_n$ of $n$ rings.

31. Let $R$ be a ring, let $S$ be a set, and let $F$ consist of all the mappings from $S$ to $R$. Define addition and multiplication in $F$ as in Example 2 of this section and show that these definitions make $F$ into a ring.

*32. Let $\mathbf{R}$ be the set of real numbers and let $H$ be the cartesian product $\mathbf{R} \times \mathbf{R} \times \mathbf{R} \times \mathbf{R}$. Show that $H$ is a noncommutative ring under the operations

$$(a, b, c, d) + (e, f, g, h) = (a + e, b + f, c + g, d + h), (a, b, c, d)(e, f, g, h) =$$

$$(ae - bf - cg - dh,\ af + be + ch - dg,\ ag - bh + ce + df,\ ah + bg - cf + de).$$

($H$ is essentially the ring of *quaternions* developed by Sir William Rowan Hamilton.)

*33. Find the invertibles of the ring $H$ of Problem 32.

---

***Emmy Noether*** *(1882–1935) One of the foremost researchers in the theory of rings in the twentieth century was also, possibly, the most important woman in the history of mathematics to date: Emmy Noether. Her father was an important mathematician at the University of Erlangen, where she studied and worked till 1916; then she went to work at Göttingen, where both Klein and Hilbert were professors. She remained at Göttingen until 1933 when she was driven out of Germany by the Nazi regime. She came to the United States and became professor of mathematics at Bryn Mawr, but only lived for a year and a half after leaving Germany.*

*Although she has been universally acknowledged as one of the great algebraists of history, she was subject during her lifetime to constant difficulties in working in a profession which until recently had not been a common one for women. Prior to the twentieth century there were few prominent women mathematicians; some of the more famous were Hypatia, a Greek mathematician murdered in Alexandria in* A.D. *415; Maria Gaetana Agnesi, the Italian mathematician for whom the famous curve, the "Sail of Agnesi" (mistranslated as "Witch of Agnesi") is named; Sonya Kovalevsky, a Russian mathematician, student of Weierstrass and professor of mathematics at the University of Stockholm.*

*Professor Noether was never given a professorship at Göttingen, although she was one of the most able mathematicians there. Hilbert once said at a meeting held to discuss a possible appointment for her: "I do not see that the sex of the candidate is an argument against her admission as Privatdozent. After all we are a university and not a bathing establishment." She did not receive the appointment, however. In 1922 she finally received the appointment of "nichtbeamteter ausserordentlicher Professor," which she held at Göttingen until she left in 1933. It was a position which involved no duties—unfortunately, it also involved no salary.*

---

## 4.2 Ring Homomorphisms and Ideals

Here we introduce the analogues in ring theory of the closely related group theory concepts of homomorphism, normal subgroup, and quotient group.

***Definition 1 Ring Homomorphism, Kernel***

*A mapping $\theta$ from a ring $R$ to a ring $R'$ such that for all $r$ and $s$ in $R$*

$$\theta(r + s) = \theta(r) + \theta(s), \qquad \theta(rs) = \theta(r)\theta(s)$$

*is a* **ring homomorphism** *from $R$ to $R'$. The* **kernel** *of $\theta$ is the subset of all $k$ in $R$ for which $\theta(k)$ is the zero of $R'$.*

In other words, a ring homomorphism is a mapping that preserves both additions and multiplications. Clearly, a ring homomorphism from $R$ to $R'$ automatically is a homomorphism from the additive group of $R$ to the additive group of $R'$.

***Definition 2 Isomorphism, Endomorphism, Automorphism***

*A bijective ring homomorphism from $R$ to $R'$ is a ring* **isomorphism**. *A ring homomorphism from $R$ to itself is a ring* **endomorphism** *of $R$. A ring isomorphism from $R$ to itself is a ring* **automorphism** *of $R$.*

In our study of groups we found that the kernel of a homomorphism from a group $G$ to a group $G'$ is always a normal subgroup in $G$. Furthermore, if $N$ is a normal subgroup in a group $G$, then there exists a homomorphism $\theta$ (the natural map) from $G$ to the quotient group $G/N$ such that the kernel of $\theta$ is $N$. Together, these two statements imply that a subset $N$ of a group $G$ is a normal subgroup in $G$ if and only if $N$ is the kernel of a homomorphism from $G$ to a group $G'$.

This motivates us to investigate the properties possessed by the kernel $K$ of a ring homomorphism $\theta$ from $R$ to $R'$. Since such a $\theta$ is a homomorphism from the additive group of $R$ to that of $R'$, $K$ must be a subgroup of the additive group of $R$. As a subset of $R$, $K$ has the properties of associativity of multiplication and distributivity. This means that $K$ will have been proved to be a subring of $R$ once we have shown closure of $K$ under multiplication. While demonstrating that $K$ is closed under multiplication, we obtain an important extra property of $K$, as follows:

Let $k$ be an element of the kernel $K$ and let $r$ be any element of $R$. Then $\theta(kr) = \theta(k)\theta(r) = 0 \cdot \theta(r) = 0$ and similarly $\theta(rk) = 0$. This means that both $kr$ and

$rk$ are in $K$. Since $r$ is any element of $R$, it may be any element of $K$. Thus we have proved that $K$ is closed under multiplication and hence is a subring of $R$.

The two following definitions help us in stating the properties of a kernel of a ring homomorphism noted above.

***Definition 3 Inside—Outside Closure***

*Let $J$ be a subset of a ring $R$ such that both $jr$ and $rj$ are in $J$ for all $j$ in $J$ and $r$ in $R$. Then we say that $J$ is* **closed under inside-outside multiplication**.

***Definition 4 Ideal in a Ring***

*Let $I$ be a subring closed under inside-outside multiplication in a ring $R$. Then $I$ is an* **ideal** *in $R$.*

The properties of a kernel found above can now be stated as follows:

***Lemma 1 Kernel of a Ring Homomorphism***

*The kernel of a ring homomorphism from $R$ to $R'$ is an ideal in $R$.*

We next show how an ideal $I$ in a ring $R$ determines a collection $\Gamma$ of cosets of $I$ in $R$ and operations of addition and multiplication that make $\Gamma$ into a ring. This will be the ring theory analogue of the construction of the quotient group $G/N$ from a group $G$ and a normal subgroup $N$ in $G$.

Under addition, an ideal $I$ in a ring $R$ is a normal subgroup in the additive group of $R$. Therefore, we use the following approach:

***Definition 5 Coset of an Ideal in a Ring***

*Let $I$ be an ideal in a ring $R$ and let $a$ be any element of $R$. Then the* **ideal coset** *for the element $a$ of the ideal $I$ in the ring $R$ is the same subset of $R$ as the coset $a + I$ considering $I$ to be a subgroup in the additive group of $R$.*

*Notation*

*The ideal coset for $a$ of $I$ in $R$ will be designated by $a + I$.*

The fact that $a + I$ might be either the coset of a subgroup or the coset of an ideal need not cause any difficulty since $a + I$ has the same elements under each interpretation.

Let $\Gamma$ be the collection of ideal cosets $a + I$ for all $a$ in $R$. We can add ideal cosets $a + I$ and $b + I$ in $\Gamma$ by using the formula

$$(a + I) + (b + I) = (a + b) + I$$

for adding them as normal subgroup cosets. It seems natural to use the formula

$$(a + I)(b + I) = ab + I \tag{M}$$

to introduce an operation of multiplication in $\Gamma$. This requires that we show that multiplication of ideal cosets is well defined by formula (M), that is, that the product is not changed by using different expressions for the factors.

So let $a + I = a' + I$ and $b + I = b' + I$. This implies that $a' = a + i$ and $b' = b + i_1$ with $i$ and $i_1$ in $I$. Now we apply formula (M) to the new expressions for the cosets and have

$$(a' + I)(b' + I) = (a'b') + I.$$

Next we note that $a'b' = (a + i)(b + i_1) = ab + i_2$, where $i_2 = ai_1 + bi + ii_1$. Since $I$ is closed under inside-outside multiplication and under addition, $i_2$ is in $I$. Hence $a'b'$ is in the same coset as $ab$ and $a'b' + I = ab + I$.

Therefore, multiplication in $\Gamma$ is well defined by (M). We leave to the reader the further details of proving the following result.

***Theorem 1 The Ring of Cosets of an Ideal***

*If $I$ is an ideal in a ring $R$, the collection $\Gamma$ of ideal cosets $a + I$ is a ring under the operations given by*

$$(a + I) + (b + I) = (a + b) + I,$$
$$(a + I)(b + I) = ab + I.$$

### *Definition 6 Quotient Ring $R/I$*

*The ring $\Gamma$ described in Theorem 1 is called the* **quotient ring** *of $R$ by $I$ and is denoted by $R/I$.*

We continue the analogy with group theory.

### *Definition 7 The Natural Map*

*Let $I$ be an ideal in a ring $R$. The mapping $\theta$ from $R$ to $R/I$ with $\theta(r) = r + I$ is called the* **natural map**; *that is, the image of an element $r$ of $R$ under the natural map is the ideal coset of $I$ to which $r$ belongs.*

### *Theorem 2 Properties of the Natural Map*

*The natural map $\theta$ from a ring $R$ to one of its quotient rings $R/I$ is a surjective ring homomorphism with $I$ as kernel.*

*Proof*

For all $a$ and $b$ in $R$,

$$\theta(a + b) = (a + b) + I = (a + I) + (b + I) = \theta(a) + \theta(b),$$

$$\theta(ab) = ab + I = (a + I)(b + I) = \theta(a)\theta(b).$$

This shows that $\theta$ is a ring homomorphism. Each coset $r + I$ in $R/I$ is the image under $\theta$ of the element $r$ of $R$; hence $\theta$ is surjective. Also, $\theta(r) = r + I$ is equal to the additive identity $I$ of $R/I$ if and only if $r$ is in $I$; thus the kernel of $\theta$ is the ideal $I$.

Theorem 2 contains a converse of Lemma 1. We restate Lemma 1 and its converse in the following form.

### *Theorem 3 Ideals and Kernels*

*A subset $I$ of a ring $R$ is an ideal in $R$ if and only if $I$ is the kernel of a ring homomorphism from $R$ to some ring $R'$.*

The next result provides a fairly easy way of showing that a subset of a ring is an ideal.

***Theorem 4 Conditions for an Ideal***

*Let $I$ be a nonempty subset of a ring $R$ and let $I$ be closed under both subtraction and inside-outside multiplication. Then $I$ is an ideal in $R$.*

***Proof***

Since closure under inside-outside multiplication implies closure under multiplication, $I$ is a subring in $R$. (See Theorem 4 of Section 4.1.) This and inside-outside closure make $I$ an ideal in $R$.

One way to obtain a subgroup $H$ in a group $G$ is to choose an element $a$ of $G$ and then let $H$ consist of all the elements $a^m$ with $m$ an integer; that is, let $H$ be the cyclic subgroup $[a]$. We next give an analogous procedure for obtaining an ideal in a commutative ring $R$.

***Theorem 5 Multiples of a Fixed Element***

*Let $a$ be a fixed element of a commutative ring $R$ and let $I$ be the subset of $R$ consisting of all the multiples $ra$ of the given element $a$ by elements $r$ of $R$. Then $I$ is an ideal in $R$.*

***Proof***

Let $sa$ and $ta$ be any two elements of $I$ and let $u$ be any element of $R$. Then $sa - ta = (s - t)a$ and $u(sa) = (us)a$ are elements of $I$. This and commutativity of $R$ imply that $I$ is closed under subtraction and under inside-outside multiplication. Also, $I$ is nonempty since it contains $0 \cdot a = 0$. Theorem 4 now tells us that $I$ is an ideal in $R$.

***Definition 8 Principal Ideal Generated by a***

*Let $a$ be a fixed element of a commutative ring $R$ and let $(a)$ consist of all products $ra$ with $r$ in $R$. Then $(a)$ is called the* **principal ideal** *generated by $a$ in $R$.*

We now have a convenient notation for an infinite number of ideals in the commutative ring $\mathbf{Z}$ of the integers. Some of these principal ideals in $\mathbf{Z}$ are $(0) = \{0\}$, $(1) = \mathbf{Z} = (-1)$,

$$(2) = (-2) = \{\ldots, -6, -4, -2, 0, 2, 4, \ldots\},$$

$$(3) = (-3) = \{\ldots, -9, -6, -3, 0, 3, 6, \ldots\}.$$

It is easily seen that for $m \neq 0$ the set of elements of the principal ideal $(m)$ is

$$\{\ldots, -3m, -2m, -m, 0, m, 2m, 3m, \ldots\},$$

the same set as that for the cyclic subgroup $[m]$ in the additive group of the integers. If $m = 0$, $(m)$ and $[m]$ each consist only of 0. Hence the sets for $(m)$ and $[m]$ are the same for every integer $m$.

Since the ring **Z** of the integers has an infinite number of ideals $I$, we can now use the concept of quotient rings to construct an infinite number of new rings $\mathbf{Z}/I$. This will give us a fertile source of concrete examples of finite rings and also will provide valuable information concerning **Z** itself. After studying polynomial rings in Chapter 5, we shall again use quotient rings to obtain new rings and additional insight into old ones.

### *Notation* *The Rings* $\mathbf{Z_m}$ *and Groups* $\mathbf{V_m}$

*For* $m = 1, 2, 3, \ldots$ *the quotient ring* $\mathbf{Z}/(m)$ *of the integers* **Z** *by the principal ideal* $(m)$ *is denoted by* $\mathbf{Z_m}$. *We use* $\mathbf{V_m}$ *to designate the multiplicative group of invertibles of* $\mathbf{Z_m}$.

The $\mathbf{Z_m}$ are called the ***modular rings*** and hence the study of these rings is sometimes called ***modular arithmetic***.

In $\mathbf{Z_m}$ let $\bar{a}$ denote the ideal coset $a + (m)$; that is, let

$$\bar{a} = \{\ldots, a - 2m, a - m, a, a + m, a + 2m, a + 3m, \ldots\}.$$

This is consistent with the previous usage of $\bar{a}$ for the subgroup coset $a + [m]$ since $a + (m)$ and $a + [m]$ have the same elements. Then the natural map $\theta$ from **Z** to $\mathbf{Z_m}$ has $\theta(a) = \bar{a}$. The following result is helpful in applying this ring homomorphism $\theta$.

### *Lemma 2* $\mathbf{Z_m}$ *and* **Z**

(a) $\bar{a} = \bar{b}$ *in* $\mathbf{Z_m}$ *if and only if* $m \mid (a - b)$ *in* **Z**.
(b) *If* $a = qm + r$ *in* **Z**, $\bar{a} = \bar{r}$; *i.e.*, $\overline{qm + r} = \bar{r}$, *in* $\mathbf{Z_m}$.
(c) $\mathbf{Z_m} = \{\bar{0}, \bar{1}, \bar{2}, \ldots, \overline{m - 1}\}$.
(d) $\overline{a + b} = \bar{a} + \bar{b}$ *and* $\overline{ab} = \bar{a} \cdot \bar{b}$ *for all integers a and b.*

*Proof*

Parts (a), (b), and (c) are restatements of Lemmas 1, 3, and 4 of Section 3.3. Using the fact that the natural map $\theta$ from $\mathbf{Z}$ to $\mathbf{Z_m}$ is a ring homomorphism, we have

$$\overline{a+b} = \theta(a+b) = \theta(a) + \theta(b) = \bar{a} + \bar{b},$$
$$\overline{ab} = \theta(ab) = \theta(a)\theta(b) = \bar{a} \cdot \bar{b}.$$

This proves (d).

*Example 1*

Make addition and multiplication tables for $\mathbf{Z}_6$.

*Solution*

Part (c) of Lemma 2 tells us that $\mathbf{Z}_6 = \{\bar{0}, \bar{1}, \bar{2}, \bar{3}, \bar{4}, \bar{5}\}$. Parts (b) and (d) of Lemma 2 facilitate addition and multiplication in $\mathbf{Z}_6$. For example,

$$\bar{4} + \bar{5} = \bar{9} = \overline{6+3} = \bar{3}, \qquad \bar{4} \cdot \bar{5} = \overline{20} = \overline{3 \cdot 6 + 2} = \bar{2} \text{ in } \mathbf{Z}_6.$$

Similarly, we obtain the information listed in the following tables:

| + | $\bar{0}$ | $\bar{1}$ | $\bar{2}$ | $\bar{3}$ | $\bar{4}$ | $\bar{5}$ |
|---|---|---|---|---|---|---|
| $\bar{0}$ | $\bar{0}$ | $\bar{1}$ | $\bar{2}$ | $\bar{3}$ | $\bar{4}$ | $\bar{5}$ |
| $\bar{1}$ | $\bar{1}$ | $\bar{2}$ | $\bar{3}$ | $\bar{4}$ | $\bar{5}$ | $\bar{0}$ |
| $\bar{2}$ | $\bar{2}$ | $\bar{3}$ | $\bar{4}$ | $\bar{5}$ | $\bar{0}$ | $\bar{1}$ |
| $\bar{3}$ | $\bar{3}$ | $\bar{4}$ | $\bar{5}$ | $\bar{0}$ | $\bar{1}$ | $\bar{2}$ |
| $\bar{4}$ | $\bar{4}$ | $\bar{5}$ | $\bar{0}$ | $\bar{1}$ | $\bar{2}$ | $\bar{3}$ |
| $\bar{5}$ | $\bar{5}$ | $\bar{0}$ | $\bar{1}$ | $\bar{2}$ | $\bar{3}$ | $\bar{4}$ |

| | $\bar{0}$ | $\bar{1}$ | $\bar{2}$ | $\bar{3}$ | $\bar{4}$ | $\bar{5}$ |
|---|---|---|---|---|---|---|
| $\bar{0}$ | $\bar{0}$ | $\bar{0}$ | $\bar{0}$ | $\bar{0}$ | $\bar{0}$ | $\bar{0}$ |
| $\bar{1}$ | $\bar{0}$ | $\bar{1}$ | $\bar{2}$ | $\bar{3}$ | $\bar{4}$ | $\bar{5}$ |
| $\bar{2}$ | $\bar{0}$ | $\bar{2}$ | $\bar{4}$ | $\bar{0}$ | $\bar{2}$ | $\bar{4}$ |
| $\bar{3}$ | $\bar{0}$ | $\bar{3}$ | $\bar{0}$ | $\bar{3}$ | $\bar{0}$ | $\bar{3}$ |
| $\bar{4}$ | $\bar{0}$ | $\bar{4}$ | $\bar{2}$ | $\bar{0}$ | $\bar{4}$ | $\bar{2}$ |
| $\bar{5}$ | $\bar{0}$ | $\bar{5}$ | $\bar{4}$ | $\bar{3}$ | $\bar{2}$ | $\bar{1}$ |

The two smallest positive primes in the integers $\mathbf{Z}$ are 2 and 3. It can be shown that all the other positive primes $p$ are in one of the two following arithmetic progressions:

$$5, 11, 17, 23, 29, 35, 41, \ldots \qquad \text{(A)}$$

$$7, 13, 19, 25, 31, 37, 43, \ldots \qquad \text{(B)}$$

[See Problem 16(v) below.] The fact that there is a ring homomorphism, the natural map, from the infinite ring $\mathbf{Z}$ to the finite ring $\mathbf{Z}_6$ can be used to obtain fairly efficient techniques for partitioning the set of integers in (A), or in (B), into a subset of composites and a subset of primes. In Example 2, which follows, we show how to do this for the progression (A) and leave the similar problem for progression (B) to the reader as Problem 17 below.

### *Example 2*

Use the multiplication table for the finite ring $\mathbf{Z}_6$ (see Example 1) to prove that an integer $c$ in the arithmetic progression

$$5, 11, 17, \ldots, 6n - 1, \ldots \tag{A}$$

is composite if and only if $c = ab$ with $a < c$, $a$ in the progression (A), and $b$ in the arithmetic progression

$$7, 13, 19, \ldots, 6n + 1, \ldots . \tag{B}$$

### *Solution*

Let $c$ be in (A). If $c = ab$ with $a < c$, $a$ in (A) and $b$ in (B), then clearly $c$ is composite.

Conversely, let $c$ be composite; that is, let $c = ab$ with $a$ and $b$ integers satisfying $1 < a < c$ and $1 < b < c$. Since $c$ is in (A) we have $\bar{c} = \bar{5}$ and so $\bar{a} \cdot \bar{b} = \overline{ab} = \bar{c} = \bar{5}$ in $\mathbf{Z}_6$. We note that $\bar{5}$ appears only twice as an entry in the multiplication table for $\mathbf{Z}_6$, namely, in the products $\bar{1} \cdot \bar{5} = \bar{5}$ and $\bar{5} \cdot \bar{1} = \bar{5}$. Hence $\bar{a} \cdot \bar{b} = \bar{5}$ implies that either $\bar{a}$ or $\bar{b}$ is $\bar{5}$ and the other is $\bar{1}$. For definiteness, let $\bar{a} = \bar{5}$ and $\bar{b} = \bar{1}$. Since $a > 1$ and $b > 1$, this means that $a$ is in (A) and $b$ is in (B).

One can easily deduce from Example 2 that the only composite numbers $c$ with $c$ in the finite arithmetic progression $5, 11, 17, \ldots, 95$ are

$$5 \cdot 7 = 35, \qquad 5 \cdot 13 = 65, \qquad 7 \cdot 11 = 77, \qquad 5 \cdot 19 = 95.$$

Hence the remaining integers

$$5, 11, 17, 23, 29, 41, 47, 53, 59, 71, 83, 89$$

in this finite progression are primes.

We next illustrate the use of the natural map from $\mathbf{Z}$ to a $\mathbf{Z}_m$ to obtain information concerning the squares of integers.

### *Example 3*

Prove that no square of an integer is present in the infinite arithmetic progression

$$2, 5, 8, 11, 14, \ldots . \tag{C}$$

*Proof*

We assume that there are integers $a$ and $b$ with $b^2 = a$ and $a$ in (C); then we show that this assumption leads to a contradiction.

All the integers in (C) are in the ideal coset $\bar{2}$ in $\mathbf{Z}_3$. Hence we use the natural map $\theta$ from $\mathbf{Z}$ to $\mathbf{Z}_3$. Since the homomorphism $\theta$ preserves the ring operations, the equation $b^2 = a$ in $\mathbf{Z}$ implies that in $\mathbf{Z}_3$

$$\bar{b}^2 = [\theta(b)]^2 = \theta(b^2) = \theta(a) = \bar{a} = \bar{2};$$

that is, $\bar{b}^2 = \bar{2}$. As $\mathbf{Z}_3$ has only three elements, we calculate

$$\bar{0}^2 = \bar{0}, \qquad \bar{1}^2 = \bar{1}, \qquad \bar{2}^2 = \bar{4} = \overline{3+1} = \bar{1}$$

and observe that $\bar{b}^2$ is not $\bar{2}$ for any $\bar{b}$ in $\mathbf{Z}_3$. Therefore, $b^2$ is not in the progression (C) for any $b$ in $\mathbf{Z}$.

In solving Example 3, we proved that there are no integers $b$ and $c$ such that $b^2 = 3c + 2$. We now apply the same technique to another equation.

*Example 4*

Prove that there are no integers $x$ and $y$ such that

$$x^2 - 17y^2 = 855. \tag{D}$$

*Proof*

We remove the influence of the term $17y^2$ in equation (D) by using the natural map $\theta$ from $\mathbf{Z}$ to $\mathbf{Z}_{17}$. Since the ring homomorphism $\theta$ preserves additions, subtractions, and multiplications, the existence of integers $x$ and $y$ satisfying (D) would imply

$$\bar{x}^2 - \overline{17}\bar{y}^2 = \overline{855}. \tag{E}$$

But $\overline{17} = \bar{0}$ and $\overline{855} = \overline{17 \cdot 50 + 5} = \bar{5}$ in $\mathbf{Z}_{17}$. Thus (E) becomes $\bar{x}^2 = \bar{5}$. Now we find all the squares in $\mathbf{Z}_{17}$; for this purpose it is convenient to represent the 17 cosets in $\mathbf{Z}_{17}$ by $\bar{0}, \pm\bar{1}, \pm\bar{2}, \ldots, \pm\bar{8}$. The table

| $\bar{x}$ | $\bar{0}$ | $\pm\bar{1}$ | $\pm\bar{2}$ | $\pm\bar{3}$ | $\pm\bar{4}$ | $\pm\bar{5}$ | $\pm\bar{6}$ | $\pm\bar{7}$ | $\pm\bar{8}$ |
|---|---|---|---|---|---|---|---|---|---|
| $\bar{x}^2$ | $\bar{0}$ | $\bar{1}$ | $\bar{4}$ | $\bar{9}$ | $\overline{16}$ | $\bar{8}$ | $\bar{2}$ | $\overline{15}$ | $\overline{13}$ |

shows that $\bar{5}$ is not the square of an element of $\mathbf{Z}_{17}$. Therefore, equation (D) has no solution in the integers.

*Example 5*

Name a ring with unity 1 which has a subring that does not contain 1.

*Solution*

The subring of the even integers does not contain the unity 1 of the integers **Z**.

*Example 6*

Find a ring $U$ with unity in which there is a subring $S$ such that $S$ does not contain the unity of $U$ but $S$ is a ring with unity.

*Solution*

Let $U$ be $\mathbf{Z}_6 = \{\bar{0}, \bar{1}, \bar{2}, \bar{3}, \bar{4}, \bar{5}\}$ and let $S$ be the principal ideal $(\bar{2}) = \{\bar{0}, \bar{2}, \bar{4}\}$. Then the unity $\bar{1}$ of $U$ is not in $S$. However, in $\mathbf{Z}_6$ (and hence in $S$)

$$\bar{4} \cdot \bar{0} = \bar{0}, \qquad \bar{4} \cdot \bar{2} = \overline{6 + 2} = \bar{2}, \qquad \bar{4} \cdot \bar{4} = \overline{2 \cdot 6 + 4} = \bar{4},$$

which shows that $\bar{4}$ is the unity of $S$.

## Problems

Problem 39 below is cited in Section 5.8.

1. Let $T$ be the subring in the real numbers consisting of all numbers of the form $a + b\sqrt{3}$ with $a$ and $b$ integers.
   (i) Use the fact that $\sqrt{3}$ is irrational (see Problem 19 of Section 1.4) to show that if $a$, $b$, $c$, and $d$ are rational numbers with $a + b\sqrt{3} = c + d\sqrt{3}$, then $a = c$ and $b = d$. Thus show that the representation $a + b\sqrt{3}$ for an element of $T$ is unique.
   (ii) For every $\alpha = a + b\sqrt{3}$ of $T$, let $\bar{\alpha} = a - b\sqrt{3}$. Let $f$ be the bijection from $T$ to $T$ given by $f(\alpha) = \bar{\alpha}$. Show that $f$ is a ring automorphism of $T$.
2. Let **R** be the ring of the real numbers and let $\theta$ be the mapping from **R** to itself with $\theta(x) = |x|$. Show that $\theta$ preserves multiplications but does not preserve additions and hence $\theta$ is not a ring homomorphism.
3. Let **Z** be the ring of the integers and let $\theta$ be the mapping from **Z** to itself with $\theta(n) = 2n$. Show that $\theta$ preserves additions but does not preserve multiplications and hence $\theta$ is not a ring homomorphism.

4. Let $\theta$ be the mapping from $\mathbf{Z}_4$ to $\mathbf{Z}_2$ with $\bar{0} \mapsto \bar{0}$, $\bar{1} \mapsto \bar{1}$, $\bar{2} \mapsto \bar{0}$, and $\bar{3} \mapsto \bar{1}$. Explain why the following are true:
   (i) $\theta(\bar{a}) = \bar{a}$.
   (ii) $\bar{a} + \bar{b} = \bar{c}$ in $\mathbf{Z}_4$ implies that $\bar{a} + \bar{b} = \bar{c}$ in $\mathbf{Z}_2$.
   (iii) $\bar{a} \cdot \bar{b} = \bar{d}$ in $\mathbf{Z}_4$ implies that $\bar{a} \cdot \bar{b} = \bar{d}$ in $\mathbf{Z}_2$.
   (iv) $\theta$ is a ring homomorphism.
5. Let $\theta$ be the mapping from $\mathbf{Z}_{15}$ to $\mathbf{Z}_3$ with $\theta(\bar{a}) = \bar{a}$.
   (i) Show that $\theta$ is a surjective ring homomorphism.
   (ii) Find the kernel of $\theta$.
6. Show that the mapping $\theta$ from $\mathbf{Z}_2$ to $\mathbf{Z}_6$ with $\theta(\bar{0}) = \bar{0}$ and $\theta(\bar{1}) = \bar{3}$ is an injective ring homomorphism.
7. Show that there is no injective ring homomorphism $\theta$ from $\mathbf{Z}_2$ to $\mathbf{Z}_4$.
8. Let $\mathbf{Z}$ and $\mathbf{Q}$ be the rings of the integers and of the rational numbers, respectively. Every rational number $r$ is uniquely expressible in the form $s/t$ with $s$ and $t$ relatively prime integers and $t$ positive. Let the mapping $\theta$ from $\mathbf{Q}$ to $\mathbf{Z}$ be given by $\theta(r) = s$. Is $\theta$ a ring homomorphism? Explain.
9. For which integer $m$ is the principal ideal $(m)$ in $\mathbf{Z}$ finite?
10. For which integers $n$ does the principal ideal $(n)$ contain all the integers? (There is a negative $n$ as well as a positive one.)
11. Let $m > 1$ so that the quotient ring $\mathbf{Z}/(m) = \mathbf{Z_m} = \{\bar{0}, \bar{1}, \ldots, \overline{m-1}\}$ has at least two elements.
    (i) What is the additive identity of $\mathbf{Z_m}$?
    (ii) Does $\mathbf{Z_m}$ have a unity? If so, which element is the unity?
    (iii) Is the additive group of $\mathbf{Z_m}$ cyclic?
    (iv) Is $\bar{0}$ an invertible in $\mathbf{Z_m}$?
12. (i) How many elements are there in $\mathbf{Z}_1 = \mathbf{Z}/(1)$?
    (ii) Does $\mathbf{Z}_1$ have a unity? If so, which element is the unity?
    (iii) Is $\bar{0}$ an invertible in $\mathbf{Z}_1$?
13. The natural map $\theta$ from $\mathbf{Z}$ to $\mathbf{Z}_{14}$ is given by

$$\theta(a) = a + (14) = \bar{a}.$$

    (i) Which integers are in the kernel $K$ of $\theta$?
    (ii) Which integers are in the complete inverse image $\theta^{-1}(\bar{1})$?
14. Answer questions (i) and (ii) of Problem 13 for the natural map $\theta$ from $\mathbf{Z}$ to $\mathbf{Z_m}$, where $m$ is an integer greater than 1.
15. Use the tables of Example 1 of this section to do the following:
    (a) Show the negative of each element of $\mathbf{Z}_6 = \{\bar{0}, \bar{1}, \bar{2}, \bar{3}, \bar{4}, \bar{5}\}$ in a table.
    (b) Show the order of each element of the additive group of $\mathbf{Z}_6$ in a table.
    (c) Which elements of $\mathbf{Z}_6$ are of the form $\bar{2} \cdot \bar{a}$ with $\bar{a}$ in $\mathbf{Z}_6$? Is $\bar{2}$ an invertible of $\mathbf{Z}_6$? Explain.

(d) Find all $\bar{a}$ and $\bar{b}$ such that $\bar{a} \cdot \bar{b} = \bar{1}$ in $\mathbf{Z}_6$.
(e) List the elements of the multiplicative group $\mathbf{V}_6$ of invertibles in $\mathbf{Z}_6$.
(f) Find all $\bar{a}$ and $\bar{b}$ in $\mathbf{Z}_6$ such that $\bar{a} \neq \bar{0}$, $\bar{b} \neq \bar{0}$, but $\overline{ab} = \bar{0}$.
(g) List the squares of the elements of $\mathbf{Z}_6$.
(h) Tabulate $y = x(x + \bar{5})$ and $z = (x + \bar{2})(x + \bar{3})$ for $x$ in $\mathbf{Z}_6$.
(i) Find all solutions of $x(x + \bar{5}) = \bar{0}$ in $\mathbf{Z}_6$.

16. Let $p$ be a prime in the integers $\mathbf{Z}$. Explain why the following are true:
(i) If $\bar{p} = \bar{2}$ in $\mathbf{Z}_6$, then $p = 2$.
(ii) If $\bar{p} = \bar{3}$ in $\mathbf{Z}_6$, then $p$ is 3 or $-3$.
(iii) If $\bar{p} = \bar{4}$ in $\mathbf{Z}_6$, then $p = -2$.
(iv) $\bar{p}$ cannot be $\bar{0}$ in $\mathbf{Z}_6$.
(v) If $p > 3$, then $\bar{p}$ must be $\bar{1}$ or $\bar{5} = -\bar{1}$ in $\mathbf{Z}_6$.

17. (i) Use Problem 15(d) to explain why a number $c$ in the set $T = \{7, 13, 19, 25, \ldots, 6n + 1, \ldots\}$ is composite if and only if $c$ is the product $ab$ of two smaller integers with either $a$ and $b$ both in $T$ or $a$ and $b$ both in the set

$$S = \{5, 11, 17, 23, \ldots, 6n - 1, \ldots\}.$$

(ii) List all the primes in the finite set $\{7, 13, 19, \ldots, 97\}$.

18. Use Problems 16 and 17 together with Example 1 to find all the primes among the first one hundred positive integers.

19. (i) Make the multiplication table for $\mathbf{Z}_5 = \{\bar{0}, \bar{1}, \bar{2}, \bar{3}, \bar{4}\}$.
(ii) Give the reciprocal (multiplicative inverse) of each of $\bar{1}, \bar{2}, \bar{3}, \bar{4}$.
(iii) State the order of the group $\mathbf{V}_5$ of invertibles of $\mathbf{Z}_5$.
(iv) Explain why $v^4 = \bar{1}$ for every $v$ in $\mathbf{V}_5$.
(v) Let $a$ be an integer that is not a multiple of 5. Explain why $\bar{a}$ is in $\mathbf{V}_5$ and hence $\bar{a}^4 = \bar{1}$.
(vi) Let $a$ be an integer that is not a multiple of 5. Explain why $5 \mid (a^4 - 1)$ in the integers $\mathbf{Z}$.
(vii) Explain why $\bar{b}^5 = \bar{b}$ for every integer $b$.
(viii) Explain why $5 \mid (b^5 - b)$ for every integer $b$.

20. Let $\mathbf{V}_9$ be the multiplicative group of invertibles of

$$\mathbf{Z}_9 = \{\bar{0}, \bar{1}, \bar{2}, \bar{3}, \bar{4}, \bar{5}, \bar{6}, \bar{7}, \bar{8}\}.$$

(i) Make the multiplication table for $\mathbf{Z}_9$.
(ii) Show that $\bar{1}, \bar{2}, \bar{4}, \bar{5}, \bar{7}, \bar{8}$ are invertibles by finding their reciprocals in $\mathbf{Z}_9$.
(iii) Explain why $\bar{0}$, $\bar{3}$, and $\bar{6}$ are not invertibles in $\mathbf{Z}_9$.
(iv) What is the order of $\mathbf{V}_9$?

(v) Explain why $\overline{a}^6 = \overline{1}$ for every $\overline{a}$ in $\mathbf{V_9}$.

(vi) Let $a$ be an integer such that $\gcd(a, 9) = 1$. Explain why $\overline{a}$ is in $\mathbf{V_9}$ and hence $\overline{a}^6 = \overline{1}$.

(vii) Let $a$ be an integer with $\gcd(a, 9) = 1$. Explain why $9 \mid (a^6 - 1)$.

21. List the elements of each of the following principal ideals in $\mathbf{Z_8}$:
(a) $(\overline{2})$; (b) $(\overline{6})$; (c) $(\overline{3})$; (d) $(\overline{5})$.

22. List the elements of each of the following principal ideals in $\mathbf{Z_7}$:
(a) $(\overline{0})$; (b) $(\overline{1})$; (c) $(\overline{2})$; (d) $(\overline{5})$.

23. (i) Explain why $(\overline{a}) = (\overline{m - a})$ for all $\overline{a}$ in $\mathbf{Z_m}$.
(ii) How many (distinct) principal ideals are there in $\mathbf{Z_{12}}$?

24. (i) Let $I$ be an ideal in a ring $U$ with unity and let $I$ contain an invertible (that is, unit) $v$ of $U$. Show that $I = U$.
(ii) Explain why $\mathbf{Z_{11}}$ has only two ideals.

25. (a) In $\mathbf{Z_{10}}$, find all $x$ such that $x^2 = x$ and show that the subring $S = (\overline{5}) = \{\overline{0}, \overline{5}\}$ has a unity even though the unity $\overline{1}$ of $\mathbf{Z_{10}}$ is not in $S$.
(b) In $\mathbf{Z_{16}}$, show that the only subrings $S$ with a unity are $\mathbf{Z_{16}}$ and the single element ring $\{\overline{0}\}$.
(c) In $\mathbf{Z_{30}}$, find all the subrings $S$ with a unity.

26. (i) Explain why a unity $u$ of a subring $S$ in a ring $R$ must satisfy $u^2 = u$.
(ii) Let $S$ be a subring in $\mathbf{Z_m}$. Prove that $S$ has a unity if and only if there exists an element $u$ with $u^2 = u$ such that $S$ is the principal ideal $(u)$.

27. Let $\theta$ be the mapping from $\mathbf{Z_{12}}$ to $\mathbf{Z_6}$ such that the image of the element $\overline{a}$ of $\mathbf{Z_{12}}$ is the element $\overline{4a}$ of $\mathbf{Z_6}$.
(i) Tabulate the mapping $\theta$.
(ii) Prove that $\theta$ is a ring homomorphism.
(iii) Show that the image set $M$ of $\theta$ is a subring in $\mathbf{Z_6}$.
(iv) Show that $\theta(\overline{1})$ is the unity of the subring $M$ but is not the unity of $\mathbf{Z_6}$.
(v) Explain why there is no homomorphism $\alpha$ from $\mathbf{Z_{12}}$ to $\mathbf{Z_6}$ for which $\alpha(\overline{1}) = \overline{2}$.

28. Let $\theta$ be a ring homomorphism from $R$ to $R'$. Let $K$ and $M$ be the kernel and image set of $\theta$, respectively. Show the following:
(a) $M$ is a subring in $R'$.
(b) The quotient ring $R/K$ is isomorphic to $M$. [See Theorem 2(f) of Section 3.3.]
(c) If $R$ has a unity 1, $\theta(1)$ is the unity of $M$ (but not necessarily the unity of $R'$).
(d) If $v$ is in the multiplicative group $V$ of invertibles in $R$, $\theta(v)$ is in the multiplicative group $V'$ of invertibles in $M$. Hence $\theta$, with its domain restricted to be just $V$, is a group homomorphism from $V$ to $V'$.

(e) If $m$ is in $M$, the complete inverse image $\theta^{-1}(m)$ is a coset of the kernel $K$.

(f) If $R$ is finite, the number of elements in $M$ is an integral divisor of the number of elements in $R$.

29. Explain why every subring, and hence every ideal, in $\mathbf{Z_m}$ is a cyclic subgroup of the additive group of $\mathbf{Z_m}$ and therefore is a principal ideal in $\mathbf{Z_m}$.

30. Prove that $(\bar{a}) = (\bar{b})$ in $\mathbf{Z_m}$ if and only if $\gcd(a, m) = \gcd(b, m)$.

31. Use Problems 29 and 30 to find all the subrings in $\mathbf{Z_{24}}$.

32. Find all the subrings in $\mathbf{Z_{36}}$.

33. (i) Prove that there is no surjective ring homomorphism from $\mathbf{Z_{30}}$ to $\mathbf{Z_7}$.
(ii) Describe a ring homomorphism from $\mathbf{Z_{30}}$ to $\mathbf{Z_7}$ that is not surjective.

34. Let $m$ and $n$ be positive integers. Show the following:
(i) If $m|n$, then $\bar{a} + \bar{b} = \bar{s}$ and $\bar{a} \cdot \bar{b} = \bar{t}$ in $\mathbf{Z_n}$ imply $\bar{a} + \bar{b} = \bar{s}$ and $\bar{a} \cdot \bar{b} = \bar{t}$ in $\mathbf{Z_m}$.
(ii) There is a surjective ring homomorphism from $\mathbf{Z_n}$ to $\mathbf{Z_m}$ if and only if $m|n$.

35. Let $n$ be a positive integer. Show that there is an injective ring homomorphism from $\mathbf{Z_2}$ to $\mathbf{Z_{2n}}$ if and only if $n$ is odd. (See Problems 6 and 7 above.)

36. Let $q$ and $q'$ be the numbers of elements in finite rings $R$ and $R'$, respectively. Let there be an injective ring homomorphism from $R$ to $R'$. Show that $q|q'$.

37. Let $n$ be a positive integer. Explain why there are at most $n + 1$ distinct squares of elements of $\mathbf{Z_{2n}}$.

38. Let $n$ be a positive integer. Explain why there are at most $n + 1$ distinct squares of elements of $\mathbf{Z_{2n+1}}$.

39. Let $I_1$ and $I_2$ be ideals in a ring $R$. Let $I$ be the subset of $R$ consisting of the elements that are in both $I_1$ and $I_2$. That is, let

$$I = I_1 \cap I_2.$$

(a) Show that $I$ is an ideal in $R$.
(b) Explain why $I$ is an ideal in $I_1$ (and in $I_2$).

40. Let $s$ and $t$ be integers and let $m = \text{lcm}[s, t]$. Show that $(s) \cap (t) = (m)$ in $\mathbf{Z}$. (See Section 1.6.)

41. Use the natural map from $\mathbf{Z}$ to $\mathbf{Z_3}$ to prove that there are no integers $x$ and $y$ satisfying

$$x^2 - 3y^2 = 992.$$

42. Prove that there are no integers $x$ and $y$ that satisfy
   (a) $x^2 + (x + 1)^2 + (x + 2)^2 = y^2$.
   (b) $x^2 + (x + 1)^2 + (x + 2)^2 + (x + 3)^2 = y^2$.

43. Prove that there are no squares of integers in the sequence

$$11, 111, 1111, 11111, \ldots$$

of integers with each digit a 1 (and at least two digits).

*44. Prove that there are no integers $x$ and $y$ satisfying

$$x^2 + 3xy - 2y^2 = 122.$$

45. Let $G = \{m + n\mathbf{i} : m, n \in \mathbf{Z}\}$ be the ring of gaussian integers. Show that $S = \{m + 2n\mathbf{i} : m, n \in \mathbf{Z}\}$ is a subring in $G$ but is not an ideal in $G$.

---

***Carl Friedrich Gauss*** *(1777–1855)* *Gauss was undeniably one of the greatest of mathematicians, if not the greatest. He started life as an astonishing child prodigy—before reaching the age of 20 he demonstrated that the 17-sided regular polygon is constructible, developed the method of least squares, and showed that every integer is the sum of three, or fewer, triangular numbers.*

*His most famous work is the* **Disquisitiones Arithmeticae** *(1801), probably the greatest work in the theory of numbers. Gauss once described mathematics as the queen of the sciences and the theory of numbers as the queen of mathematics. In the* **Disquisitiones** *he developed congruence arithmetic and proved one of the most beautiful theorems in mathematics, the law of quadratic reciprocity, a theorem known to Euler and Legendre but not proved prior to Gauss' demonstrations. Gauss struggled for a year to produce a proof of the theorem once he had discovered it—he later produced five additional proofs.*

*Gauss went on to work in many branches of mathematics as well as in physics and astronomy. He showed a reluctance to publish his results and many discoveries attributed to others were later found in Gauss' diary or in his manuscripts. Probably the most famous instance of this is the simultaneous development of non-Euclidean geometry by Bolyai and Lobachevski, a discovery made earlier by Gauss.*

*He was a professor at the University of Göttingen and never once during his life did he leave Germany. The monument to Gauss in Braunschweig has a base chosen appropriately in the form of a polygon of 17 sides.*

---

## 4.3 Congruence in $\mathbf{Z}$, The Euler and Fermat Theorems

*Algebra is generous; she often gives more than is asked of her.*

*Jean LeRond d'Alembert*

In this section we obtain some important results in the Theory of Numbers by applying group theory, especially Lagrange's Theorem, to the multiplicative groups $\mathbf{V_m}$ of the invertibles of $\mathbf{Z_m}$.

We start by characterizing the integers $a$ such that $\bar{a}$ is in $\mathbf{V_m}$.

### Lemma 1 Invertibles of $\mathbf{Z_m}$

*An $\bar{a}$ of $\mathbf{Z_m}$ is an invertible if and only if the integers $a$ and $m$ are relatively prime.*

#### Proof

First let $\gcd(a, m) = 1$. Then there are integers $h$ and $k$ such that $ah + mk = 1$. (See the corollary to Theorem 2 of Section 1.4.) Using the fact that the natural map from $\mathbf{Z}$ to $\mathbf{Z_m}$ preserves additions and multiplications, we have

$$\bar{1} = \overline{ah + mk} = \bar{a} \cdot \bar{h} + \bar{m} \cdot \bar{k}.$$

Since $\bar{m} = \bar{0}$ in $\mathbf{Z_m}$, this leads to $\bar{1} = \bar{a} \cdot \bar{h}$; thus $\bar{a}$ has a reciprocal $\bar{h}$ in $\mathbf{Z_m}$ and $\bar{a}$ is in $\mathbf{V_m}$.

Conversely, let $\bar{a}$ be an invertible in $\mathbf{Z_m}$; that is, let $\bar{a} \cdot \bar{h} = \bar{1}$ in $\mathbf{Z_m}$. Then $\overline{ah} = \bar{1}$, which implies that $m|(1 - ah)$. This in turn means that there is an integer $k$ with $1 - ah = mk$. Then $ah + mk = 1$ and it follows that $a$ and $m$ are relatively prime.

### Corollary 1

$\mathbf{V_p} = \{\bar{1}, \bar{2}, \ldots, \overline{p-1}\}$ *for all positive primes $p$.*

#### Proof

Since $\pm 1$ and $\pm p$ are the only divisors of $p$, an integer $n$ is relatively prime to $p$ if and only if $n$ is not a multiple of $p$. Hence each of $1, 2, \ldots, p - 1$ is relatively prime to $p$ and it follows from Lemma 1 that $\bar{1}, \bar{2}, \ldots, \overline{p-1}$ are in $\mathbf{V_p}$. Since $\bar{0}$ is the zero of $\mathbf{Z_p}$ and $\mathbf{Z_p}$ has more than one element, $\bar{0}$ is not an invertible.

*Corollary 2*

*An $\bar{a}$ of $\mathbf{Z_m}$ is an invertible if and only if $\bar{a}$ is a generator of the additive group of $\mathbf{Z_m}$.*

The proof of Corollary 2 is left to the reader as Problem 39 of this section.

***Example 1***

Show that $18|(a^6 - 1)$ for every integer $a$ that is relatively prime to 18.

***Solution***

Using Lemma 1, we find that the subset of invertibles of

$$\mathbf{Z_{18}} = \{\bar{0}, \bar{1}, \bar{2}, \ldots, \overline{17}\}$$

is the multiplicative group

$$\mathbf{V_{18}} = \{\bar{1}, \bar{5}, \bar{7}, \overline{11}, \overline{13}, \overline{17}\},$$

of order 6. Let $a$ and 18 be relatively prime. Then $\bar{a}$ is in $\mathbf{V_{18}}$ and Lagrange's Theorem tells us that $\bar{a}^6 = \bar{1}$. Then

$$\overline{a^6} = \bar{1} \text{ in } \mathbf{Z_{18}}$$

and hence $18|(a^6 - 1)$ in $\mathbf{Z}$.

One can translate information concerning $\mathbf{Z_m}$ into the language of the integers $\mathbf{Z}$ using the fact that

$$\bar{a} = \bar{b} \text{ in } \mathbf{Z_m} \tag{A}$$

is equivalent to

$$m|(a - b) \text{ in } \mathbf{Z}. \tag{B}$$

Clearly (A) or (B) is also equivalent to each of the following:

$a - b$ is in the principal ideal $(m)$, (C)

$a$ and $b$ are in the same coset of $(m)$, (D)

$a$ and $b$ have the same remainder in division by $m$. (E)

The theorems of this section are usually given in terms of yet another notation, which follows, for the statement in (A), (B), (C), (D), or (E).

***Notation Congruence in Z***

*Let a, b, and m be integers such that $m|(a-b)$. Then one writes*

$$a \equiv b \pmod{m},$$

*which is read as "a is congruent to b modulo m."*

If $a$ is not congruent to $b$ modulo $m$, one may write $a \not\equiv b \pmod{m}$.

We next give the definition of an important function in the Theory of Numbers.

***Notation The Euler $\phi$-Function***

*Let m be a positive integer. Then $\phi(m)$ denotes the number of integers a in $\{0, 1, 2, \ldots, m-1\}$, such that a and m are relatively prime, that is, such that $gcd(a, m) = 1$. The function $\phi$ is known as the Euler $\phi$-function, or Euler totient.*

Since an $\bar{a}$ of $\mathbf{Z_m} = \{\bar{0}, \bar{1}, \bar{2}, \ldots, \overline{m-1}\}$ is in $\mathbf{V_m}$ if and only if $a$ and $m$ are relatively prime, it is clear that $\phi(m)$ is the order of $\mathbf{V_m}$. Then it follows from Lagrange's Theorem that

$$\bar{a}^{\phi(m)} = \bar{1}$$

for every $\bar{a}$ in $\mathbf{V_m}$. Translating this into the language of the integers $\mathbf{Z}$, we have the following famous result due to Leonhard Euler. (See the biographical note after Section 2.6.)

***Theorem 1 Euler's Theorem***

*Let a and m be relatively prime integers with $m > 1$. Then*

$$m|(a^{\phi(m)} - 1),$$

*i.e., $a^{\phi(m)} \equiv 1 \pmod{m}$.*

Euler's Theorem is a generalization of the following result due to Pierre Fermat. (See the biographical note at the close of this section.)

***Theorem 2 Fermat's Theorem***

*If p is a positive prime in the integers,*

$$a^p \equiv a \ (mod\ p)$$

*for all integers a. If* $b \not\equiv 0 \ (mod\ p)$, *then*

$$b^{p-1} \equiv 1 \ (mod\ p).$$

***Proof***

Corollary 1 to Lemma 1 tells us that $\mathbf{V_p} = \{\overline{1}, \overline{2}, \ldots, \overline{p-1}\}$ and hence that the order of $\mathbf{V_p}$ is $\phi(p) = p - 1$. If $b \not\equiv 0$ (mod $p$), then $\overline{b} \neq \overline{0}$ in $\mathbf{Z_p}$ and hence $\overline{b}$ is in $\mathbf{V_p}$. Then it follows from Euler's Theorem that $b^{p-1} \equiv 1$ (mod $p$).

Now let $a$ be any integer. If $\overline{a} \neq \overline{0}$ in $\mathbf{Z_p}$, then $\overline{a}$ is in $\mathbf{V_p}$ and $\overline{a}^{p-1} = \overline{1}$ in $\mathbf{V_p}$, using Lagrange's Theorem. Multiplying both sides by $\overline{a}$ gives us $\overline{a}^p = \overline{a}$, from which it follows that $a^p \equiv a$ (mod $p$). If $\overline{a} = \overline{0}$, then $\overline{a}^p = \overline{0} = \overline{a}$ in $\mathbf{Z_m}$ and it again follows that $a^p \equiv a$ (mod $p$).

Next we use the natural map from $\mathbf{Z}$ to $\mathbf{Z_m}$ to show that congruences modulo $m$ can be added, subtracted, or multiplied to obtain new congruences.

Let $m$ be a fixed positive integer and let $\theta$ be the natural map from $\mathbf{Z}$ to $\mathbf{Z_m}$. If $\overline{a} = \overline{b}$ and $\overline{c} = \overline{d}$ in $\mathbf{Z_m}$, then, of course,

$$\overline{a} \pm \overline{c} = \overline{b} \pm \overline{d}, \qquad \overline{a} \cdot \overline{c} = \overline{b} \cdot \overline{d}. \tag{F}$$

The equations of (F) and the fact that $\theta$ is a ring homomorphism tells us that

$$\overline{a \pm c} = \overline{b \pm d}, \qquad \overline{ac} = \overline{bd}.$$

This may be translated into congruence language as follows:

***Theorem 3 Congruence Preserved under Addition, Subtraction, Multiplication***

*If* $a \equiv b \ (mod\ m)$ *and* $c \equiv d \ (mod\ m)$, *then* $a \pm c \equiv b \pm d \ (mod\ m)$ *and* $ac \equiv bd \ (mod\ m)$.

See Problem 21 of this section for some information on congruences and division.

*Example 2*

Use congruences as an aid in finding the remainder when 1453 is divided by 17.

**Solution**

$1453 = 40 \cdot 36 + 13 \equiv 6 \cdot 2 - 4 = 8 \pmod{17}$; hence the remainder is 8.

*Example 3*

Show that the square of an integer is never congruent to 2 or 3 modulo 4.

**Solution**

Modulo 4, an integer $n$ must be congruent to 0, 1, 2, or 3. Then the table

| $n \pmod 4$ | 0 | 1 | 2 | 3 |
|---|---|---|---|---|
| $n^2 \pmod 4$ | 0 | 1 | 0 | 1 |

shows that all squares of integers are congruent to either 0 or 1 and hence that no square is congruent to 2 or 3 modulo 4.

*Example 4*

Find the smallest positive integer $q$ such that

$$10^q \equiv 1 \pmod{17}.$$

**Solution**

Since 17 is a prime, the order of $\mathbf{V}_{17}$ is $17 - 1 = 16$. This and Lagrange's Theorem imply that the order of $\overline{10}$ in $\mathbf{V}_{17}$ is 2, 4, 8, or 16. Then we note that $\overline{10}^2 = \overline{15}$, $\overline{10}^4 = \overline{15}^2 = (-\overline{2})^2 = \overline{4}$, $\overline{10}^8 = \overline{4}^2 = \overline{16} = -\overline{1}$, and $\overline{10}^{16} = (-\overline{1})^2 = \overline{1}$. These calculations show that $\overline{10}$ has order 16 in $\mathbf{V}_{17}$; that is, 16 is the smallest positive integer $q$ such that $\overline{10}^q = \overline{1}$ in $\mathbf{V}_{17}$. Translating this into the language of congruences, we have 16 as the smallest $q$ with $10^q \equiv 1 \pmod{17}$.

*Example 5*

Use the fact that $\overline{10}$ has order 5 in $\mathbf{V}_{41}$ to find the unending decimal representations for 1/41 and 6/41.

*Solution*

The hypothesis tells us that $\overline{10}^5 = \overline{1}$ in $\mathbf{V}_{41}$. This implies that $41 \mid (10^5 - 1)$, i.e., $41 \mid 99999$. Divison shows that $99999 = 41 \cdot 2439$ and so $1/41 = 2439/99999$. Now the formula

$$1 + r + r^2 + r^3 + \cdots = (1 - r)^{-1}, \; |r| < 1,$$

for the sum of an infinite geometric progression, gives us

$$\begin{aligned} 1/99999 &= 1/(10^5 - 1) = (10^5 - 1)^{-1} = 10^{-5}(1 - 10^{-5})^{-1} \\ &= 0.00001(1 + 10^{-5} + 10^{-10} + 10^{-15} + \cdots) \\ &= 0.000010000100001 \ldots. \end{aligned}$$

Hence we have the decimal representations

$$1/41 = 2439/99999 = 0.024390243902439 \ldots,$$

$$6/41 = 14634/99999 = 0.1463414634 \ldots,$$

in each of which a block of five digits is repeated endlessly.

## *Problems*

Problem 38 below is cited in Section 7.3.

1. (i) List the order of each element of $\mathbf{V}_{18} = \{\overline{1}, \overline{5}, \overline{7}, \overline{11}, \overline{13}, \overline{17}\}$ in a table.
   (ii) Is $\mathbf{V}_{18}$ cyclic? If so, which elements are generators?
2. (i) Find the smallest positive integer $q$ such that $v^q = \overline{1}$ for every $v$ in $\mathbf{V}_{12}$.
   (ii) Is $\mathbf{V}_{12}$ cyclic? Explain.
3. (i) Show that $4! = 1 \cdot 2 \cdot 3 \cdot 4 \equiv -1 \pmod 5$.
   (ii) Show that $1 \cdot 2 \cdot 3 \cdot 4 \cdot 5 \cdot 6 \equiv -1 \pmod 7$.
   (iii) Translate the statement in (ii) into an equality in $\mathbf{V}_7$.
   (iv) Translate the statement in (ii) into a divisibility statement of the form "$a \mid b$ in $\mathbf{Z}$."
4. (i) Show that $3! = 1 \cdot 2 \cdot 3 \equiv 2 \pmod 4$.
   (ii) Show that $\overline{1} \cdot \overline{2} \cdot \overline{3} \cdot \overline{4} \cdot \overline{5} = \overline{0}$ in $\mathbf{Z}_6$.
   (iii) Translate the statement in (ii) into congruence language.
   (iv) Translate the statement in (ii) into a divisibility statement of the form "$a \mid b$ in $\mathbf{Z}$."
5. Show the following:
   (i) $3 \cdot 11 \equiv 1 \pmod{16}$.
   (ii) $5 \cdot 13 \equiv 1 \pmod{16}$.

(iii) $7 \cdot 9 \equiv -1 \pmod{16}$.
(iv) $15 \equiv -1 \pmod{16}$.
(v) $1 \cdot 3 \cdot 5 \cdot 7 \cdot 9 \cdot 11 \cdot 13 \cdot 15 \equiv 1 \pmod{16}$.
(vi) The product $v_1 v_2 \cdots v_8$ of the eight elements of the group $\mathbf{V_{16}}$ equals $\overline{1}$.

6. (i) Show that the product of the six elements of $\mathbf{V_{18}}$ is $-\overline{1} = \overline{17}$.
(ii) Translate the statement in (i) into congruence language.

7. Given that $a \equiv 3 \pmod 4$, use Example 3 to prove that there are no integers $x$ and $y$ such that $x^2 + y^2 = a$.

8. Given that $b \equiv 2 \pmod 4$, use Example 3 to prove that there are no integers $x$ and $y$ such that $x^2 - y^2 = b$.

9. Explain why the unity $\overline{1}$ of $\mathbf{Z_m}$ and $-\overline{1} = \overline{m-1}$ are in $\mathbf{V_m}$ for $m \geq 2$.

10. Let $m \geq 2$. Show that $\overline{m-a}$ is in $\mathbf{V_m}$ if and only if $\bar{a}$ is in $\mathbf{V_m}$.

11. Complete the following partial table for the Euler $\phi$-function:

| $m$ | 1 | 2 | 3 | 4 | 5 | 6 | 7 | 8 | 9 | 10 |
|---|---|---|---|---|---|---|---|---|---|---|
| $\phi(m)$ | 1 | 1 | | | | 2 | 6 | | | |

12. Let $\phi$ be the Euler $\phi$-function and find the following:
(i) $\phi(16)$; (ii) $\phi(32)$; (iii) $\phi(2^n)$, where $n$ is a positive integer.

13. Give a formula for $\phi(3^n)$, where $n$ is a positive integer.

14. Let $p$ be a positive prime and let $n$ be a positive integer. Give a formula for $\phi(p^n)$.

15. Find the $x$ in $\{1, 2, 3, \ldots, 20\}$ such that $10x \equiv 1 \pmod{21}$.

16. Find the $y$ in $\{1, 2, 3, \ldots, 20\}$ such that $10y \equiv 13 \pmod{21}$.

17. (i) Let $a$ and $m$ be relatively prime integers with $m > 1$. Explain why there is an $h$ in $\{1, 2, \ldots, m-1\}$ such that $ah \equiv 1 \pmod m$.
(ii) Give the analogue of (i) for the case $m = 1$.

18. Let $a$, $b$, and $m$ be integers with $m > 1$ and $\gcd(a, m) = 1$. Explain why there is a $k$ in $\{0, 1, \ldots, m-1\}$ such that

$$ak \equiv b \pmod m.$$

19. Find all elements $x$ of $\mathbf{Z_{12}}$ such that $0, x, 2 \cdot x, 3 \cdot x, \ldots, 11 \cdot x$ are distinct and hence comprise all of $\mathbf{Z_{12}}$.

20. Let $a$ and $m$ be integers with $m > 0$. Prove that

$$\overline{0}, \bar{a}, 2 \cdot \bar{a}, \ldots, (m-1) \cdot \bar{a}$$

(are distinct elements of $\mathbf{Z_m}$ and hence) comprise all of $\mathbf{Z_m}$ if and only if $\gcd(a, m) = 1$.

21. (i) Give an example of integers $a$, $b$, $c$, and $m$ such that $ab \equiv ac \pmod{m}$ but neither $b \equiv c \pmod{m}$ nor $a \equiv 0 \pmod{m}$ is true.
(ii) Prove that $ab \equiv ac \pmod{m}$ and $\gcd(a, m) = 1$ together imply $b \equiv c \pmod{m}$. (*Hint*: "Cancel" $a$, using Problem 17.)

22. Given that $a \equiv b \pmod{m}$, prove by mathematical induction that $a^n \equiv b^n \pmod{m}$ for all nonnegative integers $n$.

23. (i) Explain why $10^n \equiv 1 \pmod{9}$ for all nonnegative integers $n$.
(ii) Let $a_0, a_1, \ldots, a_d$ be integers. Explain why

$$10^d a_d + \cdots + 10a_1 + a_0 \equiv a_d + \cdots + a_1 + a_0 \pmod{9}.$$

(iii) Show that $123456789 \equiv 0 \pmod{9}$.
(iv) Explain why every number obtained by rearranging the digits of 123456789 is a multiple of 9.

24. (i) Explain why $10^n \equiv (-1)^n \pmod{11}$ for all nonnegative integers $n$.
(ii) Let $a_0, a_1, \ldots, a_d$ be integers. Explain why

$$10^d a_d + \cdots + 10a_1 + a_0 \equiv a_0 - a_1 + a_2 - \cdots + (-1)^d a_d \pmod{11}.$$

(iii) Show that $123456 \equiv 3 \pmod{11}$.
(iv) Explain why no number obtained by rearranging the digits of 123456 can be a multiple of 11.

25. For each of the following powers of 2, find the number in $\{1, 2, 3, 4\}$ to which it is congruent modulo 5.
(i) $2^6$; (ii) $2^7$; (iii) $2^8$; (iv) $2^9$; (v) $2^{10}$; (vi) $2^{777}$.

26. For each of the following values of $n$, find the number in $\{1, 3, 7, 9\}$ that $3^n$ is congruent to modulo 10.
(i) 6; (ii) 7; (iii) 8; (iv) 9; (v) 10; (vi) 555.

27. Find $x$ and $y$ in $\{1, 2, 3, \ldots, 12\}$ such that

$$3^{200} \equiv x \pmod{13}, \qquad 94^{200} \equiv y \pmod{13}.$$

28. (i) Find the last digit (units digit) of $a = 7^{123}$. That is, find the remainder in the division of $a$ by 10.
(ii) Find the last digit of $987^{123}$.

29. Let $a$ be an integer that is not an integral multiple of 7. Explain why the smallest positive integer $q$ such that $a^q \equiv 1 \pmod{7}$ must be 1, 2, 3, or 6.

30. Let $a$ be an integer with $\gcd(a, 19) = 1$. What are the possible values of the least positive integer $q$ such that $a^q \equiv 1 \pmod{19}$?

31. Let $m$ be a positive integer whose last digit (units digit) is 1, 3, 7, or 9. Explain why the following are true:
   (i) $\gcd(10, m) = 1$; that is, 10 and $m$ are relatively prime.
   (ii) There exist positive integers $s$ such that

$$10^s \equiv 1 \pmod{m}.$$

   (iii) There are integral multiples $t$ of $m$ such that every digit of $t$ is a 9.

32. (a) Find the smallest positive integer $r$ such that $17|r$ and every digit of $r$ is a 9. (See Example 4 of this section.)
   (b) Find the smallest positive integer $s$ such that $13|s$ and every digit of $s$ is a 9.
   (c) Find the smallest positive integer $t$ such that $221|t$ and every digit of $t$ is a 9.

33. (i) Show that $H = \{\overline{1}, \overline{4}\}$ and $N = \{\overline{1}, \overline{14}\}$ are normal subgroups in $\mathbf{V}_{15}$.
   (ii) Show that $\mathbf{V}_{15}/N$ is cyclic and $\mathbf{V}_{15}/H$ is not cyclic.

34. Let $N$ and $N'$ be normal subgroups in $G$ and $G'$, respectively. Show that $G/N$ and $G'/N'$ need not be isomorphic even though $G$ and $G'$ are isomorphic and $N$ and $N'$ are isomorphic. (See Problem 33.)

35. Extend the table of Problem 11 by tabulating $\phi(m)$ for $m = 11, 12, \ldots, 20$.

36. Show the following:
   (a) $\phi(1) + \phi(2) + \phi(3) + \phi(6) + \phi(9) + \phi(18) = 18$.
   (b) $\phi(1) + \phi(2) + \phi(3) + \phi(5) + \phi(6) + \phi(10) + \phi(15) + \phi(30) = 30$.

37. Show the following:
   (a) $\phi(2)\phi(9) = \phi(18) > \phi(3)\phi(6)$.
   (b) $\phi(3)\phi(4) = \phi(12) > \phi(2)\phi(6)$.
   (c) $\phi(4)\phi(9) = \phi(36) > \phi(3)\phi(12) > \phi(2)\phi(18) > [\phi(6)]^2$.

38. Explain why one could define $\phi(n)$ to be the number of integers $a$ in $\{1, 2, \ldots, n\}$ such that $\gcd(a, n) = 1$.

39. Show that an $\bar{a}$ of $\mathbf{Z}_\mathbf{m}$ is an invertible if and only if $\bar{a}$ is a generator of the additive group of $\mathbf{Z}_\mathbf{m}$ (and thus prove Corollary 2 to Lemma 1). (*Hint*: See Problems 35 and 36 of Section 2.7.)

40. Show that $\phi(n)$ is the number of generators in a cyclic group of order $n$.

41. For which integers $a$ in $\{0, 1, 2, 3, 4\}$ do there exist integers $x$ such that $x^2 \equiv a \pmod{5}$?

42. For which integers $b$ in $\{-3, -2, -1, 0, 1, 2, 3\}$ do there exist integers $y$ such that $y^2 \equiv b \pmod 7$?

43. Show that $a^m \equiv a^n \pmod 2$ for integers $a$, $m$, and $n$ with $m > 0$ and $n > 0$.

44. (a) Use the Fermat Theorem to show that $a^{2m-1} \equiv a^{2n-1} \pmod 3$ for all integers $a$ and all positive integers $m$ and $n$.
    (b) Show that $a^{2m} \equiv a^{2n} \pmod 3$ for all integers $a$ and all positive integers $m$ and $n$.

45. Let $a$, $b$, $c$, $r$, and $s$ be integers. Show the following:
    (a) If $a \equiv b \pmod{rs}$, then $a \equiv b \pmod s$.
    (b) If $a \equiv b \pmod r$, $a \equiv b \pmod s$, and $m = \text{lcm}[r, s]$, then $a \equiv b \pmod m$.
    (c) If $c \equiv a \pmod r$, $c \equiv b \pmod s$, and $d = \gcd(r, s)$, then $a \equiv b \pmod d$.

46. Show the following for all integers $a$:
    (a) $a^5 \equiv a \pmod{30}$.
    (b) $a^7 \equiv a \pmod{42}$.
    (c) $a^{13} \equiv a \pmod{2730}$.
    (d) $a^{21} \equiv a \pmod{330}$.

47. Show that $a^p \equiv a \pmod{6p}$ for all integers $a$ and all primes $p$ with $p > 3$.

48. Show that $a^p b \equiv b^p a \pmod{6p}$ for all primes $p > 3$ and all integers $a$ and $b$.

49. Explain why $a/17$ has a decimal representation in which a block of 16 digits repeats endlessly, for every positive integer $a$.

50. (i) Find the order of $\bar{2}$, $\bar{5}$, and $\overline{10}$ in $\mathbf{V}_{31}$.
    (ii) Find the smallest positive integer $m$ such that in the decimal representation for $1/31$ there is a block of $m$ digits that repeats endlessly.

51. Let $d$ and $e$ be in $\mathbf{Z}^+ = \{1, 2, 3, \ldots\}$, $n = de$, and

$$r \in \{1, 2, 3, \ldots, n\}.$$

Show that $\gcd(r, n) = e$ if and only if there is a $k$ in $\{1, 2, \ldots, d\}$ such that $r = ek$ and $\gcd(k, d) = 1$.

52. (a) Let $[a]$ be a cyclic group of order $m$, $d|m$, and $d > 0$. Prove that there are exactly $\phi(d)$ elements of order $d$ in $[a]$.
    (b) Let $d_1, d_2, \ldots, d_r$ be all the positive integral divisors of $m$. Explain why

$$\phi(d_1) + \phi(d_2) + \cdots + \phi(d_r) = m.$$

53. Prove that the number of primitive complex $n$th roots of unity is $\phi(n)$. (See Problem 24 of Section 2.7.)

***Pierre de Fermat*** *(1601-1665)* *Fermat was a lawyer by profession and served in the Parlement in Toulouse. He is probably the outstanding example of an amateur in mathematics. The word "amateur" is used only to indicate that his great mathematical work was in a field different from his profession; he certainly had the highest professional competence.*

*He discovered the basic idea of analytic geometry at least a year prior to the publication of Descartes'* **La Géométrie**. *Furthermore, in the 1630's he developed methods for calculating maxima and minima for certain curves, for determining slopes of tangent lines to curves of the form $y = f(x)$, and for determining the area under the curves $y = x^m$, m rational. Again he did not publish, so that the first published work on the calculus remains the "Nova methodus pro maximis et minimis" of Leibniz of 1684. (Newton, although he had the basic ideas of the calculus earlier, did not publish his results until after that date.)*

*Much of Fermat's work was motivated by his interest in the classics, the work on analytic geometry and calculus prompted by his study of Apollonius and Pappus. The contributions for which he is most remembered are those in the theory of numbers. He wrote in the margin of the Bachet 1621 edition of Diophantus'* **Arithmetica** *that he had a proof that for every integer $n > 2$ there exist no integer triples x, y, z satisfying $x^n + y^n = z^n$ but that there was too little room in the margin for the proof. This is the famous Fermat Theorem (or Conjecture), still unproved although verified for many values of n.*

*The "little" Fermat Theorem presented in Section 4.3 is a generalization of the special case, known to the Chinese more than 2000 years ago, that $2^p - 2$ is exactly divisible by p, for all positive primes p. The first published proof of the little Fermat Theorem was by Euler.*

*Fermat published almost nothing but instead made known his results in letters to a French priest, Marin Mersenne, who then passed them on to others. Fermat's own edition of Diophantus' text was published posthumously in Toulouse in 1670. His own works did not appear until 1679 in his* **Varia Opera Mathematica.**

## *4.4 Integral Domains*

In Section 4.1 we saw that a product of two elements of a ring is zero if one of the factors is zero. In the familiar number systems—the integers, the rational numbers, the real numbers, and the complex numbers—all of which are rings, the converse is also true; that is, a product is zero only if one of the factors is zero. In the modular ring $\mathbf{Z}_6$ this converse does not hold since $\overline{2} \cdot \overline{3} = \overline{0}$ but neither $\overline{2}$ nor $\overline{3}$ is the additive identity $\overline{0}$.

We now introduce terminology that helps us to distinguish between these two types of rings.

### *Definition 1* 0-Divisor in a Ring, Regular Element

*If $ab = 0$ with $a \neq 0$ and $b \neq 0$ in a ring $R$, $a$ and $b$ are said to be* **0-divisors** *(zero-divisors) in $R$. An element of $R$ that is not a 0-divisor is a* **regular element**.

If a ring $R$ has 0-divisors, a quadratic equation $ax^2 + bx + c = 0$ can have more than two solutions in $R$ [as one can see from Parts (h) and (i) of Problem 15 in Section 4.2].

### *Definition 2* Integral Domain

*A commutative ring $D$ with a unity $1 \neq 0$ and with no 0-divisors is called an* **integral domain**.

As one might guess from the name, the ring $\mathbf{Z}$ of the integers is an integral domain. So are the number systems $\mathbf{Q}$, $\mathbf{R}$, and $\mathbf{C}$. $\mathbf{Z}_6$ is not an integral domain since some of its elements, namely $\overline{2}$, $\overline{3}$, and $\overline{4}$, are 0-divisors.

When $m$ is composite, one can readily show that $\mathbf{Z}_m$ has 0-divisors and hence is not an integral domain. (See Problem 5 of this section.) It can also be shown that $\mathbf{Z}_p$ is an integral domain for all positive primes $p$. (See Problem 7 of this section.)

In $\mathbf{Z}_6$, $\overline{2} \cdot \overline{1} = \overline{2} \cdot \overline{4}$ but $\overline{1} \neq \overline{4}$; this shows the danger in attempting to cancel a 0-divisor. The following result indicates a situation in which we are allowed to cancel.

### *Theorem 1* Multiplicative Cancellation in Rings

*In a ring $R$, let $c$ be neither 0 nor a 0-divisor. Then either $ca = cb$ or $ac = bc$ implies that $a = b$.*

***Proof***

We deal with the case in which $ac = bc$ and leave the other case to the reader. Adding the same element $-bc$ to the given equal products, we have

$$ac + (-bc) = bc + (-bc),$$

which implies that

$$[a + (-b)]c = 0.$$

Since $c$ is neither 0 nor a 0-divisor, the other factor $a + (-b)$ must be zero. Adding $b$ to both sides of $a + (-b) = 0$ then leads to $a = b$, as desired.

**Corollary Cancellation in an Integral Domain**

*In an integral domain D, $c \neq 0$ and either $ac = bc$ or $ca = cb$ imply that $a = b$.*

**Theorem 2 A Converse of Theorem 1**

*Let c be a nonzero element of a ring R such that each of $ac = bc$ and $ca = cb$ implies that $a = b$ in R. Then c is not a 0-divisor.*

**Proof**

If $c$ were a 0-divisor, there would be an element $d \neq 0$ in $R$ such that either $cd = 0$ or $dc = 0$. We deal with the case $cd = 0$; the other case is similar. Now $cd = 0$ and $c \cdot 0 = 0$ imply that $cd = c \cdot 0$. Then the hypothesis allowing cancellation of $c$ gives us the contradiction $d = 0$, which proves the theorem.

**Corollary**

*Let D be a commutative ring with unity $1 \neq 0$ in which each of $ac = bc$ and $ca = cb$ implies that either $c = 0$ or $a = b$. Then D is an integral domain.*

Next we note a special property of the additive group of an integral domain.

**Theorem 3 Additive Order of Elements of D**

*In the additive group of an integral domain D, either all the elements except 0 have infinite order or all the nonzero elements have the same finite order.*

*Proof*

Here we denote the unity element 1 of $D$ by $e$. Let $a$ be any nonzero element of $D$. Using distributivity and mathematical induction, one can show that, for all positive integers $n$,

$$n \cdot a = a + \cdots + a = ea + \cdots + ea = (e + \cdots + e)a = (n \cdot e)a. \tag{A}$$

Let $e$ have finite order $q$ (in the additive group of $D$); then $q \cdot e = 0$ and replacing $n$ by $q$ in (A) gives us $q \cdot a = 0 \cdot a = 0$. This implies that $a$ has finite order $q'$ with $q'|q$.

Now let $a$ have finite order $q'$; then $q' \cdot a = 0$ and replacing $n$ by $q'$ in (A) leads to $0 = q' \cdot a = (q' \cdot e)a$. Since $a$ is neither 0 nor a 0-divisor, this implies that $q' \cdot e = 0$ and we see that $e$ has finite order $q$ with $q|q'$.

It follows that if either $e$ or $a$ has finite order, they both have the same finite order. This means that all nonzero elements of $D$ have the same order.

## *Definition 3 Characteristic of D*

*If the unity 1 of an integral domain D has finite order q in the additive group of D, one says that D has characteristic q. If 1 has infinite order, D has characteristic 0.*

For example, $\mathbf{Z}$ has characteristic 0, $\mathbf{Z}_2$ has characteristic 2, and $\mathbf{Z}_6$ does not have a characteristic since it is not an integral domain.

Since $1 \neq 0$ in an integral domain $D$, the characteristic of $D$ cannot be 1. It is left to the reader to show (in Problem 11 below) that the characteristic of $D$ is either 0 or a prime.

*Example 1*

Given that $4 \cdot 1 = 0$ in an integral domain $D$, show that $2 \cdot 1 = 0$.

*Solution*

Using associativity and distributivity, we have

$$\begin{aligned} 4 \cdot 1 &= 1 + 1 + 1 + 1 = (1 + 1) + (1 + 1) \\ &= 1(1 + 1) + 1(1 + 1) \\ &= (1 + 1)(1 + 1). \end{aligned}$$

Then the hypothesis $4 \cdot 1 = 0$ gives us

$$(1 + 1)(1 + 1) = 0. \tag{B}$$

Since $D$ has no 0-divisors, one of the factors on the left side of (B) must be zero. Hence $1 + 1 = 2 \cdot 1 = 0$.

The following is the natural analogue of subgroup and subring.

***Definition 4 Subdomain***

*A subdomain in an integral domain $D$ is a subset $S$ of $D$ such that $S$ is an integral domain under the operations in $D$.*

It is left to the reader, in Problem 19 below, to characterize the subrings in an integral domain $D$ that are subdomains.

## *Problems*

In this problem set, $D$ is always an integral domain.

1. Let $D = \{0, 1, c, d\}$. Show the following and then make the addition and multiplication tables for $D$:
   (i) $1 + 1 + 1 + 1 = 4 \cdot 1 = 0$.
   (ii) $D$ has characteristic 2.
   (iii) $d$, $d^2$, and $cd$ are three distinct nonzero elements of $D$.
   (iv) $\{1, c, d\}$ is a multiplicative group.
2. Let $D$ have $n$ elements and characteristic $q$. Show that $q | n$.
3. Show that $\bar{2}$ is the only 0-divisor in $\mathbf{Z_4}$.
4. Find all the 0-divisors in each of the following:
   (a) $\mathbf{Z_8}$; (b) $\mathbf{Z_9}$; (c) $\mathbf{Z_{10}}$.
5. Prove that $\mathbf{Z_m}$ is not an integral domain if $m$ is composite.
6. Let $D$ have $m$ elements and let $m$ be composite.
   (a) Use Theorem 3 of this section to prove that the additive group of $D$ is not cyclic.
   *(b) Use Cauchy's Theorem or a Sylow Theorem (see Section 2.14) to prove that $m$ is a power of a prime.
7. (i) Show that an invertible $v$ in a ring $U$ with unity is never a 0-divisor.
   (ii) Explain why $\mathbf{Z_p}$ is an integral domain for every positive prime $p$.
   (iii) What is the characteristic of $\mathbf{Z_p}$?
8. Let $a$, $b$, and $c$ be elements of a ring $R$ with $c$ neither zero nor a 0-divisor. Show that $ca = cb$ implies that $a = b$. (This is the part of the proof of Theorem 1 that was left to the reader.)

9. Prove the corollary to Theorem 2; that is, prove that if $U$ is a commutative ring with unity and $U$ has the nonzero multiplicative cancellation property, then $U$ is an integral domain.

10. Let $6 \cdot 1 = 0$ in $D$. Use the fact that $D$ has no 0-divisors to show that either $2 \cdot 1 = 0$ or $3 \cdot 1 = 0$ and hence that the characteristic of $D$ is 2 or 3.

11. Assume that it has been proved using distributivity and mathematical induction that $(m \cdot a)(n \cdot b) = mn \cdot ab$ for all positive integers $m$ and $n$ and all elements $a$ and $b$ of a ring $R$ and then use this result with $a = b = 1$ to show that the characteristic of an integral domain is either 0 or a prime in $\mathbf{Z}$.

12. Let there be $n$ distinct elements in $D = \{0, d_2, d_3, \ldots, d_n\}$. Show that:
    (i) For every $k$ with $2 \le k \le n$, the products $d_2 d_k, d_3 d_k, \ldots, d_n d_k$ are all different from one another.
    (ii) Each $d_k$ is an invertible in $D$.
    (iii) $\{d_2, d_3, \ldots, d_n\}$ is a multiplicative group.

13. Show that $x^2 - (r + s)x + rs = 0$ in $D$ if and only if $x = r$ or $x = s$.

14. (i) Show that $\mathbf{Z}_6$ has 4 elements $x$ such that $x(x + 1) = 0$.
    (ii) How many elements $x$ of an integral domain satisfy $x(x + 1) = 0$?

15. Let $D$ have characteristic 2. Show the following in $D$:
    (a) $(a + b)^2 = a^2 + b^2$.
    (b) $(a + b + c)^2 = a^2 + b^2 + c^2$.
    (c) $(a + b)^4 = a^4 + b^4$.
    (d) $(a + b + c)^8 = a^8 + b^8 + c^8$.
    (e) $(a + b)^5 = a^5 + a^4b + ab^4 + b^5$.

16. Let $D$ have characteristic 3. Show that in $D$:
    (a) $(a + b)^3 = a^3 + b^3$.
    (b) $(a + b + c)^9 = a^9 + b^9 + c^9$.
    (c) $(a + b)^5 = a^5 - a^4b + a^3b^2 + a^2b^3 - ab^4 + b^5$.

17. Prove that a subring $S$ in $D$ has a unity if and only if the unity 1 of $D$ is in $S$ or $S = \{0\}$.

18. Is the ring $G$ of gaussian integers an integral domain? ($G$ consists of the complex numbers $m + n\mathbf{i}$, with $m$ and $n$ in $\mathbf{Z}$.)

19. What conditions suffice to show that a subring $S$ in $D$ is also an integral domain, that is, for $S$ to be a subdomain?

20. Let $a$, $b$, and $p$ be integers with $p$ a positive prime. Use Problem 7(ii) to give another proof that $p|(ab)$ implies either $p|a$ or $p|b$.

21. Prove that there is no integral domain with 6 elements.

22. Let $a \neq 0$ in $D$. Show that $a = -a$ if and only if $D$ has characteristic 2.

23. Let $D$ have characteristic $q$. Let $V$ be the multiplicative group of invertibles in $D$. Let $\theta$ be the mapping from $V$ to $V$ with $\theta(v) = v^2$. Show the following:

(i) $\theta$ is a group homomorphism.
(ii) The only solutions of $x^2 = b^2$ in $D$ are $x = b$ and $x = -b$. (These are equal when $q = 2$.)
(iii) The kernel of $\theta$ is $\{1, -1\}$ if $q \neq 2$ and is $\{1\}$ if $q = 2$.
(iv) If $m$ is in the image set $M$ of $\theta$, $\theta^{-1}(m)$ has two elements if $q \neq 2$ and one element if $q = 2$.
(v) If $V$ is finite, $M$ is a subgroup in $V$ with index 2 when $q \neq 2$ and index 1 when $q = 2$.

24. Let $V$ be the multiplicative group of the invertibles in $D$. Show the following:
(i) If $x$ is in $V$ and $x^{-1} = x$, then $x = 1$ or $x = -1$. [See Problem 23(iii).]
(ii) If $V$ is finite, $V$ has odd order if and only if $D$ has characteristic 2. (*Hint*: Show that there are an even number of elements $v$ in $V$ such that $v^{-1} \neq v$.)
(iii) If $V$ is finite, it has an even number of elements $v$ such that $v^{-1} \neq v$ and the product of all these elements is 1.
(iv) If $V$ has finite order $s$, the product $v_1 v_2 \cdots v_s$ of all the elements of $V$ equals $-1$. (If the characteristic of $D$ is 2, then $-1 = 1$.)

25. Let $p$ be a positive prime in the integers $\mathbf{Z}$ and let $\mathbf{V_p} = \{\bar{1}, \bar{2}, \ldots, \overline{p-1}\}$ be the multiplicative group of the invertibles of $\mathbf{Z_p} = \mathbf{Z}/(p)$. Show the following:
(i) $\bar{1} \cdot \bar{2} \cdot \bar{3} \cdots \overline{p-2} = \bar{1}$ in $\mathbf{V_p}$. [See Problem 24(iii).]
(ii) $\bar{1} \cdot \bar{2} \cdot \bar{3} \cdots \overline{p-1} = \overline{p-1} = -\bar{1}$ in $\mathbf{V_p}$.
(iii) $(p-1)! + 1$ is a multiple of $p$ in $\mathbf{Z}$.
(iv) $(p-1)! \equiv -1 \pmod{p}$.
[This is named Wilson's Theorem after John Wilson (1741–1793). See the biographical note on Lagrange after Section 2.11.]

26. Let $p$ be a positive prime in $\mathbf{Z}$. Show that:
(i) $(p-2)! - 1$ is a multiple of $p$ in $\mathbf{Z}$.
(ii) $(p-2)! \equiv 1 \pmod{p}$.

27. Find the remainder when each of the following is divided by the prime 101.
(a) 100!; (b) 99!; *(c) 98!; *(d) 97!; *(e) 96!.

28. Find the remainder when each of the following is divided by 103:
(a) 102!; (b) 101!; *(c) 100!; *(d) 99!; *(e) 98!.

29. Explain why $(29!)^2 \equiv 1 \pmod{59}$ and $(30!)^2 \equiv -1 \pmod{61}$.

30. Let $U$ be a (not necessarily commutative) ring with unity and no 0-divisors. Do all nonzero elements of $U$ have the same order in the additive group of $U$? Explain.

31. Let $c$ be the characteristic of an integral domain $D$ and let $m$ be a positive integer such that $m \cdot x = 0$ for all $x$ in $D$. Prove that $c \mid m$.

32. Find a 0-divisor in the direct product $\mathbf{Z_2} \times \mathbf{Z_2}$. (See Problem 30 of Section 4.1 for the definition of direct product.)

33. Is the direct product $D \times D'$ of integral domains ever an integral domain? Explain.

## 4.5 Ordered Integral Domains

The integral domain **Z** has a subset, $\mathbf{Z}^+ = \{1, 2, 3, \ldots\}$, which is closed under addition and under multiplication and which has the property that **Z** is partitioned into the three subsets: $\{0\}$, $\mathbf{Z}^+$, and the set $\mathbf{Z}^- = \{-1, -2, -3, \ldots\}$ of the negatives of the elements of $\mathbf{Z}^+$. The order relation $x < y$ in **Z** can be defined in terms of this partitioning. We indicate below how this may be done in the process of considering the generalization to integral domains.

### *Definition 1 Ordered Integral Domain*

*An integral domain D is said to be* **ordered** *if it has a subset $D^+$ with the two properties that follow.*

$O_1$. **Closure.** *$D^+$ is closed under addition and multiplication.*

$O_2$. **Trichotomy.** *If a is an element of D, then one and only one of the following holds:*

*(i) $a = 0$;  (ii) a is in $D^+$;  (iii) the negative $-a$ of a is in $D^+$.*

For a given subset $D^+$ with properties $O_1$ and $O_2$, one calls the elements in $D^+$ the ***positive*** elements of $D$. Also if $-a$ is positive, $a$ is said to be ***negative***. One defines $x < y$ to mean that $y - x$ is positive, i.e., is in $D^+$. The notation $y > x$ is equivalent to $x < y$; the notation $a \le b$ (or $b \ge a$) means that either $a < b$ or $a = b$.

It is an immediate consequence of the definitions that $c > 0$ if and only if $c - 0 = c$ is positive. Similarly, $c < 0$ if and only if $c$ is negative.

We next illustrate the derivation of order properties from the ***order axioms*** $O_1$ and $O_2$.

#### *Example 1*

Let $a$, $b$, and $c$ be elements of an ordered integral domain $D$. Show that if $a < b$ and $c > 0$, then $ac < bc$.

#### *Solution*

The hypotheses $a < b$ and $c > 0$ tell us that $b - a$ and $c$ are positive. Closure of $D^+$ under multiplication then implies that $(b - a)c = bc - ac$ is positive. Finally, $ac < bc$, by definition of $x < y$.

#### *Example 2*

Show that $x^2 > 0$ for every $x \neq 0$ in an ordered integral domain $D$.

*Solution*

Let $x \neq 0$. From the trichotomy axiom $O_2$ it then follows that either $x$ is positive or $-x$ is positive. If $x$ is positive, $x^2 = x \cdot x$ is positive by closure of $D^+$ under multiplication. If $-x$ is positive, we similarly see that $x^2 = (-x)(-x)$ is positive. Hence $x^2 > 0$ whenever $x \neq 0$ and the proof is complete.

The result in Example 2 shows that $y^2 \geq 0$ for every element $y$ of an ordered integral domain $D$. Also, it follows from the example and $1 \neq 0$ that $1^2 = 1 > 0$.

In an ordered integral domain $D$, one defines the ***absolute value*** $|a|$ of an element $a$ as follows: If $a$ is positive or $a$ is 0, then $|a| = a$; while $|a|$ is the positive element $-a$ if $a$ is negative. This implies that $|x| \geq 0$, that $|x|$ is zero if and only if $x = 0$, and that $|x| > 0$ if $x$ is not zero.

Examples of ordered integral domains are the integers **Z**, the rational numbers **Q**, and the real numbers **R**. It is not possible to order the complex numbers **C**, that is, to find a subset $\mathbf{C}^+$ of **C** with the closure and trichotomy properties $O_1$ and $O_2$. (The proof of this is left to the reader as Problem 19 below.)

## *Problems*

In the problems below, $D$ always denotes an ordered integral domain and $a$, $b$, $c$, and $d$ always are elements of $D$.

1. Prove the following:
   (a) $a < b$ and $b < c$ imply that $a < c$.
   (b) $a < b$ and $c < 0$ imply that $ca > cb$.
   (c) $a < b$ if and only if $a + c < b + c$.
   (d) $a < b$ and $c < d$ imply that $a + c < b + d$.

2. By a consideration of cases, prove that $a \leq b$ and $c \leq d$ imply that $a + c \leq b + d$.

3. Let $a > 0$, $b > 0$, $c < 0$, and $d < 0$. Prove the following:
   (a) $ab > 0$.
   (b) $ac < 0$.
   (c) $cd > 0$.
   (d) $2 \cdot a > 0$ (i.e., $a + a > 0$).
   (e) $2 \cdot c < 0$.
   (f) $c < a$.

4. Show that $2 \cdot a < 2 \cdot b$ if and only if $a < b$.

5. Show the following:
   (a) $|-a| = |a|$.
   (b) $a \leq |a|$.
   (c) $-b \leq |b|$.
   (d) $-|a| \leq a$.
   (e) $a + b \leq |a| + |b|$.
   (f) $-(a + b) \leq |a| + |b|$.
   (g) $|a + b| \leq |a| + |b|$.
   (h) $|c| - |a| \leq |c - a|$.
   (i) $|c| - |b| \leq |c + b|$.
   (j) $|(|a| - |b|)| \leq |a \pm b|$.

6. Show the following:
   (a) $a^2 - 2 \cdot ab + b^2 \geq 0$.
   (b) $a^2 + b^2 \geq 2 \cdot ab$.
   (c) $a^2 + b^2 + c^2 \geq bc + ac + ab$.
   (d) $a^2 + b^2 \geq ab$.
   (e) $a^2 + b^2 \geq -ab$.

7. Given that $a \neq b$, show the following:
   (i) $a^2 - 2 \cdot ab + b^2 > 0$.
   (ii) $a^2 + b^2 > 2 \cdot ab$.
   (iii) $a^2 + b^2 + c^2 > bc + ac + ab$.
   (iv) $a^2 + b^2 > ab$.

8. Use trichotomy to show that exactly one of the following is true for fixed $b$ and $c$:
   (i) $b = c$; (ii) $b > c$; (iii) $b < c$.

9. Given that $a \geq 1$ and $b \geq 1$, show that $ab + 1 \geq a + b$.

10. Let $a > 1$, $b > 1$, and $c > 1$. Show that

$$abc + a + b + c > bc + ac + ab + 1.$$

11. Given that $a > 0$, prove that $n \cdot a > 0$ for all $n$ in $\mathbf{Z}^+$.

12. Explain why an ordered integral domain has characteristic 0.

13. Explain why an ordered integral domain must be infinite.

14. Can $\mathbf{Z_p}$ be an ordered integral domain for any prime $p$? Explain.

15. We know that the subset $\mathbf{Z}^+ = \{1, 2, 3, \ldots\}$ of the integers $\mathbf{Z}$ satisfies the order axioms $O_1$ and $O_2$. Assume that $P$ is a subset of $\mathbf{Z}$ that also has properties $O_1$ and $O_2$ and prove that $P$ must be $\mathbf{Z}^+$. (This shows that there is only one way to order $\mathbf{Z}$.)

16. Let $S$ be a subdomain in an ordered domain $D$. Describe a method of ordering $S$.

17. Prove that there is only one way to order the rational numbers $\mathbf{Q}$.

18. Prove that every ordered integral domain $D$ has a subdomain $D'$ that is isomorphic to the integers $\mathbf{Z}$.

19. Prove as follows that the complex numbers $\mathbf{C}$ cannot be ordered: Assume that a subset $\mathbf{C}^+$ has properties $O_1$ and $O_2$, show that both 1 and $-1$ are in $\mathbf{C}^+$, and note the contradiction.

*20. Prove that there is only one way to order the real numbers. (You may use the facts that every positive real number is the square of a real number and that there is a rational number between any two real numbers.)

21. Find all the real numbers $x$ such that

$$|x - 4| + |x - 7| = 6.$$

(It may be helpful to graph $y = |x - 4| + |x - 7|$.)

22. For each of the following, find all the rational numbers $x$ that satisfy the equation:

    (a) $|x - 1| + |x - 2| + |x - 3| = 5/2$.
    (b) $|x - 1| + |x - 2| + |x - 3| = 4$.
    (c) $|x - 1| + |x - 2| + |x - 3| = 1$.

23. Let $c$ be a positive composite integer. Prove that there exists a prime $p$ in $\mathbf{Z}$ with $1 < p^2 \leq c$ and $p|c$.

24. Use that fact that no one of the primes

$$2, 3, 5, 7, 11$$

is an integral divisor of 131 to prove that 131 is a prime.

25. Prove that there do not exist integers $m$ and $n$ with

$$4 < m^2 \leq n \text{ and } 2n < (m + 1)^2.$$

*26. Find the largest positive integer $n$ such that the sum of the squares of the digits of $n$ exceeds $n$.

27. (i) Find all triples of positive integers $\{x, y, z\}$ such that

$$yz + xz + xy = xyz + 1.$$

(*Hint:* Assume that $0 < x < y < z$.)

(ii) Find all triples of positive integers $\{x, y, z\}$ for which there exists a positive integer $k$ with $yz + xz + xy = kxyz + 1$.

28. Find all triples of positive integers $\{r, s, t\}$ such that simultaneously $\bar{s} \cdot \bar{t} = \bar{1}$ in $\mathbf{Z}_r$, $\bar{r} \cdot \bar{t} = \bar{1}$ in $\mathbf{Z}_s$, and $\bar{r} \cdot \bar{s} = \bar{1}$ in $\mathbf{Z}_t$.

29. Let $m = 10^{150} + 3 \cdot 10^{100} + 1$. Use the fact that $(10^{50})^3 < m < (10^{50} + 1)^3$ to explain why $m$ is not the cube of an integer.

30. (i) Show that $(a + 3)^3 < a(a + 4)(a + 6) < (a + 4)^3$ for all real $a \geq 7$.
    *(ii) Use (i) to prove that the product of six consecutive positive integers is never the cube of an integer.

31. Find all solutions of $y^2 = x^4 + 2x^3 + 2x^2 + 2x + 5$ in integers $x$ and $y$.

*32. Find all solutions in integers $x$ and $y$ of the equation

$$y^2 = x^4 + 6x^3 + 10x^2 + 15x - 13.$$

## 4.6 Fields

Let $U$ be a ring with unity in which $0 \neq 1$. By Theorem 2 of Section 4.1, we then have $0 \cdot a = 0 \neq 1$ for every $a$ in $U$. Hence the additive identity 0 has no multiplicative inverse; that is, 0 is not an invertible. This means that $U$ is not a group under multiplication.

However, the set obtained by deleting 0 from $U$ may be a multiplicative group, as is true in the number systems **Q**, **R**, and **C** and in the $\mathbf{Z_p}$ with $p$ prime. This leads us to the following:

***Definition 1 Division Ring, Field, Skew-Field***

*A* **division ring** *is a ring with unity* $1 \neq 0$ *in which every nonzero element is an invertible. A* **field** *is a commutative division ring. A* **skew-field** *is a noncommutative division ring.*

Because of the importance of fields in abstract algebra, we also give the following characterization:

***Theorem 1 Alternative Definition of a Field***

*A field $F$ may be defined as a set $\hat{F}$ with operations of addition and multiplication such that:*
(*a*) *$\hat{F}$ is an abelian group under addition.*
(*m*) *$\hat{F}$ with 0 deleted is an abelian group under multiplication.* (*This is the multiplicative group of $F$.*)
(*d*) *Multiplication is distributive over addition in $\hat{F}$.*

The fact that the definition in Theorem 1 is equivalent to the previous definition follows readily with the help of Theorem 1 of Section 4.1, which states that the invertibles in a ring with unity form a multiplicative group.

Since every nonzero element of a field is an invertible, a field has no 0-divisors. [See Problem 7(i) of Section 4.4.] It follows that every field is an integral domain. The integers **Z** furnish an example of an integral domain that is not a field.

Some infinite fields are the rational numbers **Q**, the real numbers **R**, and the complex numbers **C**. In Section 4.7 we describe the infinite skew-field of the quaternions.

The $\mathbf{Z_p}$ with $p$ prime are examples of finite fields. It is a theorem of J. H. M. Wedderburn that there are no finite skew-fields, that is, every finite division ring is a field. [For a comparatively elementary proof, see I. N. Herstein, "Wedderburn's Theorem and a Theorem of Jacobson," *American Mathematical Monthly*, 68 (1961), 249–251.]

We next define an analogue of subgroup and subring.

***Definition 2 Subfield, Extension Field***

*A subset $F$ of a field $E$ is a* **subfield** *in $E$ if $F$ is a field under the operations of $E$. If $F$ is a subfield in $E$, one also says that $E$ is an* **extension field**, *or* **superfield**, *over $F$.*

***Theorem 2 Subfield Conditions***

> *Let F be a subset with at least two elements in a field E. Then F is a subfield in E if and only if F is closed under subtraction and under division by nonzero elements.*

The proof is left to the reader as Problem 25 of this section.

Since a field or an integral domain is a ring, the definitions of ring homomorphism, isomorphism, automorphism, and endomorphism apply to fields and integral domains. Also, an ***ordered field*** is a field that is ordered as an integral domain and the ***characteristic of a field*** is its characteristic as an integral domain.

It can be shown (see Problem 26 below) that the intersection $P$ of the collection of all the subfields in a field $F$ is also a subfield in $F$; this smallest

***Table 4.1 Summary of Axioms for Groups, Rings, and Fields***

Properties

| Under Addition | Under Multiplication |
|---|---|
| 1. Closure | 6. Closure |
| 2. Associativity | 7. Associativity |
| 3. Existence of an identity | 8. Existence of an identity |
| 4. Existence of inverses | 9. Existence of inverses (for nonzero elements) |
| 5. Commutativity | 9′. Cancellation law (or no 0-divisors) |
| | 10. Commutativity |

Under Addition and Multiplication
11. Distributivity
12. $0 \neq 1$

| Algebraic Structure | Required Properties |
|---|---|
| Additive group | 1–5 |
| Multiplicative group | 6–9 |
| Ring | 1–7, 11 |
| Ring with unity | 1–8, 11 |
| Commutative ring | 1–7, 10, 11 |
| Integral domain | 1–8, 9′, 10–12 |
| Division ring | 1–9, 11, 12 |
| Skew-field | 1–9, 11, 12, and negation of 10 |
| Field | 1–12 |

subfield $P$ is called the ***prime subfield*** in $F$. It is also left to the reader to show that the prime subfield $P$ is isomorphic to the rational numbers $\mathbf{Q}$ when $F$ has characteristic 0 (see Problem 27 below) and that $P$ is isomorphic to $\mathbf{Z_p}$ when $F$ has characteristic $p$ (see Problem 28 below).

In Table 4.1 we summarize the axioms for groups and the types of rings, including fields, discussed in this text.

## *Problems*

1. Let $\mathbf{C}$ be the field of complex numbers. Every $\alpha$ in $\mathbf{C}$ is uniquely expressible as $\alpha = a + b\mathbf{i}$, with $a$ and $b$ real and $\mathbf{i}^2 = -1$. Let $\bar{\alpha} = a - b\mathbf{i}$ be the ***conjugate*** of $\alpha$ and let $|\alpha| = \sqrt{a^2 + b^2}$ be the absolute value of $\alpha$. Let $f$ be the mapping from $\mathbf{C}$ to itself with $f(\alpha) = \bar{\alpha}$. Show that:
   (i) $f$ is a bijection.
   (ii) $\overline{\alpha + \beta} = \bar{\alpha} + \bar{\beta}$, $\overline{\alpha\beta} = \bar{\alpha} \cdot \bar{\beta}$; that is, the bijection $f$ is a field automorphism of $\mathbf{C}$.
   (iii) $|\alpha|^2 = \alpha\bar{\alpha}$.
   (iv) $|\alpha\beta| = |\alpha| \cdot |\beta|$. [It may be helpful to use parts (iii) and (ii). This formula implies that the mapping $g$ with $g(\alpha) = |\alpha|$ is a homomorphism from the multiplicative group of $\mathbf{C}$ to the multiplicative group of the positive real numbers.]
2. Let $T$ be the set of all real numbers $a + b\sqrt{2}$ with $a$ and $b$ rational. Define the *conjugate* $\bar{\alpha}$ and *norm* $\|\alpha\|$ of $\alpha = a + b\sqrt{2}$ by $\bar{\alpha} = a - b\sqrt{2}$ and $\|\alpha\| = \alpha\bar{\alpha}$.
   (i) Use the fact that $\sqrt{2}$ is irrational to show that if $a + b\sqrt{2} = c + d\sqrt{2}$, then $a = c$ and $b = d$.
   (ii) Show that $T$ is a field.
   (iii) Show that the mapping $f$ with $f(\alpha) = \bar{\alpha}$ is a field automorphism of $T$.
   (iv) Show that $\|\alpha\beta\| = \|\alpha\| \cdot \|\beta\|$ for all $\alpha$ and $\beta$ in $T$.
3. Explain why a finite integral domain must be a field. (See Problem 12 of Section 4.4.)
4. Let $F$ be a finite field with $m$ elements. Explain why:
   (i) $x^{m-1} = 1$ for all nonzero $x$ in $F$. (*Hint:* Use Lagrange's Theorem.)
   (ii) $x^m - x = 0$ for all $x$ in $F$.
5. Let $F = \{0, 1, c\}$ be a field with 3 elements.
   (i) Explain why both the additive group and the multiplicative group of $F$ are cyclic.
   (ii) Explain why $F$ must have characteristic 3.
   (iii) Make the addition and multiplication tables for $F$.
   (iv) Explain why any two fields with 3 elements are isomorphic.
6. Let $F$ be a field with characteristic 3 and let $1 + 1$ be denoted by 2 in $F$. Show that $P = \{0, 1, 2\}$ is a subfield in $F$.

7. Let $F$ be a field with characteristic $p$, a prime. In $F$, let $2 \cdot 1, 3 \cdot 1, \ldots, (p-1) \cdot 1$ be denoted by $2, 3, \ldots, p-1$. Show that $P = \{0, 1, 2, \ldots, p-1\}$ is a subfield in $F$. (The field $P$ can be shown to be the prime subfield in $F$. See Problem 28 below.)
8. Let $p$ be a positive prime in $\mathbf{Z}$. Explain why any two fields with $p$ elements must be isomorphic.
9. How many automorphisms are there of $\mathbf{Z}_{23}$? Explain.
10. How many automorphisms are there of $\mathbf{Z}_p$? (Here $p$ is a positive prime.)
11. Given that $a > 0$ in an ordered field $F$, show that $a^{-1} > 0$.
12. Given that $b < 0$ in an ordered field $F$, show that $b^{-1} < 0$.
13. Given that $0 < a < b$ in an ordered field $F$, show that $0 < 1/b < 1/a$.
14. Given that $a < b < 0$ in an ordered field $F$, show that $1/b < 1/a < 0$.
15. In $\mathbf{Z}_7$, find the unique element $a$ such that $\bar{3}a = \bar{2}$.
16. Let $a$ and $b$ be elements of a field $F$ with $a \neq 0$. Explain why there is a unique $x$ of $F$ such that $ax = b$.
17. In $\mathbf{Z}_{19}$, find $x$ and $y$ satisfying the simultaneous equations

$$\begin{aligned} \bar{2}x + \bar{7}y &= \bar{4} \\ \bar{9}x + \bar{4}y &= \bar{1}. \end{aligned}$$

18. Let $a$, $b$, $c$, $d$, $h$, and $k$ be elements of a field $F$ with $ad - bc \neq 0$. Show that there exist unique elements $x$ and $y$ in $F$ satisfying the simultaneous equations

$$\begin{aligned} ax + by &= h \\ cx + dy &= k. \end{aligned}$$

19. Let $F$ be a field with 8 elements. Explain why the following hold:
    (i) The characteristic of $F$ must be 2.
    (ii) Every nonzero $a$ of $F$ satisfies $a^7 - 1 = 0$.
    (iii) Every $x$ of $F$ satisfies

$$x(x-1)(x^3 + x^2 + 1)(x^3 + x + 1) = x^8 - x = 0.$$

20. Let $F$ be a field with 9 elements. Explain why the following hold:
    (i) The characteristic of $F$ must be 3.
    (ii) Every $x$ of $F$ satisfies

$$x(x-1)(x+1)(x^2+1)(x^2+x-1)(x^2-x-1) = x^9 - x = 0.$$

21. Explain why the only ideals in a field $F$ are $\{0\}$ and $F$ itself.

22. Let $M$ be the image set of a homomorphism $\theta$ from a field $F$ to a field $F'$. Explain why either $M$ is isomorphic to $F$ or $M = \{0\}$. (*Hint:* See the preceding problem.)

23. Is $\theta(1) = 1$ necessarily true for every field homomorphism $\theta$ from $F$ to $F'$? Explain.

24. Let $\theta$ be a field isomorphism from $F$ to $F'$. Explain why the following are true:
    (i) $\theta(a) \neq 0$ for all nonzero $a$ of $F$.
    (ii) The multiplicative groups of $F$ and $F'$ are isomorphic.
    (iii) $\theta(1) = 1$.
    (iv) $\theta(a^{-1}) = [\theta(a)]^{-1}$ for every nonzero $a$ of $F$.

25. Let $F$ be a subset, with more than one element, of a field $E$. Prove that $F$ is a subfield in $E$ if and only if $F$ is closed under subtraction and under division by nonzero elements. (This is the proof of Theorem 2 that was left to the reader.)

26. Let $P$ be the intersection of all the subfields in a field $F$. Prove that $P$ is a subfield in $F$ and hence is the smallest subfield in $F$. (This smallest subfield $P$ is the prime subfield in $F$.)

27. Prove that the prime subfield $P$ in a field $F$ with characteristic 0 is isomorphic to the field $\mathbf{Q}$ of rational numbers.

28. Prove that the prime subfield $P$ in a field $F$ with characteristic $p$ is isomorphic to $\mathbf{Z}_\mathbf{p}$. (See Problems 26 and 7.)

*29. Prove that the field $\mathbf{Q}$ of rational numbers can be ordered in only one way; that is, there is only one way to select a subset $\mathbf{Q}^+$ satisfying the order axioms $O_1$ and $O_2$.

30. Prove that every ordered field has a subfield isomorphic to $\mathbf{Q}$.

31. Let $F$ be a subfield in $E$ and let $H$ consist of all the field automorphisms $\theta$ of $E$ such that $\theta(f) = f$ for all $f$ in $F$. Show that $H$ is a subgroup in the group $\mathbf{B}(E)$ of all bijections from $E$ to itself.

32. Let $R$ be a commutative ring with 0-divisors; that is, let there be nonzero elements $r$ and $s$ in $R$ with $rs = 0$. Explain why $R$ cannot be a subring in a field $F$.

33. Let $\theta$ be a ring homomorphism from a ring $U$ with unity to a field $F$. Prove that either $\theta(1) = 1$ or $\theta(u) = 0$ for all $u$ in $U$.

34. Explain why the following are true:
    (a) In a ring $U$ with unity, $\{n \cdot 1 : n \in \mathbf{Z}\}$ is a subring.
    (b) The prime subfield in a field $F$ consists of all elements of the form $(m \cdot 1)(n \cdot 1)^{-1}$, with $m$ and $n$ integers and $n \cdot 1 \neq 0$.

## 4.7 Matrices, Quaternions

*There is an astonishing imagination, even in the science of mathematics.... We repeat, there is far more imagination in the head of Archimedes than in that of Homer.*

*François Marie Arouet de Voltaire*

Here we merely sample the branch of algebra called "matrix theory." We present enough material on $m$ by $n$ matrices for our work on coding theory in Chapter 8 and then specialize to 2 by 2 matrices.

Let $R$ be a commutative ring. An ***m × n matrix*** (read as "$m$ by $n$ matrix") over $R$ is a rectangular array

$$A = \begin{pmatrix} a_{11} & a_{12} & a_{13} & \cdots & a_{1n} \\ a_{21} & a_{22} & a_{23} & \cdots & a_{2n} \\ \vdots & & & & \\ a_{m1} & a_{m2} & a_{m3} & \cdots & a_{mn} \end{pmatrix} \tag{1}$$

with $m$ rows and $n$ columns and with each entry $a_{ij}$ in $R$. For short, we write this matrix $A$ as $(a_{ij})$. If we wish to indicate its size (that is, the number of rows and number of columns), we write it as $(a_{ij})_{m \times n}$. If $m = n$, $A$ is a ***square matrix***.

### *Notation Set of m × n Matrices over R*

*Let $M(m, n, R)$ denote the set of $m \times n$ matrices over the ring $R$.*

### *Definition 1 Matrix Addition*

*Let $A = (a_{ij})$ and $B = (b_{ij})$ be in $M(m, n, R)$. Then $A + B$ is the matrix $C = (c_{ij})$ in $M(m, n, R)$ with $c_{ij} = a_{ij} + b_{ij}$ for $1 \leq i \leq m$ and $1 \leq j \leq n$.*

This definition makes $M(m, n, R)$ into an additive group isomorphic to the direct product $R \times R \times \cdots \times R$ of $mn$ copies of the additive group of $R$.

The product $AB$ is defined only when the number of columns in $A$ is the same as the number of rows in $B$; before giving that definition we need some preliminary concepts.

### *Definition 2 Transpose*

*The* **transpose** $A^T$ *of an* $m \times n$ *matrix* $A = (a_{ij})$ *is the matrix* $B = (b_{ij})$ *with* $b_{ij} = a_{ji}$ *for* $1 \leq i \leq n$ *and* $1 \leq j \leq m$.

Thus the rows of $A$ are the columns of $A^T$ (and the columns of $A$ are the rows of $A^T$).

### *Example 1*

The transpose of

$$A = \begin{pmatrix} 1 & 2 & 3 \\ 4 & 5 & 6 \end{pmatrix} \text{ is } A^T = \begin{pmatrix} 1 & 4 \\ 2 & 5 \\ 3 & 6 \end{pmatrix}.$$

Clearly, $(A^T)^T = A$ for all matrices $A$; that is, the transpose of the transpose of $A$ is the original matrix $A$.

### *Definition 3 n-Vector*

*A* $1 \times n$ *matrix over* $R$ *is also called an* **n-vector** *over* $R$. *If one separates the entries of an n-vector by commas, it becomes an element of the cartesian product* $R \times R \times \cdots \times R$ *of n copies of* $R$.

### *Definition 4 Rows and Columns*

*Let* $A = (a_{ij})$ *be the matrix in* $M(m, n, R)$ *of display* (1). *For* $1 \leq i \leq m$, *the n-vector* $V_i = (a_{i1}, a_{i2}, \ldots, a_{in})$ *is called the* **i th row** *of* $A = (a_{ij})$. *For* $1 \leq j \leq n$, *the transpose* $W_j^T$ *of the m-vector* $W_j = (a_{1j}, a_{2j}, \ldots, a_{mj})$ *is the* **j th column** *of* $A$.

### *Example 2*

Let

$$A = \begin{pmatrix} 1 & 2 & 3 \\ 6 & 5 & 4 \end{pmatrix}.$$

Then the first row is (1, 2, 3) and the second row is (6, 5, 4). The columns of $A$, in order, are $(1, 6)^T$, $(2, 5)^T$, and $(3, 4)^T$.

***Definition 5 Dot Product***

*Let $U = (u_1, u_2, \ldots, u_n)$ and $V = (v_1, v_2, \ldots, v_n)$. Then the* **dot product** *$U \cdot V$ of these two n-vectors is the sum of products*

$$u_1 v_1 + u_2 v_2 + \cdots + u_n v_n.$$

*The dot product is also called the* **scalar product**.

***Example 3***

The dot product $V_1 \cdot V_2$ of the row-vectors for the $A$ of Example 2 is $1 \cdot 6 + 2 \cdot 5 + 3 \cdot 4 = 6 + 10 + 12 = 28$.

***Definition 6 Product of Matrices of Compatible Sizes***

*Let $A = (a_{ij})_{r \times s}$ and $B = (b_{ij})_{s \times t}$ be matrices over R with the number of columns in A the same as the number of rows in B. Let $V_i$ be the ith row of A and $W_j^T$ be the jth column of B. Then the* **product** *$C = AB$ is the $r \times t$ matrix $(c_{ij})$ with the dot product $V_i \cdot W_j$ as the entry $c_{ij}$.*

***Example 4***

One can see that

$$\begin{pmatrix} 1 & 2 & 3 & 0 \\ 6 & 5 & 4 & 0 \\ 7 & 8 & 9 & 1 \end{pmatrix} \begin{pmatrix} 1 & -1 \\ 0 & 7 \\ 0 & 4 \\ 2 & 3 \end{pmatrix} = \begin{pmatrix} 1 & 25 \\ 6 & 45 \\ 9 & 88 \end{pmatrix}.$$

The entry $c_{32}$ in the product is the dot product

$$\begin{aligned} V_3 \cdot W_2 = (7, 8, 9, 1) \cdot (-1, 7, 4, 3) &= 7(-1) + 8 \cdot 7 + 9 \cdot 4 + 1 \cdot 3 \\ &= -7 + 56 + 36 + 3 = 88 \end{aligned}$$

and the other entries are calculated similarly.

Multiplication of matrices is not commutative, that is, $AB$ and $BA$ need not be equal. Moreover, $AB$ may exist without $BA$ being defined, as one sees in Example 4.

***Theorem 1 Transpose of a Product***

$$(AB)^T = B^T A^T \text{ (whenever the product } AB \text{ is defined).}$$

The proof is left to the reader.

Theorem 1 states that the transpose of a product is the product of the transposes in reverse order.

Square matrices of fixed size form a ring, that is, $M(n, n, R)$ is a ring for each positive integer $n$. The case $n = 2$ is especially useful for abstract algebra and we devote the rest of this section to that case. We abbreviate $M(2, 2, R)$ as $M(R)$. Addition and multiplication in $M(R)$ are given by

$$\begin{pmatrix} a & b \\ c & d \end{pmatrix} + \begin{pmatrix} a' & b' \\ c' & d' \end{pmatrix} = \begin{pmatrix} a + a' & b + b' \\ c + c' & d + d' \end{pmatrix}, \tag{2}$$

$$\begin{pmatrix} a & b \\ c & d \end{pmatrix} \begin{pmatrix} a' & b' \\ c' & d' \end{pmatrix} = \begin{pmatrix} aa' + bc' & ab' + bd' \\ ca' + dc' & cb' + dd' \end{pmatrix}. \tag{3}$$

Additively $M(R)$ is an abelian group since it is the direct product $R \times R \times R \times R$ of four copies of the additive group of $R$. Since a ring $R$ is closed under addition and multiplication, formula (3) shows that $M(R)$ is closed under multiplication. We omit the straightforward (but tedious) details of verifying associativity of multiplication and distributivity for $M(R)$ and thus completing the proof that $M(R)$ is a ring.

Let

$$\alpha = \begin{pmatrix} a & b \\ c & d \end{pmatrix}, \quad O = \begin{pmatrix} 0 & 0 \\ 0 & 0 \end{pmatrix}, \quad I = \begin{pmatrix} 1 & 0 \\ 0 & 1 \end{pmatrix}. \tag{4}$$

For any ring $R$, the matrix $O$ is the ***zero*** 0 of $M(R)$. If $R$ has a unity 1, the matrix $I$ is the ***unity*** 1 of $M(R)$.

***Example 5***

The following products show that $M(R)$ is not commutative when $rs \neq 0$ for some $r$ and $s$ in $R$.

$$\begin{pmatrix} 0 & r \\ 0 & 0 \end{pmatrix} \begin{pmatrix} 0 & 0 \\ s & 0 \end{pmatrix} = \begin{pmatrix} rs & 0 \\ 0 & 0 \end{pmatrix}, \quad \begin{pmatrix} 0 & 0 \\ s & 0 \end{pmatrix} \begin{pmatrix} 0 & r \\ 0 & 0 \end{pmatrix} = \begin{pmatrix} 0 & 0 \\ 0 & sr \end{pmatrix}.$$

***Definition 7 Determinant of a 2 × 2 Matrix***

*The* **determinant** *of the matrix* $\alpha$ *of* (4) *is the element* $ad - bc$ *of* $R$ *and is denoted by* det $\alpha$.

We note that the determinant of the unity matrix of display (4) is $1 \cdot 1 - 0 \cdot 0 = 1$. The mapping with $\alpha \mapsto \det \alpha$ plays an important role in matrix theory. One of its key properties is given in the following result.

***Theorem 2 Determinant of a Product***

*Let* $R$ *be a commutative ring and let* $\alpha$ *and* $\beta$ *be in* $M(R)$. *Then* $\det(\alpha\beta) = (\det \alpha)(\det \beta)$.

The straighforward proof of this theorem is left to the reader as Problem 27 of this section.

***Theorem 3 Invertible Matrices***

*Let* $U$ *be a commutative ring with unity. Then a matrix* $\alpha$ *is an invertible in* $M(U)$ *if and only if* det $\alpha$ *is an invertible in* $U$.

***Proof***

Let $\alpha\beta = 1$ in $M(U)$. Then

$$(\det \alpha)(\det \beta) = \det(\alpha\beta) = \det 1 = 1,$$

and hence det $\alpha$ is an invertible in $U$.

Conversely, let $\alpha$ be as in display (4) and let $g = \det \alpha = ad - bc$ be an invertible in $U$. Then we can let $a' = dg^{-1}$, $b' = -bg^{-1}$, $c' = -cg^{-1}$, $d' = ag^{-1}$, and

$$\alpha' = \begin{pmatrix} a' & b' \\ c' & d' \end{pmatrix}.$$

Using multiplication formula (3), we then see that $\alpha\alpha'$ and $\alpha'\alpha$ are both the unity of $M(R)$; that is, $\alpha'$ is the multiplicative inverse of $\alpha$.

We next consider some important subrings in rings of matrices.

***Theorem 4 Matric Representation of the Complex Numbers***

*Let* **R** *be the field of real numbers and let K consist of the matrices in M*(**R**) *of the form*

$$\alpha = \begin{pmatrix} a & b \\ -b & a \end{pmatrix}. \tag{5}$$

*Then K is a subring in M*(**R**) *and K is a field isomorphic to the complex numbers* **C**.

The proof of Theorem 4 is left to the reader as Problem 28 below.

***Theorem 5 The Quaternions as Matrices***

*Let* **C** *be the field of complex numbers and let H be the subset of M*(**C**) *consisting of the matrices*

$$\alpha = \begin{pmatrix} a + b\mathbf{i} & c + d\mathbf{i} \\ -c + d\mathbf{i} & a - b\mathbf{i} \end{pmatrix}. \tag{6}$$

*Then H is an infinite skew-field.*

***Partial Proof***

We leave the proof that $H$ is a noncommutative subring in $M(\mathbf{C})$ to the reader as Problem 29 below. Assuming that $H$ is a noncommutative subring, we now show it to be an infinite skew-field. Let $\alpha$ be as in (6). Then

$$\det \alpha = (a + b\mathbf{i})(a - b\mathbf{i}) - (c + d\mathbf{i})(-c + d\mathbf{i}) = a^2 + b^2 + c^2 + d^2.$$

Since squares of real numbers are never negative, this means that $\det \alpha = 0$ only when $a = b = c = d = 0$, that is, only if $\alpha$ is the zero matrix of $M(\mathbf{C})$. It now follows from Theorem 3 that every nonzero $\alpha$ in $H$ is an invertible and hence that $H$ is a skew-field. Clearly $H$ is infinite.

We next present some other notations for the elements of the field $H$ of Theorem 5. Temporarily, let the matrix $\alpha$ of (6) also be denoted by the ordered quadruple $[a, b, c, d]$. Let

$$i^* = [0, 1, 0, 0], \qquad j^* = [0, 0, 1, 0], \qquad k^* = [0, 0, 0, 1]$$

and let $a^* = [a, 0, 0, 0]$ for every real number $a$. Then it is straightforward to verify that

$$a^* + b^*i^* + c^*j^* + d^*k^* = [a, b, c, d], \tag{7}$$

and that the subset $C^*$ of $H$ consisisting of all elements of the form $a^* + b^*i^*$ is isomorphic to **C**. This isomorphism makes it possible to delete the asterisks in (7) without causing any confusion. Thus we obtain the expressions for the quaternions,

$$\alpha = a + b\mathbf{i} + c\mathbf{j} + d\mathbf{k}, \tag{8}$$

used by their originator, Sir William Rowan Hamilton. (See the biographical note after this section.) Generally we use the notation (8) for the elements of the skew-field $H$ of quaternions and consider the complex numbers **C** and the real numbers **R** to be subrings in $H$.

### *Definition 8 Conjugate and Norm of a Quaternion*

*The* **conjugate** *of* $\alpha = a + b\mathbf{i} + c\mathbf{j} + d\mathbf{k}$ *is* $\bar{\alpha} = a - b\mathbf{i} - c\mathbf{j} - d\mathbf{k}$. *The* **norm** *of* $\alpha$ *is* $\|\alpha\| = \alpha\bar{\alpha}$.

Calculation shows that $\|a + b\mathbf{i} + c\mathbf{j} + d\mathbf{k}\| = a^2 + b^2 + c^2 + d^2$; thus $\|\alpha\|$ for the $\alpha$ of (8) is the same as $\det \alpha$ for the $\alpha$ of (6). Then it follows from Theorem 2 that $\|\alpha\beta\| = \|\alpha\| \cdot \|\beta\|$.

Addition or subtraction of quaternions is easy in the form (6) and hence in the form (8). Multiplication and division of quaternions expressed as in (8) is facilitated by verifying the following:

(A) $\mathbf{i}^2 = \mathbf{j}^2 = \mathbf{k}^2 = -1$.

(B) $\mathbf{ij} = -\mathbf{ji} = \mathbf{k}$, $\mathbf{jk} = -\mathbf{kj} = \mathbf{i}$, $\mathbf{ki} = -\mathbf{ik} = \mathbf{j}$.

(C) The center of the multiplicative group (of nonzero elements) of $H$ consists of the nonzero real numbers.

(D) If $\alpha \neq 0$, then $\alpha^{-1} = \|\alpha\|^{-1}\bar{\alpha}$.

(E) $\overline{\alpha + \beta} = \bar{\alpha} + \bar{\beta}$, $\overline{\alpha\beta} = \bar{\beta} \cdot \bar{\alpha}$.

The proof of these results is left to the reader in Problems 13–20 for this section.

The mapping $f$ from $H$ to $H$ with $f(\alpha) = \bar{\alpha}$ is a bijection with $f(\alpha + \beta) = f(\alpha) + f(\beta)$ and $f(\alpha\beta) = f(\beta)f(\alpha)$; such a mapping is an ***anti-automorphism*** of the skew-field $H$.

**Theorem 6 Ring of Hamiltonian Integers**

*Let W be the subset of H consisting of the quaternions $\alpha = a + b\mathbf{i} + c\mathbf{j} + d\mathbf{k}$ with a, b, c, and d integers. Then W is a subring in H.*

**Proof**

It is easily seen that $W$ is nonempty and is closed under subtraction and multiplication. Hence $W$ is a subring in $H$.

**Theorem 7 The Quaternion Group**

*The multiplicative group of invertibles of the ring W of hamiltonian integers is*

$$\{1, -1, \mathbf{i}, -\mathbf{i}, \mathbf{j}, -\mathbf{j}, \mathbf{k}, -\mathbf{k}\}. \tag{9}$$

**Proof**

Let $\alpha = a + b\mathbf{i} + c\mathbf{j} + d\mathbf{k}$ with $a$, $b$, $c$, and $d$ integers. Then $\|\alpha\| = a^2 + b^2 + c^2 + d^2$ is a nonnegative integer. If $\alpha\beta = 1$ in $W$, then $\|\alpha\| \cdot \|\beta\| = \|\alpha\beta\| = \|1\| = 1$ in $\mathbf{Z}$. It follows from the last two statements that $\|\alpha\| = 1$. This implies that $\alpha$ is an invertible only if one of $a$, $b$, $c$, $d$ is chosen as $\pm 1$ and the others as 0, that is, only if $\alpha$ is in the set of display (9). Conversely, one sees easily that each of the eight elements of that set is an invertible in $W$.

## Problems

In this problem set $a$, $b$, $c$, and $d$ always designate real numbers.

1. Describe a noncommutative ring with 16 elements.
2. Show that there exists a noncommutative ring with $n^4$ elements for every integer $n > 1$. (The case $n = 2$ is taken up in Problem 1.)
3. (a) In $M(\mathbf{Z}_2)$, find the product

$$\begin{pmatrix} 1 & 1 \\ 1 & 1 \end{pmatrix}\begin{pmatrix} 1 & 1 \\ 1 & 1 \end{pmatrix}.$$

   (b) Does the matrix ring $M(\mathbf{Z}_2)$ have 0-divisors? Explain.

4. (a) In $M(\mathbf{Q})$, find

$$\begin{pmatrix} 1 & 0 \\ 1 & 0 \end{pmatrix}\begin{pmatrix} 0 & 0 \\ 1 & 1 \end{pmatrix}.$$

(b) Does $M(\mathbf{Q})$ have 0-divisors? Explain.

5. What are the entries of the zero matrix (additive identity) in $M(2, 3, R)$?
6. What are the entries of the additive identity $0_{m \times n}$ in $M(m, n, R)$?
7. Describe a matrix $A = (a_{ij})$ in $M(10, 10, \mathbf{Z_{10}})$ such that $A$ is not the additive identity $0_{10 \times 10}$ but $A^2 = 0_{10 \times 10}$.
8. Describe a matrix $B = (b_{ij})$ in $M(m, m, \mathbf{Z_m})$ such that $B \neq 0_{m \times m}$ and $B^2 = 0_{m \times m}$.
9. Give an example of matrices $A$, $B$, $C$ in $M(\mathbf{Q})$ with $AC = BC$, $A \neq B$, and $C$ not the zero matrix.
10. Do Problem 9 with $M(\mathbf{Q})$ replaced by $M(m, m, \mathbf{Z_m})$.
11. Let $F = \begin{pmatrix} 5 & 0 \\ 0 & 5 \end{pmatrix}$. If $A^2 = FA$ in $M(\mathbf{Q})$, must $A = F$?
12. If $AB = AC$ in $M(R)$ and $A$ is invertible, must $B = C$? Explain.
13. Verify that $\mathbf{i}^2 = \mathbf{j}^2 = \mathbf{k}^2 = -1$ in the quaternions.
14. Verify that $\mathbf{ij} = -\mathbf{ji} = \mathbf{k}$. Then use this result, Problem 13, and associativity to see that

$$\mathbf{jk} = -\mathbf{kj} = \mathbf{i}, \qquad \mathbf{ki} = -\mathbf{ik} = \mathbf{j}.$$

15. Let $\alpha = a + b\mathbf{i} + c\mathbf{j} + d\mathbf{k}$. Show the following:
    (a) $\alpha\mathbf{i} = \mathbf{i}\alpha$ if and only if $c = 0 = d$.
    (b) $\alpha\mathbf{i} = \mathbf{i}\bar{\alpha}$ if and only if $b = 0$.
16. Let $\alpha = a + b\mathbf{i} + c\mathbf{j} + d\mathbf{k}$. Give the conditions on $a$, $b$, $c$, and $d$ for each of the following to hold.
    (a) $\alpha\mathbf{j} = \mathbf{j}\alpha$; (b) $\alpha\mathbf{k} = \mathbf{k}\alpha$; (c) $\alpha\mathbf{j} = \mathbf{j}\bar{\alpha}$.
17. Show that under multiplication the quaternion $\alpha$ of Problem 15 commutes with all quaternions if and only if $\alpha$ is real, that is, if and only if $b = c = d = 0$.
18. In the skew-field $H$ of quaternions, show the following:
    (a) $\alpha^{-1} = \|\alpha\|^{-1}\bar{\alpha}$ if $\alpha \neq 0$.
    (b) $\|\bar{\alpha}\| = \|\alpha\|$.
    (c) $\|\alpha^{-1}\| = \|\alpha\|^{-1}$ if $\alpha \neq 0$.
19. Show that $\overline{\alpha + \beta} = \bar{\alpha} + \bar{\beta}$ in $H$.

20. Show that $\overline{\alpha\beta} = \overline{\beta} \cdot \overline{\alpha}$ in $H$.

21. Show that $a + b\mathbf{i} + c\mathbf{j} + d\mathbf{k} = (a + b\mathbf{i}) + (c + d\mathbf{i})\mathbf{j}$.

22. Let $z_1$, $z_2$, $w_1$, and $w_2$ be complex numbers. Show that, in $H$,

$$(z_1 + z_2\mathbf{j})(w_1 + w_2\mathbf{j}) = (z_1w_1 - z_2\overline{w}_2) + (z_1w_2 + z_2\overline{w}_1)\mathbf{j}.$$

23. In the skew-field $H$ of quaternions, show that:
    (i) $(a + b\mathbf{i} + c\mathbf{j} + d\mathbf{k})^2 = (a^2 - b^2 - c^2 - d^2) + 2ab\mathbf{i} + 2ac\mathbf{j} + 2ad\mathbf{k}$.
    (ii) $[(1 + \mathbf{i} + \mathbf{j} + \mathbf{k})/2]^2 = (-1 + \mathbf{i} + \mathbf{j} + \mathbf{k})/2$.

24. In $W$, show that $\alpha^2 = -1$ if and only if $\alpha$ is in

$$\{\mathbf{i},\ -\mathbf{i},\ \mathbf{j},\ -\mathbf{j},\ \mathbf{k},\ -\mathbf{k}\}.$$

25. Let $f$ be an antiautomorphism of the skew-field $H$ over the real numbers $\mathbf{R}$; that is, let $f$ be a bijection from $H$ to itself with

$$f(a) = a, \qquad f(\alpha + \beta) = f(\alpha) + f(\beta), \qquad f(\alpha\beta) = f(\beta)f(\alpha),$$

for all $a$ in $\mathbf{R}$ and all $\alpha$ and $\beta$ in $H$. Show the following:
    (i) $[f(\alpha)]^2 = -1$ for $\alpha$ in $\{\mathbf{i},\ -\mathbf{i},\ \mathbf{j},\ -\mathbf{j},\ \mathbf{k},\ -\mathbf{k}\}$.
    *(ii) There are at least 24 such antiautomorphisms $f$.

*26. Show that there are at least 24 automorphisms of $H$ over the real numbers $\mathbf{R}$, i.e., bijections $g$ from $H$ to $H$ with

$$g(a) = a, \qquad g(\alpha + \beta) = g(\alpha) + g(\beta), \qquad g(\alpha\beta) = g(\alpha)g(\beta),$$

for all $a$ in $\mathbf{R}$ and all $\alpha$ and $\beta$ in $H$.

27. Prove that $\det(\alpha\beta) = (\det \alpha)(\det \beta)$ in $M(R)$ (and thus prove Theorem 2).

28. Let $K$ consist of all the matrices in $M(\mathbf{R})$ of the form

$$\alpha = \begin{pmatrix} a & b \\ -b & a \end{pmatrix}.$$

Prove that $K$ is isomorphic to the field $\mathbf{C}$ of complex numbers (and thus prove Theorem 4).

29. Let $H$ be the subset of $M(\mathbf{C})$ consisting of the (quaternion) matrices

$$\alpha = \begin{pmatrix} a + bi & c + di \\ -c + di & a - bi \end{pmatrix}.$$

Prove that $H$ is a noncommutative subring in $M(\mathbf{C})$. (This is the part of Theorem 5 that was left to the reader.)

30. Show that $\|\alpha^{-1}\beta^{-1}\alpha\beta\| = 1$ for all quarternions $\alpha$ and $\beta$.
31. Show that there are six elements with order 4 in the quaternion group

$$\{1, -1, \mathbf{i}, -\mathbf{i}, \mathbf{j}, -\mathbf{j}, \mathbf{k}, -\mathbf{k}\}$$

of invertibles of the ring of hamiltonian integers and hence that this group is isomorphic to the group of Example 1 of Section 2.12 (or the group of Table 3.1 in Section 3.4).
32. Let $V$ be the multiplicative group of nonzero quaternions and let $S$ consist of all quaternions $\alpha$ with $\|\alpha\| = 1$. Show that $S$ is a subgroup in $V$.

---

***Sir William Rowan Hamilton*** *(1805–1865)* *Hamilton started as a prodigious linguist; by the age of five he read Latin, Greek, and Hebrew and he knew several oriental languages by the age of ten. He attended Trinity College, Dublin, and at the age of twenty-two was appointed Royal Astronomer of Ireland.*

*Although he made significant contributions in physics and astronomy, his name is usually associated with his attempt to generalize complex numbers to a higher dimensional system, algebraic entities he called quaternions. For ten years he worked on the problem of the definition of multiplication for such a system and finally discovered that such a multiplication would have to be noncommutative. The discovery of the relationships was made while walking along the Royal Canal in Dublin and he inscribed the basic formulas on the stone of Brougham Bridge. Today one can still locate a tablet on the bridge which reads: "Here as he walked by on the 16th of October 1843 Sir William Rowan Hamilton in a flash of genius discovered the fundamental formula for quaternion multiplication* $i^2 = j^2 = k^2 = ijk = -1$ *and cut it in a stone of this bridge." He published his full results in 1853 in his* **Lectures on Quaternions**. *The development of a noncommutative system was a remarkable breakthrough in algebra, a step comparable to the abandonment of the parallel postulate in the development of noneuclidean geometry. Present day applications make great use of a related system with noncommutative multiplication, the 3-dimensional vectors of Josiah Willard Gibbs (1839–1903), the American mathematician-physicist at Yale.*

---

## 4.8 Embedding a Ring in a Field

The elements of the field **Q** of rational numbers are of the form $ab^{-1}$, with $a$ and $b \neq 0$ in the integral domain **Z** of the integers. In this section, we generalize on this relationship between **Z** and **Q**. More specifically, for certain commutative rings $R$ we construct a field $F$ related to $R$ very much as **Q** is to **Z**.

Our first step is to note that if $R$ is a subring in a field $F$, then $R$ has no 0-divisors since $ab = 0$ and $b \neq 0$ in $F$ imply that

$$a = abb^{-1} = 0 \cdot b^{-1} = 0.$$

Hence we restrict ourselves here to rings $R$ with no 0-divisors. Also we ignore the trivial ring $\{0\}$.

***Theorem 1 Field of Fractions***

*Let $R$ be a commutative ring with at least one nonzero element and no 0-divisors. Then there is a field $F$ such that the following hold:*
*(a) $F$ has a subring $R_1$ isomorphic to $R$.*
*(b) Every element of $F$ is the quotient of elements in $R_1$.*

***Proof***

Let $P$ be the subset of the cartesian product $R \times R$ consisting of all the ordered pairs

$$(r, s), \text{ with } r \text{ and } s \text{ in } R \text{ and } s \neq 0.$$

Let $E$ be the relation in $P$ such that

$$(a, b)E(c, d) \text{ if and only if } ad = bc.$$

This relation is easily seen to be reflexive, symmetric, and transitive; thus $E$ is an equivalence relation.

Let $C(a, b)$ denote the subset of all $(r, s)$ in $P$ with

$$(a, b)E(r, s).$$

Then the family $F$ of all these equivalence classes $C(a, b)$ is a partition of $P$. We next define operations in $F$ that make $F$ into a field.

It is natural to try to define addition and multiplication in $F$ using the formulas suggested by the operations in the rational numbers, that is, to let

$$C(a, b) + C(c, d) = C(ad + bc. bd) \tag{A}$$

$$C(a, b)C(c, d) = C(ac, bd). \tag{M}$$

But this requires that the operations in $F$ be well defined by formulas (A) and (M); that is, the use of any other representation for an equivalence class $C(c, d)$ as a term in (A) or a factor in (M) should not cause a change in the sum or product. Hence we present the following preliminary result.

### *Lemma 1 Addition and Multiplication Are Well Defined*

*If* $(c, d)E(c', d')$, *then* $(ad + bc, bd)E(ad' + bc', bd')$ *and* $(ac, bd)E(ac', bd')$.

The proof is left to the reader as Problems 1 and 2 of this section.

*Continuation of the Proof of the Theorem*

We note that $(a, b)$ and $(c, d)$ play the same role in (A) and (M) since $R$ is commutative. Therefore, it is not necessary to check that the replacement of $(a, b)$ by an equivalent $(a', b')$ also would not alter the results in (A) and (M).

It is left to the reader, as Problems 1–7 of this section, to prove that $F$ is a commutative field.

Let $b$ be a fixed nonzero element of $R$ and let $\theta$ be the mapping from $R$ to $F$ with $\theta(r) = C(rb, b)$. It can be shown [see Problem 7(d) of this section] that $\theta$ is an injective ring homomorphism. This implies that the image set $R_1$ of $\theta$ is isomorphic to $R$. It is also clear that the only subfield of $F$ that contains $R_1$ is $F$ itself; hence $F$ is the smallest field containing $R_1$. One easily verifies that every $C(r, s)$ in $F$ is the quotient $C(rb, b) \div C(sb, b)$ of elements in $R_1$.

The notation $C(r, s)$ has helped us to keep in mind that each member of $F$ is an equivalence class; now we would like to replace $C(r, s)$ with symbolism carrying the connotation of a fraction or quotient. If the technique of Theorem 1 is applied to the case in which $R$ is the integral domain $\mathbf{Z}$ of the integers, the resulting field $F$ is isomorphic to the rational numbers $\mathbf{Q}$. Since each element of $\mathbf{Q}$ may be expressed as $r/s$ or $rs^{-1}$, with $r$ and $s$ in $\mathbf{Z}$ and $s \neq 0$, this motivates us to identify the original ring $R$ of Theorem 1 with its isomorphic subring $R_1$ in $F$ and to use $r/s$ or $rs^{-1}$ as a notation for $C(r, s)$. There are other good reasons for this procedure. One is that ring theory in abstract algebra is concerned with the structure of rings and not with the names for their elements. Another reason is that this is the best way to avoid the confusion of giving up an old familiar ring $R$ or having two isomorphic rings to deal with extensively. For example, we prefer not to give up the integral domain $\mathbf{Z}$ of the integers when we "embed" it in the field $\mathbf{Q}$.

## *Problems*

In Problems 1–7 below, $R$ is a commutative ring with at least two elements and no 0-divisors; also $P$, $E$, and $F$ are as described in the text above.

1. Given $(c, d)E(c', d')$, show that $(ad + bc, bd)E(ad' + bc', bd')$.
2. Given $(c, d)E(c', d')$, show that $(ac, bd)E(ac', bd')$.
3. Prove that $F$ is an abelian group under addition.
4. Prove that multiplication is associative in $F$.
5. Prove that multiplication is distributive over addition in $F$.
6. Prove that multiplication is commutative in $F$.
7. Let $b$ be a nonzero element of $R$. Show that:
   (a) $C(0, b)$ is the zero of $F$.
   (b) $C(b, b)$ is the unity of $F$.
   (c) $[C(r, b)]^{-1} = C(b, r)$ for all nonzero $r$ in $R$.
   (d) The mapping $\theta$ from $R$ to $F$ with $\theta(r) = C(rb, b)$ is an injective ring homomorphism.
8. Here let $R$ be the ring consisting of all the even integers and $F$ be its corresponding field of fractions (as defined in Theorem 1). Prove that $F$ is isomorphic to the field $\mathbf{Q}$ of rational numbers.
9. Show that a field $R$ is isomorphic with its field of fractions $F$.

## **4.9 Characterizations of* Z, Q, R, *and* C

*This therefore is Mathematics, she reminds you of the invisible forms of the soul; she gives life to her own discoveries; she awakens the mind and purifies the intellect; she brings light to our intrinsic ideas; she abolishes oblivion and ignorance which are ours by birth....*

*Proclus Diadochus*

In this section we look at the familiar number systems **Z**, **Q**, **R**, and **C** from the point of view of abstract ring and field theory.

The integers **Z** may be characterized as the smallest ordered integral domain since every ordered integral domain $D$ has a subdomain $Z'$ that is isomorphic to **Z**. (See Problem 18 of Section 4.5.)

Similarly, the field **Q** of the rational numbers is the smallest field of characteristic 0, or the smallest ordered field. (See Problems 27 and 30 of Section 4.6.)

We shall characterize the field **R** of real numbers after introducing some new terminology.

***Definition 1 Dedekind Cut***

*Let $D$ be an ordered integral domain; then a* **Dedekind Cut** *in $D$ is a partitioning of $D$ into a pair $\{A, B\}$ of subsets with the two following properties:*
(i) $A \neq \varnothing$, $B \neq \varnothing$, $A \cup B = D$, $A \cap B = \varnothing$.
(ii) *For every $a$ in $A$ and $b$ in $B$, $a < b$.*

***Definition 2 Completely Ordered Field***

*If for every Dedekind Cut $\{A, B\}$ in $D$ there is an element $c$ in $D$ that is either the greatest element in $A$ or the least element in $B$, $D$ is said to be* **completely ordered** *or* **complete**.

It can be shown that the field **Q** of the rational numbers is not complete.

Dedekind (see the biographical note after this section) gave a constructive definition of the real numbers using cuts in the rational numbers; from this definition it follows that the field **R** of the real numbers is completely ordered. Then it can be shown that in fact **R** is essentially the only completely ordered field; that is, every completely ordered field is isomorphic to **R**.

Before giving a characterization of this type for the complex numbers **C**, we indicate some relations between the additive and multiplicative groups of **C** and groups that can be constructed from the real numbers **R**.

The mapping $\theta$ with $\theta(a + b\mathbf{i}) = (a, b)$ is an isomorphism from the additive group of the complex numbers to the direct product of the additive group of the real numbers with itself. (See Problem 13 of Section 3.6.)

Let $\mathbf{R}^+$ denote the multiplicative group of the positive real numbers and let $T$ be the quotient group $\mathbf{R}/[2\pi]$ of the additive group **R** of the real numbers by its cyclic subgroup $[2\pi]$. Since every nonzero complex number has a unique expression of the form

$$r(\cos x + \mathbf{i} \sin x), \quad r > 0, 0 \leq x < 2\pi$$

and multiplication of complex numbers satisfies

$$[r_1(\cos x_1 + \mathbf{i} \sin x_1)][r_2(\cos x_2 + \mathbf{i} \sin x_2)] = r_1 r_2[\cos(x_1 + x_2) + \mathbf{i} \sin(x_1 + x_2)],$$

the mapping $f$ with

$$r(\cos x + \mathbf{i} \sin x) \overset{f}{\mapsto} (r, x)$$

is easily shown to be an isomorphism from the multiplicative group of the nonzero complex numbers to the direct product $\mathbf{R}^+ \times T$. (See Problem 16 of Section 3.6.)

We now define a phrase that allows us to characterize **C** as the smallest field with certain properties.

***Definition 3 Algebraically Closed Field***

*A field $F$ is* **algebraically closed** *if for every positive integer $d$ and any $d + 1$ elements $a_0, a_1, \ldots, a_d$ of $F$ with $a_d \neq 0$ there exists an element $s$ of $F$ such that*

$$a_d s^d + \cdots + a_1 s + a_0 = 0.$$

With terminology from the next chapter, this can be restated as: $F$ is algebraically closed if every polynomial equation of positive degree with coefficients in $F$ has a root in $F$.

The field **R** of the real numbers is not algebraically closed since the second degree polynomial equation

$$x^2 + 1 = 0$$

has no real root. (Since **R** is an ordered field, the square of a real number $r$ can never be negative and hence $r^2 \neq -1$ for all $r$ in **R**.)

In 1799, Gauss proved that the field **C** of the complex numbers is algebraically closed. (See the biographical note after Section 4.2.) This means that every polynomial equation of positive degree with complex coefficients has at least one complex root.

Hence **C** is an algebraically closed extension field (superfield) over the completely ordered field **R**. In fact, **C** may be characterized as the smallest algebraically closed extension field over a complete field.

***Richard Dedekind*** *(1831–1916) Dedekind studied at Göttingen, which between the time of Gauss and the rise of the Nazis in the 1930's was one of the greatest centers of mathematical scholarship in the world. He later taught in Zurich and Braunschweig. It is to Dedekind that we owe the development of the theory of ideals. He was, however, motivated by the work of E. E. Kummer (1810–1893) and his "ideal numbers." Kummer's work was part of an unsuccessful attempt to prove the famous Fermat Conjecture that for all integers $n > 2$ there exist no integer triples $x$, $y$, $z$ such that $x^n + y^n = z^n$. It is one of the many instances in which mathematics developed for very special purposes turns out to be of far wider interest and applicability. The work of Dedekind on rings and ideals was much further developed in the twentieth century at Göttingen by the great woman mathematician Emmy Noether (1882–1935).*

### *Review Problems*

1. In a ring $U$ with unity 1, use both left and right distributivity to expand

$$(a + b)(1 + 1)$$

and then use the results to show that commutativity of addition is a consequence of the other axioms for a ring with unity.

2. Explain why $r^2 \cdot r = r \cdot r^2$ for every $r$ of a ring $R$, even when $R$ is noncommutative.
3. Which elements of $\mathbf{Z}_{42}$ are invertibles?
4. Which elements of $\mathbf{Z}_{43}$ are invertibles?
5. Find all the primes in the finite arithmetic progression

$$101, 107, 113, 119, \ldots, 191, 197.$$

6. Find all the primes in the finite arithmetic progression

$$103, 109, 115, 121, \ldots, 193, 199.$$

7. Find the following in $\mathbf{Z}_{20}$:
   (i) All the elements $x$ such that $x^2 = x$.
   (ii) A subring $S$ with a unity $u$ different from $\bar{1}$.
8. Find all the subrings in each of the following: (a) $\mathbf{Z}_{29}$; (b) $\mathbf{Z}_{45}$.
9. Show that there is no injective homomorphism from $\mathbf{Z}_2$ to $\mathbf{Z}_8$.
10. Tabulate an injective homomorphism from $\mathbf{Z}_2$ to $\mathbf{Z}_{10}$.
11. Find the integer $x$ with both $0 < x < 41$ and $19x \equiv 22 \pmod{41}$.
12. Find the integer $y$ with both $0 < y < 144$ and $35y \equiv 42 \pmod{144}$.
13. (i) In $\mathbf{Z}_{30}$, list each invertible $v$ and its inverse $v^{-1}$ in a table.
    (ii) Show that

$$1 \cdot 7 \cdot 11 \cdot 13 \cdot 17 \cdot 19 \cdot 23 \cdot 29 \equiv 11 \cdot 19 \cdot 29 \equiv -11 \cdot 19 \equiv 1 \pmod{30}.$$

14. Let $\alpha$ be the mapping from $\mathbf{V}_{30}$ to itself with $\alpha(x) = x^3$.
    (i) Tabulate $\alpha$.
    (ii) List the elements with order 3 in $\mathbf{V}_{30}$.
    (iii) Is $\alpha$ a group homomorphism? Explain.
15. Let $\phi$ be the Euler $\phi$-function. Show the following:
    (a) $\phi(3)\phi(8) = \phi(24) = 2\phi(4)\phi(6)$.
    (b) $\phi(1) + \phi(2) + \phi(4) + \phi(5) + \phi(10) + \phi(20) = 20$.
16. (a) Show that $\phi(2)\phi(9) = \phi(18) = (3/2)\phi(3)\phi(6)$.
    (b) Show that $\phi(1) + \phi(2) + \phi(11) + \phi(22) = 22$.
17. Use $\gcd(3, 10) = 1$, $\phi(10) = 4$, and Euler's Theorem to find:
    (i) The last digit (units digit) of $3^{597}$.
    (ii) The last digit of $943^{597}$.
18. Find the last digit of $689^{1287}$.
19. Use Fermat's Theorem to show that $a^n \equiv a \pmod 7$ for all integers $a$ and for $n = 7$, 13, 19, 25, ..., that is, for $n = 6m + 1$ with $m$ a positive integer.
20. Characterize the positive integers $n$ such that $a^n \equiv a \pmod{11}$ for all integers $a$.
21. Show that $a^{19} \equiv a \pmod{798}$ for all integers $a$.
22. Show that $a^{17} \equiv a \pmod{510}$ for all integers $a$.
23. Let $p$ and $n$ be integers with $p$ a prime. Given that $p \mid n^2$, explain why both $p \mid n$ and $p^2 \mid n^2$.
24. Let $p$ and $n$ be integers with $p$ a prime. Given that $p \mid n^3$, explain why $p^3 \mid n^3$.
25. Use Problem 23 to prove that there do not exist integers $x$ and $y$ such that $25x + 10 = y^2$.
26. In each of the following, prove that there are no integers $x$ and $y$ satisfying the equation:
    (a) $9x^2 + 9x + 3 = y^2$.
    (b) $54x + 15 = y^3$.

27. Let $a$ and $m$ be integers. Show the following:
   (a) $(m - a)^2 \equiv a^2 \pmod{m}$.
   (b) $(2m - a)^2 \equiv a^2 \pmod{4m}$.
   (c) If $a^2 \equiv 6 \pmod{10}$, then $a^2 - 6 \equiv 10 \pmod{20}$.

28. (i) Show that for all integers $a$,

$$a^2 \equiv (50 - a)^2 \equiv (50 + a)^2 \equiv (100 - a)^2 \pmod{100}.$$

   (ii) Find all the numbers $b$ in $\{1, 2, 3, \ldots, 99\}$ such that $b^2 \equiv 44 \pmod{100}$.
   (iii) Find all $c$ in $\{1, 2, 3, \ldots, 99\}$ such that $c^2 \equiv 444 \pmod{1000}$.

29. Let $a$ and $m$ be integers. Prove the following:
   (i) If $a \equiv b \pmod{2m}$, then $a^2 \equiv b^2 \pmod{4m}$.
   (ii) If $a \equiv b \pmod{3m}$, then $a^3 \equiv b^3 \pmod{9m}$.
   (iii) $(2a - 1)^8 \equiv 1 \pmod{32}$.
   (iv) $(3a + 1)^{81} \equiv 1 \pmod{243}$.

30. Let $n$ be an integer. Show the following:
   (a) Either $n \equiv 0 \pmod{5}$ or $n^8 + 3n^4 \equiv 4 \pmod{10}$.
   *(b) If $b \not\equiv 1 \pmod{2}$ and $b \not\equiv 0 \pmod{10}$, then $b^4 \equiv 6 \pmod{10}$, $b^{20} \equiv 76 \pmod{100}$, and $b^{200} \equiv 376 \pmod{1000}$.

31. (i) Make the multiplication table for $\mathbf{Z_8}$.
   (ii) Show that $v^2 = \overline{1}$ for every invertible $v$ of $\mathbf{Z_8}$.
   (iii) Given that $\overline{b} \cdot \overline{c} = -\overline{1} = \overline{7}$ in $\mathbf{Z_8}$, show that $\overline{b}$ and $\overline{c}$ are invertibles in $\mathbf{Z_8}$, $\overline{b} \neq \overline{c}$, and $8 | (b + c)$ in $\mathbf{Z}$.

32. Do analogues of Problem 31(iii) in which $\mathbf{Z_8}$ is replaced by $\mathbf{Z_m}$ for each of the following values of $m$:
   (a) 2; (b) 3; (c) 4; (d) 6; (e) 12; (f) 24.

33. Let $k$ be a positive integer and let $d_1, d_2, \ldots, d_s$ be all the positive integral divisors of $24k - 1$. Use Problem 32(f) to explain why $24 | (d_1 + d_2 + \cdots + d_s)$.

34. Let $W_m$ be the set of all $v^2$ with $v$ in $\mathbf{V_m}$.
   (a) Explain why $W_m$ is a subgroup in $\mathbf{V_m}$.
   (b) Use the table in Example 4 of Section 4.2 to show that $W_{17}$ has index 2 in $\mathbf{V_{17}}$.
   (c) Find the smallest $m$ such that $W_m$ has index 4 in $\mathbf{V_m}$.

35. (i) Show that $\mathbf{V_5}$, $\mathbf{V_8}$, $\mathbf{V_{10}}$, and $\mathbf{V_{12}}$ all have the same order.
   (ii) Split the groups of (i) into two classes such that in each class any two groups are isomorphic.

36. Do as in the preceding problem with $\mathbf{V_{15}}$, $\mathbf{V_{16}}$, $\mathbf{V_{20}}$, $\mathbf{V_{24}}$, and $\mathbf{V_{30}}$.

37. Show that the group $\mathbf{V_{40}}$ is solvable by producing an appropriate chain of subgroups and indicating their relevant properties.

38. Do as in the preceding problem with $\mathbf{V_{44}}$ instead of $\mathbf{V_{40}}$.

39. Show that the multiplicative group $\mathbf{V_m}$ of invertibles in $\mathbf{Z_m}$ is cyclic for each of the following values of $m$:
    (a) 2; (b) 3; (c) 4; (d) 5; (e) 6; (f) 7; (g) 9; (h) 10;
    (i) 11; (j) 13; (k) 14; (l) 17.

40. Show that $\mathbf{V_m}$ is not cyclic for the following values of $m$:
    (a) 8; (b) 12; (c) 15; (d) 16; (e) 20.

41. (i) Tabulate the homomorphism $\theta$ from $\mathbf{V_{28}}$ to $\mathbf{V_{28}}$ given by $\theta(x) = x^2$.
    (ii) List the elements of the kernel $K$ of $\theta$.
    (iii) Tabulate the natural map $\alpha$ from $\mathbf{V_{28}}$ to $\mathbf{V_{28}}/K$. (Use letters $K, L, \ldots$ for the cosets of $K$ in $\mathbf{V_{28}}$.)
    (iv) Make a multiplication table for the quotient group $\mathbf{V_{28}}/K$.

42. (i) List the elements of $\mathbf{V_{35}}$ and thus find its order $q$.
    (ii) Find the orders of $\overline{1}, \overline{6}, \overline{29}$, and $\overline{34}$ in $\mathbf{V_{35}}$.
    (iii) Show that $\mathbf{V_{35}}$ has too many elements of order 2 to be cyclic. (How many elements of order 2 are there in a cyclic group of order $q$?)
    (iv) Show that $N = \{\overline{1}, \overline{6}, \overline{29}, \overline{34}\}$ is a normal subgroup in $\mathbf{V_{35}}$.
    (v) Show that the quotient group $\mathbf{V_{35}}/N$ is cyclic.
    (vi) Tabulate the natural map $\alpha$ from $\mathbf{V_{35}}$ to $\mathbf{V_{35}}/N$. Use symbols $N, B, C, D, \ldots$ for the cosets of $N$ in $\mathbf{V_{35}}$.)
    (vii) Show that $\mathbf{V_{35}}$ is solvable.

43. Tabulate one of the isomorphisms form $\mathbf{V_{14}}$ to $\mathbf{V_{18}}$.

44. Tabulate one of the isomorphisms from $\mathbf{V_9}$ to $\mathbf{V_{18}}$.

45. (a) Show that $\mathbf{V_8}$ is isomorphic to the direct product of two groups of order 2.
    (b) Is the same true of $\mathbf{V_{12}}$? Explain.
    (c) Is the same true of $\mathbf{V_5}$ or of $\mathbf{V_{10}}$? Explain.

46. Show that $\mathbf{V_{16}}$ is isomorphic to the direct product of two cyclic groups, with one of order 4 and the other of order 2.

47. Explain why the characteristic of an integral domain cannot be 4, 6, 8, 9, 10, 12, 14, or 15.

48. Use the fact that every abelian group of order 38 is cyclic to prove that there is no integral domain with 38 elements.

49. Let $r$, $s$, and $t$ be in an integral domain $D$. Show that

$$x^3 - (r + s + t)x^2 + (st + rt + rs)x - rst = 0$$

with $x \in D$ if and only if $x \in \{r, s, t\}$.

50. Let $D$ be an integral domain with characteristic 5. Show that in $D$:
    (i) $(a + b)^5 = a^5 + b^5$.
    (ii) $(a + b)^{25} = a^{25} + b^{25}$.

51. Show that $x^2 + 1 \geq 2 \quad x$ for all $x$ in an ordered integral domain $D$.

52. Show that $x^4 + 6 \cdot x^2y^2 + y^4 \geq 4 \cdot x^3y + 4 \cdot xy^3$ for all $x$ and $y$ in an ordered integral domain $D$.

53. Let $u$ and $v$ be in a ring $U$ with unity and let $v$ be an invertible. Show that there is a unique $x$ in $U$ such that $vx = u$.

54. Let $a$, $b$, $c$, $d$, $h$, and $k$ be in a commutative ring $U$ with unity and let $ad - bc$ be an invertible $v$. Show that there is a unique solution in $U$ for the following system of two simultaneous equations:

$$\begin{cases} ax + by = h \\ cx + dy = k. \end{cases}$$

55. For which prime $p$ is $\mathbf{Z_p}$ isomorphic to a subfield in a field $F$ with 128 elements?

56. Let $F$ be a field with 625 elements. How many elements are there in the intersection of all the subfields in $F$?

57. Show that $x^2 + x^{-2} \geq 1 + 1$ for all nonzero $x$ in an ordered field $F$.

58. Use the fact that every positive number $y$ in the real numbers $\mathbf{R}$ has square roots in $\mathbf{R}$ to show that, in $\mathbf{R}$, $y > 0$ implies that

$$y + y^{-1} \geq 2.$$

59. Find the remainder when each of the following factorials is divided by the prime 139.
(a) 138!; (b) 137!; *(c) 136!; *(d) 135!.

60. Use Wilson's Theorem to show the following:
(a) $(50!)^2 \equiv -1 \pmod{101}$.
(b) $(51!)^2 \equiv 1 \pmod{103}$.

61. Let $a$ and $b$ be in a commutative ring $R$. Given that $ab$ is neither 0 nor a 0-divisor, prove the following:
(a) $a \neq 0$ and $b \neq 0$.
(b) Neither $a$ nor $b$ is a 0-divisor.

62. Describe two ideals in the field $\mathbf{Q}$ of the rational numbers and explain why $\mathbf{Q}$ has only two ideals.

63. How many ideals are there in $\mathbf{Z_{39}}$? Explain.

64. How many ideals are there in $\mathbf{Z_{101}}$? Explain.

65. How many ring homomorphisms are there from $\mathbf{Z_{17}}$ to $\mathbf{Z_{17}}$? Explain.

66. Let $p$ be a positive prime in $\mathbf{Z}$. How many ring homomorphisms are there from $\mathbf{Z_p}$ to $\mathbf{Z_p}$? Explain.

67. Describe a homomorphism $\theta$ from $\mathbf{Z_{54}}$ to some ring $R$ such that the kernel of $\theta$ is $(\bar{3})$.

68. Let $J$ be the intersection of a collection $\Gamma$ of subrings in a ring $R$. Show the following:
    (a) $J$ is a subring in $R$.
    (b) If each subring of $\Gamma$ is an ideal, so is $J$.
    (c) If each subring of $\Gamma$ is an integral domain, so is $J$.
    (d) If each subring of $\Gamma$ is a field, so is $J$.

69. Let $R$ be a commutative ring and let $a \in R$. Explain the relationships among the following:
    (i) The principal ideal $(a)$ in $R$.
    (ii) The subring generated by $\{a\}$ in $R$, that is, the intersection $D$ of all the subrings $S$ in $R$ with $a \in S$.
    (iii) The smallest subring $J$ in $R$ with $a \in J$.

70. Let $a$ and $b$ be integers, not both zero. Let $I$ be the smallest ideal, containing both $a$ and $b$, in the ring of the integers $\mathbf{Z}$. ($I$ is the intersection of all the ideals containing $a$ and $b$.) Explain why $\gcd(a, b)$ is the unique positive generator of $I$.

## *Supplementary and Challenging Problems*

1. Let $R$ be a ring whose additive group is cyclic. Prove that $R$ is commutative (under multiplication).

2. Let $R$ be a ring with $p$ elements, where $p$ is prime. Prove that either $R$ and $\mathbf{Z_p}$ are isomorphic rings or $rs = 0$ for all $r$ and $s$ in $R$.

3. In a ring $U$ with unity $1 \neq 0$, let $ab = 1$ and let $a$ not be a 0-divisor. Prove that $ba = 1$.

4. Let $a$ be an element that is neither 0 nor a 0-divisor in a finite ring $R$. Show that $R$ has a unity and that $a$ is an invertible.

5. Let $\mathbf{V_m}$ be the multiplicative group of invertibles in $\mathbf{Z_m}$. Let $m$ be odd and $\mathbf{V_m}$ be a cyclic group $[\bar{a}]$. Prove that $\mathbf{V_{2m}}$ is cyclic and that either $\bar{a}$ or $\overline{a+m}$ is a generator of $\mathbf{V_{2m}}$. Given that $\mathbf{V_{11}} = [\bar{2}]$, find a generator for $\mathbf{V_{22}}$.

6. Given that $a \equiv 7 \pmod 8$, prove that there do not exist integers $x$, $y$, and $z$ such that $x^2 + y^2 + z^2 = a$.

7. Use the fact that no integer $x$ satisfies $x^2 \equiv 3 \pmod 4$ to show that no integer $y$ satisfies $y^2 \equiv 12 \pmod{16}$ and no integer $z$ satisfies $z^2 \equiv 4444 \pmod{10^4}$.

8. Let $a$, $b$, and $m$ be in $\mathbf{Z}$ with $m > 0$. Prove that $a \equiv b \pmod{7^m}$ implies that $a^7 \equiv b^7 \pmod{7^{m+1}}$.

9. Let $a$, $b$, and $c$ be integers with

$$bc \equiv 1 \pmod{a}, \qquad ac \equiv 1 \pmod{b}, \qquad ab \equiv 1 \pmod{c}.$$

Show that $bc + ac + ab \equiv 1 \pmod{abc}$.

10. Find positive integers $x$ and $y$, with $x$ as small as possible, such that $x^2 + (x+1)^2 + (x+2)^2 + \cdots + (x+10)^2 = y^2$.

11. Let $p$ be an odd positive prime in $\mathbf{Z}$ and let $r = (p-1)/2$. Show the following:
(a) In $\mathbf{Z}_\mathbf{p}$, $\overline{2r+1} = 0$, $\overline{-2r} = \overline{1}$, and $\overline{(p-2)}^{-1} = \overline{(-2)}^{-1} = \overline{r}$.
(b) In $\mathbf{V}_\mathbf{p}$, $\overline{1} \cdot \overline{2} \cdot \overline{3} \cdots \overline{p-3} = \overline{r}$.
(c) $(p-3)! - r$ is a multiple of $p$ in $\mathbf{Z}$.
(d) $(p-3)! \equiv r \pmod{p}$.

12. Let $p$ be a positive prime in $\mathbf{Z}$ of the form $p = 6s + 1$, with $s$ an integer. Show the following:
(a) In $\mathbf{Z}_\mathbf{p}$, $\overline{6s+1} = 0$, $\overline{-6s} = \overline{1}$, $\overline{(p-2)}^{-1} = \overline{(-2)}^{-1} = \overline{3s}$, and $\overline{(p-3)}^{-1} = \overline{(-3)}^{-1} = \overline{2s}$.
(b) In $\mathbf{V}_\mathbf{p}$, $\overline{1} \cdot \overline{2} \cdot \overline{3} \cdots \overline{p-4} = \overline{6s^2}$.
(c) $(p-4)! - 6s^2$ is a multiple of $p$ in $\mathbf{Z}$.
(d) $(p-4)! \equiv 6s^2 \pmod{p}$.

13. Let $p$ be a positive prime in $\mathbf{Z}$ of the form $p = 6t - 1$, with $t$ in $\mathbf{Z}$. Show the following:
(a) In $\mathbf{Z}_\mathbf{p}$, $\overline{(p-2)}^{-1} = \overline{-3t}$ and $\overline{(p-3)}^{-1} = \overline{-2t}$.
(b) In $\mathbf{V}_\mathbf{p}$, $\overline{1} \cdot \overline{2} \cdot \overline{3} \cdots \overline{p-4} = \overline{6t^2}$.
(c) $(p-4)! - 6t^2$ is a multiple of $p$ in $\mathbf{Z}$.
(d) $(p-4)! \equiv 6t^2 \pmod{p}$.

14. Let $p$ be a prime in the integers with $p \geq 7$. Prove that:
(a) $(p^2 - p) | [(p-2)! + p - 1]$.
(b) $(p^2 + p) | [(p-2)! - p - 1]$.

15. Prove that $[n!/(n^2 + 3n + 2)]$ is an even integer for all positive integers $n$. Here $[x]$ denotes the greatest integer in $x$, that is, the integer such that $[x] \leq x < [x] + 1$.

16. Let $S$ be the subring in the real numbers $\mathbf{R}$ consisting of all $a + b\sqrt{3}$, with $a$ and $b$ integers. Prove that $S$ is not an ideal in $\mathbf{R}$.

17. An ideal $I$ in a commutative ring $R$ is said to be a ***prime ideal*** if $a \in R$, $b \in R$, and $ab \in I$ together imply that either $a \in I$ or $b \in I$. Prove that an ideal $I$ in $R$ is prime if and only if the quotient ring $R/I$ has no 0-divisors.

18. Let $I$ be an ideal in $R$. If $a - b \in I$, we say that ***a is congruent to b modulo I*** and write $a \equiv b \bmod I$.
(a) Prove that the relation in $R$ of congruence modulo $I$ is an equivalence relation.
(b) Prove that congruence modulo $I$ is preserved under addition, subtraction, and multiplication, that is, $a \equiv b \bmod I$ and $c \equiv d \bmod I$ imply that $a \pm c \equiv b \pm d \bmod I$ and $ac \equiv bd \bmod I$.

(c) Prove that $a \equiv b \bmod I$ implies $a^n \equiv b^n \bmod I$ for all positive integers $n$.

19. Describe an infinite field with only a finite number of subfields.

20. Describe an infinite collection of subfields in the field $\mathbf{R}$ of real numbers.

21. Can an infinite integral domain $D$ that is not a field have only a finite number of subdomains? Explain.

22. Let $\mathbf{A}(R)$ be the set of automorphisms of a ring $R$. Prove that $\mathbf{A}(R)$ is a subgroup in the group $\mathbf{B}(R)$ of bijections from $R$ to $R$.

23. Define "characteristic of a group" in such a way that the previously defined characteristic of an integral domain $D$ is always equal to the characteristic of the additive group of $D$. What is the characteristic of a Klein 4-group under this definition?

24. Let $R$ be a commutative ring and let $I$ consist of all the elements $r$ of $R$ such that $r^n = 0$ for some integer $n$. Prove that:
    (i) $I$ is an ideal in $R$.
    (ii) In the quotient ring $R/I$, if $\bar{r}^n = \bar{0}$ with $n$ an integer, then $\bar{r} = \bar{0}$.

25. Let $V$ and $\mathbf{R}^+$ be the multiplicative groups of nonzero quaternions and positive real numbers, respectively. Let $S$ be the subgroup of all $\alpha$ in $V$ with $\|\alpha\| = 1$. Show that $V$ is isomorphic to the direct product $S \times \mathbf{R}^+$.

26. Let $W$ be the ring of hamiltonian integers $a + b\mathbf{i} + c\mathbf{j} + d\mathbf{k}$ (with $a, b, c, d \in \mathbf{Z}$). Let $\lambda = (1 + \mathbf{i} + \mathbf{j} + \mathbf{k})/2$ and let $W'$ consist of all $\alpha$ such that either $\alpha \in W$ or $\alpha = \beta + \lambda$ with $\beta \in W$. Show that:
    (a) $W'$ is a subring in the skew-field $H$ of quaternions.
    (b) $\|\alpha\|$ is an integer for all $\alpha$ in $W'$.

27. Show that every quaternion $\gamma$ is a sum $\gamma_1 + \gamma_2$ with $\gamma_1$ in the ring $W'$ of Problem 26 and $\|\gamma_2\| < 1$.

28. Show that every complex number $z$ is a sum $z_1 + z_2$, where $z_1$ is a gaussian integer $m + n\mathbf{i}$, with $m$ and $n$ integers, and $|z_2| \le 1/2$.

29. Let $\alpha$ and $\beta$ be in the ring $W'$ of Problem 26 and let $\beta \neq 0$. Show the following:
    (i) $\alpha = \gamma\beta$, with $\gamma$ a quaternion.
    (ii) $\alpha = \gamma_1\beta + \gamma_2\beta$, with $\gamma_1$ in $W'$ and $\|\gamma_2\| < 1$.
    (iii) $\alpha = \gamma_1\beta + \rho$, with $\gamma_1$ and $\rho$ in $W'$ and $\|\rho\| < \|\beta\|$.

30. Let $z_1$ and $z_2$ be gaussian integers with $z_2 \neq 0$. Show that there exist gaussian integers $z_3$ and $z_4$ with $z_1 = z_2 z_3 + z_4$ and $|z_4| < |z_2|$.

31. Let $L$ be a subring in the $W'$ of Problem 26 with the property that $\alpha\beta$ is in $L$ for every $\alpha$ in $W'$ and $\beta$ in $L$. (This makes $L$ a ***left ideal*** in $W'$.)
    (a) Explain why $L$ is not necessarily an ideal in $W'$.
    *(b) Show that $L$ consists of all products $\alpha\gamma$, where $\alpha$ may be any element of $W'$ and $\gamma$ is some fixed element of $L$ (that is, every left ideal in $W'$ is a ***principal left ideal***).

32. Prove that, in the ring of gaussian integers, every ideal $I$ is a principal ideal. (See Problem 31.)

33. Let $a$, $d$, and $n$ be positive integers. Prove that the infinite arithmetic progression

$$a, \quad a + d, \quad a + 2d, \quad a + 3d, \ldots$$

either has no exact $n$th powers $r^n$, of integers $r$, or it has infinitely many $n$th powers.

# 5

## *Polynomials*

Polynomial functions usually are easier to deal with than other functions. Also, there are many types of problems (for example, integration of a continuous function) in which there is an exact solution when the function involved is a polynomial and otherwise one may have to approximate the solution by approximating the given function with a polynomial.

The theory of rings is a generalization growing out of the study of the ring **Z** of the integers and of the rings of polynomials with coefficients in the number systems **Z**, **Q**, **R**, and **C**.

In this chapter, we restrict ourselves to commutative rings and always use $R$ to denote a commutative ring; $U$, a commutative ring with unity; $D$, an integral domain; and $F$, a field.

### *5.1 Polynomial Extensions of Rings*

The ring **Z** of the integers is a proper subring in the complex numbers **C**. One element of **C** that is not in **Z** is the real number $\pi = 3.14159\ldots$, the ratio of the circumference of a circle to its diameter. We ask the question: "If we adjoin $\pi$ to **Z**, what other numbers must we also include in order to have a ring?" This question is rephrased and answered in the following:

***Example 1***

What is the smallest subring in **C** that contains $\pi$ and all the integers?

***Solution***

As a start, we let $S$ designate a subring in **C** such that $S$ contains $\pi$ and all the integers. (The ring **R** of the real numbers and **C** itself are such subrings.) Since $S$ is closed under multiplication, it must contain the powers $\pi, \pi^2, \pi^3, \ldots$ of $\pi$ and also must contain all the numbers of the form $a\pi^j$, where $j$ is a nonnegative integer and $a$ is any integer. Closure of $S$ under addition then implies that $S$ contains all the numbers of the form

$$\alpha = a_d\pi^d + a_{d-1}\pi^{d-1} + \cdots + a_1\pi + a_0\,;\ a_j \text{ in } \mathbf{Z}. \tag{1}$$

An $\alpha$ of the form (1) is called a polynomial in $\pi$ with integer coefficients. The set of all such polynomials is denoted by $\mathbf{Z}[\pi]$.

We see that $\pi$ is in $\mathbf{Z}[\pi]$ by letting $d = 1$, $a_1 = 1$, and $a_0 = 0$ in (1). If we let $d = 0$ and let $a_0$ be any integer $n$, the $\alpha$ of (1) becomes $n$. Hence $\mathbf{Z}[\pi]$ contains $\pi$ and all the integers. It can also be shown that $\mathbf{Z}[\pi]$ is closed under subtraction and multiplication and hence is a subring in $\mathbf{C}$. We noted above that each $\alpha$ of $\mathbf{Z}[\pi]$ is in every subring $S$ in $\mathbf{C}$ that contains $\pi$ and all the integers. Therefore, $\mathbf{Z}[\pi]$ is the desired smallest of these subrings $S$. (See the definition in Section 3.5 of the smallest set in a collection.)

***Example 2***

What is the smallest subring in $\mathbf{C}$ that contains the complex number $\mathbf{i}$ and all the integers?

***Solution***

In a manner similar to that of Example 1, we see that the subring sought here consists of all the complex numbers of the form

$$\beta = b_d \mathbf{i}^d + b_{d-1}\mathbf{i}^{d-1} + \cdots + b_1\mathbf{i} + b_0 ;\ b_j \text{ in } \mathbf{Z}. \tag{2}$$

Such a number $\beta$ is called a polynomial in $\mathbf{i}$ with integer coefficients and the ring of all these polynomials is denoted by $\mathbf{Z}[\mathbf{i}]$.

Since $\mathbf{i}^2 = -1$, $\mathbf{i}^3 = -\mathbf{i}$, $\mathbf{i}^4 = 1$, $\mathbf{i}^5 = \mathbf{i}, \ldots$ the polynomial $\beta$ of (2) can be rewritten in the simpler form

$$\beta = a_0 + a_1\mathbf{i};\quad a_0 \text{ and } a_1 \text{ in } \mathbf{Z}. \tag{3}$$

Hence $\mathbf{Z}[\mathbf{i}]$ is the ring of gaussian integers discussed in Problem 21 of Section 4.1.

The simplification from a form such as (1) or (2), in which the number of terms is arbitrary, to a form such as (3), in which the number of terms is limited, is not possible for $\mathbf{Z}[\pi]$. The following terminology will help us discuss this difference between the nature of $\mathbf{Z}[\pi]$ and that of $\mathbf{Z}[\mathbf{i}]$.

***Definition 1 Algebraic over R***

*Let $R$ be a subring in a commutative ring $T$. An element $t$ of $T$ is said to be* **algebraic over R** *if there is a nonnegative integer $d$ and elements $a_d, a_{d-1}, \ldots, a_0$ of $R$ such that $a_d \neq 0$ and, in $T$,*

$$a_d t^d + a_{d-1}t^{d-1} + \cdots + a_1 t + a_0 = 0.$$

***Definition 2 Transcendental over R***

*Let $R$ be a subring in a commutative ring $T$. An element $t$ of $T$ is said to be* **transcendental over R** *if an equation*

$$a_d t^d + a_{d-1}t^{d-1} + \cdots + a_1 t + a_0 = 0,$$

*with the $a_j$ in $R$, implies that all the coefficients $a_d, \ldots, a_0$ are zero.*

Let $c$ be a complex number. Then $c$ is said to be ***algebraic***, or an ***algebraic number***, if $c$ is algebraic over the integers **Z**. Similarly, $c$ is ***transcendental***, or a ***transcendental number***, if $c$ is transcendental over **Z**.

Clearly every complex number is either algebraic or transcendental. One way to show that **i** is algebraic is to note that

$$a_2\mathbf{i}^2 + a_1\mathbf{i} + a_0 = 0$$

when we let $a_2 = 1$, $a_1 = 0$, and $a_0 = 1$ and that some of these integer coefficients are not zero.

In 1882 F. Lindemann established the far from trivial result that $\pi$ is transcendental. (See the biographical note at the close of this section.) We now show that this implies that a given number $\alpha$ in $\mathbf{Z}[\pi]$ has a unique expression as a polynomial of the form (1) above. To do this, we assume that $\alpha$ is expressible both as

$$\alpha = a_d\pi^d + a_{d-1}\pi^{d-1} + \cdots + a_1\pi + a_0$$

and as

$$\alpha = b_e\pi^e + b_{e-1}\pi^{e-1} + \cdots + b_1\pi + b_0$$

with the coefficients $a_d, \ldots, a_0, b_e, \ldots, b_0$ all integers. It is convenient to make $e = d$; we do this by inserting terms with zero coefficients if necessary. Then

$$0 = \alpha - \alpha = (a_d - b_d)\pi^d + \cdots + (a_1 - b_1)\pi + (a_0 - b_0). \tag{4}$$

Since $\pi$ is transcendental, it follows from (4) that $a_i - b_i = 0$ and hence $a_i = b_i$ for $0 \leq i \leq d$. This means that the expression (1) for a given $\alpha$ in $\mathbf{Z}[\pi]$ is unique.

The polynomial form (2) for a number in $\mathbf{Z}[\mathbf{i}]$ is not unique; for example,

$$2\mathbf{i}^3 + 3\mathbf{i}^2 - 8\mathbf{i} + 1 = -10\mathbf{i} - 2.$$

However, a number in $\mathbf{Z}[\mathbf{i}]$ does have a unique expression of the form $m + n\mathbf{i}$, with $m$ and $n$ integers.

Next we generalize on these two examples. Let $R$ be a subring in a commutative ring $T$ and let $t$ be an element of $T$. The notation $R[t]$ may be used to designate the smallest subring $S$ in $T$ such that $t$ and all the elements of $R$ are in $S$; one easily sees, as in the examples above, that $R[t]$ consists of all the polynomials

$$\alpha = a_d t^d + a_{d-1}t^{d-1} + \cdots + a_0 \ (a_j \text{ in } R)$$

in $t$ with coefficients in $R$.

If both $t_1$ and $t_2$ are transcendental over $R$, it can be shown that there is an isomorphism $f$ from $R[t_1]$ to $R[t_2]$ with $f(a) = a$ for all $a$ in $R$ and $f(t_1) = t_2$. That is, the structure of $R[t]$ does not depend on the choice of $t$, as long as $t$ is transcendental over $R$. Yet another way to say this is that $t$ is a "place holder" in the expression for a polynomial.

This suggests that we might be able to construct a substitute for the ring of polynomials $R[t]$ without the hypothesis that there is an element $t$ transcendental over $R$. Such a construction follows for the case in which $R$ is a commutative ring $U$ with unity. (Theorem 1 of Section 4.8 tells us that a commutative ring $R$, with at least two elements and no 0-divisors, can be embedded in a field $F$; and every field is a commutative ring with unity.)

***Theorem 1 Extension of a Commutative Ring***

*Let $U$ be a commutative ring with unity. Then there exists a ring $S^*$ in which there is a subring $U^*$ isomorphic to $U$ and an element $x^*$ that is transcendental over $U^*$.*

***Proof***

Let $S^*$ consist of the infinite sequences

$$\alpha^* = (a_0, a_1, a_2, \ldots), \quad a_j \text{ in } U,$$

in which only a finite number of the $a_j$ may be nonzero. That is, for each $\alpha^*$ in $S^*$ there is a fixed $d$ such that $a_j = 0$ for $j > d$.

Using the operations in $U[t]$ as a guide, we define addition and multiplication of sequences

$$\alpha^* = (a_0, a_1, \ldots), \qquad \beta^* = (b_0, b_1, \ldots)$$

in $S^*$ by

$$\alpha^* + \beta^* = (a_0 + b_0, a_1 + b_1, a_2 + b_2, \ldots), \tag{A}$$

$$\alpha^*\beta^* = (c_0, c_1, c_2, \ldots) \text{ with } c_k = \sum_{i+j=k} a_i b_j. \tag{M}$$

(This "sigma" notation states that $c_k$ is the sum of all the products $a_i b_j$ in which $i + j = k$.) We leave it to the reader to show that the operations (A) and (M) convert $S^*$ into a commutative ring.

Let $U^*$ be the subset of $S^*$ consisting of all the sequences $(a_0, 0, 0, \ldots)$, that is, of all

$$(a_0, a_1, a_2, \ldots), \quad a_j = 0 \text{ for } j > 0.$$

It is easily shown that $U^*$ is a subring in $S^*$, $U$ and $U^*$ are isomorphic, $x = (0, 1, 0, 0, \ldots)$ is transcendental over $U^*$, and $S^* = U^*[x]$.

The construction outlined in Theorem 1 provides partial support for the following statement:

***Assumption*** ***Existence of Transcendental Elements***

*Let $R$ be a commutative ring. Then there exists a commutative ring $T$ in which $R$ is a subring and an element $x$ in $T$ that is transcendental over $R$.*

***Definition 3*** ***Indeterminate over $R$***

*An element $x$ that is transcendental over $R$ is also called an* **indeterminate** *over $R$.*

Theorem 1 supports the above assumption in the following sense: If we have a commutative ring $U$ with unity and desire an indeterminate $x$ over $U$, we can replace $U$ with its isomorphic $U^*$ and then drop the awkward asterisks as a means of simplifying the notation. Some mathematicians refer to this process as justifiable "abuse of the language."

## Problems

1. Show that $\sqrt{2}$ is algebraic.
2. Show that $\sqrt[3]{5}$ is algebraic.
3. Let $\alpha$ be the complex number $2 + 3\mathbf{i}$. Find integers $a_1$ and $a_0$ such that $\alpha^2 + a_1\alpha + a_0 = 0$ and thus show that $\alpha$ is algebraic.
4. Show that the complex number $\beta = 7 - 5\mathbf{i}$ is algebraic.
5. Use Lindemann's result that $\pi$ is transcendental to show that $\pi^2$ is transcendental.
6. Show that $\pi + 1$ is transcendental.
7. Show that each of the following is algebraic:
   (a) $\sqrt{2} + \sqrt{3}$.

(b) $\sqrt[3]{5} + \sqrt[3]{25}$.
*(c) $\sqrt{3} + \sqrt[3]{2}$.

8. Let **Q** denote the rational numbers. Explain why every element of $\mathbf{Q}[\sqrt[3]{7}]$ can be written as

$$a_0 + a_1\sqrt[3]{7} + a_2(\sqrt[3]{7})^2,$$

with the $a_j$ rational.

9. Let **R** and **C** denote the real and the complex numbers, respectively. Let $a + b\mathbf{i}$ be a fixed complex number with $b \neq 0$. Explain why

$$\mathbf{R}[a + b\mathbf{i}] = \mathbf{R}[\mathbf{i}] = \mathbf{C}.$$

10. Let $m$ be a fixed integer. Explain why

$$\mathbf{Z}[m + \mathbf{i}] = \mathbf{Z}[\mathbf{i}].$$

11. What is the smallest subring $S$ in **C** such that 1 and $\sqrt{2}$ are in $S$?
12. What is the smallest subring $T$ of **C** such that $\sqrt{2}$ is in $T$?
13. Let **Q** be the field of rational numbers.
    (a) In $\mathbf{Q}[\sqrt{2}]$, express the reciprocal of $3 - \sqrt{2}$ in the form $h + k\sqrt{2}$, with $h$ and $k$ in **Q**.
    (b) Explain why $\mathbf{Q}[\sqrt{2}]$ is the field $T$ of Problem 2, Section 4.6.
14. (a) In $\mathbf{Q}[\sqrt{5}]$, express the reciprocal of $4 + 7\sqrt{5}$ in the form $h + k\sqrt{5}$, with $h$ and $k$ in **Q**.
    (b) Show that $\mathbf{Q}[\sqrt{5}]$ is a subfield in the real numbers **R**.

*15. Prove that $\mathbf{Z}[\sqrt{2}]$ and $\mathbf{Z}[\sqrt{3}]$ are not isomorphic rings.

---

***C. L. F. Lindemann*** *(1852–1939)* *Lindemann was a student of Karl Weierstrass and became a professor in Munich. In 1882 he published a paper in the* **Mathematische Annalen** *which showed that $e^{ix} + 1 = 0$ cannot be satisfied by an algebraic x and since Euler had proved that it was satisfied by π, π must not be algebraic. That is, π must be a transcendental number. This result settled the classical problem of whether the "circle can be squared," that is, of whether one can construct with straightedge and compass a square with the same area as a given circle. (Constructions of this type are discussed* *in Chapter 6.) It had been shown earlier by Johann Heinrich Lambert in 1770 and Adrien-Marie Legendre in 1794 that $\pi$ and $\pi^2$ were irrational, but this did not settle the classical construction problem.*

*The Lindemann result on $\pi$ followed the proof of 1873 by Charles Hermite (1822–1901) that e is transcendental. This appeared in the* **Comptes Rendus de l'Académie des Sciences, Paris.**

---

## 5.2 *Polynomials over a Commutative Ring*

Next we give a formal definition that is consistent with the discussion in the previous section.

### *Definition 1 Polynomial over R*

*Let x be an indeterminate over a commutative ring R. An expression of the form*

$$\alpha = a_d x^d + a_{d-1}x^{d-1} + \cdots + a_1 x + a_0, \tag{P}$$

*with the coefficients $a_d, \ldots, a_0$ in R, is a* **polynomial in $x$ over $R$**.

### *Definition 2 Degree of a Polynomial*

*If $a_d \neq 0$ in the expression (P), the polynomial $\alpha$ has* **degree** *d and we write* $\deg \alpha = d$.

In (P), if a coefficient $a_j = 0$, the term $a_j x^j$ may be omitted; if $a_j = 1$, the term $a_j x^j$ may be written as $x^j$. Each element $r$ of $R$ is in the form (P) with $d = 0$ and $a_0 = r$. Hence every nonzero $r$ of $R$ is a polynomial of degree 0 over $R$. Moreover, $\deg \alpha = 0$ if and only if $\alpha = r$ with $r$ a nonzero element of $R$.

The 0 of $R$ is also a polynomial over $R$; it is the only polynomial without a degree.

### *Notation R[x]*

*The set of all polynomials in x over R is denoted by R[x]. (One may read "R[x]" as "R bracket x.")*

In characterizing addition and multiplication in $R[x]$, it is convenient to express any $\alpha$ and $\beta$ in $R[x]$ as

$$\alpha = a_d x^d + a_{d-1}x^{d-1} + \cdots + a_1 x + a_0, \quad a_j \text{ in } R,$$
$$\beta = b_d x^d + b_{d-1}x^{d-1} + \cdots + b_1 x + b_0, \quad b_j \text{ in } R.$$

(We do not assume that $\deg \alpha = \deg \beta$ but merely insert terms with zero coefficients when that is needed to achieve this uniformity of appearance.)

Then the operations of addition and multiplication in $R[x]$ are given by the following

$$\alpha + \beta = c_d x^d + c_{d-1}x^{d-1} + \cdots + c_1 x + c_0, \tag{A}$$
$$\alpha\beta = e_{2d}x^{2d} + e_{2d-1}x^{2d-1} + \cdots + e_0, \tag{M}$$

where $c_j = a_j + b_j$ (for $0 \leq j \leq d$) and $e_k$ is the sum of all the products $a_i b_j$ in which $i + j = k$ (for $0 \leq k \leq 2d$).

We omit the details of showing that with these operations $R[x]$ is a commutative ring with $R$ as a subring. It is clear that the 0 of $R$ is also the zero of $R[x]$ and that the unity of $R$, if $R$ has a unity, is the unity of $R[x]$.

Since $x$ is transcendental over $R$, a polynomial

$$\alpha = a_d x^d + \cdots + a_1 x + a_0, \quad a_j \text{ in } R, \tag{P}$$

is equal to the zero polynomial if and only if all the coefficients $a_j$ are zero. As is true for polynomials in $\pi$ over $\mathbf{Z}$ (see Section 5.1), it then follows that the expression (P) for a fixed $\alpha$ in $R[x]$ is unique, except for the possible omission of terms with zeros as coefficients.

If $R$ has at least one nonzero element $r$ (that is, if $R$ has at least two elements), there are an infinite number of polynomials in $R[x]$ since, for example,

$$r, rx, rx^2, \ldots$$

are distinct polynomials in $R[x]$.

We next introduce some notation that helps us to state properties of multiplication in $R[x]$.

### *Definition 3 Leading Coefficient*

*The* **leading coefficient** $L(\alpha)$ *of a nonzero polynomial* $\alpha$ *in* $R[x]$ *is the coefficient* $a_d$ *in the expression*

$$\alpha = a_d x^d + \cdots + a_0, \quad a_j \in R, \quad a_d \neq 0;$$

*if* $\alpha$ *is the zero polynomial, its leading coefficient* $L(\alpha)$ *is the zero element of* $R$.

For example, if $\alpha$ is the polynomial $5x^3 - 7x + 8$ in $\mathbf{Q}[x]$, then $L(\alpha) = 5$.

***Definition 4 Monic Polynomial***

*If $R$ has a unity 1 and $L(\alpha) = 1$, $\alpha$ is said to be a* **monic polynomial**.

When studying polynomials in elementary algebra, one usually has the coefficients in the integral domain $\mathbf{Z}$ or one of the fields $\mathbf{Q}$, $\mathbf{R}$, and $\mathbf{C}$. Such experience leads us to expect that $\deg(\alpha\beta) = \deg \alpha + \deg \beta$ for all nonzero $\alpha$ and $\beta$, that $\alpha\beta = 0$ if and only if $\alpha$ or $\beta$ is 0, and that there is a form of unique factorization. Below, we shall see that these properties do go over to polynomials with coefficients in any field. However, they need not apply to polynomials with coefficients in a ring with 0-divisors, as we see in the following:

***Example 1***

In $Z_6[x]$, we have:

(a) $(\bar{3}x + \bar{2})(\bar{2}x + \bar{5}) = \bar{6}x^2 + (\overline{15} + \bar{4})x + \overline{10}$

$= \overline{19}x + \overline{10} = x + \bar{4},$

(b) $(\bar{3}x + \bar{3})(\bar{4}x^2 + \bar{2}) = \overline{12}x^3 + \overline{12}x^2 + \bar{6}x + \bar{6} = \bar{0},$

and

(c) $(x + \bar{1})(x + \bar{2}) = x^2 + \bar{3}x + \bar{2} = (x + \bar{4})(x + \bar{5}).$

The multiplications in Example 1 above suggest the following result.

***Lemma 1 Multiplication in*** $R[x]$

*Let $\alpha \neq 0$ and $\beta \neq 0$ in $R[x]$.*
*(a) If $\alpha\beta \neq 0$, then $\deg(\alpha\beta) \leq \deg \alpha + \deg \beta$.*
*(b) If $L(\beta)$ is not a 0-divisor in $R$, then $L(\alpha\beta) = L(\alpha)L(\beta)$, $\alpha\beta \neq 0$, and*

$$\deg(\alpha\beta) = \deg \alpha + \deg \beta.$$

***Corollary***

*In $R[x]$, a monic polynomial is never a 0-divisor.*

Lemma 1 and its corollary follow immediately from formula (M) for multiplication in $R[x]$.

Since an integral domain $D$ or a field $F$ is a ring, the results in Lemma 1 apply to $D[x]$ and $F[x]$. We also have the following improvements and additions:

### *Lemma 2 Polynomials over an Integral Domain D*

*(a)* $L(\alpha\beta) = L(\alpha)L(\beta)$ *for all* $\alpha$ *and* $\beta$ *in* $D[x]$.
*(b) If* $\alpha \neq 0$ *and* $\beta \neq 0$ *in* $D[x]$, *then* $\alpha\beta \neq 0$ *and*

$$\deg(\alpha\beta) = \deg \alpha + \deg \beta.$$

*(c)* $D[x]$ *is an integral domain.*
*(d) The invertibles of* $D[x]$ *are just the invertibles of* $D$.

***Proof***

Parts (a) and (b) follow from formula (M) for polynomial multiplication and the fact that an integral domain $D$ has no 0-divisors. Part (b) then implies that $D[x]$ has no 0-divisors. Since the unity 1 of $D$ is also the unity of $D[x]$, this means that $D[x]$ is an integral domain.

Let $\alpha$ be an invertible of $D[x]$. Then there is a $\beta$ in $D[x]$ such that $\alpha\beta = 1$. It follows that $\alpha \neq 0$ and $\beta \neq 0$. Using part (b), we have

$$\deg \alpha + \deg \beta = \deg(\alpha\beta) = \deg 1 = 0.$$

Since $\deg \alpha$ and $\deg \beta$ are nonnegative, we see that

$$\deg \alpha = \deg \beta = 0.$$

Hence we have shown that the invertible $\alpha$ of $D[x]$ is in $D$.

### *Lemma 3 The Invertibles of F[x]*

*An* $\alpha$ *of* $F[x]$ *is an invertible if and only if* $\alpha$ *is a nonzero element of* $F$.

This is an immediate consequence of Lemma 2(d) and the fact that every nonzero element of a field $F$ is an invertible.

***Example 2***

List all the first degree polynomials in $\mathbf{Z}_2[x]$.

***Solution***

We seek all polynomials $\alpha = a_1x + a_0$ with $a_1$ and $a_0$ in $\mathbf{Z}_2 = [\bar{0}, \bar{1}]$ and with $a_1 \neq \bar{0}$. Hence $a_1$ must be $\bar{1}$ and $\alpha$ must be $x$ or $x + \bar{1}$.

If one is dealing extensively with $\mathbf{Z_p}$ or $\mathbf{Z_p}[x]$, it may be convenient to drop the bars on the symbols $\bar{0}, \bar{1}, \ldots, \overline{p-1}$ for the elements of $\mathbf{Z_p}$. Also there are applications in which the polynomial $a_d x^d + a_{d-1}x^{d-1} + \cdots + a_0$ of $\mathbf{Z_p}[x]$ is represented by $a_d a_{d-1} \cdots a_0$. With these notations, the polynomials of Example 2 would be written as 10 and 11. The particular value of $p$ may be indicated by a subscript; then these polynomials would be given as $10_2$ and $11_2$.

***Definition 5 Reducible Polynomial, Irreducible Polynomial***

*A nonzero $\alpha$ is* **reducible** *in $R[x]$ if there exist $\beta$ and $\gamma$ in $R[x]$ such that $\alpha = \beta\gamma$,* deg $\beta <$ deg $\alpha$, *and* deg $\gamma <$ deg $\alpha$. *A polynomial is* **irreducible** *if it has positive degree and is not reducible.*

Briefly, a polynomial is reducible if it can be factored into polynomials with "reduced" degrees. A "reducible polynomial" is analogous to a "composite integer" since an integer is composite if and only if it is the product of two integers with smaller absolute values. We discuss another analogue in the next section.

***Example 3***

Find all the reducible polynomials of degree 2 in $\mathbf{Z}_2[x]$.

***Solution***

We saw in Example 2 that $x$ and $x + \bar{1}$ are the only first degree polynomials in $\mathbf{Z}_2[x]$. Since $\mathbf{Z}_2$ is an integral domain in $\mathbf{Z}_2[x]$ a second degree polynomial is reducible if and only if it is the product of two first degree polynomials. (See the definition of reducible polynomial and Lemma 2.) Hence the reducible polynomials of degree 2 in $\mathbf{Z}_2[x]$ are

$$x^2, \qquad (x + \bar{1})^2 = x^2 + \bar{1}, \qquad x(x + \bar{1}) = x^2 + x.$$

***Example 4***

Find all the irreducible polynomials of degree 2 in $\mathbf{Z}_2[x]$.

*Solution*

We seek polynomials of the form

$$x^2 + a_1 x + a_0,$$

with $a_1$ and $a_0$ in $\{\bar{0}, \bar{1}\}$, that are not included among the answers in Example 3. The only such polynomial is

$$x^2 + x + \bar{1}.$$

*Example 5*

Let $\alpha$ be in $R[x]$ and let $\deg \alpha = 1$. Show that $\alpha$ is irreducible.

*Solution*

Let $\alpha = \beta\gamma$ in $R[x]$. We see that $\deg \beta < \deg \alpha = 1$ and $\deg \gamma < \deg \alpha = 1$ would imply that $\deg \beta = 0 = \deg \gamma$. Then it would follow from Lemma 1(a) that

$$\deg \alpha = \deg(\beta\gamma) \leq \deg \beta + \deg \gamma = 0 + 0 = 0.$$

This would contradict the hypothesis $\deg \alpha = 1$; hence either $\deg \beta \geq \deg \alpha$ or $\deg \gamma \geq \deg \alpha$ and $\alpha$ is irreducible.

*Example 6*

Let $\alpha = \bar{2}x + \bar{1}$ in $\mathbf{Z}_6[x]$. Is $\alpha$ an invertible?

*Solution*

Let us see if there is a $\beta$ in $\mathbf{Z}_6[x]$ such that $\alpha\beta = 1$, that is, if there are $b_0, \ldots, b_d$ in $\mathbf{Z}_6$ such that

$$(\bar{2}x + \bar{1})(b_d x^d + b_{d-1}x^{d-1} + \cdots + b_0) = 1. \tag{1}$$

Expanding the left side of (1) gives us

$$\bar{2}b_d x^{d+1} + (b_d + \bar{2}b_{d-1})x^d + \cdots + (b_1 + \bar{2}b_0)x + b_0 = 1. \tag{2}$$

Equating coefficients on the two sides of (2), we find that

$$\bar{2}b_d = b_d + \bar{2}b_{d-1} = \cdots = b_1 + \bar{2}b_0 = \bar{0}, \quad b_0 = \bar{1}.$$

But $\bar{2}b_d = \bar{0}$ and $b_d \neq \bar{0}$ in $\mathbf{Z}_6$ requires that $b_d = \bar{3}$ and then

$$b_d + \bar{2}b_{d-1} = \bar{3} + \bar{2}b_{d-1} = 0$$

is not true for any $b_{d-1}$ in $\mathbf{Z}_6$. Hence $\alpha$ is not an invertible. (In this proof we tacitly assumed that $d > 0$, but it is easily seen that no $\beta$ with $\deg \beta = 0$ satisfies $\alpha\beta = 1$.)

### *Definition 6 Polynomials in $x_1, x_2, \ldots, x_n$*

*Let $R$ be a subring in a commutative ring $T$ and $x_1, x_2, \ldots, x_n$ be elements of $T$. Let $T_0 = R$ and $T_i = T_{i-1}[x_i]$ for $i = 1, 2, \ldots, n$. Then $T_n$ is the* **ring of polynomials** *in $x_1, x_2, \ldots, x_n$ over $R$ and $T_n$ is denoted by $R[x_1, x_2, \ldots, x_n]$.*

For example, $x^4 - 3xy + y^2$ is a polynomial in $x$ and $y$ over $\mathbf{Z}$.

### *Definition 7 Independent Indeterminates*

*Let $R[x_1, x_2, \ldots, x_n]$ be as in Definition 6. Then $x_1, x_2, \ldots, x_n$ are* **independent indeterminates** *over $R$ if $x_i$ is an indeterminate over $R[x_1, \ldots, x_{i-1}, x_{i+1}, \ldots, x_n]$ for $i = 1, 2, \ldots, n$.*

## *Problems*

1. (i) Explain why every nonzero polynomial in $\mathbf{Z}_2[x]$ is monic.
   (ii) List all the reducible polynomials with degree 3 in $\mathbf{Z}_2[x]$.
   (iii) List all the irreducible polynomials with degree 3 in $\mathbf{Z}_2[x]$.
2. Given that $\alpha$ and $\beta$ are monic polynomials in $U[x]$, explain why $\alpha\beta$ is monic and $\deg(\alpha\beta) = \deg \alpha + \deg \beta$.
3. In the notation introduced after Example 2 of this section, list six monic reducible polynomials with degree 2 in $\mathbf{Z}_3[x]$.
4. Explain why every first degree polynomial in $R[x]$ is irreducible.
5. How many monic polynomials with degree $d$ are there in $\mathbf{Z}_m[x]$?
6. How many polynomials with degree $d$ are there in $\mathbf{Z}_m[x]$?
7. Explain why the following are true:
   (i) A 0-divisor in $R$ is also a 0-divisor in $R[x]$.
   (ii) If $R[x]$ has 0-divisors, so does $R$.
   (iii) $R[x]$ is an integral domain if and only if $R$ is an integral domain.
   (iv) $\mathbf{Z}_m[x]$ is an integral domain if and only if the positive integer $m$ is a prime in $\mathbf{Z}$.

8. Let $\alpha$, $\beta$, and $\gamma$ be in $R[x]$ with $\alpha \neq 0$ and the leading coefficient $L(\alpha)$ not a 0-divisor in $R$. Explain why $\beta = \gamma$ follows from $\alpha\beta = \alpha\gamma$ or $\beta\alpha = \gamma\alpha$.
9. Let $\beta$ be in $U[x]$ and let the leading coefficient $L(\beta)$ be an invertible $v$ of $U$. Explain why $v^{-1}\beta$ is monic and has the same degree as $\beta$.
10. What is the only monic $\alpha$ in $U[x]$ with $\deg \alpha = 0$?
11. Can $\deg(\alpha + \beta)$ be larger than both $\deg \alpha$ and $\deg \beta$? Explain.
12. Let $\alpha$ and $\beta$ be nonzero polynomials in $R[x]$ with $\deg(\alpha + \beta)$ smaller than both $\deg \alpha$ and $\deg \beta$.
    (i) What must be the relation between $\deg \alpha$ and $\deg \beta$?
    (ii) What must be the relation between $L(\alpha)$ and $L(\beta)$?
13. Show the following:
    (a) $(x + \bar{2})(x + \bar{3}) = x(x + \bar{5})$ in $\mathbf{Z}_6[x]$.
    (b) $(x + \bar{2})(x + \bar{5}) = x(x + \bar{7})$ in $\mathbf{Z}_{10}[x]$.
    (c) $(x + \bar{3})(x + \bar{5}) = x(x + \bar{8})$ in $\mathbf{Z}_{15}[x]$.
14. Show the following:
    (a) $(x + \bar{2})^2 = x^2$ in $\mathbf{Z}_4[x]$.
    (b) $(x + \bar{4})^2 = x^2$ in $\mathbf{Z}_8[x]$.
    (c) $(x + \bar{6})^2 = x^2$ in $\mathbf{Z}_{12}[x]$.
15. Find all the factorizations of $x^2$ as a product $\alpha\beta$ of monic first degree polynomials in $\mathbf{Z}_9[x]$.
16. Show that in $\mathbf{Z}_5[x]$ there is one and only one way to factor $x^2 + x - \bar{1}$ as a product of two monic first degree polynomials.
17. In $\mathbf{Z}_2[x]$, show that

$$x(x + \bar{1})(x^3 + x + \bar{1})(x^3 + x^2 + \bar{1}) = x^8 - x.$$

18. In $\mathbf{Z}_3[x]$, show that

$$x(x - 1)(x + 1)(x^2 + 1)(x^2 + x - \bar{1})(x^2 - x - \bar{1}) = x^9 - x.$$

19. Let $D$ be an ordered integral domain. Show that $D[x]$ may be ordered by letting $D[x]^+$ consist of all $\alpha = a_d x^d + \cdots + a_0$ of $D[x]$ such that $a_d$ is in $D^+$.
20. (i) Let $D$ be an integral domain and let $D[x]$ be ordered. Explain a method of ordering $D$.
    (ii) Let $\mathbf{C}$ be the field of complex numbers. Can $\mathbf{C}[x]$ be ordered? Explain.
21. For each $\alpha = a_d x^d + a_{d-1}x^{d-1} + \cdots + a_2 x^2 + a_1 x + a_0$ in $R[x]$, let the ***derivative*** $\alpha'$ of $\alpha$ be defined to be

$$\alpha' = [d \cdot a_d]x^{d-1} + [(d - 1) \cdot a_{d-1}]x^{d-2} + \cdots + [2 \cdot a_2]x + a_1.$$

(Recall that if $n$ is a positive integer, $n \cdot a$ means the sum $a + a + \cdots + a$ of $n$ terms each of which is $a$.) Show that this definition implies that

$$(\alpha + \beta)' = \alpha' + \beta', \qquad (\alpha\beta)' = \alpha\beta' + \alpha'\beta.$$

22. Use Problem 21 to prove that $(\alpha^n)' = n \cdot \alpha^{n-1}\alpha'$ for all $\alpha$ in $R[x]$ and all nonnegative integers $n$.
23. In $U[x]$, expand $(x - 1)(a_2 x^2 + a_1 x + a_0)$ in the form

$$b_3 x^3 + b_2 x^2 + b_1 x + b_0$$

and show that $b_3 + b_2 + b_1 + b_0 = 0$.

24. In $U[x]$, expand $(x + 1)(a_3 x^3 + a_2 x^2 + a_1 x + a_0)$ in the form $b_4 x^4 + b_3 x^3 + b_2 x^2 + b_1 x + b_0$ and show that $b_4 - b_3 + b_2 - b_1 + b_0 = 0$.
25. Generalize on Problem 23.
26. Generalize on Problem 24.
27. Let $f$ be the mapping from $R[x]$ to $R$ such that the image of $a_d x^d + \cdots + a_1 x + a_0$ is $a_0$.
   (i) Show that $f$ is a surjective ring homomorphism.
   (ii) Give a generator for the kernel $K$ of $f$.
   (iii) Let $I$ be the principal ideal $(x)$ in $R[x]$. Explain why the quotient ring $R[x]/I$ is isomorphic to $R$.
28. Let $f$ be the mapping of Problem 27. Let $S$ be a subset of $R[x]$ and let $T$ consist of the images $f(\alpha)$ for all the $\alpha$ in $S$.
   (i) Prove that $T$ is a subring in $R$ if $S$ is a subring in $R[x]$.
   (ii) Prove that $T$ is an ideal in $R$ if $S$ is an ideal in $R[x]$.
29. Let $m$ be a fixed positive integer and let $a \mapsto \bar{a}$ be the natural map from $\mathbf{Z}$ to $\mathbf{Z_m}$. Let $g$ be the mapping from $\mathbf{Z}[x]$ to $\mathbf{Z_m}$ such that

$$g(a_d x^d + \cdots + a_1 x + a_0) = \bar{a}_0 = a_0 + (m).$$

Show that $g$ is a surjective ring homomorphism.

30. Let $m$ be a fixed positive integer and let $a \mapsto \bar{a}$ give the natural map from $\mathbf{Z}$ to $\mathbf{Z_m}$. Let $h$ be the mapping from $\mathbf{Z}[x]$ to $\mathbf{Z_m}[x]$ with

$$a_d x^d + \cdots + a_1 x + a_0 \overset{h}{\mapsto} \bar{a}_d x^d + \cdots + \bar{a}_1 x + \bar{a}_0.$$

   (i) Is $h$ injective?
   (ii) Is $h$ surjective?
   (iii) Is $h$ a ring homomorphism?

31. Let $p$ be a positive prime in $\mathbf{Z}$ and let $a \mapsto \bar{a}$ give the natural map from $\mathbf{Z}$ to $\mathbf{Z_p}$. Let $f$ be the mapping from $\mathbf{Z}[x]$ to $\mathbf{Z_p}[x]$ with

$$f(a_d x^d + \cdots + a_1 x + a_0) = \bar{a}_d x^d + \cdots + \bar{a}_1 x + \bar{a}_0 .$$

(i) Show that $f(\alpha) = 0$ if and only if $p | a_i$ for $0 \leq i \leq d$.
(ii) Given that $f(\beta\gamma) = 0$, show that either $f(\beta) = 0$ or $f(\gamma) = 0$.
(iii) If $\beta\gamma$ is monic, show that $f(\beta) \neq 0$ and $f(\gamma) \neq 0$.

32. Let $r \mapsto r'$ give a ring homomorphism $f$ from $R$ to $R'$. Let $g$ be the mapping from $R[x]$ to $R'[x]$ with

$$g(a_d x^d + \cdots + a_1 x + a_0) = a'_d x^d + \cdots + a'_1 x + a'_0 .$$

*(i) Prove that $g$ is a ring homomorphism.
(ii) Prove that $g$ is injective if and only if $f$ is injective.
(iii) Prove that $g$ is surjective if and only if $f$ is surjective.

33. Let $f$ be the mapping from $R[x]$ to itself with

$$a_d x^d + a_{d-1} x^{d-1} + \cdots + a_1 x + a_0 \mapsto a_d x^{2d} + a_{d-1} x^{2d-2} + \cdots + a_1 x^2 + a_0 .$$

(i) Is $f$ injective?
(ii) Is $f$ surjective?
(iii) Is $f$ a ring homomorphism?

34. Let $\alpha$ be in $R[x]$. Prove the following:
(i) There is exactly one ring homomorphism $f$ from $R[x]$ to itself with $r \mapsto r$ for all $r$ in $R$ and $x \mapsto \alpha$.
(ii) The $f$ of part (i) is surjective if and only if $\alpha = ax + b$ with $a$ and $b$ in $R$ and $a$ an invertible.

35. Explain why $R[x, y] = R[y, x]$. (See Definition 6.)

36. Explain why $R[x, y, z] = R[y, z, x] = R[y, x, z]$.

## 5.3 Divisibility in Commutative Rings

The main result of this section is the division algorithm for polynomials. First we present generalizations to an arbitrary commutative ring $R$ of some divisibility notation previously given for the ring $\mathbf{Z}$ of the integers. This material is given here rather than in Chapter 4 because the polynomial rings provide a more meaningful setting.

Let $r$ and $s$ be elements of $R$. Then $s|r$ is defined to mean that there exists an element $t$ of $R$ such that $r = st$. When $s|r$, one also says that $s$ is a ***divisor*** of $r$ in $R$ and that $r$ is a ***multiple*** of $s$ in $R$.

***Example 1***

In $\mathbf{Z}_6[x]$, let $\alpha = x + \bar{5}$ and $\beta = \bar{2}x + \bar{1}$. Show that $\beta$ is a divisor of $\alpha$ but $\alpha$ is not a divisor of $\beta$.

***Solution***

The equation

$$(\bar{2}x + \bar{1})(\bar{3}x + \bar{5}) = x + \bar{5}$$

in $\mathbf{Z}_6[x]$ shows that $\beta|\alpha$. We show that $\alpha$ is not a divisor of $\beta$ by assuming that $\beta = \alpha\gamma$, with $\gamma$ in $\mathbf{Z}_6[x]$, and obtaining a contradiction.

Since $\alpha$ is monic, it follows from Lemma 1(b) of Section 5.2 that $\beta = \alpha\gamma$ would imply

$$1 = \deg \beta = \deg \alpha + \deg \gamma = 1 + \deg \gamma$$

and hence $\deg \gamma = 0$. However, the only polynomials of degree zero in $\mathbf{Z}_6[x]$ are $\bar{1}, \bar{2}, \bar{3}, \bar{4}$, and $\bar{5}$ and it is easily seen that no one of these can be chosen as $\gamma$ so as to satisfy $\alpha\gamma = \beta$.

***Example 2***

In $\mathbf{Z}_6[x]$, let $\alpha = x + \bar{5}$ and $\beta = \bar{5}x + \bar{1}$. Show that both $\alpha|\beta$ and $\beta|\alpha$.

***Solution***

The equations

$$\beta = \bar{5}\alpha, \qquad \alpha = \bar{5}\beta$$

in $\mathbf{Z}_6[x]$ show that $\alpha|\beta$ and $\beta|\alpha$.

***Definition 1 Associates in a Commutative Ring*** $R$

*If both $a|b$ and $b|a$ in $R$, one says that $a$ and $b$ are associates of each other in $R$.*

For example, the associates of a nonzero $n$ in $\mathbf{Z}$ are $n$ and $-n$; the only associate of 0 in $\mathbf{Z}$ is 0 itself.

Example 2 above shows that $x + \overline{5}$ and $\overline{5}x + \overline{1}$ are associates in $\mathbf{Z}_6[x]$ while Example 1 shows that $x + \overline{5}$ and $\overline{2}x + \overline{1}$ are not associates in $\mathbf{Z}_6[x]$.

### *Lemma 1 Associates in an Integral Domain* $D$

*The following are all equivalent in* $D$:

(*i*) *$a$ and $b$ are associates; that is,* $a|b$ *and* $b|a$.

(*ii*) $a = vb$ (*and* $b = v^{-1}a$), *with $v$ an invertible in* $D$.

(*iii*) $(a) = (b)$; *that is, $a$ and $b$ are generators of the same principal ideal in* $D$.

#### *Proof*

Let $a|b$ and $b|a$. Then there are $u$ and $v$ in $D$ such that $b = ua$ and $a = vb$. Since $D$ is an integral domain, one may cancel a nonzero $a$ in $1 \cdot a = vua$ and so obtain $1 = vu$, which means that $v$ and $u$ are invertibles. On the other hand, $a = 0$ implies that $b = u \cdot 0 = 0$ and so $a = 1 \cdot b$ and $b = 1 \cdot a$ (with 1 an invertible).

Thus we see that whether $a = 0$ or $a \neq 0$, each of $a$ and $b$ is the other multiplied by an invertible. This shows that statement (ii) follows from statement (i). The converse follows immediately as does the equivalence of statements (i) and (iii).

### *Corollary Associates in* $D[x]$

*Polynomials $\alpha$ and $\beta$ are associates in $D[x]$ if and only if $\alpha = v\beta$ with $v$ an invertible in $D$.* If $\alpha$ and $\beta$ are associates, $\deg \alpha = \deg \beta$.

The proof is left to the reader as Problem 13(a) of this section. Note that the $D[x]$ of the corollary may be taken as the $D$ of Lemma 1.

In the rest of this section, we generalize some concepts previously introduced for the integers $\mathbf{Z}$ by applying them to general commutative rings $U$ with unity or to polynomial rings $U[x]$.

### *Definition 2 Prime in* $U$

*In a commutative ring $U$ with unity, let $p$ be an element that is not 0 and is not an invertible. Also let every divisor of $p$ be either an invertible or an associate of $p$. Then $p$ is a* **prime** *in* $U$.

***Definition 3 Composite in U***

*An element a in U is* **composite** *if a is not* 0, *not an invertible, and not a prime.*

The following examples show that our concepts of "prime polynomial" and "irreducible polynomial" are distinct analogues of "prime integer." However, in Theorem 1 below we show that they are equivalent for $F[x]$, that is, when the coefficients are in a field.

***Example 3***

Find a polynomial in $\mathbf{Z}[x]$ that is irreducible but is not a prime.

***Solution***

The polynomial $\alpha = 5x + 10$ is irreducible since $\deg \alpha = 1$. (See Example 5 of Section 5.2.) The divisor 5 of $\alpha$ is not an invertible in $\mathbf{Z}[x]$ since it is not an invertible in the integral domain $\mathbf{Z}$. Also $\deg 5 \neq \deg \alpha$ implies that 5 is not an associate of $\alpha$. Hence it follows from the definition of a prime that $\alpha$ is not a prime.

***Example 4***

Show that the monic polynomial $\alpha = x + \bar{5}$ is irreducible but is not a prime in $\mathbf{Z}_6[x]$.

***Solution***

Since $\deg \alpha = 1$, $\alpha$ is irreducible. (See Example 5 of Section 5.2.) In Example 1 above, it was shown that $\alpha$ has a divisor $\beta = \bar{2}x + 1$, which is not an associate of $\alpha$. As $\beta$ is also not an invertible (see Example 6 of Section 5.2), it follows that $\alpha$ is not a prime.

***Theorem 1 Primes and Irreducibles in*** $F[x]$

*Let F be a field. Then a polynomial $\alpha$ is prime in $F[x]$ if and only if $\alpha$ is irreducible in $F[x]$.*

***Proof***

Let $\alpha$ be an irreducible polynomial and let $\alpha = \beta\gamma$ in $F[x]$. Since $\alpha$ is irreducible, either $\deg \beta \geq \deg \alpha$ or $\deg \gamma \geq \deg \alpha$; for definiteness let $\deg \beta \geq \deg \alpha$. Then

$$\deg \alpha = \deg(\beta\gamma) = \deg \beta + \deg \gamma$$

implies that $\deg \gamma = 0$. Hence $\gamma$ is a nonzero element of $F$ and so is an invertible in $F$ and in $F[x]$. That makes $\beta$ an associate of $\alpha$ and shows that every divisor of $\alpha$ is either an invertible or an assocate of $\alpha$; that is, $\alpha$ is a prime.

Conversely, let $\alpha$ be a prime in $F[x]$. Then $\alpha \neq 0$ and $\deg \alpha > 0$ since a polynomial of degree 0 is an invertible in $F[x]$ and a prime is not an invertible. Let $\alpha = \beta\gamma$. Then the hypothesis that $\alpha$ is a prime implies that either $\beta$ is an invertible, and so of degree 0, or $\beta$ is an associate of $\alpha$, and hence $\deg \beta = \deg \alpha$. The same is true of $\gamma$. If both $\deg \beta < \deg \alpha$ and $\deg \gamma < \deg \alpha$ it would follow that $\deg \beta = 0 = \deg \gamma$ and

$$\deg \alpha = \deg \beta + \deg \gamma = 0 + 0 = 0.$$

But $\deg \alpha > 0$. Hence the assumption that $\deg \beta < \deg \alpha$ and $\deg \gamma < \deg \alpha$ is false; that is, the prime $\alpha$ is irreducible.

A ring is closed under addition, subtraction, and multiplication but may or may not contain the quotient of two given elements. In rings of polynomials, as in the ring **Z** of integers, there is a division algorithm, that is, a process of dividing $\alpha$ by $\beta$ so as to obtain a quotient $\gamma$ and a remainder $\rho$. (There are restrictions on the divisor $\beta$.) Since the steps of the algorithm are familiar from elementary algebra, we shall not describe them in detail. However, we present a theorem that has many important applications and also implicitly characterizes the algorithm.

For practical and theoretical reasons, it is convenient to start by considering the case in which $\beta$ is monic.

### *Lemma 2 Division by a Monic Polynomial*

*Let $\alpha$ and $\beta$ be in $U[x]$ with $\beta$ monic. Then there is a unique ordered pair of polynomials $\gamma$ and $\rho$ in $U[x]$, such that $\alpha = \gamma\beta + \rho$ and either $\rho = 0$ or $\deg \rho < \deg \beta$. ($\gamma$ is called the* **quotient** *and $\rho$ the* **remainder** *in the division of $\alpha$ by $\beta$.)*

#### *Proof*

We first consider the case in which $\beta|\alpha$. Then there is a $\gamma$ in $U[x]$ such that $\alpha = \gamma\beta = \gamma\beta + 0$ and we may choose this $\gamma$ as the quotient and 0 as the remainder $\rho$.

Now let $\beta$ not be a divisor of $\alpha$ and let $S$ be the set of all polynomials of the form $\alpha - \delta\beta$, with $\delta$ in $U[x]$. We see that 0 is not in $S$, since otherwise $\beta|\alpha$. Hence every polynomial in $S$ has a degree and we let $T$ be the nonempty set of all degrees of polynomials in $S$. It then follows from the Well Ordering Principle that there is a least integer $t$ in $T$. Now let $\gamma$ be a polynomial in $U[x]$ such that $\alpha - \gamma\beta$ is a polynomial $\rho$ in $S$ of this minimal degree $t$. Let $\deg \beta = d$ and

$$\rho = r_t x^t + \cdots + r_0, \quad r_t \neq 0.$$

We show that $t = \deg \rho < \deg \beta = d$ by assuming that $t \geq d$ and obtaining a contradiction. If $t \geq d$, then

$$\alpha - (\gamma + r_t x^{t-d})\beta$$

would be in $S$ and would have degree less than $t$, contradicting the minimal nature of $t$. Therefore, $\deg \rho < \deg \beta$ and the existence of a quotient and remainder with the desired properties is established.

We complete the proof by showing that the ordered pair $(\gamma, \rho)$ is unique. Let us assume that in $U[x]$

$$\alpha = \gamma\beta + \rho = \gamma'\beta + \rho'$$

with each of $\rho$ and $\rho'$ either 0 or of degree less than $d$. Then

$$(\gamma - \gamma')\beta = \rho' - \rho.$$

Since $\beta$ is monic, either $\gamma - \gamma' = 0$ or $\deg[(\gamma - \gamma')\beta] \geq \deg \beta = d$. [See part (b) of Lemma 1 in Section 5.2.] On the other hand, $\rho' - \rho$ is either 0 or of degree less than $d$. Together these statements imply that $\gamma - \gamma'$ and $\rho - \rho'$ are both zero; that is, $(\gamma, \rho)$ is unique.

We next use Lemma 2 to obtain a more general form of the division algorithm.

***Theorem 2*** ***The Division Algorithm in*** $U[x]$

*Let $\alpha$ and $\beta$ be in $U[x]$ and let $\beta$ have an invertible $v$ as its leading coefficient. Then there is a unique ordered pair $\gamma$ and $\rho$ in $U[x]$ such that $\alpha = \gamma\beta + \rho$ and either $\rho = 0$ or $\deg \rho < \deg \beta$.*

***Proof***

The polynomial $\beta' = v^{-1}\beta$ is monic and hence it follows from Lemma 2 that there are polynomials $\gamma'$ and $\rho$ in $U[x]$ with $\alpha = \gamma'\beta' + \rho$ and either $\rho = 0$ or $\deg \rho < \deg \beta' = \deg \beta$. Now we let $\gamma = v^{-1}\gamma'$. Then $\gamma' = v\gamma$ and

$$\alpha = \gamma'\beta' + \rho = (v\gamma)(v^{-1}\beta) + \rho = \gamma\beta + \rho,$$

as desired. Uniqueness is left to the reader.

***Corollary 1***

*In Theorem 2, $\beta | \alpha$ if and only if $\rho = 0$.*

### Corollary 2

*In Theorem 2, if* deg $\beta = 1$, *then* $\rho$ *is in* $U$.

In a ring $F[x]$ of polynomials over a field $F$, every nonzero $\beta$ has an invertible as its leading coefficient and hence may be used as the divisor in the division algorithm of Theorem 2. Also, in $F[x]$ the quotient $\gamma$ satisfies deg $\gamma =$ deg $\alpha$ $-$ deg $\beta <$ deg $\alpha$ when $\gamma \neq 0$ and deg $\beta > 0$.

Our base 10 notation for the integers uses the fact that every positive integer is uniquely expressible as a polynomial

$$10^s c_s + 10^{s-1} c_{s-1} + \cdots + 10c_1 + c_0, \quad c_s \neq 0,$$

with the $c_j$ integers satisfying $0 \leq c_j < 10$. (There are similar expressions for other bases.) An analogous result for polynomials over a field $F$ is given in the following:

### Theorem 3 Polynomials in $\beta$

*Let* $\beta$ *be a polynomial in* $U[x]$ *with an invertible as its leading coefficient. Then in* $U[x]$ *every nonzero* $\alpha$ *is uniquely expressible in the form*

$$\rho_s \beta^s + \cdots + \rho_1 \beta + \rho_0, \quad \rho_s \neq 0, \tag{1}$$

*where each* $\rho_j$ *either is* 0 *or has degree less than* $\beta$.

### Proof

We assume the result false. This and the Well Ordering Principle imply that among the polynomials not of the form (1) there is an $\alpha$ with least degree $d$. Theorem 2 tells us that we can let this $\alpha$ equal $\gamma\beta + \rho_0$ where either $\rho_0 = 0$ or deg $\rho_0 <$ deg $\beta$. We see that $\gamma \neq 0$ since otherwise $\alpha = \rho_0$ would be of form (1) with $s = 0$. Then $\gamma \neq 0$ implies that deg $\gamma <$ deg $\alpha$. This and the minimal nature of deg $\alpha$ mean that $\gamma$ has an expression of the form (1), which we write as

$$\gamma = \rho_t \beta^{t-1} + \cdots + \rho_2 \beta + \rho_1.$$

Then

$$\alpha = \gamma\beta + \rho_0 = \rho_t \beta^t + \cdots + \rho_1 \beta + \rho_0$$

is of form (1). This contradiction completes the proof.

## *Problems*

1. (i) In $\mathbf{Z}_{12}$, find all $\bar{a}$ such that $\bar{2}|\bar{a}$.
   (ii) In $\mathbf{Z}_{12}$, find all $\bar{b}$ such that $\bar{b}|\bar{2}$.
   (iii) In $\mathbf{Z}_{12}$, find all associates of 2.
   (iv) In $\mathbf{Z}_3[x]$, find all associates of $\bar{2}x^3 + x$.

2. Let $p$ be a positive prime in $\mathbf{Z}$.
   (i) How many invertibles are there in $\mathbf{Z}_\mathbf{p}$?
   (ii) How many associates does a nonzero $\alpha$ have in $\mathbf{Z}_\mathbf{p}[x]$?

3. Let $\mathbf{Q}$ be the field of rational numbers. In $\mathbf{Q}[x]$, find the quotient and remainder in the division of each of the following by $x - 2$:
   (i) $x^2$; (ii) $x^3$; (iii) $x^4$; (iv) $x^5$.

4. In $\mathbf{Q}[x]$ find the quotient and remainder in the division of each of the following by $x + 3$:
   (i) $x^2 - 9$; (ii) $x^3 + 27$; (iii) $x^4 - 81$.

5. In $\mathbf{Q}[x]$, let $\alpha = x^3 - 5x + 8$ and $\beta = 2x^2 + 3x - 1$. Find $\gamma$ and $\rho$ such that $\alpha = \gamma\beta + \rho$ and $\deg \rho < 2$.

6. In $\mathbf{Z}_5[x]$, let $\alpha = x^3 + \bar{3}$ and $\beta = \bar{2}x^2 + \bar{3}x + \bar{4}$. Find $\gamma$ and $\rho$ such that $\alpha = \gamma\beta + \rho$ and $\deg \rho < 2$.

7. Let $n$ be a positive integer and let $a$ be in a commutative ring $U$ with unity. In $U[x]$, find $\beta$ such that $x^n - a^n = (x - a)\beta$.

8. Let $\alpha = a_d x^d + \cdots + a_1 x + a_0$ be in $U[x]$, let $r$ be in $U$, and let $s = a_d r^d + \cdots + a_1 r + a_0$. Show that

$$(x - r)|(\alpha - s).$$

9. Let $a$ and $b$ be in $\mathbf{Q}$. Show that $(x + a + b)|(x^3 - 3abx + a^3 + b^3)$ in $\mathbf{Q}[x]$.

10. In the field $\mathbf{C}$ of complex numbers, let $w$ be any number with $w^3 = 1$. Show that in $\mathbf{C}[x]$,

$$(x + aw + bw^2)|(x^3 - 3abx + a^3 + b^3).$$

11. Let $R$ be a commutative ring. Show the following:
    (a) If $d|a$ and $d|b$ in $R$, then $d|(ah + bk)$ for all $h$ and $k$ in $R$.
    (b) Let $a = qb \pm c$ in $R$. Then $d|a$ and $d|b$ if and only if $d|b$ and $d|c$.

12. Let $U$ be a commutative ring with unity. Show that:
    (a) In $U$, $a$ is an associate of 1 if and only if $a$ is an invertible.
    (b) In $U[x]$, let $\beta$ have an invertible as its leading coefficient $L(\beta)$ and

let $\gamma$ and $\rho$ be the quotient and remainder in the division of $\alpha$ by $\beta$. If $\gamma \neq 0$, then

$$\deg \alpha = \deg \beta + \deg \gamma.$$

(c) Let $v$ be an invertible in $U$. Then in $U$, $b|a$ if and only if $b|(av)$; also $b|a$ if and only if $(bv)|a$.

13. Let $D$ be an integral domain. Show the following:
 (a) $\alpha$ and $\beta$ are associates in $D[x]$ if and only if $\alpha = v\beta$ with $v$ an invertible in $D$. Also associates in $D[x]$ have the same degree. (This proves the corollary to Lemma 1.)
 (b) In $D[x]$, a nonzero $\alpha$ has a monic associate if and only if the leading coefficient $L(\alpha)$ is an invertible. If $L(\alpha)$ is an invertible, $\alpha$ has a unique monic associate.
 (c) In $D[x]$, $\alpha$ is an associate of 1 if and only if $\alpha$ is an invertible in $D$.
 (d) If $\beta|\alpha$ in $D[x]$, then $L(\beta)|L(\alpha)$ in $D$.
 (e) In $D[x]$, if $\beta|\alpha$ and $\alpha \neq 0$, then $\deg \beta \leq \deg \alpha$.
 (f) If $\beta|\alpha$ in $D[x]$ and $\alpha$ is monic, $L(\beta)$ is an invertible in $D$.
 (g) If $\gamma|\beta$ in $D[x]$ and $\beta$ is irreducible, then $\deg \gamma$ is either 0 or $\deg \beta$.
 (h) In $D[x]$, every monic irreducible polynomial is a prime.
 (i) In $D[x]$, the only monic divisors of a monic prime $\pi$ are 1 and $\pi$.
 (j) Let $\pi$ be a prime with degree $d$ in $D[x]$. Then every divisor of $\pi$ has degree $d$ or 0.

14. Let $F$ be a field. Show the following:
 (a) A nonzero $\beta$ in $F[x]$ has exactly one monic associate.
 (b) $\alpha$ is an associate of 1 in $F[x]$ if and only if $\alpha$ is a nonzero element of $F$.
 (c) Let $I$ be a principal ideal $(\beta)$ in $F[x]$ with $\beta \neq 0$. Then there is a unique monic generator of $I$.

15. Give an example of a principal ideal in $\mathbf{Z}[x]$ that does not have a monic generator.

16. Let $U$ be a commutative ring with unity, $c \in U$, $\alpha \in U[x]$, and $\deg \alpha = d$. Explain why there exist $b_i$ in $U$ such that

$$\alpha = b_0 + b_1(x - c) + b_2(x - c)^2 + \cdots + b_d(x - c)^d.$$

17. Let $U$ be a commutative ring with unity. Let $A$ be the relation in $U$ for which $rAs$ means that $r$ and $s$ are associates of each other. Is $A$ an equivalence relation? Explain.

18. (i) Let $\alpha$ be in $D[x]$ with $\deg \alpha$ equal to 2 or 3. Explain why $\alpha$ is reducible if and only if it has a divisor of degree 1.
 (ii) Show that $x^4 + x^2 + \bar{1}$ is reducible in $\mathbf{Z}_2[x]$ even though it has no divisor of degree 1. [*Hint:* See Problem 15(b) of Section 4.4.]

19. In $D[x]$, let $\alpha = \beta\gamma$ with $\alpha$ monic. Show that $\beta$ and $\gamma$ have monic associates $\beta'$ and $\gamma'$, respectively, and that $\alpha = \beta'\gamma'$.

20. Show that a monic $\alpha$ in $D[x]$ is reducible if and only if there exist monic polynomials $\beta$ and $\gamma$ in $D[x]$ such that

$$\alpha = \beta\gamma, \qquad \deg \beta < \deg \alpha, \qquad \deg \gamma < \deg \alpha.$$

21. Use Problem 20 to show the following:
(a) $x^3 - 3x + 1$ is irreducible (and prime) in $\mathbf{Q}[x]$.
(b) $x^3 + x + \bar{1}$ is irreducible (and prime) in $\mathbf{Z}_2[x]$.

22. Use Problem 20 to find three monic irreducible polynomials of degree 2 in $\mathbf{Z}_3[x]$. (The answer to Problem 3 of Section 5.2 may be helpful.)

23. Show that the only irreducible polynomials of degree 4 in $\mathbf{Z}_2[x]$ are $x^4 + x + \bar{1}$, $x^4 + x^3 + \bar{1}$, and $x^4 + x^3 + x^2 + x + \bar{1}$.

24. Find all the irreducible polynomials of degree 5 in $\mathbf{Z}_2[x]$.

25. Let $F = \{0, 1, c, c^2\}$ be a field with four elements. (In Problem 1 of Section 4.4, we saw that $F$ has characteristic 2 and $c^2 = 1 + c$.) In $F[x]$, show the following:
(i) $x^4 - x = x(x-1)(x-c)(x-c^2)$.
(ii) $x^{16} - x = (x^4 - x)\beta$, where

$$\beta = (x^2 + cx + 1)(x^2 + c^2x + 1)(x^2 + x + c^2)(x^2 + cx + c)(x^2 + c^2x + c^2)(x^2 + x + c).$$

(iii) The six second degree factors of $\beta$ in part (ii) are all the monic irreducible polynomials of degree 2 in $F[x]$.

26. In $\mathbf{Z}_2[x]$, show the following:
(i) $x^{16} - x = x(x-\bar{1})(x^2+x+\bar{1})(x^4+x+\bar{1})(x^4+x^3+\bar{1})(x^4+x^3+x^2+x+\bar{1})$.
(ii) The factors of $x^{16} - x$ on the right side of part (i) are all the irreducible polynomials of degree 1, 2, and 4.

27. Let $F$ be a field with nine elements. Assume that in $F[x]$ the polynomial $x^{81} - x$ can be expressed as the product of all the monic irreducible polynomials of degree 1 and 2. How many of these factors are of degree 2?

28. Let $R$ be a commutative ring. Let $\alpha$ and $\beta$ be in $R[x]$ and let $\beta = b_d x^d + \cdots + b_1 x + b_0$ with $b_d \neq 0$. Show that there exist a nonnegative integer $t$ and polynomials $\gamma$ and $\rho$ in $R[x]$ such that $(b_d)^t\alpha = \gamma\beta + \rho$ and either $\rho = 0$ or $\rho$ has lower degree than $\beta$.

29. Let $p$ be a positive prime in $\mathbf{Z}$ and let $a \mapsto \bar{a}$ give the natural map from $\mathbf{Z}$ to $\mathbf{Z}_p$. Let

$$\alpha = x^d + a_{d-1}x^{d-1} + \cdots + a_1 x + a_0$$

be a monic polynomial in $\mathbf{Z}[x]$. Show the following:
(i) If $\alpha$ is reducible in $\mathbf{Z}[x]$, then

$$\bar{\alpha} = x^d + \bar{a}_{d-1}x^{d-1} + \cdots + \bar{a}_1 x + \bar{a}_0$$

is reducible in $\mathbf{Z}_p[x]$.

(ii) $x^3 + (2a + 1)x + (2b + 1)$ is irreducible in $\mathbf{Z}[x]$ for all integers $a$ and $b$.

30. Prove that $x^3 + (3a - 1)x + (3b + 1)$ is irreducible in $\mathbf{Z}[x]$, for all integers $a$ and $b$.

## 5.4 Polynomial Functions

Associated with each polynomial

$$\alpha = a_d x^d + \cdots + a_1 x + a_0, \quad a_j \text{ in } R,$$

over a commutative ring $R$ is the mapping $\alpha^*$ from $R$ to $R$ for which the image of an $r$ in $R$ is

$$a_d r^d + \cdots + a_1 r + a_0. \tag{1}$$

Such a mapping $\alpha^*$ is called a ***polynomial function***. We let $R^*$ represent the set of $\alpha^*$ for all $\alpha$ in $R[x]$.

Instead of designating the image of $r$ under $\alpha^*$ given in display (1) as $\alpha^*(r)$, we shall use the simpler notation $\alpha(r)$. The sum $\sigma^*$ and product $\pi^*$ of polynomial functions $\alpha^*$ and $\beta^*$ are defined, as in calculus, by

$$\sigma(r) = \alpha(r) + \beta(r), \qquad \pi(r) = \alpha(r)\beta(r). \tag{2}$$

Formulas (2) are equivalent to

$$(\alpha + \beta)^* = \alpha^* + \beta^*, \qquad (\alpha\beta)^* = \alpha^*\beta^*. \tag{3}$$

The operations defined in (2), or (3), make $R^*$ into a commutative ring. (The details are tedious but straightforward and are left to the reader. See Problem 31 of Section 4.1.)

It also follows from (3) that the surjection $f$ from $R[x]$ to $R^*$ with $\alpha \mapsto \alpha^*$ is a ring homomorphism. This homomorphism need not be injective, as we see in the following example.

### *Example 1*

In $\mathbf{Z}_3[x]$, let $\alpha = x^3 + x$ and $\beta = \bar{2}x$. Show that $\alpha^* = \beta^*$ (although $\alpha \neq \beta$) and hence that the homomorphism $f$ with $\alpha \mapsto \alpha^*$ is not an isomorphism from $\mathbf{Z}_3[x]$ to $\mathbf{Z}_3^*$.

***Solution***

Since $\mathbf{Z}_3$ has only three elements, we solve our problem easily by noting that under either $\alpha^*$ or $\beta^*$ we have $\bar{0} \mapsto \bar{0}$, $\bar{1} \mapsto \bar{2}$, and $\bar{2} \mapsto \bar{1}$.

Let $R$ be a finite ring with $m$ elements. Since there are $m^m$ mappings from $R$ to $R$ (see Problem 10 of Section 3.1), there are at most $m^m$ polynomial functions in $R^*$. If $m > 1$, there are an infinite number of polynomials in $R[x]$, and hence no mapping from $R[x]$ to $R^*$ can be injective.

We show below that the homomorphism $f$ from $D[x]$ to $D^*$ with $\alpha \mapsto \alpha^*$ is an injection, and hence an isomorphism, for all infinite integral domains $D$. (See Corollary 2 to Theorem 3.)

***Definition 1 Zero of a Function, Root of an Equation***

*Let $R$ be a subring of a commutative ring $S$ and let $\alpha$ be in $R[x]$. An element $s$ of $S$, such that $\alpha(s) = 0$, is a* **zero** *in $S$ of the function $\alpha^*$ and of the polynomial $\alpha$. Also, $s$ is a* **root** *of the polynomial equation $\alpha(x) = 0$.*

We next present some results linking zeros and first degree divisors of a polynomial over a ring $U$ with unity.

***Lemma 1***

*For each $r$ in $U$ and each positive integer $n$,*

$$(x - r) \mid (x^n - r^n) \text{ in } U[x].$$

***Proof***

This follows from the easily verified factorization

$$x^n - r^n = (x - r)(x^{n-1} + rx^{n-2} + \cdots + r^{n-2}x + r^{n-1}).$$

***Lemma 2***

*Let $\alpha$ be in $U[x]$ and let $r$ be in $U$. Then, in $U[x]$,*

$$(x - r) \mid [\alpha - \alpha(r)].$$

***Proof***

The desired result is a consequence of Lemma 1 and the following:

$$\begin{aligned}\alpha - \alpha(r) &= (a_d x^d + \cdots + a_0) - (a_d r^d + \cdots + a_0)\\ &= a_d(x^d - r^d) + \cdots + a_1(x - r).\end{aligned}$$

### *Theorem 1 The Remainder Theorem*

*Let $r$ be in $U$. Then in $U[x]$ the remainder in the division of $\alpha$ by $x - r$ is $\alpha(r)$.*

***Proof***

Lemma 2 tells us that $\alpha - \alpha(r) = (x - r)\gamma$ with $\gamma$ in $U[x]$. Transposing, we have

$$\alpha = (x - r)\gamma + \alpha(r). \tag{4}$$

Since $\alpha(r)$ is either 0 or of degree 0, it is the unique remainder in the division of $\alpha$ by the first degree $x - r$.

### *Theorem 2 The Factor Theorem*

*Let $r$ be in $U$ and $\alpha$ be in $U[x]$. Then $r$ is a root of $\alpha(x) = 0$, that is, a zero of $\alpha$, if and only if $x - r$ is a divisor of $\alpha$ in $U[x]$.*

***Proof***

If $r$ is a zero of $\alpha$, that is, if $\alpha(r) = 0$, it follows immediately from (4) that $(x - r)|\alpha$. If $(x - r)|\alpha$, then (4) implies that

$$(x - r)|\alpha(r)$$

and this requires that $\alpha(r) = 0$, since a polynomial of degree 0 is not a multiple of a monic polynomial of degree 1.

For polynomials over an integral domain $D$, we have the following result.

### *Theorem 3 Linear Factors*

*Let $\alpha$ be in $D[x]$ and let $r_1, r_2, \ldots, r_n$ be $n$ distinct zeros of $\alpha$ in $D$. Then there is a $\gamma$ in $D[x]$ such that*

$$\alpha = (x - r_1)(x - r_2) \cdots (x - r_n)\gamma.$$

***Outline of Proof***

It follows from the Factor Theorem that $\alpha = (x - r_1)\beta$, with $\beta$ in $D[x]$. Then

$$0 = \alpha(r_i) = (r_i - r_1)\beta(r_i).$$

For $2 \leq i \leq n$, $r_i - r_1 \neq 0$, since the zeros $r_i$ are distinct. The fact that $D$ has no 0-divisors then implies that, for $i > 1$, $\beta(r_i) = 0$; that is, $r_i$ is a zero of $\beta$. An inductive argument may be used to complete a formal proof of this theorem. Informally, one gives $\beta$ the treatment given to $\alpha$ above and continues the process until $n$ first degree divisors have been factored out of $\alpha$.

### Corollary 1

*If $\alpha$ has $n$ distinct zeros and $\alpha \neq 0$, $\deg \alpha \geq n$. Hence a nonzero polynomial $\alpha$ has a finite number of zeros.*

### Corollary 2

*If $D$ is infinite and $\beta \neq \gamma$ in $D[x]$, then $\beta^* \neq \gamma^*$ in $D^*$; that is, the ring homomorphism $\alpha \mapsto \alpha^*$ is an injection, and hence is an isomorphism.*

The proofs of the corollaries are left to the reader as Problems 11 and 12 of this section.

Next we consider multiple roots of polynomial equations.

### Definition 2 Multiple Zero (or Root)

*Let $r$ be an element of $U$ and let $m$ be a positive integer. If $(x - r)^m$ is a divisor of $\alpha$ and $(x - r)^{m+1}$ is not a divisor of $\alpha$ in $U[x]$, one says that $r$ is a zero of $\alpha$ [or a root of $\alpha(x) = 0$] with* **multiplicity** *$m$.*

A zero with multiplicity 1 is also called a ***simple zero.***

### Theorem 4 Test for Multiplicity

*Let $m$ be a positive integer and let $\alpha = (x - r)^m\beta$ in $U[x]$. If $r$ is not a zero of $\beta$, then $r$ is a zero of $\alpha$ with multiplicity $m$.*

#### Proof

The hypothesis tells us that $(x - r)^m$ is a divisor of $\alpha$; hence we need only show that $(x - r)^{m+1}$ is not a divisor of $\alpha$. We do this by assuming that $\alpha = (x - r)^{m+1}\gamma$ in $U[x]$ and obtaining a contradiction.

This assumption leads to $(x-r)^m\beta = (x-r)^{m+1}\gamma$. The corollary to Lemma 1 of Section 5.2 tells us that the monic polynomial $x - r$ is not a 0-divisor in $U[x]$. Hence it may be cancelled $m$ times in the last equation, thus giving us $\beta = (x-r)\gamma$. Then the Factor Theorem implies that $r$ is a zero of $\beta$. This contradicts the hypothesis and completes the proof.

***Example 2***

Show that 1 is a zero with multiplicity 2 of the polynomial $\alpha = x^3 - 3x + 2$ in $\mathbf{Q}[x]$.

***Solution***

We divide $\alpha$ by $(x-1)^2$ and find that

$$\alpha = (x-1)^2(x+2).$$

Since 1 is not a zero of $x + 2$, it follows from Theorem 4 that 1 is a zero with multiplicity 2 of $\alpha$.

The two following results are needed in our impossibility proofs, in Section 6.3, concerning angle-trisection and other classical problems on constructions with compass and straightedge. Also, Theorem 5 implies that the non-real zeros of a polynomial with real coefficients occur in pairs of complex conjugates.

### *Theorem 5 Pairs of Conjugate Roots*

*Let $F$ be a subfield in a field $E$. Let $t$ be an element of $E$ such that $t$ is not in $F$ but $t^2$ is in $F$. If $r = a + bt$ is a zero in $E$ of an $\alpha$ of $F[x]$, so is $\bar{r} = a - bt$.*

***Proof***

Let $t^2 = s$ and let $\beta = (x-r)(x-\bar{r}) = [x-(a+bt)][x-(a-bt)] = x^2 - 2ax + (a^2 - sb^2)$. Note that $\beta \in F[x]$. Let $\gamma$ and $\rho$ be the quotient and remainder, in $F[x]$, in the division of $\alpha$ by $\beta$.

In $E[x]$, we have $(x-r)|\beta$ by definition of $\beta$ and $(x-r)|\alpha$ using the hypothesis that $r$ is a zero of $\alpha$ and the Factor Theorem. It follows from these facts and $\rho = \alpha - \gamma\beta$ that $(x-r)|\rho$ in $E[x]$. Since deg $\beta = 2$, $\rho$ does not have degree greater than 1. Hence $(x-r)|\rho$ implies that $\rho = c(x-r)$, with $c$ in $E$. Since $\rho$ is in $F[x]$, its coefficients $c$ and $-cr$ are in $F$.

We now consider two cases. If $c \neq 0$, then $r = -c^{-1}(-cr)$ and so $r$ is in $F$. Since $t$ is not in $F$, $r = a + bt$ can only be in $F$ if $b = 0$; this would imply that $\bar{r} = a - bt = a = r$ is a zero of $\alpha$.

If $c = 0$, then $\rho = 0$, $\beta|\alpha$, and hence $(x-\bar{r})|\alpha$. It follows that, in this case also, $\bar{r}$ is a zero of $\alpha$.

### *Corollary Complex Zeros of a Real Polynomial*

*Let $a + b\mathbf{i}$ be a zero in $\mathbf{C}$ of an $\alpha$ in $\mathbf{R}[x]$. Then $a - b\mathbf{i}$ is also a zero of $\alpha$.*

The proof of the corollary is left to the reader as Problem 19 of this section.

### *Theorem 6 Zeros of Cubics over Quadratic Extensions*

*Let $K$ be an extension field over $F$. Let $t \in K$, $t^2 \in F$, and $E = \{a + bt: a, b \in F\}$. Then $E$ is an extension field over $F$ such that any polynomial $\alpha$ of degree 3 in $F[x]$ with a zero in $E$ also has a zero in $F$.*

*Proof*

If $t$ is in $F$, then $E = F$ and the desired result holds. So let $t$ not be in $F$. One sees from

$$\frac{1}{a + bt} = \frac{a - bt}{(a + bt)(a - bt)} = \frac{a}{a^2 - b^2t^2} - \frac{b}{a^2 - b^2t^2}t$$

that an element $a + bt$ has as its inverse in $E$ the element $c + dt$ with $c = a/(a^2 - b^2t^2)$ and $d = -b/(a^2 - b^2t^2)$ unless the denominator $a^2 - b^2t^2 = 0$. But $a^2 - b^2t^2 = 0$ implies that $b = 0$ and then $a = 0$ and $a + bt = 0$ since $b \neq 0$ and $a^2 - b^2t^2 = 0$ lead to the contradiction that $t = \pm a/b$ is in $F$. Hence every nonzero element of $E$ has an inverse in $E$. It is now easy to see that $E$ is a field with $F$ as a subfield.

Let $r = a + bt$ be a zero in $E$ of a polynomial $\alpha$ with degree 3 in $F[x]$. Now it follows from Theorem 5 above that $\bar{r} = a - bt$ is also a zero of $\alpha$ in $E$. If $b = 0$, $r = a$ is the desired zero in $F$ of $\alpha$. If $b \neq 0$, then $\bar{r} \neq r$ and we can factor $\alpha$ in $E[x]$ as

$$\alpha = (x - r)(x - \bar{r})\beta.$$

Since $\deg \alpha = 3$, we have $\deg \beta = 1$; that is, $\beta = hx + k$ with $h$ and $k$ in $F$ and $h \neq 0$. Let $u = -kh^{-1}$. Then $u$ is a zero of $\beta$ and hence is a zero of $\alpha$. Also,

$$\alpha = h(x - r)(x - \bar{r})(x - u) = c_3x^3 + c_2x^2 + c_1x + c_0$$

with the $c_i$ in $F$ and $c_3 \neq 0$. Expanding the left side and equating coefficients, we find that $c_3 = h$ and

$$c_2 = -h(r + \bar{r} + u) = -c_3(a + bt + a - bt + u) = -c_3(2 \cdot a + u).$$

Solving for $u$ gives us $u = -c_2c_3^{-1} - 2 \cdot a$. Since $a$, $c_2$, and $c_3$ are in $F$, this tells us that the zero $u$ of $\alpha$ is in $F$, as desired.

## *Problems*

Problem 20 below is cited in Section 5.9. Problems 37 and 38 are cited in Section 7.4.

1. In $\mathbf{Q}[x]$, find the remainder in the division of $x^3 + 2x^2 - 5x - 7$ by $x - 4$ directly; then check using the Remainder Theorem.
2. In $\mathbf{Q}[x]$, find the remainder in the division of $4x^5 + 5x^4 - 1$ by $x + 1$ directly; then check using the Remainder Theorem.
3. Use the Remainder Theorem to show that $x - 1$ is a factor of $a_d x^d + \cdots + a_1 x + a_0$ in $U[x]$ if and only if $a_d + \cdots + a_1 + a_0 = 0$ in $U$.
4. Use the Remainder Theorem to find a necessary and sufficient condition on the coefficients $a_i$ for $x + 1$ to be a factor of $a_d x^d + \cdots + a_1 x + a_0$.
5. In $\mathbf{Z}_5[x]$, let $\alpha = x^6 + \bar{2}x^5 - \bar{2}x$ and $\beta = x^2$. Tabulate $\alpha(r)$ and $\beta(r)$ for all $r$ in $\mathbf{Z}_5$ and thus show that $\alpha^* = \beta^*$ in $\mathbf{Z}_5^*$ (even though $\alpha \neq \beta$ in $\mathbf{Z}_5[x]$).
6. Let $\alpha = x^7 - x$ in $\mathbf{Z}_7[x]$. Tabulate $\alpha(r)$ for all $r$ in $\mathbf{Z}_7$ and thus show that $\alpha^*$ is the zero function of $\mathbf{Z}_7^*$.
7. Let $F$ be a finite field with $m$ elements. Explain why every element of $F$ is a zero of $\alpha = x^m - x$ and hence $\alpha^*$ is the zero function of $F^*$.
8. Explain why $x^m - x + 1$ has no zero in a field $F$ with $m$ elements and hence has no first degree divisor in $F[x]$.
9. (i) Show that $(x - \bar{2})(x - \bar{3}) = x(x - \bar{5})$ in $\mathbf{Z}_6[x]$.
   (ii) How many zeros does $x^2 - \bar{5}x$ have in $\mathbf{Z}_6$?
10. How many zeros does $x^2 - \bar{7}x$ have in $\mathbf{Z}_{12}$?
11. Prove that an $\alpha$ with degree $d$ in $D[x]$ has at most $d$ zeros in the integral domain $D$ (and thus prove Corollary 1 to Theorem 3 above).
12. Let $D$ be an infinite integral domain. Prove that $\alpha \mapsto \alpha^*$ gives a ring isomorphism from $D[x]$ to $D^*$. (This proves Corollary 2 to Theorem 3 above.)
13. Let $\beta(h_i) = \gamma(h_i)$ for $i = 0, 1, \ldots, t$, where the $h_i$ are $t + 1$ distinct elements of an integral domain $D$ and $\beta \neq \gamma$ in $D[x]$. Explain why at least one of $\beta$ and $\gamma$ must have degree greater than $t$.
14. Let $s_1, s_2, \ldots, s_d$ be $d$ distinct zeros in an integral domain $D$ of an $\alpha$ with degree $d$ in $D[x]$.
   (i) Use Theorem 3 above to show that
   $$\alpha = a(x - s_1)(x - s_2) \cdots (x - s_d),$$
   where $a$ is a nonzero element of $D$.
   (ii) Explain why $a = 1$ in part (i) if and only if $\alpha$ is monic.
15. Let $F = \{b_1, b_2, \ldots, b_m\}$ be a field with $m$ elements and let $h$ be the ring

homomorphism from $F[x]$ to $F^*$ with $\alpha \mapsto \alpha^*$. Let $K$ be the kernel of $h$; that is, let $K$ consist of all the $\alpha$ of $F[x]$ such that $\alpha^*$ is the zero function of $F^*$. Let

$$\beta = (x - b_1)(x - b_2) \cdots (x - b_m).$$

Prove the following:

(i) $\deg \alpha \geq m$ for every nonzero $\alpha$ in $K$.
(ii) $\alpha^* = \rho^*$ in $F^*$ if and only if $\alpha - \rho$ is in $K$.
(iii) Let $\alpha = \gamma\beta + \rho$ in $F[x]$. Then $\alpha^* = \rho^*$ in $F^*$ and hence $\alpha$ is in $K$ if and only if $\rho$ is in $K$.
(iv) $\beta | \alpha$ for every $\alpha$ in $K$; that is, $K = (\beta)$.
(v) $\alpha^* = \rho^*$ in $F^*$ if and only if $\beta | (\alpha - \rho)$ in $F[x]$.
(vi) For every $\alpha$ in $F[x]$, there is a $\rho$ in $F[x]$ such that $\alpha^* = \rho^*$ and with either $\deg \rho < m$ or $\rho = 0$.

16. Let $p$ be a positive prime in $\mathbf{Z}$ and let $h$ be the ring homomorphism from $\mathbf{Z_p}[x]$ to $\mathbf{Z_p^*}$ with $\alpha \mapsto \alpha^*$. Let $K$ be the kernel of $h$ and let

$$\beta = x(x - \bar{1})(x - \bar{2}) \cdots (x - \overline{p-1}).$$

Explain why the following are true:

(i) $K = (\beta)$.
(ii) $\beta = x^p - x$.
(iii) $(x - \bar{1})(x - \bar{2}) \cdots (x - \overline{p-1}) = x^{p-1} - 1$.
(iv) $\bar{1} \cdot \bar{2} \cdot \bar{3} \cdots \overline{p-1} = -\bar{1}$ in $\mathbf{Z_p}$. [It may help to consider two cases: (a) $p = 2$, and (b) $p$ an odd prime.]
(v) $(p-1)! + 1$ is a multiple of $p$ in $\mathbf{Z}$.
(vi) $(p-1)! \equiv -1 \pmod{p}$. (This is known as Wilson's Theorem. See Problem 25 of Section 4.4 for an alternative proof.)

17. In each part below, $\alpha$ is in $F[x]$ and $s$ is a zero of $\alpha$. Find the multiplicity $m$ of the zero $s$.

(a) $\alpha = 3x^4 - 4x^3 + 1$, $s = 1$, $F = \mathbf{Q}$.
(b) $\alpha = x^3 + x^2 - 2x - 2$, $s = -\sqrt{2}$, $F = \mathbf{R}$.
(c) $\alpha = x^5 - x^4 + 2x^3 - 2x^2 + x - 1$, $s = \mathbf{i}$, $F = \mathbf{C}$.
(d) $\alpha = x^9 + \bar{1}$, $s = \bar{2}$, $F = \mathbf{Z_3}$.
(e) $\alpha = x^{10} - \bar{2}x^5 + \bar{1}$, $s = \bar{1}$, $F = \mathbf{Z_5}$.
(f) $\alpha = nx^{n+1} - (n+1)x^n + 1$, $s = 1$, $F = \mathbf{Q}$. Here $n$ is a positive integer.

*18. In $F[x]$, let $\alpha = \pi_1\pi_2 \cdots \pi_d$, with $\deg \pi_i = 1$ for each $i$. Prove that $\alpha$ has no zero with multiplicity greater than 1 if and only if $\gcd(\alpha, \alpha') = 1$. (Here $\alpha'$ is the derivative of $\alpha$ defined in Problem 21 of Section 5.2.)

19. Let $a + b\mathbf{i}$ (with $a$ and $b$ real) be a zero of a polynomial $\alpha$ with real coefficients. Show that $a - b\mathbf{i}$ is a zero of $\alpha$ (and thus prove the corollary to Theorem 5 of this section).

20. (i) Let $\alpha \in F[x]$ and $\deg \alpha \in \{2, 3\}$. Explain why $\alpha$ is reducible (and composite) in $F[x]$ if and only if $\alpha$ has at least one zero in $F$.
(ii) Show that $\alpha = x^4 + x^2 + \overline{1}$ is reducible in $\mathbf{Z}_2[x]$ even though $\alpha$ has no zero in $\mathbf{Z}_2$. (See Problem 18 of Section 5.3.)

21. Use Problems 8 and 20(i) to prove that $x^3 - x + \overline{1}$ is irreducible (and prime) in $\mathbf{Z}_3[x]$.

22. Use Problem 20(i) to find all the monic irreducible polynomials of degree 3 in $\mathbf{Z}_3[x]$.

23. In a field $F$, let $a \neq 0$. In $F[x]$, let

$$a(x - r_1)(x - r_2)(x - r_3) = ax^3 + bx^2 + cx + d.$$

Show the following:
(a) $r_1 + r_2 + r_3 = -b/a$.
(b) $r_2 r_3 + r_1 r_3 + r_1 r_2 = c/a$.
(c) $r_1 r_2 r_3 = -d/a$.
(d) $r_1^2 + r_2^2 + r_3^2 = (b^2 - 2 \cdot ac)/a^2$.

24. Let $U$ be a commutative ring with unity. In $U[x]$, let

$$(x - r_1)(x - r_2)(x - r_3)(x - r_4) = x^4 + a_3 x^3 + a_2 x^2 + a_1 x + a_0.$$

(a) Show that $r_1 + r_2 + r_3 + r_4 = -a_3$.
(b) Show that $r_1 r_2 r_3 r_4 = a_0$.
(c) Express each of $a_1$ and $a_2$ in terms of the $r_i$.

25. Let $U$ be a commutative ring with unity. In $U[x]$, let

$$(x - r_1)(x - r_2) \cdots (x - r_d) = x^d + a_{d-1}x^{d-1} + \cdots + a_1 x + a_0.$$

(a) Express $r_1 + r_2 + \cdots + r_d$ in terms of the $a_i$.
(b) Express $r_1 r_2 \cdots r_d$ in terms of the $a_i$.

26. Let $a_d \neq 0$ in a field $F$. In $F[x]$, let

$$a_d(x - r_1)(x - r_2) \cdots (x - r_d) = a_d x^d + \cdots + a_1 x + a_0.$$

(a) Express $r_1 + r_2 + \cdots + r_d$ in terms of the $a_i$.
(b) Express $r_1 r_2 \cdots r_d$ in terms of the $a_i$.
(c) Show that $r_1^2 + r_2^2 + \cdots + r_d^2 = (a_{d-1}^2 - 2 \cdot a_d a_{d-2})/a_d^2$.

27. In $\mathbf{Q}[x]$, let

$$(x - 1)(x - 2)(x - 3) \cdots (x - d) = x^d - s_1 x^{d-1} + s_2 x^{d-2} - \cdots + (-1)^d s_d.$$

Show the following:

(a) $s_1 = d(d+1)/2$.
(b) $s_d = 1 \cdot 2 \cdot 3 \cdots d = d!$
*(c) $2s_2 = (1^3 + 2^3 + \cdots + d^3) - (1^2 + 2^2 + \cdots + d^2)$.

28. In $F[x]$, let

$$ax^2 + bx + c = a(x - r)(x - s).$$

For all nonnegative integers $n$, let $h_n = r^n + s^n$. Show the following:
(i) $ar^{n+2} + br^{n+1} + cr^n = 0$.
(ii) $as^{n+2} + bs^{n+1} + cs^n = 0$.
(iii) $ah_{n+2} + bh_{n+1} + ch_n = 0$.

29. Using the notation of Problem 28, express $h_5$ in terms of $a$, $b$, and $c$, assuming that $a \neq 0$.

30. In $F[x]$, let $(x - r)(x - s)(x - t) = x^3 + ux^2 + vx + w$. Express $r^5 + s^5 + t^5$ in terms of $u$, $v$, and $w$. Use the answer to Problem 29 as a partial check (by letting $t = 0 = w$, $a = 1$, $b = u$, and $c = v$).

31. Let $\mathbf{R}$ and $\mathbf{C} = \mathbf{R}[\mathbf{i}]$ be the real and complex fields, respectively. Let $\alpha = x^4 - 2x^3 + x^2$ and $\beta = x^2 + 1$.
(i) Find the remainder $\rho$ in the division of $\alpha$ by $\beta$ in $R[x]$.
(ii) Considering $\alpha$ to be in $\mathbf{C}[x]$, evaluate $\alpha(\mathbf{i})$ in the form $a + b\mathbf{i}$.
(iii) Explain the relation between the answers to parts (i) and (ii).

32. Let $\mathbf{R}$ be the field of real numbers. Describe an easy method for finding the remainder in the division of an $\alpha$ in $\mathbf{R}[x]$ by $x^2 + 1$. (*Hint*: See the preceding problem.)

33. Let $\mathbf{R}$ and $\mathbf{C}$ be the real and complex fields, respectively. Let $I$ consist of all $\alpha$ in $\mathbf{R}[x]$ such that $\mathbf{i}$ is a zero in $\mathbf{C}$ of $\alpha$. Prove that $I$ is the principal ideal $(x^2 + 1)$.

*34. Let $\mathbf{Q}$ and $\mathbf{R}$ be the rational and real fields, respectively. Let $K$ consist of all $\alpha$ in $\mathbf{Q}[x]$ such that $\sqrt{2} + \sqrt{3}$ is a zero in $R$ of $\alpha$. Prove that $K$ is a principal ideal and find a monic generator of $K$.

35. In $\mathbf{Q}[x]$, let $\alpha = 3x^4 - 7x^3 - 1$. Let $c$ be in $\mathbf{Q}$ and

$$\beta = \alpha + c(x + 2)(x + 1)x(x - 1)(x - 2).$$

(i) Explain why $\alpha(r) = \beta(r)$ for $r = -2, -1, 0, 1$, and $2$.
(ii) Given that $\beta(3) = 13$, find $c$.

36. In $\mathbf{Q}[x]$, let $\alpha = 2x^3 - 3x^2 + 5$. Find a polynomial $\beta$ in $\mathbf{Q}[x]$ with $\deg \beta = 4$; $\beta(r) = \alpha(r)$ for $r = -2, -1, 0$, and $1$; and $\beta(2) = -4$. (See Problem 35.)

37. Let $V$ be the multiplicative group of invertibles in an integral domain $D$. Let ord $V$ be an even integer $2t$ and $W$ be the subset $\{v^2 : v \in V\}$ of squares in $V$. Show the following:
(i) Each $v$ of $V$ is a root of $x^{2t} - 1 = 0$.

(ii) Of the $2t$ elements of $V$, $t$ elements are roots of $x^t - 1 = 0$ and the other $t$ elements are roots of $x^t + 1 = 0$.
(iii) There are $t$ elements in $W$. [See Problem 23(v) of Section 4.4.]
(iv) An element of $V$ is in $W$ if and only if it is a root of $x^t - 1 = 0$.
(v) $-1$ is in $W$ if and only if $t$ is even.
(vi) If $w$ is in $W$, then $-w$ is in $W$ if and only if $t$ is even.

38. Let $p$ be an odd positive prime in $\mathbf{Z}$ and let $W_p$ be the subset of squares in the multiplicative group $\mathbf{V_p}$ of invertibles in $\mathbf{Z_p} = \mathbf{Z}/(p)$. Show the following:
(i) $-1$ is in $W_p$ if and only if $p \equiv 1 \pmod 4$.
(ii) If $w$ is in $W_p$, then $-w$ is in $W_p$ if and only if $p \equiv 1 \pmod 4$.

## 5.5 *Integral and Rational Roots*

Here we present methods of determining all the integers and all the rational numbers that are roots of a polynomial equation $\alpha(x) = 0$ in which the coefficients are integers or rational numbers.

If the polynomial equation under consideration has rational coefficients, it can be converted, without changing the roots, to one that has integer coefficients by multiplying both sides of the equation by a common denominator for the original coefficients. Hence we let $\alpha$ be in $\mathbf{Z}[x]$; that is, we let

$$\alpha = a_d x^d + a_{d-1}x^{d-1} + \cdots + a_0, \quad a_i \text{ in } \mathbf{Z}.$$

We first seek a means of discovering the integers, if any, that are zeros of $\alpha$. We note that $r$ is a zero of $\alpha$ if and only if

$$a_d r^d + a_{d-1}r^{d-1} + \cdots + a_1 r + a_0 = 0$$

or

$$r(a_d r^{d-1} + a_{d-1}r^{d-2} + \cdots + a_1) = -a_0. \tag{1}$$

If $r$ is an integer, equation (1) implies that $r \mid a_0$. We now formally restate this result.

***Theorem 1 Integer Roots***

*The only integers that may be zeros of a polynomial*

$$\alpha = a_d x^d + a_{d-1}x^{d-1} + \cdots + a_0,$$

*with integer coefficients $a_i$, are the integral divisors of $a_0$.*

If $a_0 \neq 0$, there are only a finite number of integral divisors $d$ of $a_0$ and one can check to see which are zeros of $\alpha$ in a finite number of steps. If $a_0 = 0$, one modifies the procedure slightly, as indicated in the following example.

*Example 1*

Find all the integers that are zeros of

$$\alpha = x^7 + x^6 - 5x^5 + x^4 - 6x^3.$$

*Solution*

We note that $\alpha = x^3\beta$, where

$$\beta = x^4 + x^3 - 5x^2 + x - 6.$$

Clearly 0 is a zero of $\alpha$, with multiplicity 3, and all the other zeros of $\alpha$ are zeros of $\beta$. Theorem 1 tells us that the set of integers that are zeros of $\beta$ is a subset of $\{1, -1, 2, -2, 3, -3, 6, -6\}$, the set of integral divisors of $-6$. Evaluating $\beta(d)$ for each of these divisors $d$ of $a_0$, we find that $\beta(2) = 0$, $\beta(-3) = 0$, and $\beta(d) \neq 0$ for the other choices of $d$. Hence the integral zeros of $\alpha$ are 0, 2, and $-3$.

Next we take up the question of rational zeros of a polynomial $\alpha$ in $\mathbf{Z}[x]$. Let

$$\alpha = a_d x^d + a_{d-1}x^{d-1} + \cdots + a_0, \quad a_i \text{ in } \mathbf{Z}.$$

A rational number $r$ is expressible as $s/t$ where $s$ and $t$ are relatively prime integers and $t \neq 0$. Then $r$ is a zero of $\alpha$ if and only if

$$(a_d s^d/t^d) + (a_{d-1}s^{d-1}/t^{d-1}) + \cdots + (a_1 s/t) + a_0 = 0,$$

which is equivalent to each of the following:

$$a_d s^d + a_{d-1}s^{d-1}t + \cdots + a_1 st^{d-1} + a_0 t^d = 0,$$

$$a_d s^d = -t(a_{d-1}s^{d-1} + \cdots + a_1 st^{d-2} + a_0 t^{d-1}), \tag{2}$$

$$a_0 t^d = -s(a_d s^{d-1} + a_{d-1}s^{d-2}t + \cdots + a_1 t^{d-1}). \tag{3}$$

From (2) we see that $t\,|\,(a_d s^d)$. Since $t$ and $s$ are relatively prime, it follows that $t$ and $s^d$ are relatively prime and then that $t\,|\,a_d$. (See Problems 10 and 9 of Section 1.6.) Similarly, equation (3) implies that $s\,|\,a_0$. This leads to the following result.

***Theorem 2*** ***Rational Roots***

*The only possibilities for rational zeros of a polynomial*

$$a_d x^d + \cdots + a_1 x + a_0,$$

*with integer coefficients* $a_i$, *are the numbers of the form* $s/t$, *where* $s$ *and* $t$ *are integers,* $s \mid a_0$, *and* $t \mid a_d$. *[One may restrict* $s$ *and* $t$ *so that they also satisfy* $t > 0$ *and* $gcd(s, t) = 1$.*]*

***Example 2***

Show that the polynomial $\alpha = 8x^3 - 6x - 1$ is irreducible in $\mathbf{Q}[x]$.

***Solution***

Since $\deg \alpha = 3$, $\alpha$ is reducible if and only if it has a first degree factor $ax + b$ and therefore a rational zero $-b/a$. Hence it will suffice to show that $\alpha$ has no rational zero.

Using Theorem 2 and the fact that $\alpha$ is also in $\mathbf{Z}[x]$, we see that the only possibilities for rational zeros are of the form $s/t$ with $s$ an integral divisor of 1 and $t$ an integral divisor of 8; thus the only candidates are 1, $-1$, 1/2, $-1/2$, 1/4, $-1/4$, 1/8, and $-1/8$. Trial shows that no one of these rational numbers is a zero of $\alpha$. Therefore, $\alpha$ has no rational roots. Hence $\alpha$ has no factor of degree 1 in $\mathbf{Q}[x]$ and is irreducible.

***Example 3***

Find all the rational roots of $\alpha = 0$, where

$$\alpha = 6x^4 - 31x^3 + 25x^2 + 33x + 7,$$

and then find the other roots of this equation.

***Solution***

The possibilities for rational roots are the rational numbers $s/t$ with $s \mid 7$ and $t \mid 6$; these numbers are $\pm 1$, $\pm 7$, $\pm 1/2$, $\pm 7/2$, $\pm 1/3$, $\pm 7/3$, $\pm 1/6$, and $\pm 7/6$. We substitute each of them into the given equation. The first root that we find is 7/2; this root and the Factor Theorem tell us that $x - (7/2)$ is a factor of $\alpha$ in $\mathbf{Q}[x]$. It is more convenient to use its associate $\beta = 2x - 7$ in seeking the complementary factor $\gamma$ such that $\alpha = \beta\gamma$; then division by $\beta$ leads to

$$\gamma = 3x^3 - 5x^2 - 5x - 1.$$

The roots of $\alpha = 0$ are 7/2 and the roots of $\gamma = 0$. The possibilities for rational zeros of $\gamma$

are $\pm 1$ and $\pm 1/3$; trial shows that $-1/3$ is a zero. Then $3x + 1$ is a factor of $\gamma$ and division shows that

$$\gamma = (3x + 1)(x^2 - 2x - 1).$$

The zeros of $x^2 - 2x - 1$ are $1 \pm \sqrt{2}$. Hence the zeros of $\alpha$ are

$$\frac{7}{2}, -\frac{1}{3}, 1 + \sqrt{2}, 1 - \sqrt{2}.$$

## *Problems*

1. Find an integer root of $x^3 + 3x^2 - 2 = 0$ and then use the Factor Theorem as an aid in finding the other roots in the field **C** of complex numbers.
2. Do as in Problem 1 for $x^3 - 6x + 4 = 0$.
3. Find all the roots of $x^4 + 2x^3 - x^2 + 4x - 6 = 0$ in **C**.
4. Find all the roots in **C** of

$$x^5 - 8x^4 + 15x^3 + 8x^2 - 64x + 120 = 0.$$

5. Find all the roots in **C** of each of the following equations:
   (a) $3x^3 + 4x^2 - 21x + 10 = 0$.
   (b) $12x^3 - 11x^2 - 13x + 10 = 0$.
   (c) $x^4 - x^3 - 3x^2 + 5x - 2 = 0$.
   (d) $81x^5 - 54x^4 + 3x^2 - 2x = 0$.
   (e) $x^4 + 3x^2 + 2 = 0$.
6. Find all the rational zeros of each of the following polynomials:
   (a) $2x^4 + 5x^3 - 5x^2 + 7x - 3$.
   (b) $x^5 - 6x^4 + 3x^3 - 3x^2 + 2x - 12$.
7. What are the possibilities for integral roots of

$$5x^4 + ax^3 + bx^2 + cx - 21 = 0,$$

   given that $a$, $b$, and $c$ are integers?
8. What are the possibilities for rational roots of the equation in the previous problem?
9. Given that $a$ and $b$ are integers, is it possible for $35/21$ to be a root of $9x^3 + ax^2 + bx - 10 = 0$? Explain.
10. Let $a$, $b$, and $c$ be in **Z** and let $33/18$ be a root of

$$12x^3 + ax^2 + bx + c = 0.$$

Explain why $11|c$ and why it is not necessarily true that $33|c$.

11. Let $r$ be a rational zero of a monic

$$\alpha = x^d + a_{d-1}x^{d-1} + \cdots + a_0, \quad a_i \in \mathbf{Z}.$$

Explain why $r$ must be an integer.

12. Let $\alpha = a_d x^d + \cdots + a_1 x$ be in $\mathbf{Z}[x]$. (Note that $a_0 = 0$.) State a necessary condition for an integer $n$ to be a zero of $\alpha$.

13. Let $s$ and $t$ be relatively prime integers. State a necessary condition for the rational number $s/t$ to be a zero of the $\alpha$ of the preceding problem.

14. Let $a \in \mathbf{Z}$ and $\alpha = x^3 + x^2 + 2ax - 21$. Explain why $n^3 + n^2$ is even for every $n$ in $\mathbf{Z}$ and use this to show that $\alpha$ has no integral zero. Can $\alpha$ have a rational zero? Explain.

15. Show the following:
   (i) $x^5 - 7 = 0$ has no rational roots.
   (ii) $\sqrt[5]{7}$ is not a rational number.

16. For each of the following polynomials $\alpha$, show that $\alpha$ has no zero in $\mathbf{Q}$ and hence that $\alpha$ is irreducible in $\mathbf{Q}[x]$:
   (a) $x^3 - 3x + 1$.
   (b) $x^3 + x^2 - 2x - 1$.
   (c) $8x^3 + 4x^2 - 4x - 1$.

17. Let

$$\alpha = (x - \sqrt{2} - \sqrt{3})(x - \sqrt{2} + \sqrt{3})(x + \sqrt{2} - \sqrt{3})(x + \sqrt{2} + \sqrt{3}).$$

Show the following:
   (i) $\alpha$ is expressible as $x^4 + a_3 x^3 + a_2 x^2 + a_1 x + a_0$, with the $a_i$ integers.
   (ii) $\alpha$ has no rational zeros.
   (iii) $\sqrt{2} + \sqrt{3}$ is irrational.

18. Prove that $\sqrt{5} - \sqrt{3}$ is irrational. (*Hint*: See Problem 17.)

19. Let $\alpha$ be in $\mathbf{Z}[x]$ and let $r$, $s$, and $m$ be integers. Given that $r \equiv s \pmod{m}$, explain why $\alpha(r) \equiv \alpha(s) \pmod{m}$.

20. Let $\alpha$ be a polynomial in $\mathbf{Z}[x]$ such that $\alpha(0)$ and $\alpha(1)$ are both odd. Show that $\alpha$ has no zeros in $\mathbf{Z}$.

21. Let $\alpha$ be in $\mathbf{Z}[x]$ and let $k$ be a positive integer such that $\alpha(j)$ is not an integral multiple of $k$ for $j = 0, 1, 2, \ldots, k - 1$. Show that $\alpha$ has no zeros in $\mathbf{Z}$.

22. Let $\alpha$ be in $\mathbf{Z}[x]$ and let $\alpha(0)$, $\alpha(1)$, and the leading coefficient $L(\alpha)$ all be odd. Show that $\alpha = 0$ has no rational roots.

## 5.6 Polynomial Fitting, Finite Differences

*Algebra is the intellectual instrument which has been created for rendering clear the quantitative aspect of the world.*

*A. N. Whitehead*

*One cannot escape the feeling that these mathematical formulae have an independent existence and an intelligence of their own, that they are wiser than we are, wiser even than their discoverers, that we get more out of them than we originally put into them.*

*Heinrich Hertz*

One of the most important applications of polynomials is to problems involving the fitting of an analytical formula to a finite table of values of a function. This arises in interpolation or extrapolation, approximating definite integrals, numerical solution of differential equations, and so on.

We now present one of Newton's techniques for fitting a table with a polynomial function. (See the biographical note on Newton after this section.)

Suppose that we are given a table,

| $x$ | $h_0$ | $h_1$ | $h_2$ | $\dots$ | $h_t$ |
|---|---|---|---|---|---|
| $y$ | $k_0$ | $k_1$ | $k_2$ | $\dots$ | $k_t$ |

in which the $h_i$ and $k_i$ are elements of a field $F$. We seek a polynomial $\alpha$, of lowest possible degree, in $F[x]$ such that

$$\alpha(h_i) = k_i \text{ for } i = 0, 1, \dots, t. \tag{1}$$

We assume that the $h_i$ are $t + 1$ distinct elements of $F$, since otherwise there are conditions in (1) that are either redundant or contradictory.

Practically and theoretically, there are great advantages in looking for the polynomial $\alpha$ in the form

$$\alpha = c_0 + c_1(x - h_0) + c_2(x - h_0)(x - h_1) + \cdots + c_t[(x - h_0)(x - h_1) \cdots (x - h_{t-1})], \quad (2)$$

with the $c_i$ elements of $F$.

***Theorem 1*** ***Newton's Polynomial Fitting Formula***

*Let $h_0, h_1, \ldots, h_t$ be $t + 1$ distinct elements of a field $F$ and let $k_0, k_1, \ldots, k_t$ be in $F$. Then in $F[x]$ there is a unique polynomial $\alpha$ of the form*

$$\alpha = a_t x^t + a_{t-1}x^{t-1} + \cdots + a_0 \quad (3)$$

*[or of the form (2)] such that*

$$\alpha(h_i) = k_i \text{ for } i = 0, 1, \ldots, t. \quad (4)$$

***Proof***

We start by showing that the conditions in (4) determine one and only one sequence $c_0, c_1, \ldots, c_t$ of coefficients in (2). When evaluating $\alpha(h_i)$, for a given $i$, all the products in (2) that have $x - h_i$ as a factor contribute zero. More specifically, we note that (2) and (4) imply that

$$\alpha(h_0) = c_0, \qquad \alpha(h_1) = c_0 + c_1(h_1 - h_0),$$
$$\alpha(h_2) = c_0 + c_1(h_2 - h_0) + c_2(h_2 - h_0)(h_2 - h_1),$$

and so forth. Then $\alpha(h_i) = k_i$ for $0 \le i \le t$ if and only if

$$c_0 = k_0, \qquad c_1 = (k_1 - c_0)/(h_1 - h_0),$$

and so forth. More precisely, we know that $c_0$ must equal $k_0$ and if we assume that $c_0, c_1, \ldots, c_{j-1}$ have been determined, then $c_j = u_j/v_j$, where

$$u_j = k_j - c_0 - c_1(h_j - h_0) - c_2(h_j - h_0)(h_j - h_1) - \cdots - c_{j-1}[(h_j - h_0)(h_j - h_1) \cdots (h_j - h_{j-2})]$$

and

$$v_j = (h_j - h_0)(h_j - h_1)(h_j - h_2) \cdots (h_j - h_{j-1}).$$

It is easily seen that either $\alpha = 0$ or $\deg \alpha \le t$, with $\deg \alpha$ the largest value of $j$ for which $c_j \neq 0$.

The work above shows that the form (2) is unique for an $\alpha$ satisfying (4). To show uniqueness of the form (3), we let $\alpha$ satisfy (3) and (4) and let $\beta = b_d x^d + \cdots + b_0$ have $\beta(h_i) = k_i$ for $0 \le i \le t$. Also, let $\gamma = \alpha - \beta$. Then

$$\gamma(h_i) = \alpha(h_i) - \beta(h_i) = k_i - k_i = 0,$$

and so the $h_i$ are $t + 1$ distinct zeros of $\gamma$. It follows from Corollary 1 to Theorem 3 of Section 5.4 that either $\gamma = 0$ or deg $\gamma \geq t + 1$. But the difference of two polynomials of the form (3) is either 0 or of degree at most $t$. Hence $\gamma = 0$ and $\alpha = \beta$. This proves uniqueness.

### Corollary 1

*If $F = \{a_1, \ldots, a_m\}$ is a finite field, every mapping from $F$ to $F$ is a polynomial function in $F^*$, and so $F^*$ has $m^m$ elements.*

#### Proof

This follows from the Polynomial Fitting Theorem and the fact that a mapping $\theta$ from $F$ to $F$ is completely determined by the values of $\theta(a_1), \ldots, \theta(a_m)$.

#### Example 1

Find the polynomial $\alpha$ of lowest possible degree in $\mathbf{Q}[x]$ such that the equation $y = \alpha(x)$ fits the data in the table:

| $x$ | 0 | 1 | 2 | 3 | 4 | 5 |
|---|---|---|---|---|---|---|
| $y$ | 0 | 0 | 1 | 7 | 22 | 50 |

#### Solution

Theorem 1 tells us that the degree of $\alpha$ is 5 or less and that we may seek $\alpha$ in the form

$$\alpha = a + bx + cx(x - 1) + \cdots + fx(x - 1)(x - 2)(x - 3)(x - 4).$$

Using the fact that $\alpha(0) = 0$, we see that $a = 0$. Then $\alpha(1) = 0$ leads to $b = 0$. Next $\alpha(2) = 1$ implies that $1 = 2 \cdot 1 \cdot c$, or $c = 1/2$. Using these values of $a$, $b$, and $c$ and $\alpha(3) = 7$ gives us

$$7 = \tfrac{1}{2} \cdot 3 \cdot 2 + 3 \cdot 2 \cdot 1 \cdot d \quad \text{or} \quad d = \tfrac{2}{3}.$$

Similarly, one finds that $e = 0$ and $f = 0$. Hence

$$\begin{aligned}\alpha &= \tfrac{1}{2}x(x - 1) + \tfrac{2}{3}x(x - 1)(x - 2) \\ &= \tfrac{1}{6}x(x - 1)(4x - 5).\end{aligned}$$

No polynomial of lower degree fits the data since there is only one polynomial with degree 5 or less that meets the conditions.

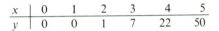

*Example 2*

Give a precise interpretation of the following question and answer it: "What value of $y$ should be associated with $x = 7/2$ so as to conform best with the table of Example 1?"

***Solution***

We interpret this problem as asking for the value of $\alpha(7/2)$, where $\alpha$ is the unique polynomial of degree 5 or less that fits the data of Example 1; that is, $\alpha$ is the polynomial

$$x(x-1)(4x-5)/6$$

found in that example. With this interpretation, the answer is

$$\alpha(7/2) = (7/2)(5/2)(9)/6 = 105/8.$$

*Example 3*

Let $y = f(x)$ have values as given in the table of Example 1. Describe a method of approximating $I = \int_0^5 f(x)\,dx$.

***Solution***

We let $\alpha$ be the polynomial $x(x-1)(4x-5)/6$ found in Example 1 and use $\int_0^5 \alpha(x)\,dx$ to approximate $I$.

In the remainder of this section, we let $F$ be a field with characteristic 0; then the unity 1 of $F$ generates an infinite cyclic subgroup in the additive group of $F$. We use a notation that allows us to think of this subgroup [1] as being $\mathbf{Z}$; specifically, we use the integer $n$ as a symbol for $n \cdot 1$ in $F$.

If $n$ and $k$ are integers with $0 < k \leq n$, the binomial coefficient "$n$ choose $k$" is given by

$$\binom{n}{k} = \frac{n(n-1)(n-2)\cdots(n-k+1)}{1 \cdot 2 \cdots k}. \tag{5}$$

For fixed $k$, the right side of (5) is a polynomial in $n$; this motivates the following introduction of a useful sequence of polynomials:

***Definition 1 Binomial Polynomials***

*For each positive integer $k$, the $k$th* **binomial polynomial** *is*

$$B_k = \frac{1}{k!}[x(x-1)(x-2)\cdots(x-k+1)].$$

*The zeroth binomial polynomial is* $B_0 = 1$.

Let $k > 0$; then it is easily seen that for $n = 0, 1, 2, \ldots, k - 1$ we have $B_k(n) = 0$ while for $n = k, k + 1, k + 2, \ldots$ the value of $B_k(n)$ is the binomial coefficient $\binom{n}{k} = \dfrac{n!}{k!(n-k)!}$.

Applying Theorem 1, or the technique of its proof, one easily obtains the following result.

***Theorem 2 The Binomial Representation***

*A polynomial* $\alpha$ *of degree* $d$ *in* $F[x]$ *is uniquely expressible as*

$$\alpha = b_0 + b_1 B_1 + b_2 B_2 + \cdots + b_d B_d, \quad b_i \in F, \quad b_d \neq 0,$$

*where the* $B_i$ *are the binomial polynomials.*

Explicit formulas for the coefficients $b_i$ in this representation are given in Problem 23(i) of this section. Also see Problems 25 and 26 for applications to polynomial fitting.

It is well known that the binomial coefficients $\binom{n}{k}$ satisfy

$$\binom{n}{k} + \binom{n}{k+1} = \binom{n+1}{k+1} \quad \text{or} \quad \binom{n}{k} = \binom{n+1}{k+1} - \binom{n}{k+1}.$$

In the notation of the binomial polynomials, the second of these becomes

$$B_k(n) = B_{k+1}(n+1) - B_{k+1}(n).$$

After we introduce some terminology, this formula will be generalized in Lemma 1 below.

***Notation Composition of Polynomials***

*Let* $R$ *be a commutative ring. Let* $\alpha$ *and* $\beta$ *be in* $R[x]$, *where*

$$\alpha = a_d x^d + a_{d-1} x^{d-1} + \cdots + a_1 x + a_0.$$

*Then* $\alpha(\beta)$ *denotes the polynomial*

$$a_d \beta^d + a_{d-1} \beta^{d-1} + \cdots + a_1 \beta + a_0.$$

In this notation, $\alpha(x)$ is the result of replacing $x$ by $x$ in $\alpha$; that is, $\alpha(x)$ is another notation for $\alpha$.

***Notation*** ***The Difference Operator Δ***

*The difference $\Delta\alpha$ of a polynomial $\alpha$ in $U[x]$ is given by*

$$\Delta\alpha = \alpha(x + 1) - \alpha. \tag{6}$$

We note that $\Delta$ is a mapping from $U[x]$ to itself and that the image of $\alpha$ under $\Delta$ is written as $\Delta\alpha$ rather than as $\Delta(\alpha)$. Since $\alpha(x) = \alpha$, formula (6) may be rewritten as

$$\Delta\alpha(x) = \alpha(x + 1) - \alpha(x).$$

***Lemma 1*** ***Differencing a Binomial Polynomial***

$\Delta B_k = B_{k-1}$ *for* $k > 0$.

***Proof***

We see that

$$\begin{aligned}
\Delta B_k &= B_k(x + 1) - B_k(x) \\
&= [(x + 1)x(x - 1)(x - 2) \cdots (x - k + 2)/k!] - [x(x - 1)(x - 2) \cdots (x - k + 1)/k!] \\
&= [(x + 1) - (x - k + 1)][x(x - 1)(x - 2) \cdots (x - k + 2)/k!] \\
&= kx(x - 1)(x - 2) \cdots (x - k + 2)/k! \\
&= x(x - 1)(x - 2) \cdots (x - k + 2)/(k - 1)! \\
&= B_{k-1}.
\end{aligned}$$

***Theorem 3*** ***Differencing Using the Binomial Representation***

*Let $\alpha = b_0 + b_1B_1 + \cdots + b_dB_d$, with $b_i$ in $F$ and with $B_i$ the ith binomial polynomial. If $d > 0$,*

$$\Delta\alpha = b_1 + b_2B_1 + b_3B_2 + \cdots + b_dB_{d-1}.$$

*Proof*

One easily sees that

$$\begin{aligned}\Delta\alpha &= \Delta[b_0 + b_1 B_1 + \cdots + b_d B_d] \\ &= \Delta b_0 + \Delta(b_1 B_1) + \cdots + \Delta(b_d B_d) \\ &= 0 + b_1 \Delta B_1 + b_2 \Delta B_2 + \cdots + b_d \Delta B_d \\ &= b_1 + b_2 B_1 + \cdots + b_d B_{d-1},\end{aligned}$$

using Lemma 1.

### *Corollary*

*If* $\deg \alpha = d > 0$, *then* $\deg(\Delta\alpha) = d - 1$. *If* $\alpha = 0$ *or* $\deg \alpha = 0$ (*that is, if* $\alpha$ *is in* $F$), *then* $\Delta\alpha = 0$.

### *Lemma 2 Summing Values of a Binomial Polynomial*

*For any* $a \in F$ *and any positive integers* $k$ *and* $n$,

$$B_{k-1}(a) + B_{k-1}(a+1) + \cdots + B_{k-1}(a+n) = B_k(a+n+1) - B_k(a). \qquad (7)$$

*Proof*

Lemma 1 tells us that $B_{k-1} = \Delta B_k$; hence the left side of (7) can be rewritten as

$$\begin{aligned}&\Delta B_k(a) + \Delta B_k(a+1) + \cdots + \Delta B_k(a+n) \\ &\qquad = [B_k(a+1) - B_k(a)] + [B_k(a+2) - B_k(a+1)] + \cdots + [B_k(a+n+1) - B_k(a+n)].\end{aligned}$$

When the terms that appear with both plus and minus signs are cancelled, the only remaining terms are

$$-B_k(a) + B_k(a+n+1).$$

This equals the right side of (7), and so the lemma is proved.

We now use Theorem 2 and Lemma 2 to obtain the following result.

### *Theorem 4 Summing Values of a Polynomial*

*Let $a \in F$, $\alpha \in F[x]$, and $\deg \alpha = d$. Then there is a $\beta$ of degree $d + 1$ in $F[x]$ such that*

$$\alpha(a) + \alpha(a + 1) + \cdots + \alpha(a + n) = \beta(n). \tag{8}$$

***Proof***

Since $B_0 = 1$, Theorem 2 tells us that we can write

$$\alpha = b_0 B_0 + b_1 B_1 + \cdots + b_d B_d, \quad b_d \neq 0. \tag{9}$$

Now let

$$\gamma = b_0 B_1 + b_1 B_2 + \cdots + b_d B_{d+1}.$$

We show below that the desired $\beta$ is $\gamma(x + a + 1) - \gamma(a)$.

Let $S$ be the sum on the left side of (8). Using (9) to express $\alpha(j)$ in terms of the $B_i(j)$, we find that

$$S = b_0[B_0(a) + B_0(a + 1) + \cdots + B_0(a + n)] + \cdots + b_d[B_d(a) + B_d(a + 1) + \cdots + B_d(a + n)].$$

Then Lemma 2 helps us to convert this to

$$\begin{aligned} S &= b_0[B_1(a + n + 1) - B_1(a)] + b_1[B_2(a + n + 1) - B_2(a)] + \cdots \\ &\quad + b_d[B_{d+1}(a + n + 1) - B_{d+1}(a)] \\ &= [b_0 B_1(a + n + 1) + b_1 B_2(a + n + 1) + \cdots + b_d B_{d+1}(a + n + 1)] \\ &\quad - [b_0 B_1(a) + b_1 B_2(a) + \cdots + b_d B_{d+1}(a)] \\ &= \gamma(a + n + 1) - \gamma(a). \end{aligned}$$

Hence $S = \beta(n)$, where $\beta(x) = \gamma(x + a + 1) - \gamma(a)$, as desired.

***Example 4***

Find a closed form for the sum

$$C_n = 0^2 + 1^2 + 2^2 + \cdots + n^2.$$

***Solution***

We note that

$$C_n = \alpha(0) + \alpha(1) + \alpha(2) + \cdots + \alpha(n)$$

if $\alpha$ is chosen as $x^2$. Then it follows from Theorem 4 that $C_n = \beta(n)$ for some $\beta$ of degree 3. Hence we may look for $\beta$ in the form

$$\beta = a + bx + cx(x-1) + dx(x-1)(x-2). \tag{10}$$

One easily calculates that $\beta(0) = C_0 = 0$, $\beta(1) = C_1 = 1$, $\beta(2) = 1 + 4 = 5$, and $\beta(3) = 1 + 4 + 9 = 14$.

Successively letting $x = 0, 1, 2$, and 3 in (10), one then finds that $a = 0$, $b = 1$, $c = 3/2$, and $d = 1/3$. Hence

$$\beta = x + \left(\frac{3}{2}\right)x(x-1) + \left(\frac{1}{3}\right)x(x-1)(x-2).$$

Expanding and collecting like terms, we have

$$\beta = \frac{2x^3 + 3x^2 + x}{6}.$$

Therefore,

$$1^2 + 2^2 + 3^2 + \cdots + n^2 = \frac{2n^3 + 3n^2 + n}{6}.$$

The Maclaurin Series Expansion of calculus for a polynomial $f(x)$ of degree $d$ is the finite sum

$$f(x) = f(0) + f'(0)x + (1/2!)f''(0)x^2 + \cdots + (1/d!)f^{(d)}(0)x^d.$$

The analogue of this formula, with derivatives replaced by differences and powers $x^n$ replaced by the binomial polynomials $\mathbf{B_n}$, is in Problem 23 of this section.

## *Problems*

1. Find the polynomial $\alpha$ of degree 4 in $\mathbf{Q}[x]$ such that 1, 2, 3, and 5 are zeros of $\alpha$ and $\alpha(4) = -18$.
2. Find the polynomial $\alpha$ of degree 3 in $\mathbf{Z}_7[x]$ such that $\bar{1}$, $\bar{3}$, and $\bar{4}$ are zeros of $\alpha$ and $\alpha(\bar{0}) = \bar{4}$.
3. Find the polynomial $\beta$ of least degree in $\mathbf{Z}_5[x]$ such that $\beta(\bar{0}) = \bar{1}$, $\beta(\bar{1}) = \bar{0}$, $\beta(\bar{2}) = \bar{3}$, $\beta(\bar{3}) = \bar{2}$, and $\beta(\bar{4}) = \bar{4}$.
4. Find the polynomial $\beta$ of least degree in $\mathbf{Z}[x]$ such that $\beta(-1) = -1$, $\beta(0) = -3$, $\beta(2) = 5$, and $\beta(5) = 47$.

5. Prove that there is no $\alpha$ of degree 3 in $\mathbf{Q}[x]$ to fit the table:

| $x$ | 1 | 2 | 3 | 4 | 5 | 6 |
|---|---|---|---|---|---|---|
| $\alpha(x)$ | 1 | 2 | 6 | 20 | 76 | 312 |

6. Prove that $\deg \beta > 3$ for every $\beta$ in $\mathbf{Q}[x]$ that fits the table:

| $x$ | 0 | 1 | 2 | 3 | 4 |
|---|---|---|---|---|---|
| $\beta(x)$ | 6 | 5 | 6 | 13 | 32 |

7. Let $\alpha = x^3 - 3x^2 + 2x$. Express

$$\alpha(0) + \alpha(1) + \alpha(2) + \cdots + \alpha(n)$$

in the form $\beta(n)$ with $\beta$ in $\mathbf{Q}[x]$.

8. Let $\alpha = x^4 - 6x^3 + 11x^2 - 6x$. Express

$$\alpha(0) + \alpha(1) + \alpha(2) + \cdots + \alpha(n)$$

as a polynomial in $n$ with coefficients in $\mathbf{Q}$.

9. Express the following sum as $\beta(n)$ with $\beta$ in $\mathbf{Q}[x]$:

$$1^3 + 2^3 + 3^3 + \cdots + n^3.$$

10. Express the following sum as $\beta(n)$ with $\beta$ in $\mathbf{Q}[x]$:

$$1^4 + 2^4 + 3^4 + \cdots + n^4.$$

11. Let $A_n$ be the maximum number of regions into which a plane can be cut up by $n$ straight lines.
    (i) Find $A_n$ for $n = 1, 2, 3,$ and 4.
    (ii) Prove that $A_n = A_{n-1} + n$ for $n > 1$.
    (iii) Express $A_n$ as a polynomial in $n$.

12. Let $B_n$ be the maximum number of regions into which 3-space can be decomposed by $n$ planes.
    (i) Find $B_n$ for $n = 1, 2, 3,$ and 4.
    *(ii) Prove that $B_n = B_{n-1} + (n^2 - n + 2)/2$ for $n > 1$.
    (iii) Find a closed form for $B_n$; that is, express $B_n$ as a polynomial in $n$.

13. Let $d$ be a nonnegative integer. For $k = 0, 1, \ldots, d$ let $\alpha_k$ be either 0 or a polynomial of degree $k$ in $F[x]$ and let

$$\alpha = n^d\alpha_0 + n^{d-1}\alpha_1 + n^{d-2}\alpha_2 + \cdots + n\alpha_{d-1} + \alpha_d.$$

Prove that there is a polynomial $\beta$ in $F[x]$ with

$$\alpha(0) + \alpha(1) + \cdots + \alpha(n) = \beta(n)$$

and either $\beta = 0$ or $\deg \beta \le d + 1$.

14. Let

$$A_n = 0 \cdot n + 1 \cdot (n-1) + 2(n-2) + 3(n-3) + \cdots + (n-1) \cdot 1 + n \cdot 0.$$

(i) Explain why there is a polynomial $\beta$, of degree at most 3, in $\mathbf{Q}[x]$ such that $A_n = \beta(n)$.
(ii) Find a closed form for $A_n$. (*Hint:* Find $\beta$.)

15. For all nonnegative integers $n$, prove that

$$0^2 \cdot n + 1^2 \cdot (n-1) + 2^2 \cdot (n-2) + \cdots + (n-1)^2 \cdot 1 + n^2 \cdot 0 = n^2(n^2-1)/12.$$

16. Find a closed form for

$$0^2 \cdot n^2 + 1^2 \cdot (n-1)^2 + 2^2 \cdot (n-2)^2 + \cdots + (n-1)^2 \cdot 1^2 + n^2 \cdot 0^2.$$

17. In $\mathbf{Q}[x]$, let $\alpha = a_3 x^3 + a_2 x^2 + a_1 x + a_0$ with $a_3 \neq 0$. Let $h = a_2/3a_3$. Prove that

$$\alpha(x-h) = a_3 x^3 + b_1 x + b_0$$

with $b_1$ and $b_0$ in $\mathbf{Q}$.

18. In $\mathbf{Q}[x]$, let $\alpha = a_d x^d + a_{d-1} x^{d-1} + \cdots + a_1 x + a_0$ with $a_d \neq 0$. Find an $h$ such that the coefficient of $x^{d-1}$ in $\alpha(x-h)$ is 0.

19. Let the $n$th difference operator $\Delta^n$ be defined inductively for all positive integers $n$ by $\Delta^1\alpha = \Delta\alpha$,

$$\Delta^2\alpha = \Delta(\Delta\alpha) = \Delta[\alpha(x+1) - \alpha] = [\alpha(x+2) - \alpha(x+1)] - [\alpha(x+1) - \alpha]$$
$$= \alpha(x+2) - 2\alpha(x+1) + \alpha, \ldots,$$

$\Delta^{m+1}\alpha = \Delta(\Delta^m\alpha)$. Let the binomial coefficient $\binom{n}{k} = n!/k!(n-k)!$. Prove that

$$\Delta^n\alpha = \binom{n}{n}\alpha(x+n) - \binom{n}{n-1}\alpha(x+n-1) + \binom{n}{n-2}\alpha(x+n-2) - \cdots + (-1)^n\binom{n}{0}\alpha(x).$$

20. Let $\deg \alpha = d > 0$. Let $\Delta^n$ be as in the preceding problem. Prove that:
    (a) $\deg(\Delta^n \alpha) = d - n$ for $1 \le n \le d$.
    (b) $\Delta^d \alpha = d!L(\alpha)$, where $L(\alpha)$ is the leading coefficient of $\alpha$.
    (c) $\Delta^n \alpha = 0$ for $n > d$.

21. Let $k$ be a positive integer, let $\binom{n}{k} = n!/k!(n-k)!$, and let

    $B_k = x(x-1)\cdots(x-k+1)/k!$ Show that:
    (a) $B_k(n) = 0$ for $n = 0, 1, 2, \ldots, k-1$.
    (b) $B_k(n) = \binom{n}{k}$ for $n$ an integer and $n \ge k$.
    (c) $B_k(n) = (-1)^k \binom{k-n-1}{-n-1}$ for negative integers $n$.

22. Explain why $B_k(n)$ must be an integer for all integers $n$ and all nonnegative integers $k$. (You may use the fact that $\binom{n}{k}$ is an integer for $k$ and $n$ integers with $0 \le k \le n$.)

23. Let $\alpha = b_0 + b_1 B_1 + b_2 B_2 + \cdots + b_d B_d$. Show that:
    (i) $\alpha(0) = b_0$ and $\Delta^j \alpha(0) = b_j$ for $j = 1, 2, \ldots, d$.
    (ii) $b_j = \binom{j}{j}\alpha(j) - \binom{j}{j-1}\alpha(j-1) + \binom{j}{j-2}\alpha(j-2) - \cdots + (-1)^j \binom{j}{0}\alpha(0)$.
    (iii) If $\alpha(n)$ is an integer for $n = 0, 1, \ldots, d$, then $b_j$ is an integer for $j = 0, 1, \ldots, d$.
    (iv) If $\alpha(n)$ is an integer for $n = 0, 1, \ldots, d$ or $b_j$ is an integer for $j = 0, 1, \ldots, d$, then $\alpha(n)$ is an integer for all integers $n$.

24. Express $\alpha = (x^4 - 2x^3 + 11x^2 + 14x)/24$ in the form

$$\alpha = b_0 + b_1 B_1 + b_2 B_2 + b_3 B_3 + b_4 B_4$$

    and also prove that $\alpha(n)$ is an integer for all integers $n$.

25. Given that $\deg \beta = 3$, complete the following table and then use the column for $x = 0$ and Problem 23(i) to express $\beta$ in terms of the binomial polynomials $B_1$, $B_2$, and $B_3$:

| $x$ | 0 | 1 | 2 | 3 |
|---|---|---|---|---|
| $\beta(x)$ | 1 | 1 | 7 | 25 |
| $\Delta\beta(x)$ | 0 | 6 | 18 | |
| $\Delta^2\beta(x)$ | 6 | | | |
| $\Delta^3\beta(x)$ | | | | |

26. Given that deg $\alpha = 4$, use the table that follows to find $\Delta^j\alpha(0)$ for $j = 0, 1, 2, 3, 4$ and then use Problem 23(i) to express $\alpha$ in terms of the binomial polynomials $B_k$.

| $x$ | 0 | 1 | 2 | 3 | 4 |
|---|---|---|---|---|---|
| $\alpha(x)$ | 0 | 0 | 7 | 39 | 126 |

27. Let $c$ be in $F$ and $\alpha$ be in $F[x]$ with deg $\alpha = d$. Let $\alpha'$ be the derivative of $\alpha$, $\alpha^{(2)}$ be the derivative of $\alpha'$, ..., and $\alpha^{(d)}$ be the derivative of $\alpha^{(d-1)}$. Show that

$$\alpha = \alpha(c) + (x - c)\alpha'(c) + (x - c)^2\alpha^{(2)}(c)/2! + \cdots + (x - c)^d\alpha^{(d)}(c)/d!.$$

(The definition of derivative is in Problem 21 of Section 5.2. It may be helpful to see Problem 16 of Section 5.3.)

28. Use the notation of the preceding problem and let $k$ be an integer with $1 \leq k \leq d$. Prove that $(x - c)^k | \alpha$ if and only if

$$\alpha(c) = \alpha'(c) = \alpha^{(2)}(c) = \cdots = \alpha^{(k-1)}(c) = 0.$$

29. Let $c$ be a zero with multiplicity $m$ of an $\alpha$ in $F[x]$. Prove that:
(a) If $m = 1$, $c$ is not a zero of $\alpha'$ (the derivative of $\alpha$).
(b) If $m > 1$, $c$ is a zero of $\alpha'$ with multiplicity $m - 1$.

30. Let $\mathbf{C}$ be the field of complex numbers. In $\mathbf{C}[x]$, let

$$\alpha = x^3 + 3x^2 + 6x + 6.$$

Prove that each zero of $\alpha$ has multiplicity 1.

31. Let $\alpha \in \mathbf{Z}[x]$, $h \in \mathbf{Z}$, and $\alpha(h) = \alpha(h + 1) = \alpha(h + 2) = 0$. Show that $6 | \alpha(n)$ for all integers $n$.

32. Let $\alpha \in \mathbf{Z}[x]$, $h \in \mathbf{Z}$, $k \in \mathbf{Z}^+$, and

$$0 = \alpha(h) = \alpha(h + 1) = \alpha(h + 2) = \cdots = \alpha(h + k - 1).$$

Show that $k! | \alpha(n)$ for all integers $n$.

---

***Sir Isaac Newton*** *(1642–1727) Nicolaus Copernicus, a Polish astronomer, published a book on the heliocentric theory in 1543. Tycho Brahe, a Dane, carefully recorded volumes of accurate observations of the positions of planets* *with the times of the observations. Galileo Galilei, an Italian, refined the telescope and saw Jupiter's orbiting moons. Johannes Kepler, a German, patiently tried pattern after pattern until he fitted Brahe's data into three laws of planetary motion which state that the path of a planet is an ellipse with the sun at a focus, the line segment joining the planet and sun sweeps out equal areas in equal times, and the squares of the times of revolution of any two planets are proportional to the cubes of their distances to the sun.*

*Newton, an Englishman, went far beyond Kepler's tremendous feat. He created calculus and differential equations and used them to show that Kepler's laws were a consequence of Newton's universal law concerning the gravitational attraction between any two particles.*

*One particular result of Newton's work was that the path of a rocket to the moon or another body in space could be charted accurately. A general consequence is that the last three centuries have been a time in which mathematics has been applied to an extremely wide range of problems.*

*From an inauspicious background Newton rose to become the towering figure of British science during his lifetime and one of the greatest men of science of all time. Contemporary portraits, busts, and medallions show him in almost god-like poses, one late portrait making him appear somewhat like a Roman senator. Recent studies, notably the biography by Manuel, have focused on some flaws in his character, such as lack of generosity in failing to concede the accomplishments of his contemporaries and in refusing to recognize ideas of his successors when they did not agree with his own. He controlled the Royal Society for many years and, partly because of his appointment as Director of the Royal Mint, he wielded considerable power. Nevertheless, his reputation as a great scientist is not seriously tarnished.*

*His greatest work was the* **Philosophiae Naturalis Principia Mathematica** *(1687). It contains some of his most startling discoveries concerning the applications of mathematics to physical problems. Laplace said that the* **Principia,** *because of its profound and original ideas, was assured "a lasting pre-eminence over all other productions of the human mind." Some indication of the importance of this book might be the fact that the first edition is presently priced at about $10,000 on the rare book market.*

*His* **Arithmetica Universalis** *(1707) contained many original results in the theory of algebraic equations and his* **Method of Fluxions** *(1736) concerns not only an explanation of his discoveries in calculus but also his work on the approximation of roots of equations.*

*The question of whether Newton or Leibniz invented the calculus caused heated debate on both sides of the English Channel for many years. Newton's first ideas on the calculus apparently date from about 1665, but his first writings on the subject were the* **De analysi per aequationes numero terminorum infinitas,** *written in 1669 and first published in 1711, and his* **Methodus fluxionum et serierum infinitorum**, *written in 1671 and first published in English translation in*

*1736 and in the original Latin in 1742. Calculus methods were employed in the* **Principia**, *in 1687. So that although he may have had the ideas of calculus prior to Leibniz, nevertheless Leibniz published first in 1684 with his "Nova methodus pro maximis et minimis" in the* **Acta Eruditorum**. *Furthermore, it is generally conceded today that both men discovered the calculus independently. This was not accepted generally during the height of the conflict between the two men and their followers. Newton's friends accused Leibniz of plagiarism. In 1687 Newton had generously acknowledged in the first edition of the* **Principia** *that Leibniz had discovered a method similar to his. Following the development of the dispute over priority, Newton deleted this reference in the third edition of 1726. The conflict had the unfortunate effect of forcing mathematicians into two camps, the followers of Newton and the followers of Leibniz. British scientists used the notation of Newton, which is inferior to that of Leibniz, and following the deaths of Newton and Leibniz, mathematicians on the continent, using the notation of Leibniz, far surpassed the British in mathematical advances. The Leibniz notation is essentially that used in calculus courses today; that of Newton is reserved for certain applications in physics.*

*Whereas Leibniz was primarily a philosopher who created great mathematics as well, Newton was primarily a physicist who created the mathematics he needed. His* **Opticks** *(1704) is another great milestone in the history of science. Newton, who died a powerful and wealthy man, much honored in Britain and elsewhere, is buried in Westminster Abbey in a massive marble tomb prominently displayed in the nave. Leibniz died neglected; it is reported that his funeral was attended by a single mourner.*

---

## 5.7 *Equations of Degree 2, 3, and 4*

For every positive integer $n$, a nonzero complex number

$$r(\cos a + \mathbf{i} \sin a), \quad a \in \mathbf{R},\ r \in \mathbf{R}^+$$

has exactly $n$ $n$th roots given by

$$\sqrt[n]{r}\{\cos[(a + 2k\pi)/n] + \mathbf{i} \sin[(a + 2k\pi)/n]\};\ k = 0, 1, \ldots, n - 1,$$

where $\sqrt[n]{r}$ denotes the unique positive $n$th root of $r$. The only $n$th root of 0 is 0 itself. (These facts are part of the well known Theorem of DeMoivre; they also follow readily from Theorem 2 of Section 2.6.)

Hence the field **C** of complex numbers is closed under extraction of $n$th roots as well as under the four rational operations (addition, subtraction, multiplication, and division by nonzero elements.)

***Definition 1 Solution in Radicals***

*A polynomial equation $\alpha = 0$ is solvable in radicals if each root is expressible in terms of the coefficients using only the rational operations and extraction of $n$th roots (with each $n$ in $\mathbf{Z}^+$).*

This section deals with methods for solution in radicals of polynomial equations with degree 2, 3, and 4. We have previously mentioned that it is not possible for such a method to apply to all equations with degree 5 or any higher degree. (See the biographical notes on Tartaglia, Ferrari, Ruffini, Abel, and Galois and Section 5.12 below.)

In the rest of this section, $F$ always designates a field that is closed under extraction of $n$th roots and has characteristic different from 2 and from 3. The methods given apply to all such fields $F$ and hence to the complex numbers **C**. Since the characteristic of $F$ is not 2 or 3, we have $2 \cdot 1 \neq 0$ and $3 \cdot 1 \neq 0$ in $F$. Then $4 \cdot 1 = (1 + 1)(1 + 1) \neq 0$ since $F$ has no 0-divisors. In $F$, we designate these nonzero elements $2 \cdot 1$, $3 \cdot 1$, and $4 \cdot 1$ by 2, 3, and 4, respectively.

If $f$ is in $F$, the roots of

$$x^d + f = 0 \tag{1}$$

are the $d$th roots of $-f$; hence (1) is solvable in radicals. One approach to solving a general monic equation

$$x^d + a_{d-1}x^{d-1} + \cdots + a_1 x + a_0 = 0, \quad a_i \in F, \tag{2}$$

is to seek techniques for reducing (2) to (1). Such techniques are possible for $d \in \{2, 3, 4\}$ but we shall take only the first step in this direction.

For $d$ in $\{2, 3, 4\}$, $h = a_{d-1}d^{-1}$ denotes an element of $F$. Then the substitution

$$x = y - h = y - a_{d-1}d^{-1}$$

in equation (2) leads to an equation

$$y^d + b_{d-2}y^{d-2} + \cdots + b_1 y + b_0, \quad b_i \in F \tag{3}$$

in which the coefficient of $y^{d-1}$ turns out to be 0. If the roots of (3) can be found, one subtracts $h$ from each of them to obtain the roots of (2).

*Example 1*

Find all roots of $x^2 + 6x + 13 = 0$ in $\mathbf{C}$.

*Solution*

Here $d = 2$, $h = 6/2 = 3$, and the substitution $x = y - 3$ leads to

$$\begin{aligned} (y-3)^2 + 6(y-3) + 13 &= 0 \\ (y^2 - 6y + 9) + (6y - 18) + 13 &= 0 \\ y^2 + 4 &= 0 \\ y^2 &= -4. \end{aligned}$$

Hence $y = \pm 2\mathbf{i}$ and $x = y - 3 = -3 \pm 2\mathbf{i}$.

*Example 2*

In $F$, solve $ax^2 + bx + c = 0$, $a \neq 0$.

*Solution*

Applying the technique of Example 1 to

$$x^2 + (b/a)x + (c/a) = 0,$$

we substitute $x = y - (b/2a)$ and find, after a few steps, that $y = \pm\sqrt{b^2 - 4ac}/2a$. Then

$$x = y - \left(\frac{b}{2a}\right) = \frac{-b \pm \sqrt{b^2 - 4ac}}{2a}.$$

Next we turn to monic cubic equations

$$x^3 + a_2 x^2 + a_1 x + a_0 = 0.$$

The substitution $x = y - (a_2/3)$ leads to an equation of the form

$$y^3 + py + q = 0. \tag{4}$$

Since it took about 2000 years to go from solving quadratics to solving cubics, one should not be surprised that at this point we have to pull a rabbit out of a hat. Our approach is to factor the left side of (4) into first degree polynomials with the help of the factorization

$$y^3 - 3rsy + (r^3 + s^3) = (y + r + s)(y + wr + w^2s)(y + w^2r + ws), \tag{5}$$

where $w = (-1 + \sqrt{-3})/2$. [In **C**, one can rewrite $w$ as $(-1 + \mathbf{i}\sqrt{3})/2$.] The factorization in (5) is easily verified by expanding the right side and using the fact that $w^2 + w + 1 = 0$.

Now we have to choose $r$ and $s$ in (5) so as to make the left side of (5) the same as the left side of (4). That is, we want $r$ and $s$ to satisfy

$$-3rs = p \quad \text{and} \quad r^3 + s^3 = q.$$

The first of these equations tells us that $s = -p/3r$; substituting this in the second equation leads to

$$\begin{aligned} r^3 + (-p/3r)^3 &= q \\ r^3 - (p^3/27r^3) &= q \\ r^6 - (p^3/27) &= qr^3 \\ (r^3)^2 - qr^3 - (p^3/27) &= 0. \end{aligned}$$

Now the quadratic formula (given in Example 2) leads to

$$r^3 = [q \pm \sqrt{q^2 + (4p^3/27)}]/2. \tag{6}$$

We do not need all the solutions for $r$ and $s$; one pair will enable us to factor the left side of (4). Hence we let $r$ be a cube root of the right side of (6), with one of the two choices in the $\pm$ sign. Then we let $s = -p/3r$ and equate to zero the first degree factors on the right side of (5). In this way we find the roots of (4) in the form

$$-(r + s), \qquad -(wr + w^2s), \qquad -(w^2r + ws). \tag{7}$$

***Example 3***

Find all the roots of $x^3 - 3x + 1 = 0$ in **C**.

***Solution***

The technique used above on the general cubic tells us that we shall have the factorization

$$x^3 - 3x + 1 = (x + r + s)(x + wr + w^2 s)(x + w^2 r + ws)$$

if $r$ and $s$ satisfy

$$-3rs = -3, \; r^3 + s^3 = 1.$$

Substituting $s = -3/(-3r) = 1/r$ in $r^3 + s^3 = 1$ leads to

$$\begin{aligned} r^3 + (1/r^3) &= 1 \\ (r^3)^2 + 1 &= r^3 \\ (r^3)^2 - r^3 + 1 &= 0. \end{aligned}$$

This quadratic for $r^3$ is satisfied by $r^3 = (1 + \sqrt{1 - 4})/2$. Hence we may let

$$r = \sqrt[3]{(1 + \mathbf{i}\sqrt{3})/2}, \quad s = 1/r$$

and then the roots of $x^3 - 3x + 1 = 0$ are given by the expressions in (7).

Finally we consider fourth degree polynomial equations

$$x^4 + ax^3 + bx^2 + cx + d = 0. \tag{8}$$

Here we do not make the substitution $x = y - (a/4)$, which would knock out the $y^3$ term. Instead, our approach is to factor the left side of (8) into quadratics. The first step is to attempt to rewrite (8) in the form

$$(x^2 + hx + k)^2 - (ux + v)^2 = 0. \tag{9}$$

Then (9) can be put in the desired form

$$[(x^2 + hx + k) + (ux + v)][(x^2 + hx + k) - (ux + v)] = 0 \tag{10}$$

and solved by two applications of the quadratic formula.

What is necessary to convert (8) into (9)? Expanding the left side of (9) and collecting like terms, we see that (8) goes into (9) if we can choose

$h$, $k$, $u$, and $v$ so as to satisfy the simultaneous conditions

$$a = 2h, \quad b = h^2 + 2k - u^2, \quad c = 2hk - 2uv, \quad d = k^2 - v^2.$$

The first of these conditions shows that we should let $h = a/2$. We substitute $h = a/2$ in the other three conditions and rewrite them as

$$u^2 = (a^2/4) + 2k - b, \qquad 2uv = ak - c, \qquad v^2 = k^2 - d. \tag{11}$$

Since $4u^2v^2 - (2uv)^2 = 0$, the conditions in (11) can only be satisfied if

$$4[(a^2/4) + 2k - b][k^2 - d] - [ak - c]^2 = 0.$$

Simplifying this equation, we find that $k$ has to satisfy

$$8k^3 - 4bk^2 + (2ac - 8d)k + (4bd - a^2d - c^2) = 0. \tag{12}$$

Equation (12) is called the ***resolvent cubic*** for the quartic (8).

Now one can show that the quartic (8) factors as in (10) if one chooses $k$ to be a root of (12) and lets

$$h = a/2, \qquad u = \sqrt{h^2 + 2k - b}, \qquad v = \sqrt{k^2 - d}. \tag{13}$$

The details are left to the reader.

One may note that this solution for the quartic equation involves solutions of a cubic and two quadratics.

***Example 4***

Let $\alpha = x^4 + 2x^3 + 3x^2 + 2x + 2$. Find all the roots of $\alpha = 0$ in the field $\mathbf{C}$ of complex numbers.

***Solution***

First we try the easy (when it works) method of seeking rational roots using Theorem 2 of Section 5.5. That result tells us that the only possibilities for rational roots are the integers $-2$, $-1$, 1, and 2. Substitution shows that no one of these integers is a zero of $\alpha$; hence we turn to the method for general quartics.

The resolvent cubic equation (12) for our given $\alpha$ is

$$2k^3 - 3k^2 - 2k + 3 = 0.$$

To solve $\alpha = 0$, we need only one root of this cubic. Hence we hope for a rational root. One of the possibilities for rational roots is 3/2, and we are fortunate that it actually checks out to be a root; so we let $k = 3/2$.

In equation (10) above, $h$ is half of the coefficient of $x^3$ in $\alpha$; thus $h = 2/2 = 1$. Using $h = 1$ and $k = 3/2$ in (13), we calculate $u = \sqrt{1 + 3 - 3} = 1$ and $v = \sqrt{(9/4) - 2} = 1/2$. Then we substitute these values for $h$, $k$, $u$, and $v$ in (10) and $\alpha = 0$ is converted to

$$(x^2 + 1)(x^2 + 2x + 2) = 0.$$

Equating each of these quadratic factors to zero and solving the two quadratic equations that result, we find that the set of zeros of $\alpha = 0$ is $\{\mathbf{i}, -\mathbf{i}, -1 + \mathbf{i}, -1 - \mathbf{i}\}$.

In elementary algebra, the discriminant of the monic quadratic equation

$$x^2 + bx + c = 0 \tag{14}$$

is $D = b^2 - 4c$. It is left to the reader as Problem 5 below, to show that $D$ equals the square of the difference of the roots of equation (14) and that, when $b$ and $c$ are real, one obtains information concerning the roots by knowing whether $D = 0$, $D > 0$, or $D < 0$. We now generalize to monic polynomials of arbitrary degree that can be factored in the form

$$\alpha = (x - r_1)(x - r_2) \cdots (x - r_d). \tag{15}$$

***Definition 2 Discriminant***

*The* **discriminant** *of the* $\alpha$ *of* (15) *is*

$$\left[\prod_{1 \le i < j \le d} (r_i - r_j)\right]^2,$$

*that is, the square of the product of the differences of the roots of* $\alpha = 0$.

For example, the discriminant of $(x - r_1)(x - r_2)(x - r_3)$ is

$$[(r_1 - r_2)(r_1 - r_3)(r_2 - r_3)]^2.$$

Discriminants of monic cubics are dealt with in Problems 7–10 below.

## Problems

1. Let $a$ and $b$ be real numbers with $b \neq 0$. Let $r = \sqrt{a^2 + b^2}$. Show that the square roots of $a + b\mathbf{i}$ are

$$\pm[\sqrt{(r+a)/2} + \mathbf{i}b/2\sqrt{(r+a)/2}].$$

2. Let $a$ and $b$ be real numbers with $b \neq 0$. Let $r = \sqrt{a^2 + b^2}$. Show that the square roots of $a + b\mathbf{i}$ are

$$\pm[(b/2\sqrt{(r-a)/2}) + \mathbf{i}\sqrt{(r-a)/2}].$$

3. Solve $z^2 - 2\mathbf{i}\sqrt{7}\,z - 24\mathbf{i} = 0$ for $z$ in the form $a + b\mathbf{i}$. [The result in Problem 1 (or Problem 2) may be helpful.]

4. Find the roots of $x^2 + 2\sqrt{5}\,x - 12\mathbf{i} = 0$ in the form $a + b\mathbf{i}$.

5. Let $x^2 + bx + c = (x - r)(x - s)$ in $\mathbf{C}[x]$. Let $D = (r - s)^2$. Show the following:
   (i) $D = b^2 - 4c$.
   (ii) $D = 0$ if and only if $r = s$.
   In the parts which follow, let $b$ and $c$ be real.
   (iii) $D > 0$ if and only if $r$ and $s$ are real and distinct.
   (iv) $D < 0$ if and only if $r$ and $s$ are nonreal complex conjugates.

6. For each of the following monic quadratics $\alpha$, use the discriminant tests of Problem 5 to see whether the roots of $\alpha = 0$ are equal, real and distinct, or nonreal complex conjugates:
   (a) $x^2 + 3x + 1$.
   (b) $x^2 + 3x + 3$.
   (c) $x^2 + 14x + 49$.

7. Let $x^3 + px + q = (x - r)(x - s)(x - t)$ in $\mathbf{C}[x]$. Let $D = [(r - s)(r - t)(s - t)]^2$. Show that:
   (i) $D = -(4p^3 + 27q^2)$.
   (ii) $D = 0$ if and only if two of $r$, $s$, and $t$ are equal.
   In the parts which follow, let $p$ and $q$ be real.
   (iii) $D > 0$ if and only if $r$, $s$, and $t$ are real and distinct.
   (iv) $D < 0$ if and only if one of $r$, $s$, and $t$ is real and the other two are nonreal complex conjugates.

8. Let $\alpha = x^3 + ax^2 + bx + c$ in $\mathbf{C}[x]$, let $h = a/3$, and $\alpha(x - h) = \beta$. Show that:
   (i) $\beta$ is of the form $x^3 + px + q$.
   (ii) The discriminants of $\alpha$ and $\beta$ are equal.

9. Use the result in Problem 7(i) to find the discriminant of each of the following monic cubics and then classify the roots as in parts (ii)–(iv) of Problem 7:
   (a) $x^3 - x - 1 = 0$.
   (b) $x^3 - 12x + 16 = 0$.
   (c) $x^3 - 5x + 2 = 0$.

10. Use the result in Problem 8 to find the discriminant of

$$x^3 - 6x^2 + 7x + 9 = 0$$

and classify the roots as in parts (ii)–(iv) of Problem 7.

11. Find all the roots in **C** of $x^3 + 3x - 1 = 0$.

12. Find all the roots in **C** of $x^3 + 3x - 4 = 0$.

13. Use the triple-angle formula $\cos(3\theta) = 4\cos^3\theta - 3\cos\theta$ to show that $\cos 10°$, $\cos 110°$, and $\cos 130°$ are the three roots of

$$4x^3 - 3x - (\sqrt{3}/2) = 0.$$

14. Express the roots of $4x^3 - 3x = 1/2$ in terms of cosines. (See the preceding problem.)

15. Find all the zeros of $\alpha$ in **C**, for each of the following $\alpha$'s:
   (a) $x^4 - 4x^3 - 6x^2 - 12x + 9$.
   (b) $x^4 - 10x^3 + 26x^2 - 5x - 2$.
   (c) $x^4 + 4x^3 + 6x^2 + 4x + 1$.

16. Find all the zeros of $\alpha$ in **C**, for each of the following $\alpha$'s:
   (a) $x^4 - x^3 + x^2 - 3x + 2$.
   (b) $x^4 + x^3 - 3x^2 - 5x - 2$.

17. Show that $c^2 = da^2$, given that, in $\mathbf{C}[x]$,

$$x^4 + ax^3 + bx^2 + cx + d = (x^2 + sx + p)(x^2 + tx + p).$$

18. Let $\alpha = x^4 + ax^3 + bx^2 + cx + d$ in $\mathbf{C}[x]$ with $c^2 = da^2 \neq 0$. Show that the roots of $\alpha = 0$, when suitably numbered, satisfy $r_1 r_2 = r_3 r_4$.

19. Let $\alpha = x^4 + ax^3 + bx^2 + cx + d$ in $\mathbf{C}[x]$ with $4ab = a^3 + 8c$. Show that the roots of $\alpha = 0$, when suitably numbered, satisfy $r_1 - r_2 = r_3 - r_4$.

20. Let $\alpha = x^4 + x^3 - x^2 - x + 1$. Factor $\alpha$ as in Problem 17 and thus find all the zeros of $\alpha$ in **C**.

21. In $\mathbf{C}[x]$, let

$$x^3 - ax^2 + bx - c = (x - r)(x - s)(x - t)$$

and let $N_k = r^k + s^k + t^k$. Show that:
(i) $N_0 = 3$ if $c \neq 0$.
(ii) $N_1 = a$.
(iii) $N_2 = a^2 - 2b$.
(iv) $N_{k+3} - aN_{k+2} + bN_{k+1} - cN_k = 0$ for $k = 0, 1, 2, \ldots$.

22. In $\mathbf{C}[x]$, let

$$x^4 - ax^3 + bx^2 - cx + d = (x - r)(x - s)(x - t)(x - u)$$

and let $N_k = r^k + s^k + t^k + u^k$. Show that:

(i) $N_0 = 4$ if $d \neq 0$.
(ii) $N_1 = a$.
(iii) $N_2 = a^2 - 2b$.
(iv) $N_3 = a^3 - 3ab + 3c$.
(v) $N_{k+4} - aN_{k+3} + bN_{k+2} - cN_{k+1} + dN_k = 0$ for $k = 0, 1, 2, \ldots$.

(The results in Problems 21 and 22 are special cases of identities, known as *Newton's Formulas*, on the sums of powers of the roots of a polynomial equation. The general case is given in Problem 30 of the Supplementary and Challenging Problems at the close of this chapter.)

## 5.8 *Ideals in* $F[x]$

In this section we note some properties of ideals in the integral domain $F[x]$, of polynomials over a field $F$, that are closely analogous to those for ideals in the integral domain $\mathbf{Z}$.

In $F[x]$ every nonzero $\beta$ has an invertible as its leading coefficient. Therefore, the only restriction on the divisor $\beta$ in the division algorithm is $\beta \neq 0$. Below this is used to show that $F[x]$ shares with $\mathbf{Z}$ (but not with $\mathbf{Z}[x]$) the property that every ideal is a principal ideal. Then principal ideals help us to define the greatest common divisor and the least common multiple of polynomials in $F[x]$.

The division algorithm is also the basis for the proof of unique factorization into primes in $F[x]$ (see Section 5.9) and for the proof that $F[x]/(\beta)$ is a field whenever $\beta$ is a prime (see Section 5.11).

We now go back to general commutative rings $U$ with unity.

### *Lemma 1 Linear Combinations in U*

*Let a and b be fixed elements of a commutative ring U with unity and let I consist of all the linear combinations*

$$ax + by,$$

*with x and y in U. Then I is an ideal in U and both a and b are in I. (I is the smallest ideal containing both a and b in the ring U.)*

***Proof***

As in the Linear Combinations Theorem for **Z** in Section 1.4, one easily sees that $a$ and $b$ are in $I$ and that $I$ is closed under subtraction. It is also clear that $I$ is closed under inside-outside multiplication. Hence $I$ is an ideal.

Every ideal in the ring **Z** of the integers is a principal ideal. We next see that this is not true in $\mathbf{Z}[x]$.

***Example 1***

Let $I$ consist of all the polynomials

$$\alpha = a_d x^d + \cdots + a_0, \quad a_j \text{ in } \mathbf{Z},$$

such that $7|a_0$ in **Z**. Show that $I$ is an ideal, but not a principal ideal, in $\mathbf{Z}[x]$.

***Solution***

We see readily that $\alpha$ is in $I$ if and only if $\alpha = 7\beta + x\gamma$, with $\beta$ and $\gamma$ in $\mathbf{Z}[x]$. It then follows from the Linear Combinations Lemma that $I$ is an ideal and that both 7 and $x$ are in $I$.

Let us assume that $I$ is a principal ideal $(\delta)$. This implies that $\delta|7$ and $\delta|x$. Since **Z** is an integral domain, it follows from $\delta|7$ that $\deg \delta \leq \deg 7 = 0$ and hence that $\deg \delta = 0$. However, the only polynomials of degree 0 in $I$ are 7 and $-7$. But $I \neq (7) = (-7)$ since $x$ is not a multiple of 7 (or $-7$) in $\mathbf{Z}[x]$. This contradiction shows that $I$ is not a principal ideal.

Let $\alpha$ be a polynomial in a principal ideal $(\beta)$ in $F[x]$ and let $\beta \neq 0$. Then $\beta|\alpha$ and $\deg \beta \leq \deg \alpha$. This shows that if $I$ is an ideal in $F[x]$ and $I \neq \{0\}$, a possible generator of $I$ must be sought among the nonzero polynomials of least degree in $I$. The following result tells us that such a search will always be successful.

***Theorem 1*** $F[x]$ ***Is a Principal Ideal Domain***

*Every ideal I in F[x] is a principal ideal. If $I \neq \{0\}$, a nonzero $\beta$ in I is a generator of I if and only if $\deg \beta \leq \deg \alpha$ for all nonzero $\alpha$ in I; also, I has a unique monic generator.*

***Proof***

The ideal $\{0\}$ is the principal ideal generated by 0; hence we may assume that there are nonzero polynomials in $I$. Then it follows from the Well Ordering Principle that there is a $\beta$ in $I$ with $\deg \beta \leq \deg \alpha$ for all nonzero $\alpha$ in $I$.

Using the division algorithm, we note that for any $\alpha$ in $I$ there exist $\gamma$ and $\rho$ in $F[x]$ such that $\alpha = \gamma\beta + \rho$ and either $\rho = 0$ or $\deg \rho < \deg \beta$. Then $\rho = \alpha - \gamma\beta$ is in $I$, since an ideal is closed under linear combinations. The inequality $\deg \rho < \deg \beta$ would contradict the minimal nature of the degree of $\beta$; hence $\rho = 0$ and $\beta|\alpha$. This shows that $I = (\beta)$.

All generators of $I$ are associates of each other and just one of these associates is monic; thus the monic generator is unique.

### Theorem 2 Common Divisors

*Let $\alpha$ and $\beta$ be in $F[x]$ with either $\alpha \neq 0$ or $\beta \neq 0$. Then there is a unique monic $\delta$ in $F[x]$ such that both of the following hold:*
*(i) $\delta|\alpha$ and $\delta|\beta$.*
*(ii) If $\gamma|\alpha$ and $\gamma|\beta$, then $\gamma|\delta$.*
*This unique $\delta$ is expressible as $\delta = \xi\alpha + \eta\beta$ with $\xi$ and $\eta$ in $F[x]$.*

***Proof***

Let $I$ consist of all linear combinations

$$\xi\alpha + \eta\beta \tag{1}$$

with $\xi$ and $\eta$ in $F[x]$. Then $I$ is an ideal and both $\alpha$ and $\beta$ are in $I$ by Lemma 1. Since either $\alpha \neq 0$ or $\beta \neq 0$, we have $I \neq \{0\}$. It follows from Theorem 1 that $I$ is a principal ideal and has a unique monic generator, which we designate as $\delta$.

Since $\alpha$ and $\beta$ are in $(\delta)$, both $\delta|\alpha$ and $\delta|\beta$; i.e., we have proved part (i). Also, $\delta$ is of the form (1) since $\delta$ is in the set $I$. Then $\gamma|\alpha$ and $\gamma|\beta$ together imply $\gamma|(\xi\alpha + \eta\beta)$, that is, $\gamma|\delta$. This establishes part (ii).

To show uniqueness, we let each of $\delta$ and $\delta'$ be a monic polynomial satisfying (i) and (ii). It follows that both $\delta|\delta'$ and $\delta'|\delta$. But monic associates are equal; hence $\delta = \delta'$ as desired.

### Definition 1 Greatest Common Divisor

*If $\alpha$, $\beta$, and $\delta$ are as in Theorem 2, $\delta = gcd(\alpha, \beta)$. Also, $gcd(0, 0) = 0$.*

Given $\alpha$ and $\beta$, one can find their greatest common divisor $\delta$ and the coefficients $\xi$ and $\eta$ in the linear combination

$$\delta = \xi\alpha + \eta\beta$$

by the euclidean algorithm for polynomials over $F$, which is analogous to the algorithm described for the integers in Section 1.5.

***Definition 2 Relatively Prime Polynomials***

*If* $gcd(\alpha, \beta) = 1$, $\alpha$ *and* $\beta$ *are* **relatively prime**.

The following result is easily obtained from Theorem 2 and the two definitions above.

***Theorem 3 Linear Combinations***

*Let* $\alpha$, $\beta$, *and* $\gamma$ *be in* $F[x]$. *Then:*
(*a*) $\gamma$ *is expressible as* $\xi\alpha + \eta\beta$ *with* $\xi$ *and* $\eta$ *in* $F[x]$ *if and only if* $gcd(\alpha, \beta)|\gamma$.
(*b*) $\alpha$ *and* $\beta$ *are relatively prime if and only if* $1 = \xi\alpha + \eta\beta$, *with* $\xi$ *and* $\eta$ *in* $F[x]$.

The proof is left to the reader.

Now let $\pi$ be a prime in $F[x]$. Then the only monic divisors of $\pi$ are 1 and the unique monic associate $\pi_1$ of $\pi$. Hence $\gcd(\alpha, \pi) = \pi_1$ when $\pi|\alpha$ and $\gcd(\alpha, \pi) = 1$ when $\alpha$ is not a multiple of $\pi$.

***Lemma 2 Analogue of Euclid's Lemma***

*In* $F[x]$, *let* $\pi$ *be a prime and let* $\pi|(\alpha\beta)$. *Then either* $\pi|\alpha$ *or* $\pi|\beta$.

The proof is similar to that of Euclid's Lemma in Section 1.7 and is left to the reader as Problem 7 of this section.

Let $\alpha$ and $\beta$ be in $F[x]$. Then the common multiples of $\alpha$ and $\beta$ are the elements common to the principal ideals $(\alpha)$ and $(\beta)$, that is, the elements of the intersection

$$I = (\alpha) \cap (\beta).$$

The intersection of ideals is also an ideal (see Problem 39 of Section 4.2). If $\alpha \neq 0$ and $\beta \neq 0$, $I$ has nonzero elements and it follows from Theorem 1 above that $I$ is a principal ideal with a unique monic generator $\mu$; it is natural to define this $\mu$ to be the ***least common multiple*** of $\alpha$ and $\beta$, which we denote by $\text{lcm}[\alpha, \beta]$. For completeness, we let $\text{lcm}[\alpha, 0] = 0 = \text{lcm}[0, \beta]$.

The following result establishes a mapping from any extension field $E$ over $F$ to the family of ideals $I$ in $F[x]$ such that $I$ has an irreducible generator.

***Theorem 4 Polynomials with a Fixed Zero***

*Let $F$ be a subfield in $E$, let $s$ be an element of $E$, and let $I$ consist of all $\alpha$ in $F[x]$ such that $s$ is a zero of $\alpha$. Then $I$ is a principal ideal. If $I \neq \{0\}$, the unique monic generator $\mu$ of $I$ is irreducible.*

***Proof***

$I$ is not empty since 0 is in $I$. Let $\alpha$ and $\beta$ be in $I$ and let $\gamma$ be in $F[x]$. Then it is clear that $\alpha - \beta$ and $\gamma\alpha$ are in $I$. That is, $I$ is a nonempty set closed under subtraction and inside-outside multiplication and so $I$ is an ideal. If $I \neq \{0\}$, it follows from Theorem 1 above that $I$ is a principal ideal and that its unique monic generator $\mu$ is also the unique monic polynomial of least degree in $I$. If $\mu$ were reducible in $F[x]$, we would have

$$\mu = \mu_1\mu_2, \quad \deg \mu_1 < \deg \mu, \quad \deg \mu_2 < \deg \mu$$

and $\mu_1(s)\mu_2(s) = \mu(s) = 0$. It would follow that $s$ is a zero of either $\mu_1$ or $\mu_2$, contradicting the minimal nature of the degree of $\mu$. This contradiction proves that $\mu$ is irreducible.

***Corollary***

*Let $\alpha$ and a prime $\pi$ be in $F[x]$. In an extension field $E$ over $F$, let $s$ be a zero of both $\alpha$ and $\pi$. Then $\pi | \alpha$ in $F[x]$.*

The proof is left to the reader as Problem 8 below.

## *Problems*

1. In $\mathbf{Z}_2[x]$, let $\alpha = x^5 + x^4 + \bar{1}$ and $\beta = x^5 + x + \bar{1}$. Find $\delta = \gcd(\alpha, \beta)$ and $\xi$ and $\eta$ in $\mathbf{Z}_2[x]$ such that

$$\delta = \xi\alpha + \eta\beta, \quad \text{with } \deg \eta \leq 2.$$

2. In $\mathbf{Q}[x]$, let $\alpha = x^5 + 2x^4 + 2x^3 + 2x^2 + x + 1$ and $\beta = x^4 - x$. Find $\delta = \gcd(\alpha, \beta)$ and $\xi$ and $\eta$ such that

$$\delta = \xi\alpha + \eta\beta, \quad \text{with } \deg \xi < 3.$$

3. Explain why distinct monic primes $\pi$ and $\pi'$ in $F[x]$ must be relatively prime.

4. Show with an example that distinct primes in $F[x]$ need not be relatively prime.
5. Explain why $x - a$ and $x - b$ are relatively prime in $F[x]$ if and only if $a \neq b$.
6. In $F[x]$, prove that $\gcd(\alpha, \beta) = 1$, $\alpha|\gamma$, and $\beta|\gamma$ together imply $(\alpha\beta)|\gamma$.
7. In $F[x]$, let $\pi$ be a prime and let $\pi|(\alpha\beta)$. Prove that either $\pi|\alpha$ or $\pi|\beta$, and thus prove Lemma 2 above. (Also, see Euclid's Lemma in Section 1.7.)
8. Let $\alpha$ and a prime $\pi$ be in $F[x]$. Let $\alpha$ and $\pi$ have a common zero in an extension field $E$ over $F$. Show that $\pi|\alpha$ (and thus prove the corollary to Theorem 4 above).
9. Let $\alpha = x + 1$ and $I$ be the principal ideal $(\alpha^2)$ in $\mathbf{Q}[x]$. Prove that the quotient ring $\mathbf{Q}[x]/I$ is not a field.
10. Show that the quotient ring $\mathbf{Z}[x]/(x^2 + 1)$ is isomorphic to the ring of gaussian integers $m + n\mathbf{i}$ (with $m$ and $n$ in $\mathbf{Z}$).
11. Let $s$ be a fixed element of a commutative ring $R$ and let $h$ be the mapping from $R[x]$ to $R$ with $\alpha \mapsto \alpha(s)$.
    (i) Explain why $h$ is a ring homomorphism.
    (ii) What is the relation between $s$ and the kernel $K$ of $h$?
12. Let $h$ be the mapping from $U[x]$ to $U$ with
    $$a_d x^d + \cdots + a_1 x + a_0 \mapsto a_d + \cdots + a_1 + a_0 .$$
    (i) Explain why $h$ is a ring homomorphism.
    (ii) Find the monic generator for the kernel of $h$.
13. Show that there exists an infinite field of characteristic $p$ for every positive prime $p$ in $\mathbf{Z}$.
14. Let $x$ and $y$ be independent indeterminates over $\mathbf{Q}$. Let $I$ consist of all linear combinations $x\alpha + y\beta$, with $\alpha$ and $\beta$ in $\mathbf{Q}[x, y]$. Show that $I$ is an ideal in $\mathbf{Q}[x, y]$ but is not a principal ideal. (See Definitions 6 and 7 of Section 5.2.)

## 5.9 *Factorizations of Polynomials*

This section deals with analogues for the polynomial rings of the factorization theorems for the integers in Section 1.7.

### *Theorem 1 Factorization into Irreducible Polynomials*

*Let R be a commutative ring. Then in R[x] every $\alpha$ of positive degree is expressible in the form*

$$\alpha = \beta_1\beta_2 \cdots \beta_n, \tag{I}$$

*where the $\beta_j$ are irreducible.*

***Proof***

We assume the result false and obtain a contradiction. Since the degree of a polynomial is a nonnegative integer, this assumption and the Well Ordering Principle imply that in the nonempty subset of $R[x]$ consisting of the polynomials that are neither irreducible nor the product of a finite number of irreducibles, there is an $\alpha$ with the minimum degree $d$ for this subset.

Since this $\alpha$ is reducible, $\alpha = \alpha_1\alpha_2$ with $0 < \deg \alpha_1 < d$ and $0 < \deg d_2 < d$. The minimal nature of $d$ then implies that both $\alpha_1$ and $\alpha_2$ are expressible in the form (I); that is, we have

$$\alpha_1 = \beta_1 \cdots \beta_s, \qquad \alpha_2 = \beta_{s+1} \cdots \beta_t,$$

with the $\beta_j$ irreducible polynomials in $R[x]$. Then

$$\alpha = \alpha_1\alpha_2 = \beta_1 \cdots \beta_s\beta_{s+1} \cdots \beta_t$$

is the desired contradiction and the theorem is proved.

So far, the analogy between factorization into primes in **Z** and factorization into irreducible polynomials in $R[x]$ has been fairly close. When we consider uniqueness, some obstacles appear, as is seen in the following:

***Example 1***

In $\mathbf{Z}_6[x]$, $x(x + \bar{5}) = (x + \bar{2})(x + \bar{3})$.

***Example 2***

In $\mathbf{Z}_5[x]$, $(\bar{2}x + 1)(\bar{4}x + \bar{3}) = (x + \bar{3})(\bar{3}x + \bar{1})$.

Since all polynomials of degree 1 are irreducible, these examples show that a given polynomial may be factored into irreducible polynomials in more than one way. However, the factors in Example 1 are not primes in $\mathbf{Z}_6[x]$. (See Example 4 of Section 5.3, which shows that $x + \bar{5}$ is not a prime in $\mathbf{Z}_6[x]$.)

As a step toward resolving the lack of uniqueness in Example 2 we recall that, technically speaking, factorization of an integer into primes in **Z** is not unique; for example,

$$21 = 3 \cdot 7 = (-3)(-7).$$

But there is unique factorization into *positive* primes for an integer greater than 1.

In the ring $F[x]$ of polynomials over a field $F$, the irreducible polynomials are the same as the primes and the analogue of a positive prime in **Z** is a monic prime polynomial. By restricting the ring $R$ of coefficients to be a field $F$, we eliminate Example 1 from consideration since $\mathbf{Z}_6$ is not a field.

We introduce uniqueness into Example 2 by factoring out the leading coefficients from the first degree factors and thus bringing monic primes into the picture. Then both sides of the equation in Example 2 become

$$\overline{3}(x + \overline{3})(x + \overline{2}).$$

We now present the general result.

***Theorem 2 Unique Factorization in*** $F[x]$

*Every polynomial $\alpha$ of positive degree in $F[x]$ is expressible in the form*

$$\alpha = v\pi_1^{n_1}\pi_2^{n_2}\cdots\pi_r^{n_r}, \tag{F}$$

*where $v$ is a nonzero element of $F$, the $\pi_j$ are distinct monic primes in $F[x]$, and the $n_j$ are positive integers. This factorization is unique (except for the order of appearance of the powers of the $\pi_j$). If $\alpha$ is monic, $v = 1$ in display (F).*

The proof is similar to that for unique factorization in the integers and is left to the reader.

Let **R** and **C** denote the fields of real and complex numbers. In $\mathbf{R}[x]$ and $\mathbf{C}[x]$ the above unique factorization results can be made more explicit using the theorem of Gauss, stated in Section 4.9, that $C$ is algebraically closed. We next restate Gauss' Theorem in the following form:

***Theorem 3 Fundamental Theorem of Algebra***

*If $\alpha$ is a polynomial of positive degree in $\mathbf{C}[x]$, $\alpha = 0$ has a root in* **C**.

A proof requires advanced techniques of analysis or topology; we assume this result without proof.

The Fundamental Theorem of Algebra and the Factor Theorem together imply the following result.

***Theorem 4 Primes in*** $\mathbf{C}[x]$

*The primes (or irreducible polynomials) in* $\mathbf{C}[x]$ *are just the polynomials of degree 1.*

Using the Unique Factorization Theorem for polynomials and Theorem 4, we obtain:

***Theorem 5 Linear Factorization***

*Let* $\alpha$ *be a polynomial of positive degree* $d$ *in* $\mathbf{C}[x]$. *Then* $\alpha$ *has* $d$ *zeros* $r_1, r_2, \ldots, r_d$ *(not necessarily distinct) in* $\mathbf{C}$ *and*

$$\alpha = a(x - r_1)(x - r_2) \cdots (x - r_d), \tag{1}$$

*with* $a \neq 0$ *in* $\mathbf{C}$.

Since the real numbers $\mathbf{R}$ form a subfield in $\mathbf{C}$, a polynomial $\alpha$ in $\mathbf{R}[x]$ is also in $\mathbf{C}[x]$ and so is expressible in the form (1) if $\deg \alpha > 0$. Some or all of the zeros $r_j$ in (1) may be real; the others come in pairs of complex conjugates. (See the corollary to Theorem 5 of Section 5.4.) We use this fact to characterize unique factorization in $\mathbf{R}[x]$.

***Theorem 6 Primes in*** $\mathbf{R}[x]$

*The primes (or irreducible polynomials) of* $\mathbf{R}[x]$ *are all the first degree polynomials and those second degree polynomials* $ax^2 + bx + c$ *(*$a$, $b$, *and* $c$ *real) with* $b^2 - 4ac < 0$.

***Proof***

A polynomial $\alpha$ of positive degree in $\mathbf{R}[x]$ has a complex zero $r$ by the Fundamental Theorem. If $r$ is real, $x - r$ divides $\alpha$ in $\mathbf{R}[x]$, while if $r = h + k\mathbf{i}$ is not real, $x^2 - (r + \bar{r})x + r\bar{r}$ divides $\alpha$ in $\mathbf{R}[x]$, where $\bar{r}$ is the complex conjugate $h - k\mathbf{i}$ of $r$. Hence an irreducible $\alpha$ must be of degree 1 or 2.

Let $\alpha = ax^2 + bx + c$. Then the (not necessarily distinct) zeros of $\alpha$ in $\mathbf{C}$ are

$$r_1 = \frac{-b + \sqrt{b^2 - 4ac}}{2}, \qquad r_2 = \frac{-b - \sqrt{b^2 - 4ac}}{2}.$$

A polynomial of degree 2 over the real numbers is composite if and only if it has a real root. [See Problem 20(i) of Section 5.4.] Hence $\alpha$ is prime in $\mathbf{R}[x]$ if and only if the roots $r_1$ and $r_2$ are not real; this is true if and only if $b^2 - 4ac < 0$.

***Theorem 7*** ***Unique Factorization in*** $\mathbf{R}[x]$

*Every* $\alpha$ *of positive degree in* $\mathbf{R}[x]$ *is expressible as*

$$\alpha = a(\pi_1)^{n_1}(\pi_2)^{n_2} \cdots (\pi_m)^{n_m}, \quad a \text{ in } \mathbf{R}, \tag{2}$$

*where the* $\pi_j$ *are distinct monic irreducible polynomials of degree 1 or 2 in* $\mathbf{R}[x]$. *The factorization* (2) *is unique except for the order of appearance of the* $(\pi_j)^{n_j}$.

***Proof***

This is an immediate consequence of Theorems 6 and 2 above.

## Problems

1. Explain why the following are true in $\mathbf{Z_{31}}[x]$:
   (i) There are 31 monic polynomials of degree 1.
   (ii) There are 961 monic polynomials of degree 2.
   (iii) Of the 961 monic second degree polynomials, 496 are reducible (and composite) and 465 are irreducible (and prime).
2. Do the analogue of Problem 1 for $\mathbf{Z_p}[x]$, where $p$ is a prime in $\mathbf{Z}$.
3. Let $\alpha_1$, $\alpha_2$, and $\alpha_3$ denote distinct monic polynomials of degree 1 and let $\beta$ denote a monic prime of degree 2 in $\mathbf{Z_{31}}[x]$. Explain why the following are true in $\mathbf{Z_{31}}[x]$:
   (i) There are 4495 polynomials of the form $\alpha_1\alpha_2\alpha_3$.
   (ii) There are 930 polynomials of the form $\alpha_1^2\alpha_2$.
   (iii) There are 31 polynomials of the form $\alpha_1^3$.
   (iv) There are 14415 polynomials of the form $\alpha_1\beta$.
   (v) There are 9920 monic primes of degree 3.
4. Do the analogue of Problem 3 for $\mathbf{Z_p}[x]$, where $p$ is a prime in $\mathbf{Z}^+$.
5. Express each of the following irreducible polynomials as a product $\alpha\beta$ with $\alpha$ and $\beta$ polynomials of degree 1 in $\mathbf{Z_6}[x]$:
   (a) $x$; (b) $x + \overline{2}$; (c) $x + \overline{3}$.
6. Find first degree polynomials $\alpha$ and $\beta$ in $\mathbf{Z_{10}}[x]$ such that $\alpha\beta = x + \overline{3}$.
7. In $F[x]$, explain why $\gcd(\alpha, \beta) = 1$ if and only if there is no prime $\pi$ with both $\pi | \alpha$ and $\pi | \beta$.
8. In $F[x]$, explain why $\gcd(\alpha_1\alpha_2 \cdots \alpha_m, \beta_1\beta_2 \cdots \beta_n) = 1$ if and only if $\gcd((\alpha_i, \beta_j) = 1$ for all $i$ and $j$ with $1 \leq i \leq m$ and $1 \leq j \leq n$.
9. Let $\pi_1, \pi_2, \ldots, \pi_n$ be $n$ distinct monic primes in $F[x]$ and let $r_1, r_2, \ldots, r_n$ be

(not necessarily distinct) positive integers. Explain why

$$\gcd(\pi_1^{r_1}, \pi_2^{r_2}\pi_3^{r_3} \cdots \pi_n^{r_n}) = 1.$$

10. Let $a_1, a_2, \ldots, a_n$ be $n$ distinct elements of a field $F$ and let $r_1, r_2, \ldots, r_n$ be (not necessarily distinct) positive integers. Explain why $(x - a_1)^{r_1}$ and $(x - a_2)^{r_2}(x - a_3)^{r_3} \cdots (x - a_n)^{r_n}$ are relatively prime.

## 5.10 Partial Fractions

If $F$ is a field, we can apply the construction outlined in Section 4.8 to the polynomial ring $F[x]$ and thus obtain the field of all quotients $\alpha/\beta$, with $\alpha$ and $\beta$ in $F[x]$ and $\beta \neq 0$.

### *Notation Field of Quotients of Polynomials*

*If $x$ is an indeterminate over a field $F$, the field*

$$\{\alpha/\beta : \alpha, \beta \in F[x], \beta \neq 0\}$$

*is denoted by $F(x)$.*

It can be shown that the field $F(x)$ is essentially (in the sense of isomorphism) the smallest field in which $x$ and all the elements of $F$ are present. This justifies our calling $F(x)$ the ***field extension*** of $F$ by $x$. In Section 5.11, field extensions of $F$ by finite sets of elements (that may be algebraic or transcendental over $F$) will be discussed.

Unique factorization into primes in the polynomial ring $F[x]$ enables one to express any quotient of polynomials in $F(x)$ uniquely as a finite sum of special elements to be described below. This representation has important applications, one of which is in the calculus topic of integration of rational functions.

### *Definition 1 Prime Power Denominator Term*

*A* **ppd-term** *is an element $\gamma/\pi^k$ of $F(x)$ with $k$ a positive integer, $\pi$ a monic prime in $F[x]$, and* $\deg \gamma < \deg \pi$.

Using the lemmas that follow, we shall show that every quotient of polynomials $\alpha/\beta$ of $F(x)$ is the sum of a polynomial in $F[x]$ and a finite number of ppd-terms.

***Lemma 1***

*Let $\alpha/\beta$ be in $F(x)$ and let $\beta = \beta_1\beta_2$ with* $\gcd(\beta_1, \beta_2) = 1$. *Then there exist $\alpha_1$ and $\alpha_2$ in $F[x]$ such that*

$$\frac{\alpha}{\beta} = \frac{\alpha_1}{\beta_1} + \frac{\alpha_2}{\beta_2}.$$

***Proof***

Since $\beta_1$ and $\beta_2$ are relatively prime, every $\alpha$ of $F[x]$ is a linear combination $\alpha_1\beta_2 + \alpha_2\beta_1$, with $\alpha_1$ and $\alpha_2$ in $F[x]$. Then

$$\frac{\alpha}{\beta} = \frac{\alpha_1\beta_2 + \alpha_2\beta_1}{\beta_1\beta_2} = \frac{\alpha_1}{\beta_1} + \frac{\alpha_2}{\beta_2}.$$

***Lemma 2***

*Let $\beta$ be a polynomial of positive degree in $F[x]$ and let $s$ be a positive integer. Then an element $\alpha/\beta^s$ of $F(x)$ is expressible as*

$$\frac{\alpha}{\beta^s} = \gamma_0 + \frac{\gamma_1}{\beta} + \frac{\gamma_2}{\beta^2} + \cdots + \frac{\gamma_s}{\beta^s} \tag{1}$$

*with $\gamma_i$ in $F[x]$ and either $\gamma_i = 0$ or* $\deg \gamma_i < \deg \beta$ *for $0 \le i \le s$.*

***Proof***

Theorem 3 of Section 5.3 tells us that there exist $\gamma_i$ in $F[x]$ such that either $\gamma_i = 0$ or $\deg \gamma_i < \deg \beta$ and

$$\alpha = \gamma_0\beta^s + \gamma_1\beta^{s-1} + \cdots + \gamma_{s-1}\beta + \gamma_s. \tag{2}$$

Division of both sides of equation (2) by $\beta^s$ leads to the desired equation (1).

We are now ready for the main theorem of this section.

***Theorem 1 Partial Fraction Decomposition***

*Every $\alpha/\beta$ of $F(x)$ is expressible as a sum of a polynomial in $F[x]$ and a finite number of ppd-terms $\gamma/\pi^m$ (with $\pi$ a monic prime and* $\deg \gamma < \deg \pi$).

*Proof*

If $\deg \beta = 0$, $\alpha/\beta$ is a polynomial. Hence let $\deg \beta > 0$. The leading coefficient of $\beta$ is a nonzero element of the field $F$ and can be cancelled from both the numerator and denominator of $\alpha/\beta$; thus, we may assume that $\beta$ is monic.

Then it follows from the Unique Factorization Theorem of Section 5.9 that

$$\beta = (\pi_1)^{n_1}(\pi_2)^{n_2} \cdots (\pi_r)^{n_r},$$

with the $\pi_i$ distinct monic primes in $F[x]$.

If $r > 1$, we let $\beta_1 = (\pi_1)^{n_1}$ and $\beta' = (\pi_2)^{n_2} \cdots (\pi_r)^{n_r}$. Then $\beta_1$ and $\beta'$ are relatively prime and Lemma 1 tells us that $\alpha/\beta = (\alpha_1/\beta_1) + (\alpha'/\beta')$, with $\alpha_1$ and $\alpha'$ in $F[x]$. If $r > 2$, $\alpha'/\beta'$ is broken up similarly. Proceeding in this way, we express $\alpha/\beta$ in the form

$$\frac{\alpha}{\beta} = \frac{\alpha_1}{\pi^{n_1}} + \frac{\alpha_2}{\pi^{n_2}} + \cdots + \frac{\alpha_r}{\pi^{n_r}}. \tag{3}$$

Then each term on the right side of (3) can be expressed as the sum of the form shown in (1) and this gives us the desired sum for $\alpha/\beta$.

In applications of this partial fraction decomposition (for example, to integration theory), the field $F$ of coefficients is usually the field $\mathbf{R}$ of real numbers or the field $\mathbf{C}$ of complex numbers. Therefore, we make some further observations for these fields.

Theorem 4 of Section 5.9 tells us that $\deg \pi = 1$ for every prime $\pi$ in $\mathbf{C}[x]$ and hence in every ppd-term $\gamma/\pi^m$ of $\mathbf{C}(x)$. Similarly, we see from Theorem 6 of Section 5.9 that $\deg \pi$ is 1 or 2 for every prime $\pi$ in $\mathbf{R}[x]$ and in each ppd-term $\gamma/\pi^m$ of $\mathbf{R}(x)$.

Hence every ppd-term of $\mathbf{R}(x)$ is of one of the forms

$$a(x + b)^{-m}, \qquad (ax + b)(x^2 + cx + d)^{-m} \tag{4}$$

with $a$, $b$, $c$, and $d$ real numbers and $m$ a positive integer. In calculus texts it is shown that an integral of any function of one of these forms can be expressed in terms of polynomials, logarithms, and inverse trigonometric functions. It then follows from the partial fraction representation (Theorem 1) that an integral of each $\alpha/\beta$ of $\mathbf{R}(x)$ is expressible in this manner.

## Problems

1. In $\mathbf{Q}(x)$, express each of the following as a finite sum of ppd-terms:
   (a) $(5x + 7)/(x - 2)(x + 3)$.
   (b) $(2x^3 - 5x^2 - 8)/(x + 1)^2(x^2 + 4)$.

2. Express $x^5/(x^2+1)(x^2+9)$ as the sum of a polynomial in $\mathbf{Q}[x]$ and a finite number of ppd-terms in $\mathbf{Q}(x)$.
3. Express $(x^3+\bar{2}x)/(x-\bar{1})(x+\bar{1})$ as a sum of a polynomial in $\mathbf{Z}_5[x]$ and a finite number of ppd-terms in $\mathbf{Z}_5(x)$.
4. In $\mathbf{Z}_7(x)$, express $(\bar{2}x+\bar{1})/(x^2+\bar{3})(x^2+\bar{5})$ as a sum of ppd-terms.
5. Let $r$ and $s$ be distinct elements of a field $F$. In $F(x)$, show that

$$[(x-r)(x-s)]^{-1}=[(r-s)(x-r)]^{-1}+[(s-r)(x-s)]^{-1}.$$

6. Let $r$, $s$, and $t$ be three distinct elements of a field $F$. In $F(x)$, show that

$$[(x-r)(x-s)(x-t)]^{-1}=a(x-r)^{-1}+b(x-s)^{-1}+c(x-t)^{-1},$$

where $a=[(r-s)(r-t)]^{-1}$, $b=[(s-r)(s-t)]^{-1}$, and $c=[(t-r)(t-s)]^{-1}$.

7. Generalize on Problems 5 and 6.

*8. Let $\alpha(x)=(x-r_1)(x-r_2)\cdots(x-r_n)$, with the $r_i$ distinct elements of a field $F$, and let $\alpha'(x)$ be the derivative of $\alpha(x)$. (See Problem 21 of Section 5.2.) In the field $F(x)$, show the following:

$$\frac{1}{\alpha(x)}=\frac{1}{\alpha'(r_1)(x-r_1)}+\frac{1}{\alpha'(r_2)(x-r_2)}+\cdots+\frac{1}{\alpha'(r_n)(x-r_n)}.$$

*9. Let $\alpha(x)$ and $\alpha'(x)$ be as in Problem 8. Show that

$$\frac{\alpha'(x)}{\alpha(x)}=\frac{1}{x-r_1}+\frac{1}{x-r_2}+\cdots+\frac{1}{x-r_n}.$$

10. For each $a$ in $\{1, 5, 7, 11, 13\}$, show that

$$\frac{a}{72}=\frac{b}{2}+\frac{c}{4}+\frac{d}{8}+\frac{e}{3}+\frac{f}{9}$$

with $|b|$, $|c|$, and $|d|$ in $\{0, 1\}$; and $|e|$ and $|f|$ in $\{0, 1, 2\}$.

11. Define an analogue for the rational numbers $\mathbf{Q}$ of a ppd-term in $F(x)$ and then state an analogue of Theorem 1 above in the form of a result on partial fraction decomposition of rational numbers.

## 5.11 Extension Fields

*The mathematician, carried along on his flood of symbols, dealing apparently with purely formal truths, may still reach results of endless importance for our description of the physical universe.*

Karl Pearson

In every field $E$, there is a smallest subfield $P$, the prime subfield. We know all the prime subfields since $P$ is isomorphic to the rational numbers $\mathbf{Q}$ when $E$ has characteristic 0 and $P$ is isomorphic to $\mathbf{Z_p}$ when $E$ has prime characteristic $p$. This suggests that an approach to studying fields is to investigate methods of extending a given field $F$ into a larger one.

There are two methods that use the integral domain $F[x]$ of polynomials over $F$. One is to apply the technique of Section 4.8 to $F[x]$ and thus obtain the field $F(x)$ of polynomial fractions $\alpha/\beta$. Another method of extending $F$ is to apply to $F[x]$ the technique used to construct the finite fields $\mathbf{Z_p}$ from the integers $\mathbf{Z}$. Specifically, we prove below that when $\pi$ is a prime in $F[x]$, the quotient ring $E = F[x]/(\pi)$ is a field having a subfield isomorphic to $F$.

Given a polynomial $\alpha$ of positive degree in $F[x]$, the latter technique can be used to extend $F$ so that $\alpha$ factors into first degree polynomials over the new field.

**Theorem 1** ***Quotient Rings*** $F[x]/(\beta)$

*Let $\beta$ be a polynomial of degree $d$ in $F[x]$.*

(a) *The elements of $E = F[x]/(\beta)$ are the cosets $\bar{\rho} = \rho + (\beta)$ with either* $\deg \rho < d$ *or* $\rho = 0$.

(b) *If $F$ has $q$ elements, $E$ has $q^d$ elements.*

(c) *If $\beta$ is a prime, $E$ is a field that has a subfield $\bar{F}$ isomorphic to $F$.*

***Proof***

For each $\alpha$ in $F[x]$ let $\bar{\alpha}$ denote the ideal coset $\alpha + (\beta)$. The division algorithm gives us $\alpha = \gamma\beta + \rho$ with either $\deg \rho < \deg \beta = d$ or $\rho = 0$. Since the natural map $\alpha \mapsto \bar{\alpha}$ preserves additions and multiplications,

$$\bar{\alpha} = \overline{\gamma\beta + \rho} = \bar{\gamma}\bar{\beta} + \bar{\rho}.$$

Since $\overline{\beta} = \overline{0}$ in $E$, this means that $\overline{\alpha} = \overline{\rho}$ with

$$\rho = r_{d-1}x^{d-1} + \cdots + r_1 x + r_0, \quad r_i \text{ in } F. \tag{1}$$

The representation (1) is unique since $\overline{\rho} = \overline{\rho}_1$ if and only if $\beta|(\rho - \rho_1)$ and the difference $\rho - \rho_1$ of two polynomials of the form (1) cannot be a multiple of $\beta$ unless $\rho - \rho_1 = 0$.

If $F$ is a finite field with $q$ elements, there are $q$ choices for each of the $d$ coefficients $r_{d-1}, \ldots, r_0$ in (1) and hence $q^d$ elements (cosets $\overline{\rho}$) in the ring $E$.

Now let $\beta$ be a prime. If $\overline{\alpha} \neq 0$, $\alpha$ is not a multiple of the prime $\beta$ and it follows that $\gcd(\alpha, \beta) = 1$. Then

$$\xi\alpha + \eta\beta = 1, \tag{2}$$

with $\xi$ and $\eta$ in $F[x]$. Applying the natural map, we have

$$\overline{\xi}\overline{\alpha} + \overline{\eta}\overline{\beta} = \overline{1}.$$

But $\overline{\beta} = \overline{0}$ and so $\overline{\xi}\overline{\alpha} = \overline{1}$. Therefore, every nonzero $\overline{\alpha}$ of $E$ is an invertible and $E$ is a field.

Now let $a$ be in $F$. Then $a$ is also in $F[x]$ and $\overline{a}$ is in $E$. Let $\overline{F}$ consist of the $\overline{a}$ for all $a$ in $F$. It is easily seen that $\overline{F}$ is a subfield in $E$ and that the mapping $\theta$ with $a \mapsto \overline{a}$ is an isomorphism from $F$ to $\overline{F}$. (We note that $\theta$ is the modification of the natural map from $F[x]$ to $E$ in which the domain is restricted to the subset $F$ of $F[x]$.)

### *Example 1*

Let $\mathbf{R}$ be the field of real numbers, $\beta = x^2 + 1$, and $E = \mathbf{R}[x]/(\beta)$. Explain why $E$ is isomorphic to the complex numbers $\mathbf{C}$.

### *Solution*

Theorem 1 tells us that the elements of $E$ are the cosets

$$\overline{\rho} = \overline{a + bx} = \overline{a} + \overline{b}\overline{x},$$

with $a$ and $b$ any real numbers. Since $\overline{\beta} = \overline{0}$,

$$\overline{x^2 + 1} = (\overline{x})^2 + \overline{1} = \overline{0}. \tag{3}$$

With the help of this equation, it is easy to see that the bijection $\theta$ from $E$ to $\mathbf{C}$ with

$$\theta(\overline{a} + \overline{b}\overline{x}) = a + b\mathbf{i}$$

preserves additions and multiplications and hence is an isomorphism.

The polynomial equation $x^2 + 1 = 0$ does not have a root in the real numbers $\mathbf{R}$. Even if we did not know of the complex numbers, we could produce a field $E = \mathbf{R}[x]/(x^2 + 1)$ in which the coset $\bar{x} = x + (x^2 + 1)$ is a root. [See equation (3) in Example 1.] The polynomial $x^2 + 1$ is a prime in $\mathbf{R}[x]$ but in $E[x]$ it factors as a product $(x - \bar{x})(x + \bar{x})$ of first degree polynomials.

We now generalize on this example. Let $F$ be any field and let $\alpha$ be any polynomial of positive degree in $F[x]$. If $\alpha$ does not have a zero in $F$, we can choose a prime divisor $\pi$ of $\alpha$ in $F[x]$ and construct the field $E = F[x]/(\pi)$, in which there is a subfield $F_1$ isomorphic to $F$. To avoid complicating the terminology, we act as if $F_1$ were $F$; that is, we assume that $F$ is a subfield in $E$.

Then the coset $\bar{x} = x + (\pi)$ is a zero of $\pi$, and hence of $\alpha$, in $E[x]$. The Factor Theorem then tells us that $\alpha = (x - \bar{x})\beta$, with $\beta$ in $E[x]$. If $\beta$ does not have a zero in $E$ (or, equivalently, a first degree factor in $E[x]$) and $\deg \beta > 0$, one can give $\beta$ the treatment accorded $\alpha$. Continuing in this way, one shows that an extension of $F$ exists over which $\alpha$ factors into first degree polynomials.

We now introduce some terminology applying to extension fields.

### *Definition* *Field Extension*

*Let $F$ be a subfield in $E$ and let $s_1, s_2, \ldots, s_n$ be elements of $E$. Then $F(s_1, s_2, \ldots, s_n)$ denotes the smallest subfield $S$ in $E$ such that each $s_i$ is in $S$ and each element of $F$ is in $S$. The field $S$ is called the* **field extension** *of $F$ by the $s_i$ in $E$.*

If $x$ is an element of $E$ which is transcendental over $F$ (that is, if $x$ is an indeterminate over $F$), then $F(x)$ consists of all polynomial fractions $\alpha/\beta$, with $\alpha$ and $\beta \neq 0$ in $F[x]$; such an $F(x)$ is called a ***simple transcendental extension*** of $F$.

If $s$ is algebraic over $F$, we show below that the polynomial ring $F[s]$ is a field and hence is the same as the field extension $F(s)$; such an $F(s)$ is a ***simple algebraic extension*** of $F$.

If an $\alpha$ of $F[x]$ is the product

$$a(x - s_1)(x - s_2) \cdots (x - s_n),$$

with $a$ in $F$ and the $s_i$ in $E$, the algebraic extension $F(s_1, s_2, \ldots, s_n)$ is called a ***splitting field*** (or a ***root field***) over $F$ for $\alpha$. It can be shown that any two splitting fields over $F$ for $\alpha$ are isomorphic.

By definition, an element $s$ of an extension $E$ of $F$ is algebraic over $F$ if and only if $s$ is a zero of a nonzero polynomial $\alpha$ in $F[x]$. If $s$ is algebraic over $F$, it follows from Theorem 4 of Section 5.8 that there is a unique monic irreducible $\pi$ in $F[x]$ with the property that $s$ is a zero in $E$ of an $\alpha$ in $F[x]$ if and only if $\pi \mid \alpha$; this $\pi$ is called the ***minimal polynomial for s over F***. If $\deg \pi = d$, one says that $s$ is ***algebraic of degree d over F***.

### *Theorem 2 Simple Algebraic Extension*

*Let $s$ be algebraic of degree $d$ over $F$ and let $\pi$ be its minimal polynomial. Then:*
*(a) $F[s]$ is isomorphic to $F[x]/(\pi)$.*
*(b) The polynomial extension $F[s]$ is the same as the field extension $F(s)$.*
*(c) Every element of $F(s)$ is of the form*

$$a_{d-1}s^{d-1} + \cdots + a_1 s + a_0,\ a_i \text{ in } F.$$

***Proof***

The elements of $F[s]$ are the $\alpha(s)$ for all $\alpha$ in $F[x]$. Clearly, $\alpha_1(s) = \alpha_2(s)$ if and only if $s$ is a zero of $\alpha_1 - \alpha_2$. By Theorem 4 of Section 5.8, $s$ is a zero of $\alpha_1 - \alpha_2$ if and only if $\pi|(\alpha_1 - \alpha_2)$, that is, if and only if $\alpha_1$ and $\alpha_2$ are in the same coset of $(\pi)$ in $F[x]$.

Then the mapping $f$ from $F[s]$ to $F[x]/(\pi)$ with $\alpha(s) \mapsto \bar{\alpha} = \alpha + (\pi)$ is easily seen to be a bijection that preserves additions and multiplications; thus $f$ is a ring isomorphism. Since $\pi$ is a prime in $F[x]$, Theorem 1 tells us that $F[x]/(\pi)$ is a field. Hence the isomorphic $F[s]$ is also a field. This makes $F[s]$ the smallest extension field $F(s)$ of $F$ in which $s$ is present.

Part (c) follows from the fact that every $\alpha(s)$ of $F[s]$ may be rewritten as $\rho(s)$, where $\rho$ is the remainder in the division of $\alpha$ by $\pi$.

### *Corollary 1 Simple Algebraic Extension of a Finite Field*

*Let $F$ have $q$ elements, let $s$ be algebraic of degree $n$ over $F$, and let $m = q^n$. Then $F[s]$ has $m$ elements and the minimal polynomial for $s$ over $F$ is a divisor of $x^m - x$.*

### *Corollary 2 Number of Elements in a Finite Field*

*Let $F$ have $m$ elements and characteristic $p$. Then $m = p^n$ with $n$ a positive integer.*

### *Corollary 3 Isomorphic Algebraic Extensions*

*Let $s$ and $t$ (be algebraic and) have the same minimal polynomial over $F$. Then $F[s]$ and $F[t]$ are isomorphic.*

The proofs of these corollaries are left to the reader as Problems 5–7 of this section.

Corollary 2 to Theorem 2 states that a finite field has $p^n$ elements, where $p$ and $n$ are positive integers and $p$ is a prime. Theorem 1 tells us how to construct such a field, if we can find an irreducible polynomial $\pi$ of degree $n$ in $\mathbf{Z}_p[x]$. Corollary 1 indicates that such a $\pi$ should be sought among the factors of $x^m - x$, where $m = p^n$.

It can be proved that for every positive prime $p$ in $\mathbf{Z}$ there exist prime polynomials of all positive degrees in $\mathbf{Z}_p[x]$ and hence fields with $p^n$ elements for all positive integers $n$. [See G. J. Simmons: "The Number of Irreducible Polynomials of Degree $n$ over $GF(p)$," *The American Mathematical Monthly*, 77 (1970), 743–745.]

It can also be proved that any two finite fields with the same number of elements are isomorphic. A finite field is called a ***Galois field.*** If a Galois field has $p^n$ elements, it is designated as $GF(p^n)$.

We now illustrate the construction of Galois fields.

*Example 2*

Construct a field with 8 elements and discuss some of its properties.

*Solution*

Since $8 = 2^3$, we start with a prime field $F = \{0, 1\}$ of characteristic 2 and look for an irreducible polynomial of degree 3 in $F[x]$. One sees that $\pi = x^3 + x + 1$ has no zeros in $F$ by noting that $\pi(0) = 1 = \pi(1)$. It follows that $\pi$ has no first degree divisors in $F[x]$. This and the fact that $\deg \pi = 3$ imply that $\pi$ is irreducible.

Using Theorem 1, one now sees that $E = F[x]/(\pi)$ is a field with $2^3 = 8$ elements. We choose a notation for the elements of $E$ that allows us to consider $E$ to be an extension of $F$. Specifically, we let the elements $\overline{0}$, $\overline{1}$, and $\overline{x}$ of $E$ be denoted by 0, 1, and $c$, respectively. Then the set for $E$ is

$$\{0, 1, c, c + 1, c^2, c^2 + 1, c^2 + c, c^2 + c + 1\}.$$

Addition is performed easily in $E$ using the fact that $y + y = 0$ for all $y$ in $E$; that is, $E$ has characteristic 2. For example,

$$(c^2 + c) + (c^2 + 1) = c + 1.$$

Since $\overline{\pi} = 0$ in $E$, $c^3 + c + 1 = 0$; that is, $c$ is a zero in $E$ of $\pi$. The multiplicative group $V$ of $E$ has 7 elements. Since 7 is a prime, $V$ is cyclic with $c$ one of its 6 generators. Using the fact that $c^3 = c + 1$, we express the elements of $V$ as powers of $c$ in the following table:

| $n$ | 0 | 1 | 2 | 3 | 4 | 5 | 6 |
|---|---|---|---|---|---|---|---|
| $c^n$ | 1 | $c$ | $c^2$ | $c + 1$ | $c^2 + c$ | $c^2 + c + 1$ | $c^2 + 1$ |

Lagrange's Theorem tells us that $c^j$ is a zero of $x^7 - 1$ for $j = 0, 1, \ldots, 6$. Then it follows from Theorem 3 of Section 5.4 that in $E[x]$ we have

$$x^7 - 1 = (x - 1)(x - c)(x - c^2)(x - c^3)(x - c^4)(x - c^5)(x - c^6). \tag{4}$$

Since $\pi$ is the minimal polynomial for $c$ over $F$ and $x^7 - 1$ is a polynomial in $F[x]$ with $c$ as a zero, it follows that $\pi$ is a divisor of $x^7 - 1$. (See the corollary to Theorem 4 of Section 5.8.) In fact,

$$x^7 - 1 = (x - 1)(x^3 + x + 1)(x^3 + x^2 + 1).$$

This and the factorization (4) imply that $c^j$, for $1 \leq j \leq 6$, is a zero of either $\pi$ or $\pi' = x^3 + x^2 + 1$. Substitution shows that $c$, $c^2$, and $c^4$ are zeros of $\pi$ while $c^3$, $c^5$, and $c^6$ are zeros of $\pi'$.

## *Problems*

1. In $\mathbf{Z}_3[x]$, let $\beta = x^3 + \bar{2}x + 1$ and $I = (\beta)$.
   (i) Show that $\beta$ has no zero in $\mathbf{Z}_3 = \{\bar{0}, \bar{1}, \bar{2}\}$.
   (ii) Explain why $\mathbf{Z}_3[x]/I$ is a field $F$.
   (iii) How many elements are there in the $F$ of (ii)?
2. Find a monic prime $\pi$ in $\mathbf{Q}[x]$ such that $\mathbf{Q}[x]/(\pi)$ is isomorphic to the field extension $\mathbf{Q}(\sqrt{5})$.
3. Describe the construction of a field with 25 elements.
4. For which integers $m$ in $\{1, 2, 3, \ldots, 20\}$ is there
   (i) a field with $m$ elements?
   (ii) an integral domain with $m$ elements?
5. Let $s$ be algebraic of degree $n$ over a field $F$ with $q$ elements. Prove that $F[s]$ has $q^n$ elements and that the minimal polynomial for $s$ over $F$ is a divisor of $x^m - x$, where $m = q^n$. (This is the proof of Corollary 1 to Theorem 2.)
6. Prove that the number of elements in a finite field $F$ is of the form $p^n$, where $p$ is the characteristic of $F$ and $n$ is a positive integer (and thus prove Corollary 2 to Theorem 2).
7. Let $s$ and $t$ have the same minimal polynomial over $F$. Prove that $F[s]$ and $F[t]$ are isomorphic (and thus prove Corollary 3 to Theorem 2).
8. Let $s$ be a zero in an extension field $E$ over $F$ of a monic prime $\pi$ in $F[x]$. Show that $\pi$ is the minimal polynomial for $s$ over $F$.
9. Let $F$ be a field with characteristic 5 and let $\theta$ be the mapping from $F$ to itself with $\theta(x) = x^5$. Show the following:
   (a) $\theta$ is an endomorphism (that is, a homomorphism from $F$ to itself).
   (b) If $F$ is finite, $\theta$ is an automorphism and every element of $F$ has a unique fifth root in $F$.
10. Do the analogue of Problem 9 in which $F$ is a field of characteristic $p$ and $\theta(x) = x^p$.

11. Let $s$ and $t$ be complex numbers. Use the fact that in $\mathbf{C}[x]$

$$(x + s + t) \mid (x^3 - 3stx + s^3 + t^3)$$

to find the minimal polynomial for $\sqrt[3]{2} + \sqrt[3]{4}$ over $\mathbf{Q}$.

12. Find the minimal polynomial for $\sqrt[3]{5} + \sqrt[3]{25}$ over $\mathbf{Q}$.

13. Show that the minimal polynomial for $\sqrt{7} + \mathbf{i}$ is:
(i) $x^2 - 2\sqrt{7}x + 8$ over the real numbers $\mathbf{R}$.
(ii) $x^4 - 12x^2 + 64$ over the rational numbers $\mathbf{Q}$.

14. Find the minimal polynomial for $\sqrt{7} + \mathbf{i}$ over $\mathbf{Q}(\mathbf{i})$.

15. Let $V$ be the multiplicative group of a field $F$ with 9 elements. Show that $V$ is cyclic.

16. Let $P$ be the prime subfield in a finite field $F$ with $m$ elements. Explain why $F$ is the splitting field over $P$ of $x^m - x$.

17. Let $\alpha$ be in $F[x]$ and let $\alpha'$ be the derivative of $\alpha$. Prove that $\gcd(\alpha, \alpha') = 1$ if and only if there is no extension field $E$ over $F$ in which $\alpha$ has a zero with multiplicity greater than 1. (See Problem 29 of Section 5.6.)

18. Show that no one of the following polynomials in $\mathbf{C}[x]$ has a zero of multiplicity greater than 1. (*Hint*: See the previous problem.)
(i) $x^2 + 2x + 2$.
(ii) $x^3 + 3x^2 + 6x + 6$.
(iii) $x^4 + 4x^3 + 12x^2 + 24x + 24$.
(iv) $x^5 + 5x^4 + 20x^3 + 60x^2 + 120x + 120$.

19. Show that $x^3 - 3x^2 + 6x - 6$ does not have a zero with multiplicity greater than 1.

20. Generalize on Problems 18 and 19.

*21. Let $F$ be a finite field with 256 elements. How many monic irreducible polynomials of degree 8 are there with coefficients in $F$?

## *5.12 *Automorphisms of E over F, Insolvability of a Quintic*

*Mathematicians are like Frenchmen: whatever you say to them they translate into their own language, and forthwith it is something entirely different.*

*Johann Wolfgang von Goethe*

Galois reduced the problem of deciding whether a given polynomial equation can be solved in radicals to that of testing whether a group of permutations of the roots of the equation is a solvable group. Here we present the first small steps of this process. We also give an equation of the fifth degree which is not solvable in radicals together with a number of theorems, some without proof, which imply this impossibility result.

We start by considering some special automorphisms of an extension field $E$ over $F$.

### *Definition 1 Automorphism of* E *over* F

*Let $F$ be a subfield in $E$. An* **automorphism of E over F** *is an automorphism $\theta$ of $E$ such that $\theta(a) = a$ for all $a$ in $F$.*

### *Theorem 1 Automorphisms over* F *Send Roots to Roots*

*Let $\theta$ be an automorphism of $E$ over $F$ and let $r$ be a zero in $E$ of an $\alpha$ in $F[x]$. Then the image of $r$ under $\theta$ is also a zero of $\alpha$.*

#### *Proof*

Let the image of an element $b$ of $E$ under $\theta$ be denoted by $\bar{b}$. Let

$$\alpha = a_d x^d + \cdots + a_1 x + a_0, \quad a_i \text{ in } F.$$

Since $\theta$ preserves additions and multiplications, applying $\theta$ to the given equation

$$a_d r^d + \cdots + a_1 r + a_0 = 0$$

leads to

$$\bar{a}_d \bar{r}^d + \cdots + \bar{a}_1 \bar{r} + \bar{a}_0 = \bar{0}.$$

Since $\theta$ leaves each element of $F$ invariant, we then have

$$a_d \bar{r}^d + \cdots + a_1 \bar{r} + a_0 = 0,$$

that is, $\alpha(\bar{r}) = 0$ as desired.

Let $E$ be a splitting field for an $\alpha$ of positive degree in $F[x]$. In $E[x]$, let

$$\alpha = a(x - s_1)(x - s_2) \cdots (x - s_n),$$

where $a$ is in $F$ and the $s_i$ are $n$ distinct elements of $E$. Let $\theta$ be an automorphism of $E$ over $F$. Then it follows from Theorem 1 that

$$s_i \mapsto \theta(s_i), \qquad i = 1, 2, \ldots, n$$

is a permutation on the set $\{s_1, s_2, \ldots, s_n\}$. This leads us to the following result.

**Theorem 2** ***Galois Group of a Polynomial***

*Let $E$ be a splitting field for an $\alpha$ of degree $n$ in $F[x]$. Let $\alpha$ have $n$ distinct zeros in $E$. Then the automorphisms of $E$ over $F$ form a group $G$ isomorphic to a subgroup $G'$ in $\mathbf{S}_n$. (The operation in $G$ is composition of mappings.)*

The proof is left to the reader as Problem 1 of this section.

The group $G'$ of Theorem 2 is called the ***Galois group*** for $\alpha$ over $F$.

***Example 1***

Show that the Galois group $G$ for $\pi = x^3 + x + 1$ over $F = \{0, 1\}$ is a cyclic group of order 3.

***Solution***

In Example 2 of Section 5.11, we saw that $\pi$ has a splitting field $E = \{0, 1, c, c^2, \ldots, c^6\}$ and that, in $E[x]$,

$$\pi = (x - c)(x - c^2)(x - c^4).$$

Let $\theta$ be an isomorphism of $E$ over $F$; then

$$\theta(0) = 0, \ \theta(c^k) = [\theta(c)]^k \quad \text{for } 0 \le k \le 6.$$

This shows that $\theta$ is completely determined by the value of $\theta(c)$. Theorem 1 tells us that $\theta(c)$ must be in the set $\{c, c^2, c^4\}$ of zeros of $\pi$ in $E$.

Let $\alpha$ be the mapping from $E$ to $E$ with

$$\alpha(0) = 0, \ \alpha(c^k) = c^{2k} \quad \text{for } 0 \le k \le 6.$$

It is now straightforward to verify that $\alpha$, $\alpha^2$, and the identity automorphism $\varepsilon$ are the 3 automorphisms of $E$ over $F$. (This is left to the reader.) Hence the Galois group of $\pi$ is cyclic and of order 3 since it is isomorphic to the automorphism group $\{\varepsilon, \alpha, \alpha^2\}$.

We next state, but do not prove, an extremely difficult result that has greatly influenced abstract algebra.

***Theorem 3*** ***A Galois Theorem***

*Let $F$ be a field of characteristic 0 and let $\alpha \in F[x]$. Then $\alpha = 0$ is solvable in radicals if and only if the Galois group of $\alpha$ over $F$ is a solvable group.*

It can be shown that the Galois group of $x^5 - 4x^3 - 2$ over the rational numbers $\mathbf{Q}$ is the symmetric group $\mathbf{S_5}$. (This is not proved here. A proof is indicated in Exercise 8 of Chapter IX in Chih-Han Sah, *Abstract Algebra*, New York: Academic Press, 1967.)

Assuming this fact, it follows from Theorem 3 above and Theorem 4 of Section 2.13 that

$$x^5 - 4x^3 - 2 = 0$$

is not solvable in radicals; that is, its roots cannot be expressed in terms of the coefficients using only the operations of addition, subtraction, multiplication, division, and extraction of $n$th roots (with $n$ a positive integer).

## *Problems*

1. Let $E$ be a splitting field for an $\alpha$ of degree $n$ in $F[x]$. Prove that the automorphisms of $E$ over $F$ form a group $G$ isomorphic to a subgroup $G'$ in $\mathbf{S_n}$ (and thus prove Theorem 2).
2. Show that Theorem 3 is true when $\deg \alpha \leq 4$, using appropriate references to previous results in this text (for example, Problem 20 in the Supplementary and Challenging Problems for Chapter 2).
3. How many automorphisms are there of
   (i) $\mathbf{Q}(\sqrt{2})$ over $\mathbf{Q}$? (ii) $\mathbf{Q}(\sqrt[3]{2})$ over $\mathbf{Q}$?
4. Explain why every automorphism $\theta$ of a field $F$ is an automorphism of $F$ over its prime subfield $P$.
5. Find the order of the Galois group of $x^3 - 5$ over
   (i) the rational numbers $\mathbf{Q}$; (ii) the real numbers $\mathbf{R}$.

6. Find the order of the Galois group of $x^8 - x$ over $\mathbf{Z}_2$.
7. Let $s$ be a zero in $E = F(s)$ of a prime $\pi$ in $F[x]$. Prove that there are at least as many automorphisms of $E$ over $F$ as there are zeros of $\pi$ in $E$.

---

## *Review Problems*

1. Show that $\sqrt{17} + 4\mathbf{i}$ is algebraic (over $\mathbf{Z}$).
2. Show that every complex number $a + b\mathbf{i}$ is algebraic over the field $\mathbf{R}$ of real numbers.
3. Explain why every element of $\mathbf{Q}[\sqrt[3]{5}]$ can be expressed in the form $a\sqrt[3]{25} + b\sqrt[3]{5} + c$, with $a$, $b$, and $c$ in $\mathbf{Q}$.
4. List the 10 monic primes of degree 2 in $\mathbf{Z}_5[x]$.
5. Expand $(x + \bar{2})(x + \bar{1})x(x - \bar{1})(x - \bar{2})$ in $\mathbf{Z}_5[x]$.
6. In $\mathbf{Z}_{11}[x]$, express $x^{10} - \bar{1}$ as a product of monic first degree polynomials.
7. Show that $(x^2 + x + 1) | (x^6 - 2x^3 + 1)$ in $\mathbf{Z}[x]$.
8. Let $\alpha = x^2 + 5$ and $\beta = 2x^2 + 10$.
   (i) Are $\alpha$ and $\beta$ associates in $\mathbf{Q}[x]$? Explain.
   (ii) Are $\alpha$ and $\beta$ associates in $\mathbf{Z}[x]$? Explain.
9. Find the quotient and remainder in the division of
   (i) $x^6 + 13x^5 + 5x^4 + 52x^3 + 3x^2 + 3x - 13$ by $x^2 + 4$ in $\mathbf{Z}[x]$.
   (ii) $x^6 + \bar{3}x^5 + \bar{2}x^3 + \bar{3}x^2 + \bar{3}x + \bar{2}$ by $x^2 - \bar{1}$ in $\mathbf{Z}_5[x]$.
10. Let $c$ and $k$ be complex numbers and let $\bar{c}$ and $\bar{k}$ be their complex conjugates. Show that the expansion of

$$(x^2 + cx + k)(x^2 + \bar{c}x + \bar{k})$$

    has all real coefficients.
11. How many roots are there of $x^2 - x = 0$
    (i) in $\mathbf{Z}_{30}$? (ii) in $\mathbf{Z}_{29}$?
12. How many zeros are there of $x^3 - x$
    (i) in $\mathbf{Z}_{30}$? (ii) in $\mathbf{Z}_{31}$?
13. Expand $(x^2 - \bar{2})^7$ in $\mathbf{Z}_7[x]$.
14. Expand $(x^5 + x^4 + \bar{1})^{81}$ in $\mathbf{Z}_3[x]$.
15. Let $M$ consist of all $\alpha$ in $\mathbf{Z}_{17}[x]$ such that $\alpha^* = 0$ in $\mathbf{Z}_{17}^*$. Explain why $M$ is an ideal in $\mathbf{Z}_{17}[x]$ and name the monic generator of $M$.
16. Show that $x^{11} - x^2 - x + \bar{2}$ has no first degree factor in $\mathbf{Z}_{11}[x]$.
17. Find a rational root of $2x^3 - 5x^2 + 18x - 45 = 0$ and then find the other roots.

18. Give the complete set of possibilities for rational roots of

$$91x^3 + bx^2 + cx + 77 = 0,$$

where $b$ and $c$ are integers.

19. Find the $\alpha$ with $\deg \alpha \leq 3$ in $\mathbf{Z}_7[x]$ that fits the table:

| $x$ | $\bar{1}$ | $\bar{4}$ | $\bar{5}$ | $\bar{6}$ |
|---|---|---|---|---|
| $\alpha(x)$ | $\bar{5}$ | $\bar{6}$ | $\bar{4}$ | $\bar{1}$ |

20. Find the $\beta$ in $\mathbf{Q}[x]$ such that

$$1 \cdot 2^2 + 2 \cdot 3^2 + 3 \cdot 4^2 + \cdots + n(n + 1)^2 = \beta(n).$$

21. Let $\alpha = x^3 - x^2 - 8x + 12$. Express $\alpha$ in the form

$$a + b(x - 2) + c(x - 2)^2 + d(x - 2)^3,$$

with $a$, $b$, $c$, and $d$ in $\mathbf{Q}$ and thus find the multiplicity of 2 as a zero of $\alpha$.

22. Show that 1 is a zero of $nx^{n+2} - (2n + 1)x^{n+1} + (n + 1)x^n + x - 1$ with multiplicity 3 for all positive integers $n$.

23. (i) Show that $x^3 + x^2 - 2x - 1$ has no rational zeros.
(ii) Find all the complex zeros of this cubic.

24. Find all the roots in $\mathbf{C}$ of each of the following:
(a) $x^4 - 13x^2 + 36 = 0$.
(b) $x^4 + 4x^3 + 4x^2 - 4x - 5 = 0$.
(c) $x^4 - 4x^3 + 8x + 20 = 0$.

25. Let $r$ be a fixed element of a commutative ring $U$ with unity and let $f$ be the mapping from $U[x]$ to $U$ with $\alpha \mapsto \alpha(r)$.
(i) Explain why $f$ is a ring homomorphism.
(ii) Is $r$ a zero of every $\alpha$ in the kernel $K$ of $f$?
(iii) Name the monic generator of the kernel $K$.

26. Which pair from the following three polynomials of $\mathbf{Q}[x]$ are relatively prime?
(i) $x^2 - 9$; (ii) $x - 3$; (iii) $x + 3$.

27. How many monic second degree primes are there in $\mathbf{Z}_{29}[x]$?

28. Express $x^3/(x + \bar{1})(x + \bar{3})$ as the sum of a polynomial and a finite number of ppd-terms in $\mathbf{Z}_7(x)$; that is, find the partial fraction decomposition of this quotient of polynomials.

29. Is there a field with 99 elements? Explain.

30. Is there a field with 49 elements? Explain.

31. Let $F$ be a finite field with $n$ elements. Let $\pi$ be a prime of degree $d$ in $F[x]$ and let $m = n^d$. Prove that $\pi|(x^m - x)$ in $F[x]$.
32. Let $\alpha$ and $\pi$ be in $F[x]$, with $\pi$ a prime. Prove that $\pi|\alpha$ in $F[x]$ if and only if there exists an extension field $E$ over $F$ in which $\alpha$ and $\pi$ have a common zero.
33. Let $F$ be a field with 81 elements. Is every automorphism $\theta$ of $F$ an automorphism of $F$ over its prime subfield $P = \{0, 1, 1 + 1\}$? Explain.
34. Is every automorphism $\theta$ of the complex numbers **C** an automorphism of **C** over **Q**? Explain.

## *Supplementary and Challenging Problems*

1. Let $z$ and $w$ be complex numbers with $z$ transcendental (over **Z**). Prove the following:
   (i) There is exactly one homomorphism $\theta$ from $\mathbf{Z}[z]$ to $\mathbf{Z}[w]$ with $\theta(1) = 1$ and $\theta(z) = w$.
   (ii) The $\theta$ of (i) is an isomorphism if and only if $w$ is transcendental (over **Z**).
2. Let $\alpha$ and $\beta$ be in $\mathbf{Q}[x]$. In $\mathbf{C}[x]$, let

$$\alpha = (x - r)(x - s), \qquad \beta = (x - a)(x - b)(x - c).$$

   Prove the following:
   (i) $r - a$ is algebraic over **Q** (and hence over **Z**).
   (ii) $ra$ is algebraic over **Q**.
   (iii) $1/a$ is algebraic over **Q** if $a \neq 0$.
3. Let $A$ be the set of all complex numbers $a$ that are algebraic (over **Z**). Prove that $A$ is a subfield in the complex numbers **C**.
4. (i) Let $n$ be an integer and let $\beta|\alpha$ in $\mathbf{Z}[x]$. Show that $\beta(n)|\alpha(n)$ in **Z**.
   (ii) Find all the integers $m$ such that $x^2 - x + m$ is a divisor of $x^{13} + x + 90$ in $\mathbf{Z}[x]$.
5. Let $L_n$ be the $n$th Lucas number defined by

$$L_1 = 1, \qquad L_2 = 3, \qquad L_n = L_{n-1} + L_{n-2} \quad \text{for } n = 3, 4, 5, \ldots.$$

   Prove that, for all positive integers $n$, $x^2 - x - 1$ is a divisor of $x^{2n} - L_n x^n + (-1)^n$ in $\mathbf{Z}[x]$.
6. Let $\alpha = u_0 x^d + u_1 x^{d-1} + \cdots + u_d$ and $\beta = v_0 x^{d-1} + v_1 x^{d-2} + \cdots + v_{d-1}$ in $F[x]$ with the coefficients satisfying

$$v_j = u_0 + u_1 + \cdots + u_j \quad \text{for } 0 \leq j < d.$$

   Also let $\alpha$ and $\beta$ have a common zero in an extension field $E$ over $F$. Prove that $\alpha = \beta\gamma$ with $\gamma$ in $F[x]$ and find $\gamma$.

7. (i) Let $p\alpha = \beta\gamma$ with $p$ a prime in $\mathbf{Z}$ and $\alpha$, $\beta$, and $\gamma$ in $\mathbf{Z}[x]$. Also, let $\gamma$ have a coefficient that is not a multiple of $p$. Prove that every coefficient of $\beta$ is a multiple of $p$.
   (ii) Prove that $\alpha$ is reducible in $\mathbf{Z}[x]$ if and only if it is reducible (and composite) in $\mathbf{Q}[x]$.

8. Let $\alpha = a_d x^d + \cdots + a_1 x + a_0$ be in $\mathbf{Z}[x]$. Let $p$ be a prime in $\mathbf{Z}$ and let

$$a_d \not\equiv 0 \pmod{p}, \qquad a_j \equiv 0 \pmod{p} \text{ for } 0 \le j < d,$$
$$a_0 \not\equiv 0 \pmod{p^2}.$$

   Prove that $\alpha$ is irreducible (and a prime) in $\mathbf{Q}[x]$. (This is known as ***Eisenstein's Irreducibility Criterion.***)

9. Prove that $x^5 - 4x^3 - 2$ is a prime in $\mathbf{Q}[x]$.

10. (i) Prove that $x^4 + 5x^3 + 10x^2 + 10x + 5$ is a prime in $\mathbf{Q}[x]$.
    (ii) Use the substitution $x = y + 1$ to prove that

$$x^4 + x^3 + x^2 + x + 1$$

   is a prime in $\mathbf{Q}[x]$.

11. Let $m$, $n$, and $p$ be positive integers with $p$ prime. Let $\alpha_n$ be the complex number $\cos(2\pi/n) + \mathbf{i}\sin(2\pi/n)$. The minimal polynomial $C_n(x)$ for $\alpha_n$ over $\mathbf{Q}$ is called the $n$th ***cyclotomic polynomial***. Prove the following:
    (a) $C_n(x)$ is the minimal polynomial for each primitive $n$th root of unity, that is, for each

$$\cos(2k\pi/n) + \mathbf{i}\sin(2k\pi/n) \quad \text{with } \gcd(k, n) = 1.$$

    The degree of $C_n(x)$ is $\phi(n)$, where $\phi$ is the Euler $\phi$-function.
    (b) $x^n - 1$ is the product of the $C_d(x)$ for all positive integral divisors $d$ of $n$.
    (c) $C_p(x) = x^{p-1} + x^{p-2} + \cdots + x + 1$. (Recall that $p$ is a prime.)
    (d) If $p \mid n$, then $C_{pn}(x) = C_n(x^p)$.
    (e) If $p$ is not an integral divisor of $n$, then

$$C_{pn}(x)C_n(x) = C_n(x^p).$$

    *(f) If $p$ is not an integral divisor of $n$, then

$$C_{pn}(x) = C_n(\alpha_p x)C_n(\alpha_p^2 x)C_n(\alpha_p^3 x) \cdots C_n(\alpha_p^{p-1} x)$$

    and $C_{pn}(x)C_n(x) = C_n(x^p) = D(p, n)$, where $D(p, n)$ is the determinant of the $p$ by $p$ matrix $(a_{ij})$ in which $a_{ij}$ is the sum of all the terms $c_k x^k$ of $C_n(x)$ for which $k \equiv i - j + 1 \pmod{p}$.
    (g) $C_1(1) = 0$, $C_n(1) = p$ if $n$ is a power $p^m$, and $C_n(1) = 1$ if $n$ is neither 1 nor a power of a prime.
    *(h) Each coefficient of $C_n(x)$ is in $\{-1, 0, 1\}$ if $n$ is not a multiple of three distinct odd primes.

12. Prove that $x^{n-1} + x^{n-2} + \cdots + x + 1$ is a prime in $\mathbf{Q}[x]$ if and only if $n$ is a prime in $\mathbf{Z}^+$.

13. In $\mathbf{Z}_2[x]$, let $\beta_n = x^{n-1} + x^{n-2} + \cdots + x + \bar{1}$.
 (i) Prove that $\beta_n$ is reducible whenever $n$ is a composite integer.
 (ii) Show that $\beta_p$ may or may not be reducible when $p$ is a prime in $\mathbf{Z}$.

14. Let $\beta = (x^3 - 1)(x^2 - 1)(x - 1)$ and $\alpha_n = (x^{n+2} - 1)(x^{n+1} - 1)(x^n - 1)$. Prove that $\beta | \alpha_n$ in $\mathbf{Z}[x]$ for all positive integers $n$. Generalize.

15. Let $p = 2q + 1$, where $q$ is an odd positive integer and $p$ is a prime. Prove that $q! \equiv \pm 1 \pmod{p}$.

16. Let $p$ and $d$ be positive integers with $p$ a prime and $d | (p - 1)$. Show that there are at most $\phi(d)$ elements with order $d$ in the multiplicative group $\mathbf{V_p}$ of invertibles in $\mathbf{Z_p}$. [Here $\phi$ is the Euler $\phi$-function. It can be shown that there are exactly $\phi(d)$ elements with order $d$ in $\mathbf{V_p}$. See Problem 5 of the Supplementary and Challenging Problems for Chapter 7.]

17. Let $r = \cos(2\pi/n) + \mathbf{i} \sin(2\pi/n)$, where $n$ is an integer greater than 1. Show the following:
 (i) $x^n - 1 = (x - 1)(x - r) \cdots (x - r^{n-1})$ in $\mathbf{C}[x]$.
 (ii) $1 + r + r^2 + \cdots + r^{n-1} = 0$.
 (iii) $1 \cdot r \cdot r^2 \cdots r^{n-1} = (-1)^{n-1}$.
 (iv) Let $p_n = (1 + r)(1 + r^2) \cdots (1 + r^{n-1})$. Then $p_n = 0$ when $n$ is even and $p_n = 1$ when $n$ is odd.
 (v) $\{1, r, r^2, \ldots, r^{n-1}\}$ is the only subgroup with order $n$ in the multiplicative group of the nonzero complex numbers.

18. Let $r_1, r_2, \ldots, r_n$ be the $n$ $n$th roots of 1 in $\mathbf{C}$. Evaluate

$$\prod_{i<j} (r_i - r_j)^2,$$

that is, the product of the squares of the differences of the $n$th roots of unity.

19. Let $\alpha = a + b(x + 3) + c(x + 3)(x - 4)$, where $a$, $b$, and $c$ are in $\mathbf{Q}$. Find $a$, $b$, and $c$ given that, in $\mathbf{Q}[x]$,

$$(x + 3) | (\alpha - 6), \ (x - 4) | (\alpha + 7), \text{ and } (x - 6) | (\alpha - 10).$$

Note that here the parentheses are symbols of aggregation and do not denote principal ideals.

20. Find the smallest positive integer $a$ such that there is a polynomial $ax^2 + bx + c$ in $\mathbf{Z}[x]$ with real zeros $r$ and $s$ satisfying $0 < r < s < 1$.

21. For every positive integer $n$, let

$$(x + 1)^n(x^2 - 1.998x + 1) = x^{n+2} + a_{n+1}x^{n+1} + a_n x^n + \cdots + a_0.$$

Find the smallest $n$ such that $a_j > 0$ for $0 < j < n + 1$.

22. Let $F$ be an extension field over $\mathbf{Q}$ and let $n$ denote $n \cdot 1$ in $F$. In $F[x]$, let

$$x_a^{(d)} = (x + a)(x + a - 1)(x + a - 2) \cdots (x + a - d + 1).$$

Prove that for every $\alpha$ of degree $d$ in $F[x]$, there are $c_i$ in $F$ such that

$$\alpha = c_0 x_0^{(d)} + c_1 x_1^{(d)} + \cdots + c_d x_d^{(d)}.$$

23. In $\mathbf{Q}[x]$, let $\alpha = x^4 + x^3 - x^2 - x + 1$.
(i) Show that if $r$ is a zero in $\mathbf{C}$ of $\alpha$, so is $-1/r$.
(ii) Find the monic second degree $\beta$ in $\mathbf{Q}[x]$ such that $s = r - r^{-1}$ is a zero of $\beta$ when $r$ is a zero of $\alpha$.
(iii) Find all the roots of $\alpha = 0$ in $\mathbf{C}$.

24. Generalize on parts (i) and (ii) of Problem 23.

25. Let $\alpha = x^4 + ax^3 + bx^2 + cx + d$ be in $\mathbf{C}[x]$. Use the substitution $x = \sqrt{y}$ to show that the square of a root of $\alpha = 0$ satisfies

$$y^4 + (2b - a^2)y^3 + (b^2 + 2d - 2ac)y^2 + (2bd - c^2)y + d^2 = 0.$$

26. Let $\alpha = x^3 + ax^2 + bx + c$ be in $\mathbf{C}[x]$. Find a monic cubic equation satisfied by the cubes of the roots of $\alpha = 0$.

27. Assume the fact that if $r_1, r_2, \ldots, r_n$ are nonnegative real numbers and

$$(x - r_1)(x - r_2) \cdots (x - r_n) = x^n - s_1 x^{n-1} + s_2 x^{n-2} - \cdots + (-1)^n s_n,$$

then $s_1/n \geq \sqrt[n]{s_n}$ and use this result to prove that the roots of

$$x^n + a_1 x^{n-1} + a_2 x^{n-2} + \cdots + a_{n-1} x + a_n = 0$$

cannot all be real when each $a_i$ is in $\{-1, 1\}$ and $n > 3$.

28. Prove that there is no choice of $k$ as a real number such that

$$x^4 + x^3 - x^2 + kx + 1 = 0$$

has 4 real roots.

29. Let $\mathbf{C}$ be the field of complex numbers. Let $\alpha$ and $\beta$ be in $\mathbf{C}[x]$ and have the same set of zeros in $\mathbf{C}$ (but possibly with different multiplicities). Let the same be true for the polynomials $\alpha + 1$ and $\beta + 1$. Prove that $\alpha = \beta$.

30. Let $F$ be an extension field over $\mathbf{Q}$. In $F[x]$, let

$$(x - r_1)(x - r_2) \cdots (x - r_n) = x^n - s_1 x^{n-1} + s_2 x^{n-2} - \cdots + (-1)^n s_n.$$

Also let $N_k = (r_1)^k + (r_2)^k + \cdots + (r_n)^k$. Show that:
(i) $N_0 = n$ if $s_n \neq 0$.

(ii) $N_k - s_1 N_{k-1} + s_2 N_{k-2} - \cdots + (-1)^{k-1} s_{k-1} N_1 + (-1)^k s_k k = 0$ for $k = 1, 2, \ldots, n-1$.

(iii) $N_k - s_1 N_{k-1} + s_2 N_{k-2} - \cdots + (-1)^n s_n N_{k-n} = 0$ for $k = n, n+1, \ldots$.

(These results are known as the ***Newton formulas*** on the sums of the powers of the roots of a polynomial equation.)

31. Let $w = (-1 + \mathbf{i}\sqrt{3})/2$. Find all the invertibles in $\mathbf{Z}[w]$.

32. Find all the invertibles in $\mathbf{Z}[\mathbf{i}\sqrt{5}]$.

33. In the integral domain $\mathbf{Z}[\mathbf{i}]$ of the gaussian integers, let

$$\alpha_1 \alpha_2 \cdots \alpha_m = \beta_1 \beta_2 \cdots \beta_n,$$

with the $\alpha$'s and $\beta$'s primes in $\mathbf{Z}[\mathbf{i}]$. Prove the following:

(i) $\alpha_1$ has an associate among the $\beta$'s.

(ii) $m = n$.

(iii) There is a bijection (one-to-one onto mapping) $f$ from $\{\alpha_1, \ldots, \alpha_m\}$ to $\{\beta_1, \ldots, \beta_m\}$ such that $\alpha_i$ and $f(\alpha_i)$ are associates for $1 \leq i \leq m$.

(This is an analogue for $\mathbf{Z}[\mathbf{i}]$ of the unique factorization theorem for $\mathbf{Z}$ given in Theorem 2 of Section 1.7.)

34. Let $D = \mathbf{Z}[\mathbf{i}\sqrt{5}]$. For all $\alpha = a + b\mathbf{i}\sqrt{5}$ (with $a$ and $b$ in $\mathbf{Z}$) in $D$, let $\bar{\alpha} = a - b\mathbf{i}\sqrt{5}$ and $\|\alpha\| = \alpha\bar{\alpha}$. Prove the following:

(i) $\|\alpha\beta\| = \|\alpha\| \cdot \|\beta\|$ in $D$.

(ii) $D$ is an integral domain.

(iii) $6 = 2 \cdot 3 = (1 + \mathbf{i}\sqrt{5})(1 - \mathbf{i}\sqrt{5})$ in $D$.

(iv) 2, 3, $1 + \mathbf{i}\sqrt{5}$, and $1 - \mathbf{i}\sqrt{5}$ are primes in $D$.

(v) No two of the primes in part (iv) are associates. (This shows that $\mathbf{Z}[\mathbf{i}\sqrt{5}]$ does not have the type of unique factorization into primes described for $\mathbf{Z}[\mathbf{i}]$ in the preceding problem.)

35. Let $F$ be a finite field with $m$ elements and let $q$ be a positive prime in $\mathbf{Z}$. Show that there are exactly $(m^q - m)/q$ monic prime polynomials of degree $q$ in $F[x]$.

36. Let $\theta$ be a mapping from $\mathbf{C}$ to $\mathbf{C}$ with $\Delta^2\theta(x) = 2\theta(x)$, i.e.,

$$\theta(x+2) - 2\theta(x+1) + \theta(x) = 2\theta(x).$$

Explain why there cannot be any polynomial $\alpha$ in $\mathbf{C}[x]$ such that the polynomial function $\alpha^* = \theta$.

37. For each integer $s > 1$, let $d_s$ be the lowest degree of a monic polynomial $\alpha$ in $\mathbf{Z}[x]$ with the property that $s \mid \alpha(n)$ for all integers $n$. Prove that $d_s$ is the smallest positive integer $k$ such that $s \mid k!$.

38. Find all the primes $\pi = m + n\mathbf{i}$ in the gaussian integers $\mathbf{Z}[\mathbf{i}]$ such that $\pi$ is an associate of its conjugate $\bar{\pi} = m - n\mathbf{i}$.

39. Find a positive prime $p$ in $\mathbf{Z}$ and a prime $\pi = m + n\mathbf{i}$ in $\mathbf{Z}[\mathbf{i}]$ such that $\pi^2 \mid p$ in $\mathbf{Z}[\mathbf{i}]$.

# 6

## *Euclidean Constructions*

*Indeed, when in the course of a mathematical investigation we encounter a problem or conjecture a theorem, our minds will not rest until the problem is exhaustively solved and the theorem rigorously proved; or else, until we have found the reasons which made success impossible and, hence, failure unavoidable. Thus, the proofs of the impossibility of certain solutions play a predominant role in modern mathematics; the search for an answer to such questions has often led to the discovery of newer and more fruitful fields of endeavour.*

*David Hilbert*

Students of euclidean geometry learn how to perform various constructions, including the bisection of a given angle, with straightedge and compass. On the other hand, a few euclidean construction problems, including angle trisection, had intrigued people for more than 2000 years without anyone producing a valid solution.

In the early part of the nineteenth century, mathematicians began to prove that certain unsolved problems were in that state because people were looking for things that do not, and will not ever, exist. They proved that no formula which merely involves the rational operations and extraction of roots can ever be found for solving general fifth degree polynomial equations. Then they used similar algebraic methods to prove trisection of a general angle and some other classical euclidean constructions impossible to achieve.

But this has not stopped the "angle trisectors." Mathematics departments and libraries still receive unsolicited manuscripts and privately printed books that claim to contain a solution.

This wasted effort is due to several misconceptions. One is the mistaken notion that the problem is still unsolved (thus permitting a prospective solver to gain undying fame in its solution), while in actual fact the problem has been solved by the proof that a valid construction is not possible.

There is also lack of understanding of what is allowed in a euclidean construction. This topic is part of euclidean geometry in which lines have no thickness and points have neither length nor breadth nor depth. What is sought is not just an excellent approximation to a geometrical figure but a theoretical method for obtaining the exact figure. Of course there are constructions yielding a figure so close to the desired one that the eye cannot perceive the difference; these are not acceptable euclidean constructions. What is wanted is a method of construction that would do the job exactly with the allowable tools if they were used perfectly. This problem is not in applied science but in conceptual mathematics.

We shall describe the problem in technical terms in the next section.

## 6.1 Closure under Euclidean Constructions

> *"When I use a word," Humpty Dumpty said in a rather scornful tone, "it means just what I choose it to mean—neither more nor less."*
>
> *Charles L. Dodgson (Lewis Carroll)*

We deal with a fixed plane, which we call *the plane*. We assume that we are given two fixed points $\pi_0$ and $\pi_1$ and that the distance between them is the unit of distance. We use "line" to mean "straight line."

Let $\Gamma$ be a collection of points, lines, and circles of the plane. $\Gamma$ is said to be *closed under euclidean constructions* if the following three conditions are met:

(a) The line through any two distinct points of $\Gamma$ is in $\Gamma$.

(b) Let $\alpha$, $\beta$, and $\gamma$ be points of $\Gamma$ with $\alpha \neq \beta$. Let $r$ be the distance between $\alpha$ and $\beta$. Then the circle with center at $\gamma$ and radius $r$ is in $\Gamma$.

(c) If $\alpha$ is a point of intersection of two distinct lines of $\Gamma$, or of two distinct circles of $\Gamma$, or of a line and a circle of $\Gamma$, then $\alpha$ is in $\Gamma$.

If $\Gamma$ satisfies these three conditions, it is clear that $\Gamma$ contains every point, line, and circle that might be obtained from those in $\Gamma$ by constructions with straightedge and compass.

In Section 6.3 below, we shall describe a $\Gamma$ that is closed under euclidean constructions (constructions with straightedge and compass) and in which there is a pair of lines forming a 60° angle. If it were possible to trisect every angle with euclidean constructions, there would have to be a pair of lines making a 20° angle in $\Gamma$, since $\Gamma$ is closed under euclidean constructions. However, we shall prove that no lines of the specific $\Gamma$ of Section 6.3 form a 20° angle; and thus prove that a general method for trisecting all angles with straightedge and compass is impossible.

The same technique will be used to prove that some other classical constructions cannot be performed with straightedge and compass. Our advantage over the great mathematicians of ancient times is that now we can make use of analytic geometry and the theory of field extensions.

We start by listing the operations on lengths that have been known to be possible since Euclid's time. In elementary euclidean geometry one learns that if given lengths of 1, $a$, and $b$, then one can construct with straightedge and compass lengths of

$$a + b, \quad |a - b|, \quad ab, \quad a/b, \quad \sqrt{a}. \tag{1}$$

Figure 6.1 illustrates the construction of lengths $a + b$ and $a - b$ from given lengths $a$ and $b$ (with $a > b$). The techniques for constructing lengths $ab$ and $a/b$ using given lengths 1, $a$, and $b$ are indicated in Figure 6.2. The mean proportional construction to obtain $\sqrt{a}$ from 1 and $a$ is shown in Figure 6.3.

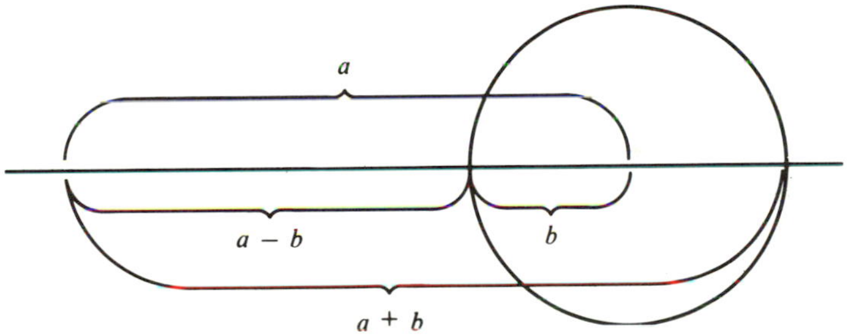


*Figure 6.1*

Since we intend to use both positive and negative numbers as coordinates (with the sign an indication of direction), we may replace $|a - b|$ with $a - b$ in (1) above. Thus we are motivated to consider subsets $F$, with at least two elements, in the real numbers $\mathbf{R}$ such that $F$ is closed under addition, subtraction, multiplication, division by nonzero numbers, and extraction of square roots of nonnegative numbers. A subset $F$ with these properties is a subfield in $\mathbf{R}$ such that $\sqrt{a}$ is in $F$ whenever $a$ is in $F$ and $a \geq 0$.

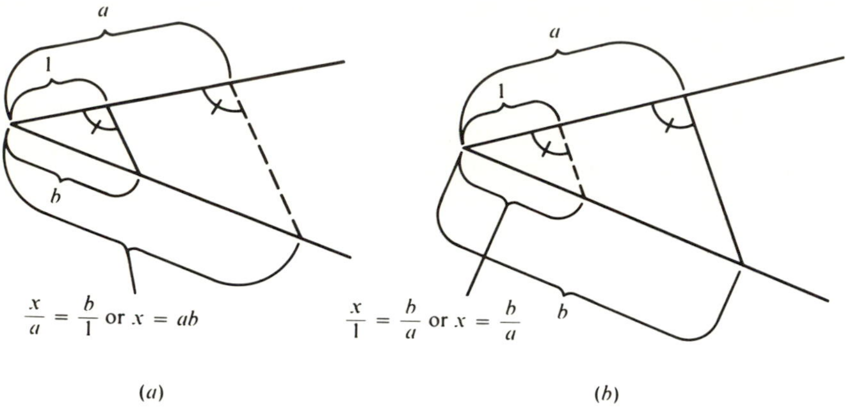


*Figure 6.2*

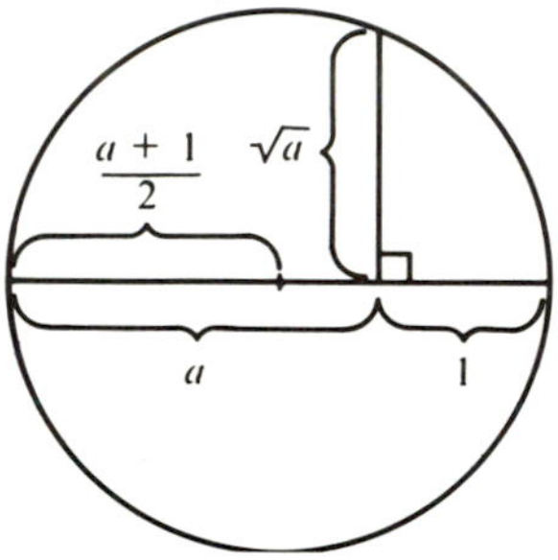


*Figure 6.3*

There are such subfields since the field **R** of all the real numbers has this property. For our present purposes, the smallest $F$ of this nature is most suitable; it will be characterized in the next section.

## 6.2 Closure under Square Roots

The field **Q** of the rational numbers is the prime subfield in the real numbers **R**; this means that **Q** is contained in every subfield in **R**. We know that $\sqrt{a}$ need not be in **Q** when $a$ is a positive rational number since $\sqrt{2}$, $\sqrt{3}$, $\sqrt{5}$, and so on are not in **Q**.

We seek the smallest extension $F$ over $\mathbf{Q}$ such that $F$ is closed under extraction of square roots of positive numbers in $F$. If $a$ is a positive rational number, not only $\sqrt{a}$ but all the numbers in the field extension $\mathbf{Q}(\sqrt{a})$ must be in $F$. If $b$ is a positive real number in $Q_1 = \mathbf{Q}(\sqrt{a})$, we similarly see that every number in $Q_1(\sqrt{b}) = \mathbf{Q}(\sqrt{a}, \sqrt{b})$ must be in $F$.

This motivates us to consider finite chains

$$Q_0, Q_1, Q_2, \ldots, Q_s, \tag{1}$$

such that $Q_0 = \mathbf{Q}$ and $Q_i = Q_{i-1}(r_i)$, with $r_i$ a square root of a positive number in $Q_{i-1}$, for $i = 1, 2, \ldots, s$. We call the end result $Q_s = \mathbf{Q}(r_1, r_2, \ldots, r_s)$, of such a chain, a ***multiquadratic extension*** over $\mathbf{Q}$.

***Example 1***

Find a multiquadratic extension $E = \mathbf{Q}(a, b, c)$ over $\mathbf{Q}$ such that all four roots of $x^4 - 8x^2 + 5 = 0$ are in $E$.

***Solution***

Substituting $u$ for $x^2$ converts the given equation into the quadratic $u^2 - 8u + 5 = 0$, which has $4 + \sqrt{11}$ and $4 - \sqrt{11}$ as its solutions. Then $x = \pm\sqrt{u}$ leads to the four solutions for $x$:

$$x_1 = \sqrt{4 + \sqrt{11}}, \quad x_2 = -x_1, \quad x_3 = \sqrt{4 - \sqrt{11}}, \quad x_4 = -x_3.$$

Now we let $a = \sqrt{11}$, $b = x_1$, and $c = x_3$. Then $a^2$, $b^2$, and $c^2$ are all positive real numbers with $a^2$ in $\mathbf{Q}$, $b^2$ in $\mathbf{Q}(a)$, and $c^2$ in $\mathbf{Q}(a, b)$. Thus $E = \mathbf{Q}(a, b, c)$ is a multiquadratic extension. It is also clear that all four roots $x_i$ are in $E$.

***Notation***

*In the rest of this chapter, $F$ denotes the set of all real numbers in all multiquadratic extensions $E$ over $\mathbf{Q}$; that is, $F$ is the union of all such fields $E$.*

Next we show that this $F$ is a field that is closed under extraction of square roots of positive numbers.

Let $h$ and $k$ be in $F$. Then $h$ is in a multiquadratic extension $Q_s = \mathbf{Q}(a_1, a_2, \ldots, a_s)$ with each $a_{i+1}^2$ in $\mathbf{Q}(a_1, \ldots, a_i)$ and $k$ is in a multiquadratic extension $Q_t = \mathbf{Q}(b_1, b_2, \ldots, b_t)$ with each $b_{j+1}^2$ in $\mathbf{Q}(b_1, \ldots, b_j)$. Then $h - k$ and $hk^{-1}$ (if $k \neq 0$) are in $F$ since they are in the multiquadratic extension field

$$\mathbf{Q}(a_1, \ldots, a_s, b_1, \ldots, b_t).$$

Since $F$ is nonempty, it now follows from Theorem 2 of Section 4.6 that $F$ is a field.

Also, if $g$ is a positive number in $F$, then $g$ is in a multiquadratic extension $\mathbf{Q}(a_1, \ldots, a_s)$. Hence $\sqrt{g}$ is in $\mathbf{Q}(a_1, \ldots, a_s, \sqrt{g})$ and therefore is in $F$. This means that $F$ is closed under extraction of square roots of positive elements.

## Problems

Here $F$ always denotes the special field that is the union of all the multiquadratic extensions of the field $\mathbf{Q}$ of rational numbers.

1. Show that $\cos(\pi/3)$ is in $\mathbf{Q}$, $\cos(\pi/4)$ is in $\mathbf{Q}(\sqrt{2})$, and hence both $\cos(\pi/3)$ and $\cos(\pi/4)$ are in the field $F$.
2. Use the fact that the field $F$ is closed under extraction of square roots of positive numbers to show that $\sin x$ is in $F$ if and only if $\cos x$ is in $F$.
3. Given that $\cos x$ and $\cos y$ are in $F$, show that $\cos(x + y)$ and $\cos(x - y)$ are in $F$.
4. Given that $\cos(2x)$ is in $F$, show that $\cos x$ is in $F$.
5. Show that $\cos(\pi/12)$ is in $F$.
6. Show that $\sin(\pi/24)$ is in $F$.
7. Let $c = \cos(2\pi/5)$, $s = \sin(2\pi/5)$, and $z = c + \mathbf{i}s$. Show the following:
   (i) $z^5 - 1 = 0$, $z \neq 1$.
   (ii) $z^4 + z^3 + z^2 + z + 1 = 0$.
   (iii) $z^2 + z + 1 + z^{-1} + z^{-2} = 0$.
   (iv) $z^{-1} = c - \mathbf{i}s$, $z + z^{-1} = 2c$.
   (v) $z^2 + z^{-2} = 4c^2 - 2$.
   (vi) $4c^2 + 2c - 1 = 0$.
   (vii) $\cos(2\pi/5) = c = (-1 + \sqrt{5})/4$.
   (viii) $\cos(2\pi/5)$ is in the field $F$.
8. Show that $\cos x$ is in $F$ for each of the following:
   (i) $x = \pi/15$; (ii) $x = \pi/30$; (iii) $x = \pi/60$.
9. Show that all four roots of $x^4 - 6x^2 + 2 = 0$ are in a multiquadratic extension $E = \mathbf{Q}(a, b, c)$ over $\mathbf{Q}$.

## 6.3 Constructible Points, Lines, and Circles

*Circles to square and cubes to double*
*would give a man excessive trouble.*

*Matthew Prior*

Let $\pi_0$ and $\pi_1$ be the two fixed points of the plane used in Section 6.1 to determine the unit of length. We set up a cartesian coordinate system in the plane with $\pi_0$ as the *origin*, the line through $\pi_0$ and $\pi_1$ as the *x-axis*, the perpendicular to the $x$-axis at the origin as the *y-axis*, and the distance between $\pi_0$ and $\pi_1$ as the common unit of length on both axes.

Then every point of the plane is uniquely represented by its coordinates $(x, y)$, an ordered pair of real numbers. We shall sometimes use the ordered pair $(x, y)$ as a name for the point.

We are now ready to define the collection $\Gamma$ (of points, lines, and circles of the plane) that will enable us to prove certain classical constructions to be impossible.

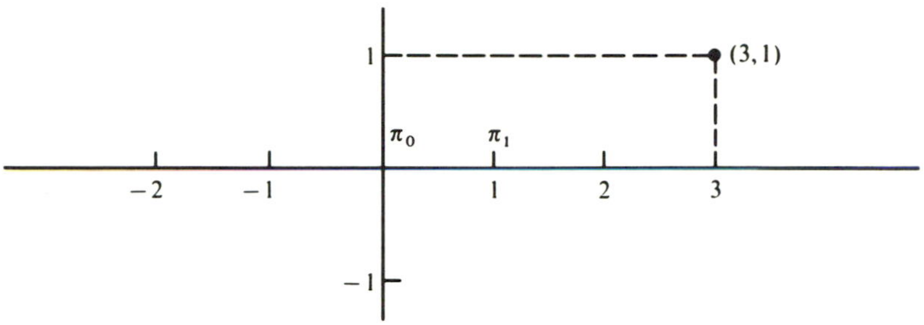


*Figure 6.4*

The condition for a point $(u, v)$ of the plane to be in $\Gamma$ is that each of its coordinates $u$ and $v$ is in the field $F$ of the preceding section. A line is in $\Gamma$ if and only if it has an equation $ax + by = c$ with each of $a$, $b$, and $c$ in $F$, and either $a$ or $b$ not zero. A circle is in $\Gamma$ if and only if its radius $r$ and the coordinates of its center $(h, k)$ are in $F$. Such a circle has an equation of the form

$$x^2 + y^2 - rx - sy = t$$

with $r = 2h$, $s = 2k$, and $t = r^2 - h^2 - k^2$ (and hence $r$, $s$, and $t$ in $F$).

Next we give a number of lemmas which together show that this collection $\Gamma$ is closed under euclidean constructions.

### *Lemma 1*

*The line through two distinct points in $\Gamma$ is also in $\Gamma$.*

#### *Proof*

Let the points have coordinates $(u_1, v_1)$ and $(u_2, v_2)$. Since the two points are distinct, either $u_1 - u_2 \neq 0$ or $v_1 - v_2 \neq 0$. Then the line through the two points has

$$(v_1 - v_2)(x - u_2) - (u_1 - u_2)(y - v_2) = 0$$

as an equation. This can be put in the form $ax + by = c$ with $a = v_1 - v_2$, $b = -(u_1 - u_2)$, and $c = v_1u_2 - u_1v_2$. Since $F$ is a field, we see that $a$, $b$, and $c$ are in $F$ and hence the line is in $\Gamma$.

### *Lemma 2*

*The point of intersection of two nonparallel lines in $\Gamma$ is also in $\Gamma$.*

#### *Proof*

Let the lines have equations

$$a_1x + b_1y = c_1, \qquad a_2x + b_2y = c_2,$$

with $a_1$, $b_1$, $c_1$, $a_2$, $b_2$, and $c_2$ in the field $F$. Let $d = a_1b_2 - a_2b_1$. Since the lines are not parallel, $d \neq 0$. Solving the simultaneous linear equations, we find that the point of intersection is $(u, v)$ with

$$u = (c_1b_2 - c_2b_1)/d, \qquad v = (a_1c_2 - a_2c_1)/d.$$

Now $u$ and $v$ are in the field $F$ and the point $(u, v)$ is in $\Gamma$.

### *Lemma 3*

*A point of intersection of a circle and line of $\Gamma$ is also in $\Gamma$.*

#### *Proof*

Let the equation of the circle be

$$x^2 + y^2 - rx - sy = t \tag{1}$$

and that of the line be $ax + by = c$ with $r$, $s$, $t$, $a$, $b$, and $c$ in $F$ and either $a \neq 0$ or $b \neq 0$. For definiteness, let us assume that $a \neq 0$.

Then the $y$-coordinates of the points of intersection are found by solving $ax + by = c$ for $x$ in terms of $y$, that is, as

$$\frac{c - by}{a}, \tag{2}$$

then substituting this for $x$ in (1), and finally solving the resulting second degree equation for $y$. The operations that have to be performed are additions, subtractions, multiplications, divisions by nonzero numbers, and extraction of a square root (of a nonnegative number when there is at least one actual point of intersection). Since $F$ is closed under all these operations, the $y$ for an intersection is in $F$. Substituting such a $y$ in (2), we obtain the corresponding $x$ coordinate and see that it is also in $F$. Hence the point of intersection is in $\Gamma$.

### *Lemma 4*

*A point of intersection of two circles of $\Gamma$ is also in $\Gamma$.*

***Proof***

Let the circles have equations

$$x^2 + y^2 - r_i x - s_i y = t_i, \quad i = 1 \text{ and } 2.$$

Subtracting corresponding sides of these equations, we see that an intersection also satisfies

$$(r_1 - r_2)x + (s_1 - s_2)y = t_2 - t_1. \tag{3}$$

Since intersecting circles cannot have the same center, either $r_1 - r_2 \neq 0$ or $s_1 - s_2 \neq 0$. Hence (3) is the equation of a line. Then a point of intersection of the two circles is a point of intersection of this line and either one of the circles and thus is in $\Gamma$ by Lemma 3.

Summing up these four lemmas we have the following.

### *Theorem 1 Closure under Euclidean Constructions*

*$\Gamma$ is closed under euclidean constructions.*

We need a few more lemmas.

### *Lemma 5*

*There are lines in $\Gamma$ that form a 60° angle.*

***Proof***

Clearly $\sqrt{3}$ is in $F$ and hence the lines with

$$y = 0, \qquad \sqrt{3}x - y = 0$$

as equations are in $\Gamma$. These lines form a 60° angle.

### Lemma 6

*If $a$ is an angle made by two lines of $\Gamma$, $\cos a$ is in $F$. (The converse is left to the reader as Problem 1 of this section.)*

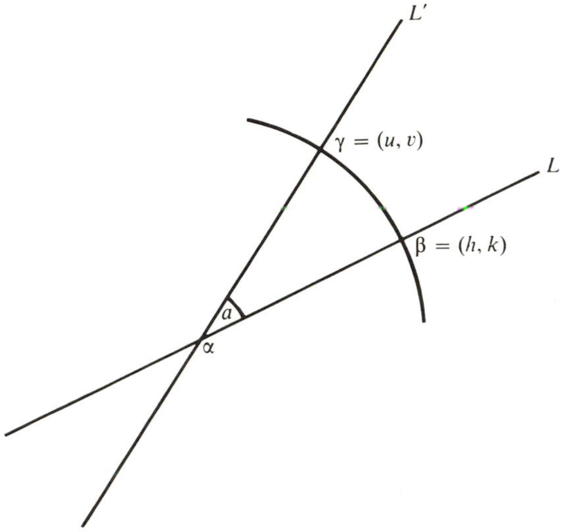


*Figure 6.5*

### Proof

Let lines $L$ and $L'$ be in $\Gamma$ and make an angle of $a$ at their point of intersection $\alpha$. This implies that $\alpha$ is in $\Gamma$. Then it follows that the circle with $\alpha$ as center and 1 as radius is in $\Gamma$. Hence the intersections $\beta$ and $\gamma$ of this circle with the lines $L$ and $L'$ are in $\Gamma$. Let $\beta$ and $\gamma$ have $(h, k)$ and $(u, v)$ as coordinates. Then $h$, $k$, $u$, and $v$ are in the field $F$.

Also line segments $\alpha\beta$, $\alpha\gamma$, and $\beta\gamma$ have 1, 1, and $\sqrt{(h-u)^2+(k-v)^2}$ as their lengths. Using the Law of Cosines, one finds that

$$\cos a = \frac{1^2 + 1^2 - (h-u)^2 - (k-v)^2}{2 \cdot 1 \cdot 1}.$$

Since this expression for $\cos a$ involves operations under which the field $F$ is closed, $\cos a$ is in $F$.

### Lemma 7

*Let $p$, $q$, and $r$ be rational numbers and let*

$$\alpha = x^3 + px^2 + qx + r.$$

*If $\alpha = 0$ has no root in $\mathbf{Q}$, it has no root in $F$.*

*Proof*

We give an indirect proof. That is, we assume that $\alpha = 0$ has a root in $F$ but not in $\mathbf{Q}$ and then show that this leads to a contradiction. From the assumption that there is a root of $\alpha = 0$ in $F$, it follows that there is a root in a multiquadratic extension $\mathbf{Q}(a_1, a_2, \ldots, a_n)$.

Let $Q_0 = \mathbf{Q}$ and $Q_i = \mathbf{Q}(a_1, \ldots, a_i)$ for $1 \leq i \leq n$. Also let $h$ be the smallest integer in $\{0, 1, \ldots, n\}$ such that there is a root of $\alpha = 0$ in $Q_h$. By hypothesis, $h \neq 0$.

Since $Q_h = Q_{h-1}(a_h)$ with $a_h^2$ in $Q_{h-1}$, Theorem 6 of Section 5.4 tells us that if $\alpha = 0$ has a root in $Q_h$ it has a root in $Q_{h-1}$. This contradicts the minimal nature of $h$ and completes the proof.

### Lemma 8

*No lines of $\Gamma$ make a $20^\circ$ angle.*

*Proof*

Let $s = \cos 20^\circ$ and $\alpha = 8x^3 - 6x - 1$. Substituting $\theta = 20^\circ$ in the triple-angle formula

$$4\cos^3\theta - 3\cos\theta = \cos(3\theta),$$

we obtain $4s^3 - 3s = \cos 60^\circ = 1/2$. Then $8s^3 - 6s - 1 = 0$; thus $s$ is a root of $\alpha = 0$.

In Example 2 of Section 5.5, we showed that $\alpha = 0$ has no rational root. Then Lemma 7 above tells us that $\alpha = 0$ has no root in the special field $F$ of this chapter. This means that the root $\cos 20^\circ$ of $\alpha = 0$ is not in $F$. Finally, it follows from Lemma 6 above that no lines of $\Gamma$ make a $20^\circ$ angle.

### Theorem 2 No General Trisection Technique

*There is no general method for trisecting angles by euclidean constructions.*

*Proof*

The collection $\Gamma$ of points, lines, and circles is closed under euclidean constructions and has lines making a $60^\circ$ angle. (See Lemma 5.) If there were a method for trisecting all angles using euclidean constructions, a $20^\circ$ angle would exist in $\Gamma$.

But Lemma 8 states that $\Gamma$ has no $20^\circ$ angle. This proves the theorem.

Another classical euclidean construction problem is that of "duplication of a cube." In this problem, one is given a length $s$ and is asked to construct a length $x$ such that a cube with edges of length $x$ will have twice the volume of a cube with edges of length $s$, that is, with $x^3 = 2s^3$.

According to a legend about this problem of doubling the volume of a cube (sometimes called the ***Delian problem***), the ancient Athenians were advised by the oracle at Delos that a plague would end when they constructed a new altar to Apollo in the form of a cube with double the volume of the old altar, that was a cube. They doubled the edge, which multiplied the volume by 8 instead of 2. This and other false constructions failed to end the plague; it continued to run rampant through the population and provided a degree of motivation unusual in mathematical problems.

Despite this alleged motivation, the great ancient Greek geometers could not solve the problem. We shall soon see that they had to fail because there is no general method for obtaining, with euclidean constructions, a length $x$ such that $x^3 = 2s^3$, where $s$ is a given length. First we need the following preliminary result.

### *Lemma 9*

*The number $\sqrt[3]{2}$ is not in the special field $F$.*

#### *Proof*

It follows easily from unique factorization of positive integers as products of primes that $\sqrt[3]{2}$ is not a rational number. [See Problem 13(b) of Section 1.7.] Hence $\alpha = x^3 - 2$ has no zero in $\mathbf{Q}$. Then Lemma 7 tells us that $\alpha$ has no zero in $F$; hence $\sqrt[3]{2}$ is not in $F$.

Now we assume that the given cube has sides of unit length. Then the cube sought in the Delian problem has edges of length $x$ satisfying $x^3 = 2 \cdot 1^3$, that is, $x = \sqrt[3]{2}$.

If such a cube were constructible with straightedge and compass, the vertices of a cube with edges of length $\sqrt[3]{2}$ would be points of our special set $\Gamma$. Then we would have

$$\sqrt[3]{2} = \sqrt{(a-c)^2 + (b-d)^2} \tag{4}$$

with $(a, b)$ and $(c, d)$ coordinates of points of $\Gamma$ and hence $a$, $b$, $c$, and $d$ in $F$. Since $F$ is a field closed under extraction of square roots of nonnegative numbers, equation (4) would imply that $\sqrt[3]{2}$ is in $F$. Since this would contradict Lemma 9, there is no solution to the Delian problem of "duplicating a cube" within the rules of euclidean constructions.

## *Problems*

1. Given that $\cos u$ is in the field $F$, show the following:
   (i) The point $(\cos u, \sin u)$ is in the collection $\Gamma$.
   (ii) In $\Gamma$, there is a line which makes an angle $u$ with the $x$-axis.

2. Explain why $\cos u$ is in $F$ if and only if $\cos(\pi - u)$ is in $F$.
3. Show that in $\Gamma$ there are two lines forming an angle of $12°$ (that is, $\pi/15$).
4. Show that in $\Gamma$ there are two lines forming an angle of $3°$.
5. Show that $\Gamma$ does not contain two lines forming an angle of $1°$.
6. Show that $\Gamma$ does not contain two lines forming an angle of $2°$.
7. Show that $\cos 40°$ is not in $F$.
8. Show that $\cos 140°$ is not in $F$.
9. Let $s$ be a positive real number in $F$. Show that in $\Gamma$ there exist five points which are the vertices of a regular pentagon with sides of length $s$. That is, show that a regular pentagon with sides of given length $s$ can be obtained by euclidean constructions. [See Problem 7 (viii) of Section 6.2.]
10. Show that in $\Gamma$ there do not exist nine points which are the vertices of a regular polygon (with nine sides), that is, a regular 9-gon is not constructible with straightedge and compass.
11. Show that $x^3 + x^2 - 2x - 1 = 0$ has no roots in $F$.
12. Show that $8y^3 + 4y^2 - 4y - 1 = 0$ has no roots in $F$.
13. Let $c = \cos(2\pi/7)$, $s = \sin(2\pi/7)$, and $z = c + \mathbf{i}s$. Show that:
    (i) $z^7 - 1 = 0, \quad z \neq 1$.
    (ii) $z^6 + z^5 + z^4 + z^3 + z^2 + z + 1 = 0$.
    (iii) $z^3 + z^2 + z + 1 + z^{-1} + z^{-2} + z^{-3} = 0$.
    (iv) $z^{-1} = c - is$, $z + z^{-1} = 2c$.
    (v) $z^2 + z^{-2} = 4c^2 - 2$, $z^3 + z^{-3} = 8c^3 - 6c$.
    (vi) $8c^3 + 4c^2 - 4c - 1 = 0$.
    (vii) $\cos(2\pi/7)$ is not in the field $F$. (See Problem 12.)
14. Show that in $\Gamma$ there do not exist seven points that are the vertices of a regular polygon, that is, a regular 7-gon is not constructible with straightedge and compass. (See the preceding problem.)

---

***Pierre Laurent Wantzel*** *(1814–1848)* *Wantzel was the first to provide a rigorous proof of the impossibility of the trisection of general angles by straightedge and compass. He was répétiteur at the École Polytechnique in Paris and in 1845 provided a further proof of the insolvability of the quintic equation, a proof that, however, in parts resembled those of Ruffini and Abel. He, like his contemporaries Galois and Abel, died young.*

---

## Review Problems

1. Show that $\cos(\pi/20)$ is in the special field $F$.
2. Show that $\cos(19\pi/20)$ is in $F$.
3. Let $\cos a$ be in $F$ and let $n$ be a positive integer. Show that $\cos(na)$ is in $F$.
4. Let $\cos a$ be in $F$ and let $n$ be a positive integer. Show that $\cos(a/2^n)$ is in $F$.
5. Show that $\cos(\pi/7)$ is not in $F$.
6. Show that $\cos(\pi/180)$ is not in $F$.
7. Show that a regular polygon with 40 sides, each of given length, is constructible with straightedge and compass.
8. Given a unit length, can one construct with straightedge and compass a regular 14-gon, with sides of unit length? Explain.

## Supplementary and Challenging Problems

1. Prove that a regular polygon with 17 sides, each of given length, is constructible with straightedge and compass.
2. Prove that one cannot construct with straightedge and compass a regular 11-gon having sides of a given length.

# 7

## *More on the Integers*

*Mathematics is the queen of the sciences and number theory the queen of mathematics.*

*Carl Friedrich Gauss*

In this chapter we present some topics in the Theory of Numbers that are related to our previous work but were not required in our study of groups, rings, and fields.

### 7.1 Simultaneous Congruences— The Chinese Remainder Theorem

It seems safe to assume that the positive integers formed the first number system used by human beings. Therefore, it should not be surprising that certain types of mathematical puzzles, in which the answers are positive integers, go back to the beginnings of history. We present an example of such a problem (dressed in words of our times).

***Example 1***

A child who can only count to 10 is given a batch of marbles for his birthday. He arranges them in columns each having 9 marbles and finds that he has 7 left over. So he arranges the entire batch in columns of 7 and ends up with 6 in an incomplete column. Finally he tries columns of 5; this leaves him with a remainder of 2. Find the two smallest positive integers which could be the total number of marbles in his batch.

***Solution***

We seek the first two numbers common to all three arithmetic progressions:

$$7, \quad 16, \quad 25, \quad 34, \quad \ldots; \tag{1}$$

$$6, \quad 13, \quad 20, \quad 27, \quad \ldots; \tag{2}$$

$$2, \quad 7, \quad 12, \quad 17, \quad \ldots. \tag{3}$$

We see that 7 appears in both (1) and (3). It can also be seen that the other terms common to these two progressions are

$$7 + 45, \qquad 7 + 90, \qquad 7 + 135, \qquad \ldots .$$

Hence we seek the first two common terms of the two progressions

$$6, \quad 13, \quad 20, \quad 27, \quad \ldots; \tag{2}$$

$$7, \quad 52, \quad 97, \quad 142, \quad \ldots . \tag{4}$$

We test the first few numbers in (4) to see if any have a remainder of 6 when divided by 7 and find that 97 is a solution to our problem. It is then fairly easy to see that the next solution is $97 + 7 \cdot 45 = 412$.

In the congruence notation, the problem of Example 1 could be stated as follows: Find the two smallest positive integers satisfying the three simultaneous congruences

$$x \equiv 7 \pmod 9, \qquad x \equiv 6 \pmod 7, \qquad x \equiv 2 \pmod 5.$$

***Definition 1 Linear Congruence***

*If a, b, and m are fixed integers, the congruence $ax \equiv b \pmod m$ is said to be* **linear**; *if $a = 1$, then the linear congruence is* **monic**.

The results that follow enable one to tackle systems of any finite number of simultaneous congruences in which the one unknown $x$ appears only to the first power.

***Lemma 1 Making a Linear Congruence Monic***

(*a*) *A congruence*

$$ax \equiv c \pmod b \tag{5}$$

*has solutions in* **Z** *if and only if* $(a, b) \mid c$.

(*b*) *If the set of solutions of* (5) *in* **Z** *is nonempty, it is also the set of solutions of a congruence*

$$x \equiv k \pmod s. \tag{6}$$

*Proof*

Part (a) is a restatement of Theorem 3 of Section 1.4.

If an integer $x$ satisfies (5), there is an integer $y$ such that

$$ax + by = c \tag{7}$$

and $d = \gcd(a, b)$ is an integral divisor of $c$. Then we can divide each of $a$, $b$, and $c$ by $d$ in (7) and obtain

$$rx + sy = t,$$

with $r$ and $s$ relatively prime. Hence (5) is equivalent to

$$rx \equiv t \pmod{s}, \qquad \gcd(r, s) = 1. \tag{8}$$

Translating (8) into the language appropriate to $\mathbf{Z_s} = \mathbf{Z}/(s)$ gives us

$$\bar{r} \cdot \bar{x} = \bar{t} \text{ in } \mathbf{Z_s}. \tag{9}$$

Since $\gcd(r, s) = 1$, $\bar{r}$ has a reciprocal in $\mathbf{Z_s}$ and (9) is equivalent to

$$\bar{x} = (\bar{r})^{-1}\bar{t} \text{ in } \mathbf{Z_s}. \tag{10}$$

Letting $k$ be an integer such that $\bar{k} = (\bar{r})^{-1}\bar{t}$ in $\mathbf{Z_s}$, we can translate (10) back into the desired form (6).

Because of the result in Lemma 1(b), we restrict our consideration of linear congruences in what follows to those that are monic, that is, of the form $x \equiv k \pmod{s}$.

***Theorem 1 Simultaneous Congruences***

*In* $\mathbf{Z}$, *let* $d = \gcd(s, t)$ *and* $m = \operatorname{lcm}[s, t]$.

(*a*) *There is an integer* $x$ *satisfying the two simultaneous congruences*

$$\begin{aligned} x &\equiv a \pmod{s} \\ x &\equiv b \pmod{t}. \end{aligned} \tag{11}$$

*if and only if* $a \equiv b \pmod{d}$.

(*b*) *If* $x = c$ *satisfies the congruences* (11), *the set of all solutions of* (11) *in* $\mathbf{Z}$ *is the same as the set of all solutions of* $x \equiv c \pmod{m}$ *in* $\mathbf{Z}$.

*Proof*

Let $c$ be a solution of (11). Then

$$a + hs = c = b + kt$$

for some integers $h$ and $k$. Hence

$$b - a = hs - kt.$$

Now it follows from Theorem 3 of Section 1.4 that $b - a$ is an integral multiple of $d = \gcd(s, t)$; that is, $a \equiv b \pmod{d}$.

Conversely, if $a \equiv b \pmod{d}$ it follows from Theorem 3 of Section 1.4 that there are integers $u$ and $v$ such that $b - a = us + vt$. Then $b - vt = a + us$ and $x = a + us$ is a solution of both congruences in (11). Hence part (a) is proved.

We start the proof of part (b) by letting $c$ be a solution of the system (11). Then $x$ is a solution of (11) if and only if $x \equiv c \pmod{s}$ and $x \equiv c \pmod{t}$, that is, if and only if both $s|(x - c)$ and $t|(x - c)$. This means that $x$ is a solution of (11) if and only if $x - c$ is a common multiple of $s$ and $t$. Since the common multiples of $s$ and $t$ are the multiples of the least common multiple $m$ (see Section 1.6), $x$ is a solution of (11) if and only if $m|(x - c)$, that is, if and only if $x \equiv c \pmod{m}$.

***Example 2***

Are there any solutions of the system of two simultaneous congruences

$$x \equiv 11 \pmod{15}, \qquad x \equiv 4 \pmod{21}?$$

***Solution***

We observe that $\gcd(15, 21) = 3$ and that 11 is not congruent to 4 modulo 3. Therefore, it follows from Theorem 1(a) that no integer $x$ satisfies both congruences.

***Example 3***

Find a single monic linear congruence $x \equiv c \pmod{m}$ that is equivalent to the system of two simultaneous congruences

$$x \equiv 14 \pmod{15}, \qquad x \equiv 5 \pmod{21}. \tag{A}$$

***Solution***

We use the technique of the proof of Theorem 1. An integer $x$ satisfying both congruences of (A) exists if and only if there exist integers $y$ and $z$ such that

$$x = 14 + 15y = 5 + 21z. \tag{B}$$

The second equality in (B) shows that we have to solve

$$21z - 15y = 14 - 5 = 9$$

for integers $y$ and $z$. The euclidean algorithm of Section 1.5 provides a constructive method for doing this. However, the coefficients here are fairly small, so we can solve the equation by systematic trial.

First we divide by $3 = \gcd(21, 15)$ and convert the equation into

$$7z - 5y = 3 \quad \text{or} \quad z = (5y + 3)/7.$$

We try 0, 1, 2, ... for $y$ until we find that $y = 5$ yields an integer, namely 4, for $z$. Substituting $y = 5$ or $z = 4$ in (B) results in $x = 89$ as a solution of the simultaneous congruences (A). Since $\text{lcm}[15, 21] = 105$, Theorem 1(b) now tells us that the system (A) is equivalent to the single congruence $x \equiv 89 \pmod{105}$.

**Theorem 2** ***Relatively Prime Moduli***

*Let $s$ and $t$ be relatively prime integers. Then the system of simultaneous congruences*

$$\begin{aligned} x &\equiv a \pmod{s} \\ x &\equiv b \pmod{t} \end{aligned} \tag{12}$$

*always has solutions. If $c$ is one solution, the solutions of* (12) *are the solutions of*

$$x \equiv c \pmod{st}. \tag{13}$$

***Proof***

We are given that $\gcd(s, t) = 1$. Since any two integers $a$ and $b$ are congruent modulo 1, Theorem 1(a) tells us that there is a solution $c$ of (12).

Using Theorem 1 of Section 1.6, we see that $\gcd(s, t) = 1$ implies $\text{lcm}[s, t] = |st|$. Now it follows from Theorem 1(b) above that $x$ satisfies (12) if and only if

$$x \equiv c \pmod{|st|}. \tag{14}$$

Since integers are congruent modulo $-st$ if and only if they are congruent modulo $st$, (14) and (13) are equivalent.

The solving of simultaneous congruences

$$x \equiv a \pmod{s}, \qquad x \equiv b \pmod{t}$$

involves the solution in integers $y$ and $z$ of

$$sy - tz = b - a.$$

We saw in Section 1.5 that such an equation can be solved with the euclidean algorithm. We extend this technique to systems with any finite number of linear congruences after introducing some terminology.

### *Definition 2 Set of Pairwise Relatively Prime Integers*

*The integers of a set S are* **pairwise relatively prime** *if* $\gcd(m, n) = 1$ *for all distinct m and n in S.*

### *Theorem 3 Chinese Remainder Theorem*

*Let $m_1. m_2, \ldots, m_r$ be pairwise relatively prime. Then the system of simultaneous congruences*

$$x \equiv a_1 \pmod{m_1}, \ldots, \quad x \equiv a_r \pmod{m_r} \tag{15}$$

*always has solutions. If c is one solution, the set of solutions of* (15) *is the set of solutions of* $x \equiv c \pmod{m_1 m_2 \cdots m_r}$.

#### *Outline of Proof*

One uses Theorem 2 to replace two of the congruences of (15) with a single congruence. This process is continued until one has reduced the system to one congruence. (This is an inductive argument.)

It is left to the reader to show that the moduli are always relatively prime in the reduction process.

#### *Example 4*

Use Euclid's Theorem that there exist an infinite number of positive primes in **Z** to prove that there are a million consecutive positive integers

$$x, \quad x + 1, \quad x + 2, \quad \ldots, \quad x + 999999,$$

each of which is an integral multiple of the cube of a prime.

#### *Solution*

Let the first million positive primes be

$$p_1, \quad p_2, \quad \ldots, \quad p_{1000000}.$$

Then we seek a positive integer $x$ that satisfies the million simultaneous congruences.

$$x \equiv 0 \pmod{2^3}, \quad x \equiv -1 \pmod{3^3}, \quad \ldots, \quad x \equiv -999999 \pmod{p_{1000000}^3}.$$

Since $\gcd(p^3, q^3) = 1$ for distinct primes $p$ and $q$, the existence of such an $x$ is guaranteed by the Chinese Remainder Theorem. (The most modern electronic computer should have considerable difficulty in finding the smallest positive solution.)

## *Problems*

1. Find a congruence $rx \equiv t \pmod{s}$, with $\gcd(r, s) = 1$, which has the same solutions as $56x \equiv 259 \pmod{987}$.
2. Find an integer $c$ with $0 < c < 21$ such that $8x \equiv 15 \pmod{21}$ has the same solutions as $x \equiv c \pmod{21}$.
3. Find the two smallest positive integers $x$ satisfying the simultaneous congruences
$$x \equiv 13 \pmod{14}, \qquad x \equiv 14 \pmod{15}.$$
4. Find the smallest positive integer and the largest negative integer satisfying the simultaneous congruences $3x \equiv 4 \pmod{5}$ and $8x \equiv 15 \pmod{21}$.
5. Find the two smallest positive integers having remainders 1, 2, 3, 4, and 5 when divided by 2, 3, 4, 5, and 6, respectively.
6. Find the two smallest positive integers having remainders 1, 2, 3, 4, and 5 when divided by 13, 14, 15, 17, and 19, respectively.
7. Find the smallest positive integer $n$ satisfying the simultaneous conditions $4|n$, $9|(n + 1)$, and $25|(n + 2)$.
8. An integer is called ***square-free*** if it is not an integral multiple of the square of a prime. Do there exist one million consecutive positive integers no one of which is square-free? Justify your answer.
9. Let $x$ and $y$ be integers. Show that $19|(3x + 5y)$ if and only if $19|(13x + 9y)$.
10. Let $x$ and $y$ be integers. Find an integer $c$ with $0 < c < 13$ such that $13|(5x - 8y)$ if and only if $13|(7x + cy)$.
11. (i) Find $a$ and $r$ in $\mathbf{Z}^+$ such that $y \equiv a \pmod{r}$ if and only if simultaneously $y \equiv 0 \pmod{7}$ and $y + 1 \equiv 0 \pmod{11}$.
    (ii) Find $b$ and $s$ in $\mathbf{Z}^+$ such that $x \equiv b \pmod{s}$ if and only if simultaneously $x \equiv 0 \pmod{14}$ and $x + 2 \equiv 0 \pmod{22}$.
    (iii) Find $c$ and $t$ in $\mathbf{Z}^+$ such that $x \equiv c \pmod{t}$ if and only if simultaneously $x \equiv 0 \pmod{14}$, $x + 1 \equiv 0 \pmod{195}$, $x + 2 \equiv 0 \pmod{22}$.

*12. Explain why there must exist an integer $y$ satisfying the simultaneous congruences $y \equiv 0 \pmod{10}$, $y + 1 \equiv 0 \pmod{231}$, and $y + 2 \equiv 0 \pmod{26}$.

*13. Explain why there exist integers $x$ and $y$ such that $\gcd(x+i, y+j) > 1$ for all $i$ and $j$ in $\{0, 1, 2\}$.

*14. Prove that for every positive integer $n$ there exist integers $x$ and $y$ such that $\gcd(x+i, y+j) > 1$ for all $i$ and $j$ in $\{0, 1, 2, \ldots, n\}$.

## 7.2 More on Mathematical Induction

*How often have I said to you that when you have eliminated the impossible, whatever remains, however improbable, must be the truth.*

*Sir Arthur Conan Doyle*

Below we extend the problem set on mathematical induction of Section 1.1. First we present a result that is frequently used as a substitute for the Well Ordering Principle. This theorem is the basis for a form of mathematical induction that is more direct than that usually needed with well ordering proofs; it does not require the assumption that the desired statement is false, that is, a proof by contradiction.

**Theorem 1 Strong Induction**

*If $S$ is a subset of $\mathbf{Z}^+ = \{1, 2, 3, \ldots\}$ such that 1 is in $S$ and $k+1$ is in $S$ whenever $1, 2, 3, \ldots, k$ are in $S$, then $S = \mathbf{Z}^+$.*

***Proof***

Let $T$ consist of the integers of $\mathbf{Z}^+$ that are not in $S$. We assume that $T$ is nonempty and seek a contradiction. Under this assumption, it follows from the Well Ordering Principle that $T$ has a least integer $t$.

Since 1 is in $S$, $t \neq 1$. The fact that $t$ is the least integer in $T$ means that $1, 2, \ldots, t-1$ are in $S$. The hypothesis then implies that $(t-1)+1 = t$ is in $S$. Hence $t$ is in both $S$ and $T$, which contradicts the definition of $T$. This contradiction shows that $T$ is empty; that is, $S = \mathbf{Z}^+$.

The reader may find any one of the inductive techniques, given in Section 1.1 or here, to be most convenient for a given problem in the set which follows.

*Problems*

Some of the following problems do not require an inductive proof.

1. Prove that there exists an infinite sequence of positive primes

$$p_1, p_2, p_3, \ldots$$

in $\mathbf{Z}$ such that $p_{n+1} > p_n$ for all positive integers $n$.

2. In a round robin tennis tournament, each player had exactly one match with every other contestant and there were no ties. Prove that it is possible to list the players' names in a column so that each contestant beat the one listed immediately below him (except that the last one listed need not have won any match).

3. In the "Hit One Hundred" game the two players alternate in announcing numbers. The first number must be in $\{1, 2, 3\}$. Then each player must announce a number that is 1, 2, or 3 more than the last number given by his opponent. A player wins if he calls out the number 100. Prove that the second player can always win if he uses perfect strategy.

4. Generalize on the game of Problem 3.

5. State a generalized version of strong induction analogous to Theorems 2 and 3 of Section 1.1.

6. Prove that every integer from $-(3^{n+1} - 1)/2$ through $(3^{n+1} - 1)/2$ is uniquely expressible in the form

$$a_0 + 3a_1 + 3^2a_2 + \cdots + 3^na_n,$$

with each $a_i$ in $\{-1, 0, 1\}$.

7. Let a sequence $u_0, u_1, u_2, \ldots$ be defined by $u_0 = 1$ and $u_{n+1} = 4u_n - 1$. Prove that $u_n = (2 \cdot 4^n + 1)/3$.

8. Let a sequence $v_0, v_1, \ldots$ be defined by $v_0 = 0$, $v_1 = 1$, and $v_n = v_{n-1} - v_{n-2}$ for $n \geq 2$. Prove that $v_{n+3} = -v_n$ and $v_{n+6} = v_n$ for all $n \geq 0$.

9. For every $n$ in $\mathbf{Z}^+$, let $n! = 1 \cdot 2 \cdot 3 \cdots n$. Let $\binom{n}{k} = n!/[k!\,(n-k)!]$ for integers $k$ and $n$ with $0 \leq k \leq n$. Prove that:

(a) $\binom{n}{k-1} + \binom{n}{k} = \binom{n+1}{k}$ for $1 \leq k \leq n$.

(b) $(1 + x)^n = \binom{n}{0} + \binom{n}{1}x + \binom{n}{2}x^2 + \cdots + \binom{n}{n}x^n$.

(c) $\binom{n}{k}$ is the number of subsets with $k$ elements in a set with $n$ elements.

10. For $k = 1, 2, \ldots, n$ let

$$u^{(k)} = \frac{d^k u}{dx^k} \quad \text{and} \quad v^{(k)} = \frac{d^k v}{dx^k}$$

exist. Prove ***Leibniz' Rule***, that is,

$$\frac{d^n(uv)}{dx^n} = \binom{n}{0} uv^{(n)} + \binom{n}{1} u^{(1)}v^{(n-1)} + \binom{n}{2} u^{(2)}v^{(n-2)} + \cdots + \binom{n}{n} u^{(n)}v.$$

11. Let $F_1, F_2, \ldots$ be the Fibonacci sequence $1, 1, 2, 3, \ldots$. Prove that

$$\binom{n}{0} F_m + \binom{n}{1} F_{m+1} + \binom{n}{2} F_{m+2} + \cdots + \binom{n}{n} F_{m+n} = F_{m+2n}.$$

12. Prove that the number of terms in $(x_1 + x_2 + \cdots + x_k)^n$ is

$$\binom{n+k-1}{k-1}.$$

13. Let $S$ be a set $\{p_1, p_2, \ldots, p_k\}$ of $k$ distinct positive primes in $\mathbf{Z}$. Let $A(k, n)$ be the number of positive integers $s$ such that $s$ is a product $q_1 q_2 \cdots q_n$ of $n$ (not necessarily distinct) primes with each $q_i$ in $S$. Find a simple closed form for $A(k, n)$ and prove it correct for all positive integers $k$ and $n$.

14. Prove the following:
    (a) $1 + 2 + 3 + \cdots + n = n(n+1)/2$.
    (b) $1 \cdot 2 + 2 \cdot 3 + 3 \cdot 4 + \cdots + n(n+1) = n(n+1)(n+2)/3$.
    (c) $1 \cdot n + 2(n-1) + 3(n-2) + \cdots + (n-1) \cdot 2 + n \cdot 1 = n(n+1)(n+2)/6$.
    (d) $1^3 + 2^3 + 3^3 + \cdots + n^3 = (1 + 2 + 3 + \cdots + n)^2$.

15. A permutation $\theta$ in $\mathbf{S_n}$ is a ***derangement*** if $\theta(x) \neq x$ for all $x$ in $\{1, 2, \ldots, n\}$. For $n \geq 1$, let $d_n$ be the number of derangements in $\mathbf{S_n}$. Also let $d_0 = 1$. Prove that
    (a) $d_n = (n-1)(d_{n-1} + d_{n-2})$ for $n \geq 2$.
    (b) $d_n = nd_{n-1} + (-1)^n$ for $n \geq 1$.
    (c) $d_n = n!\left[\frac{1}{0!} - \frac{1}{1!} + \frac{1}{2!} - \cdots + (-1)^n \frac{1}{n!}\right]$.
    (d) $\binom{n}{0} d_0 + \binom{n}{1} d_1 + \binom{n}{2} d_2 + \cdots + \binom{n}{n} d_n = n!$.

(e) $n!\binom{n}{n} - (n-1)!\binom{n}{n-1} + (n-2)!\binom{n}{n-2} - \cdots + (-1)^n 0!\binom{n}{0} = d_n$.

(f) $\binom{n}{1}d_{n-1} + 2\binom{n}{2}d_{n-2} + 3\binom{n}{3}d_{n-3} + \cdots + n\binom{n}{n}d_0 = n!$.

16. Prove that all the subsets of a set $X$ with $n$ elements can be numbered as $\beta_1, \beta_2, \ldots, \beta_m$ (with $m = 2^n$) such that $\beta_1$ is the empty set, $\beta_m$ has a single element, and for $1 \le i < m$ the set $\beta_{i+1}$ is obtained either by adjoining a single element of $X$ to $\beta_i$ or by deleting a single element from $\beta_i$.

17. Let $A_n$ be the maximum number of regions into which a plane can be cut by $n + 5$ straight lines, of which 5 are parallel to each other. Find and prove a simple closed form for $A_n$.

18. Let $B(m, n)$ be the maximum number of 3-dimensional regions into which 3-space can be cut using $m + n$ planes of which $m$ are parallel to each other. Find and prove a closed form for $B(m, n)$.

19. Let $U$ be a noncommutative ring with unity. Let $r$ and $v$ be in $U$ with $v$ an invertible. Prove that there exists an element $s$ in $U$ such that $vr^n = s^n v$ for all positive integers $n$.

20. Prove the following for all integers $n$; it may sometimes be helpful to give a proof first for all nonnegative $n$:
    (a) $n^5 + 4n \equiv 0 \pmod{5}$.
    (b) $n(n^2 + 1)(n^2 + 4) \equiv 0 \pmod{10}$.
    (c) $n^5 - 5n^3 + 4n \equiv 0 \pmod{120}$.
    (d) $n^5 + 11n^3 + 4n \equiv 0 \pmod{8}$.

21. Prove that $n^5 - 5n^3 + 4n \equiv 0 \pmod{360}$ for all integers $n$ such that $n \not\equiv 0 \pmod{3}$.

22. Let $a$ be in $\mathbf{Z}$, let $n$ be in $\mathbf{Z}^+$, and let $m = 2^n$.
    (i) Prove that $(2a - 1)^m \equiv 1 \pmod{4m}$.
    (ii) Explain why the multiplicative group $\mathbf{V}_{4m}$, of invertibles in $\mathbf{Z}_{4m}$, is not cyclic.

23. Let $a$ be in $\mathbf{Z}$, let $n$ be in $\mathbf{Z}^+$, and let $m = 3^n$. Prove that:
    (i) $(3a \pm 1)^m \equiv \pm 1 \pmod{3m}$. (The $\pm$ signs must be chosen to be the same.)
    (ii) Let $q = (3m - 1)/2$. Then there do not exist $x_1, x_2, \ldots, x_{q-1}$ in $\mathbf{Z}$ such that

$$x_1^m + x_2^m + \cdots + x_{q-1}^m = 3ma \pm q.$$

24. Let $D$ be an ordered integral domain. Prove the following:
    (i) If $0 < a < 1$ in $D$, then $0 < a^{n+1} < a^n < 1$ for all $n$ in $\mathbf{Z}^+$.
    (ii) If $1 < b$ in $D$, then $1 < b^n < b^{n+1}$ for all $n$ in $\mathbf{Z}^+$.
    (iii) If $v$ is an invertible in $D$ and the multiplicative order of $v$ is finite, $v$ must be in $\{-1, 1\}$.

25. Let $D$ be an ordered integral domain. Given that $1 + a \geq 0$ in $D$, prove that $(1 + a)^n \geq 1 + n \cdot a$ for all $n$ in $\mathbf{Z}^+$.

26. Prove that $3^n \geq n^3$ for all $n$ in $\mathbf{Z}^+$.

27. Prove that $2n\sqrt{n}/3 < \sqrt{1} + \sqrt{2} + \sqrt{3} + \cdots + \sqrt{n}$.

28. Prove that $\sqrt{1} + \sqrt{2} + \cdots + \sqrt{n} < (4n + 3)\sqrt{n}/6$.

29. Let $F_n$ be the $n$th Fibonacci number and let $R_n = F_{n+1}/F_n$. Prove that $1 \leq R_{2n-1} < R_{2n+1} < R_{2n+2} < R_{2n}$.

30. Let $F_n$ and $L_n$ be the $n$th Fibonacci and $n$th Lucas numbers, respectively. Let $a = (1 + \sqrt{5})/2$ and $b = (1 - \sqrt{5})/2$. Prove the following:
   (i) $F_n = (a^n - b^n)/(a - b)$.
   (ii) $L_n = a^n + b^n$.
   (iii) $a^n = (L_n + F_n\sqrt{5})/2$.
   (iv) $b^n = (L_n - F_n\sqrt{5})/2$.
   (v) $a^n = aF_n + F_{n-1}$ for $n \geq 2$.

31. Let $F$ be a field with $m$ elements. Let $d$ be a fixed positive integer. How many distinct polynomials $\alpha$ are there such that

$$\alpha = (x - r_1)(x - r_2)(x - r_3) \cdots (x - r_d)$$

with the $r_i$ (not necessarily distinct) elements of $F$?

32. Complete the proof of the Chinese Remainder Theorem (Theorem 3 of Section 7.1).

33. Let $x = t - t^{-1}$ and $y_n = t^n - t^{-n}$ for all positive integers $n$. Prove that $y_n$ is a polynomial in $x$ with rational coefficients if and only if $n$ is odd.

34. Let $u_1, u_2, u_3, \ldots$ be defined by $u_1 = 2$ and $u_{n+1} = u_n(u_n - 1) + 1$ for $n \geq 1$. In $\mathbf{Q}[x]$ let

$$(x - u_1)(x - u_2) \cdots (x_1 - u_n) = x^n - s_{n-1}x^{n-1} + \cdots + (-1)^{n-1}s_1x + (-1)^n s_0.$$

Prove that $s_1 = s_0 - 1$.

35. Let the $u_i$ be as in Problem 34. Let $v_i = u_i$ for $1 \leq i < n$ and $v_n = u_n - 2$. In $\mathbf{Q}[x]$, let

$$(x - v_1)(x - v_2) \cdots (x - v_n) = x^n - t_{n-1}x^{n-1} + \cdots + (-1)^{n-1}t_1x + (-1)^n t_0.$$

Prove that $t_1 = t_0 + 1$.

36. Let $n$ be a positive integer; $m = 2^n$; and $a_1, a_2. \ldots, a_m$ be positive elements of an ordered integral domain $D$. Prove that

$$m^m \cdot (a_1a_2 \cdots a_m) \leq (a_1 + a_2 + \cdots + a_m)^m$$

with equality holding if and only if $a_1 = a_2 = \cdots = a_m$.

37. Let $b_1, b_2, \ldots, b_n$ be positive elements of an ordered integral domain $D$. Prove that

$$n^n \cdot (b_1 b_2 \cdots b_n) \leq (b_1 + b_2 + \cdots + b_n)^n,$$

with equality holding if and only if $b_1 = b_2 = \cdots = b_n$.

38. Let $a_1, a_2, \ldots, a_n$ be nonnegative real numbers;

$$A = (a_1 + a_2 + \cdots + a_n)/n; \quad \text{and} \quad G = \sqrt[n]{a_1 a_2 \cdots a_n}.$$

Prove that $A \geq G$ and that $A = G$ if and only if

$$a_1 = a_2 = \cdots = a_n.$$

(This is known as the ***Inequality on the Means.***)

39. Let $a_1, a_2, \ldots, a_n$ be positive real numbers,

$$G = \sqrt[n]{a_1 a_2 \cdots a_n}, \quad \text{and} \quad H = \frac{n}{\dfrac{1}{a_1} + \dfrac{1}{a_2} + \cdots + \dfrac{1}{a_n}}.$$

Prove that $G \geq H$ and that $G = H$ if and only if

$$a_1 = a_2 = \cdots = a_n.$$

40. Let $n > 1$; let $r_1, r_2, \ldots, r_n$ be nonnegative real numbers; and let

$$(x - r_1)(x - r_2) \cdots (x - r_n)$$
$$= x^n - s_{n-1}x^{n-1} + s_{n-2}x^{n-2} - \cdots + (-1)^{n-1}s_1 x + (-1)^n s_0.$$

Prove that $s_{n-j}s_j \geq \binom{n}{j}^2 s_0$ for $0 < j < n$ $\left[\text{where } \binom{n}{j} \text{ is the binomial coefficient } n(n-1)\cdots(n-j+1)/1 \cdot 2 \cdots j\right]$.

## *7.3 Some Number Theoretic Functions*

Here we derive explicit formulas for some of the more frequently encountered functions from the Theory of Numbers.

Let $\phi$ be the Euler $\phi$-function; then $\phi(n)$ is the number of integers in $\{1, 2, \ldots, n\}$ that are relatively prime to $n$. (See Problem 38 of Section 4.3.) We now define two other number-theoretic functions (and later in this section introduce one more, the partition function).

### *Notation The $\sigma$ and $\tau$ Functions*

*Given a positive integer n, $\sigma(n)$ denotes the sum of the positive integral divisors of n and $\tau(n)$ denotes the number of such divisors.*

Let us first consider the problem of evaluating $\phi(n)$.

#### *Example 1*

Calculate $\phi(28)$.

#### *Solution*

The set of positive integers less than or equal to 28 and relatively prime to 28 is

$$A = \{1, 3, 5, 9, 11, 13, 15, 17, 19, 23, 25, 27\}.$$

Let us count these numbers in a way that will suggest a method to handle the general case. Since $28 = 2^2 \cdot 7$, a number is in $A$ if and only if it is in $X = \{1, 2, \ldots, 28\}$ and is not an integral multiple of 2 or of 7. In $X$ there are 28/2 multiples of 2 and 28/7 multiples of 7.

When we delete the multiples of 2 and the multiples of 7 from $X$, are we left with $28 - 14 - 4$ integers in $A$? No, since our tally of the integers deleted has counted the integers 14 and 28 twice, once as a multiple of 2 and once as a multiple of 7. We add on a correction term and have

$$\begin{aligned}\phi(28) &= 28 - \frac{28}{2} - \frac{28}{7} + \frac{28}{2 \cdot 7} \\ &= 28\left(1 - \frac{1}{2} - \frac{1}{7} + \frac{1}{2 \cdot 7}\right) \\ &= 28\left(1 - \frac{1}{2}\right)\left(1 - \frac{1}{7}\right) = 28 \cdot \frac{1}{2} \cdot \frac{6}{7} = 12.\end{aligned}$$

Next we take up the general case. In the following, we let the standard factorization of $n$ be

$$n = (p_1)^{r_1}(p_2)^{r_2} \cdots (p_m)^{r_m}. \tag{S}$$

Note that here $p_1, p_2, \ldots, p_m$ are distinct positive primes.

***Theorem 1 Euler $\phi$-Function***

*Let the standard factorization of n be as in* (S). *Then*

$$\phi(n) = n\left(1 - \frac{1}{p_1}\right)\left(1 - \frac{1}{p_2}\right)\cdots\left(1 - \frac{1}{p_m}\right).$$

***Proof***

As in the example above, we first count all the positive integers less than or equal to $n$ and subtract away the number of multiples of $p_1$, of $p_2$, of $p_3$, and so on to $p_m$, since these are clearly the integers not relatively prime to $n$. But we have deleted multiples of $p_1p_2$ twice from our list, for example, once as multiples of $p_1$ and once as multiples of $p_2$. We must, therefore, add back to our sum the number of multiples of $p_1p_2$, $p_1p_3$, $p_2p_3$, and so forth. Now we consider multiples of $p_1p_2p_3$ and other such triple products. They have been deleted three times (once for each prime factor) and put back in three times (once for each combination of two prime factors). But they must be deleted so we must subtract away the number of multiples of three prime factors. We continue this process and obtain the following value for $\phi(n)$:

$$\begin{aligned}\phi(n) &= n - \sum_{1\le i\le m} \frac{n}{p_i} + \sum_{1\le i<j\le m} \frac{n}{p_ip_j} - \sum_{1\le i<j<k\le m} \frac{n}{p_ip_jp_k} + \cdots + (-1)^m \frac{n}{p_1p_2\cdots p_m} \\ &= n\left(1 - \frac{1}{p_1}\right)\left(1 - \frac{1}{p_2}\right)\cdots\left(1 - \frac{1}{p_m}\right).\end{aligned}$$

The technique of proof above, of counting the elements of a set by alternately deleting from and adding to a set, is known as ***Sylvester's principle*** or the "inclusion-exclusion" principle. (See the biographical note on Sylvester at the close of this section.)

***Theorem 2 $\sigma$-Function***

*If the standard factorization of n is*

$$n = (p_1)^{r_1}(p_2)^{r_2}\cdots(p_m)^{r_m},$$

*then*

$$\sigma(n) = \frac{p_1^{r_1+1} - 1}{p_1 - 1}\cdot\frac{p_2^{r_2+1} - 1}{p_2 - 1}\cdots\frac{p_m^{r_m+1} - 1}{p_m - 1}.$$

*Proof*

Let us consider the product

$$(1 + p_1 + \cdots + p_1^{r_1})(1 + p_2 + \cdots p_2^{r_2}) \cdots (1 + p_m + \cdots + p_m^{r_m}). \quad \text{(P)}$$

When multiplied out this product will consist of all possible terms of the form

$$p_1^{s_1} p_2^{s_2} \cdots p_m^{s_m},$$

where $0 \leq s_i \leq r_i$ for $i = 1, 2, \ldots, m$ and each term will appear only once. But such products are, in fact, the divisors of $n$ and hence the product (P) is the sum of all the divisors of $n$. Using the formula for the sum of a geometric progression to sum each factor, we have the formula for $\sigma(n)$ given above.

***Theorem 3*** ***τ-Function***

*If the standard factorization of n is*

$$n = (p_1)^{r_1}(p_2)^{r_2} \cdots (p_m)^{r_m},$$

*then*

$$\tau(n) = (r_1 + 1)(r_2 + 1) \cdots (r_m + 1).$$

*Proof*

In the multiplied out expansion of the product (P) given in Theorem 2, if each term (that is, positive integral divisor of $n$) is replaced by 1, then the sum of the divisors is changed into the number of divisors. Thus $\tau(n)$ may be calculated by letting $p_1 = p_2 = \cdots = p_m = 1$ in (P); thus $\tau(n) = (r_1 + 1)(r_2 + 1) \cdots (r_m + 1)$.

Each of the functions $\phi$, $\sigma$, and $\tau$ has the following property.

***Definition 1*** ***Multiplicative Function***

*A function f from* $\mathbf{Z}^+$ *to* $\mathbf{Z}^+$ *is said to be* **multiplicative** *if* $f(rs) = f(r)f(s)$ *whenever* $gcd(r, s) = 1$.

The reader is asked in Problems 8, 24, and 32 of this section to show that each of $\phi$, $\sigma$, and $\tau$ is multiplicative.

We next use the $\sigma$-function to introduce a concept that is involved in a famous unsolved problem.

***Definition 2 Perfect Numbers***

*A positive integer n is perfect if* $\sigma(n) = 2n$.

One easily sees that 6 is perfect since $\sigma(6) = 1 + 2 + 3 + 6 = 12 = 2 \cdot 6$. The next perfect integer is 28. Each of these perfect numbers is of the form $2^s p$, where $s$ and $p$ are positive integers with $p$ prime and

$$p = 1 + 2 + 2^2 + \cdots + 2^s = 2^{s+1} - 1.$$

It is left to the reader, as Problem 19 of this section, to prove that all numbers of this form are perfect. The much harder converse—that all even perfect numbers are of this form—was proved by Euler. It has been conjectured that no odd perfect numbers exist, but no proof or counterexample has been obtained yet. It is known that an odd perfect number would have to be greater than $10^{36}$, so one should not expect to find a counterexample easily.

A somewhat related concept is given in the following:

***Definition 3 Amicable Pair***

*Positive integers a and b form an* **amicable pair** *if*

$$\sigma(a) = a + b = \sigma(b).$$

In Problem 21 of this section, the reader is asked to show that 220 and 284 form an amicable pair.

In Section 2.13 we used the calculation of the number of cycles and products of disjoint cycles of various types to obtain important results concerning the symmetric group $\mathbf{S_5}$. One may note that every permutation in $\mathbf{S_5}$ is representable in one of the following ways:

$$(abcde),\ (abcd)(e),\ (abc)(de),\ (abc)(d)(e),\ (ab)(cd)(e),\ (ab)(c)(d)(e),\ (a)(b)(c)(d)(e),$$

where $a$, $b$, $c$, $d$, and $e$ are distinct integers in $\{1, 2, 3, 4, 5\}$.

These types of representations correspond to the ways in which 5 can be written as a positive integer or a sum of positive integers:

$$5,\ 4 + 1,\ 3 + 2,\ 3 + 1 + 1,\ 2 + 2 + 1,\ 2 + 1 + 1 + 1,\ 1 + 1 + 1 + 1 + 1. \quad (1)$$

We see that there are seven such partitions of the positive integer 5.

In classifying the permutations in $\mathbf{S_n}$, and in other applications, it is helpful to be able to generalize from the integer 5 to a general positive integer $n$, as we now do.

***Definition 4 The Partition Function p***

*An expression for a positive integer n as (a positive integer or) a sum of positive integers will be called a* **partition of n**. *The number of such partitions, where the order is not taken into account, is denoted by p(n), and p is called the* **(unordered) partition function**.

We note that $3 + 1 + 1$, $1 + 3 + 1$, and $1 + 1 + 3$ are all considered to be the same partition of 5 and that 5 itself is a partition of 5. Display (1) above shows that $p(5) = 7$.

Euler examined this partition function in his *Introductio in Analysin Infinitorum* (1748) and obtained the following remarkable result.

***Theorem 4 Generating Function for p***

$$\frac{1}{(1-x)(1-x^2)(1-x^3)(1-x^4)\cdots} = p(0) + p(1)x + p(2)x^2 + \cdots.$$

This means that if the function on the left is expanded as a power series in $x$, we can read off the values of $p(n)$ as the coefficients of the powers of $x$ in the series; such a function is called a *generating function.* [For uniformity we let $p(0)$ be 1.]

We do not prove Theorem 4 or even define precisely what is meant by the infinite product in the denominator of the generating function but instead give an intuitive argument to show that the result is plausible. If we were to expand

$$\frac{1}{1-x}\cdot\frac{1}{1-x^2}\cdot\frac{1}{1-x^3}\cdot\frac{1}{1-x^4}\cdots$$

factor by factor into series of powers of $x$, we would have

$$(1 + x + x^2 + \cdots)(1 + x^2 + x^4 + \cdots)(1 + x^3 + x^6 + \cdots)(1 + x^4 + x^8 + \cdots)\cdots$$
$$= (1 + x^1 + x^{1+1} + \cdots)(1 + x^2 + x^{2+2} + \cdots)(1 + x^3 + x^{3+3} + \cdots)\cdots. \quad (2)$$

For example, what is the coefficient of $x^5$ when this is multiplied out and like terms collected? First we note that we can ignore series $(1 + x^m + x^{2m} + \cdots)$ in which $m > 5$. Then the following table shows all the ways of selecting terms $x^j$ from the series that are the first five factors in (2) so as to obtain $x^5$ in the expansion.

| $x^5$ term in expansion | Term chosen from 1st factor | 2nd factor | 3rd factor | 4th factor | 5th factor |
|---|---|---|---|---|---|
| $x^5$ | 1 | 1 | 1 | 1 | $x^5$ |
| $x^{1+4}$ | $x^1$ | 1 | 1 | $x^4$ | 1 |
| $x^{2+3}$ | 1 | $x^2$ | $x^3$ | 1 | 1 |
| $x^{1+1+3}$ | $x^{1+1}$ | 1 | $x^3$ | 1 | 1 |
| $x^{1+2+2}$ | $x^1$ | $x^{2+2}$ | 1 | 1 | 1 |
| $x^{1+1+1+2}$ | $x^{1+1+1}$ | $x^2$ | 1 | 1 | 1 |
| $x^{1+1+1+1+1}$ | $x^{1+1+1+1+1}$ | 1 | 1 | 1 | 1 |

Thus we see that $x^5$ appears precisely 7 times in the product; the coefficient of $x^5$ is therefore 7, which is $p(5)$.

The values of $p(n)$ can also be calculated from a recursion formula, which we now give without proof:

$$\begin{aligned} p(n) = {} & p(n-1) + p(n-2) - p(n-5) - p(n-7) \\ & + p(n-12) + p(n-15) - p(n-22) - p(n-26) \\ & + \cdots \\ & + (-1)^k p\left(n - \frac{k(3k-1)}{2}\right) + (-1)^k p\left(n - \frac{k(3k+1)}{2}\right) + \cdots \end{aligned}$$

Neither of these results makes the calculation of $p(n)$ very easy; however, P. A. MacMahon used the recursion formula to calculate that $p(200) = 3{,}972{,}999{,}029{,}388$. It is evident that $p(n)$ grows rather rapidly! In 1917 G. H. Hardy and S. Ramanujan discovered an explicit formula for $p(n)$; however, it is very complicated and of limited utility, being mainly useful for large $n$. Readers interested in pursuing this matter further should read lecture VIII of G. H. Hardy's book *Ramanujan: Twelve Lectures on Subjects Suggested by His Life and Work*, Cambridge: University Press, 1940. In this lecture Hardy also gave the value of $p(14031)$, a formidable number indeed. In 1958, tables of $p(n)$ were published for $n \leq 1000$. (See the biographical notes on Hardy and Ramanujan at the close of this section.)

## *Problems*

1. Calculate the following:
   (a) $\phi(60)$; (b) $\sigma(728)$; (c) $\tau(728)$.
2. Calculate the following:
   (a) $\phi(360)$; (b) $\sigma(7280)$; (c) $\tau(7280)$.

3. Find all the positive integers $n$ such that:
   (a) $\phi(n) = 4$; (b) $\phi(n) = 6$.
4. Find all the positive integers $n$ such that:
   (a) $\phi(n) = 8$; (b) $\phi(n) = 10$.
5. Explain why $\phi(n) = \phi(2n)$ for all odd positive integers $n$.
6. Show the following:
   (i) If $\gcd(a, n) = 1$, then $\gcd(n - a, n) = 1$.
   (ii) For $n \geq 2$, the sum of all the integers $a$ such that both $1 \leq a \leq n$ and $\gcd(a, n) = 1$ is equal to $n\phi(n)/2$.
7. Let $p, q, \ldots, r$ be distinct positive primes and let $a, b, \ldots, c$ be any positive integers. Explain why

$$\phi(p^a q^b \cdots r^c) = \phi(p^a)\phi(q^b) \cdots \phi(r^c).$$

8. Prove that $\phi$ is a multiplicative function; that is, prove that $\phi(rs) = \phi(r)\phi(s)$ whenever $r$ and $s$ are relatively prime positive integers.
9. Let $a$ and $b$ be positive integers and let $d = \gcd(a, b)$. Prove that $\phi(ab) = d\phi(a)\phi(b)/\phi(d)$.
10. Let $p$ and $n$ be positive integers with $p$ prime. Prove that:
    (a) $p|n$ implies that $(p - 1)|\phi(n)$.
    (b) $(p - 1)|\phi(n)$ does not imply that $p|n$.
11. Let $d$ and $n$ be positive integers with $d|n$. Show that $\phi(d)|\phi(n)$.
12. (i) Find the smallest positive prime $p$ with $7|\phi(p)$.
    (ii) Prove that there is no positive integer $n$ with $\phi(n) = 14$.
13. (i) Let $\phi(n) = 48$, $p$ be a positive prime, and $p|n$. Prove that $p$ is 2, 3, 5, 7, or 13.
    (ii) Given that $\phi(n) = 48$, prove that $n$ must be an integral multiple of 7 or 9 or 13.
    (iii) Find the ten positive integers $n$ satisfying $\phi(n) = 48$.

*14. How many rational numbers are expressible as $a/b$ with $a$ and $b$ integers satisfying both $0 \leq a < b \leq n$ and $\gcd(a, b) = 1$?

15. Show that, for every positive integer $n$,

$$\frac{\sigma(n)}{n} = \sum_{d|n} \frac{1}{d},$$

where $\sum_{d|n}$ means the sum over all positive integral divisors $d$ of $n$. $\left[\text{For example, } \sum_{d|6} \frac{1}{d} = \frac{1}{1} + \frac{1}{2} + \frac{1}{3} + \frac{1}{6} = \frac{12}{6} = \frac{\sigma(6)}{6}.\right]$

16. Let $\sigma_2(n)$ denote the sum of the squares of the positive integral divisors of $n$; that is, let $\sigma_2(n) = \sum_{d|n} d^2$. Also let the standard factorization of $n$ be

$$n = (p_1)^{r_1}(p_2)^{r_2} \cdots (p_m)^{r_m}.$$

Show that

$$\sigma_2(n) = \sigma(n) \prod_{i=1}^{m} \frac{(p_i)^{r_i+1} + 1}{p_i + 1}.$$

17. Prove that 6 and 28 are the two smallest perfect numbers.
18. Show that 496 is a perfect number.
19. Let $s$ be a positive integer such that $2^{s+1} - 1$ is a prime $p$. Prove that $2^s p$ is perfect.
20. Given that $n$ is perfect, show that $\sum_{d|n} \frac{1}{d} = 2$.
21. Show that 220 and 284 are an amicable pair.
22. (i) Find $\sigma(1184) - 1184$.
    (ii) Find $\sigma(1210) - 1210$.
    (iii) Find an amicable pair of positive integers whose difference is 26.
23. Let $p, q, \ldots, r$ be distinct positive primes and let $a, b, \ldots, c$ be positive integers. Explain why

$$\sigma(p^a q^b \cdots r^c) = \sigma(p^a)\sigma(q^b) \cdots \sigma(r^c).$$

24. Prove that $\sigma$ is multiplicative; that is, prove that $\sigma(mn) = \sigma(m)\sigma(n)$ whenever $m$ and $n$ are relatively prime positive integers.
25. (i) Tabulate $\tau(n)$ for $n = 1, 2, 3, \ldots, 20$.
    (ii) For which numbers $n$ in $\{1, 2, 3, \ldots, 20\}$ is $\tau(n)$ odd?
26. Make a conjecture concerning the positive integers $n$ such that $\tau(n)$ is odd and then prove that the conjecture is correct.
27. Characterize the positive integers $n$ such that $\tau(n)$ is a prime $p$.
28. Find the smallest positive integer $n$ such that $\tau(n) = 5$.
29. Find the smallest positive integer $n$ such that $\tau(n) = 10$.
30. (a) Find the smallest positive integer that has at least 10 positive integral divisors.
    (b) Find the smallest positive integer with exactly 28 positive integral divisors.
31. Do the analogue of Problems 7 and 23 for the $\tau$-function.

32. Prove that $\tau$ is multiplicative.
33. Tabulate the partition function $p(n)$ for $n = 1, 2, \ldots, 8$.
34. Classify the permutations in $\mathbf{S}_6$ according to the type of the standard form and count the number of each type. Check using ord $\mathbf{S}_6 = 6! = 720$.

***James Joseph Sylvester*** (*1814–1897*) *Sylvester was born in London and was educated in England at St. John's College, Cambridge. He spent a considerable amount of time in the United States, however. In 1841 he was appointed professor of mathematics at the University of Virginia but soon returned to England. He again took a position in the United States in 1876 at the newly established Johns Hopkins University in Baltimore and there served as first editor of the* **American Journal of Mathematics**. *In 1884 he became Savilian Professor of Geometry at Oxford.*

*Between his two sojourns in the United States he worked with Cayley in England and with him developed much of their important work on the theory of invariants. He was the first to use a number of words currently common in mathematics: invariant, Jacobian, discriminant, Hessian.*

***Godfrey Harold Hardy*** (*1877–1947*) *G. H. Hardy was Savilian Professor of Geometry at Oxford and Sadeirian Professor of Pure Mathematics at Cambridge. He was certainly one of the greatest mathematicians of the twentieth century and made, along with his principal collaborators, S. Ramanujan and J. E. Littlewood, profound contributions to several general areas of number theory.*

*Much could be written about Hardy but we shall refer the reader to Hardy's own account of his life in mathematics,* **A Mathematician's Apology** (*Cambridge: University Press, 1969*), *and an account of another of Hardy's collaborators, G. Polya, "Some Mathematicians I Have Known"* [**The American Mathematical Monthly,** *76* (*1969*), *no. 7, 746–753*].

*No mathematician should ever allow himself to forget that mathematics, more than any other art or science, is a young man's game. ... Galois died at twenty-one, Abel at twenty-seven, Ramanujan at thirty-three, Riemann at forty. ... I do not know of an instance of a major mathematical advance initiated by a man past fifty.*

*G. H. Hardy*

---

***Srinivasa Ramanujan*** (*1887–1920*) *Ramanujan was born in the Tanjore district of India and received only a minimal formal education in mathematics. He nevertheless had a phenomenal talent for discovering profound and complex formulas, many of which he was able to prove, or at least prove to the satisfaction of other mathematicians of the day. He corresponded with Hardy, who made arrangements for him to move to England. He arrived in England in 1914 and proceeded to fill in gaps in his understanding of function theory and modern concepts of what constituted proof.*

*He became ill in 1917 and died in 1920, only six years after arriving in England. During those final years he continued to make dramatic new discoveries and it was during this period, much of which he spent in the hospital, that he gave us one of the most famous anecdotes passed along in mathematics classrooms. On one visit to Ramanujan in the hospital Hardy mentioned that he had arrived in taxi number 1729. Ramanujan immediately commented that 1729 was indeed an interesting number since*

$$1729 = 9^3 + 10^3 = 1^3 + 12^3$$

*is the smallest positive integer that can be represented in two different ways as the sum of two cubes!*

---

## 7.4 *Quadratic Residues*

*The moving power of mathematical invention is not reasoning but imagination.*

*Augustus de Morgan*

In the Theory of Numbers, an integer $a$ is called a ***quadratic residue*** modulo a positive integer $m$ if $\gcd(a, m) = 1$ and there exist integers $x$ that satisfy

$$x^2 \equiv a \pmod{m}. \tag{1}$$

If $\gcd(a, m) = 1$ and there are no integer solutions of the congruence (1), $a$ is a ***quadratic nonresidue*** modulo $m$.

If $a \equiv b \pmod{m}$, then $b$ is a quadratic residue modulo $m$ if and only if $a$ is a quadratic residue; similarly for nonresidues. Hence we may restrict ourselves to integers $a$ satisfying $0 < a < m$. We also note that $a$ is a quadratic residue modulo $m$ if and only if there is an integer $x$ such that $\bar{x}^2 = \bar{a}$ in the multiplicative group $\mathbf{V_m}$ of invertibles of $\mathbf{Z_m} = \mathbf{Z}/(\mathrm{m})$.

For small positive integers $m$, one can easily find the quadratic residues modulo $m$ by squaring $m$ consecutive integers, as one can see in Examples 3 and 4 of Section 4.2 and Example 3 of Section 4.3. (In those examples, we used quadratic nonresidues to prove nonexistence of integer solutions for certain quadratic equations.)

Let us examine the pattern formed by the quadratic residues among $1, 2, 3, \ldots, p - 1$ for some small odd primes $p$. In Table 7.1, an entry "$R$" means that $n$ is a quadratic residue modulo $p$ while an entry "—" means that it is not.

***Table 7.1***

| $p$ \ $n$ | 1 | 2 | 3 | 4 | 5 | 6 | 7 | 8 | 9 | 10 | 11 | 12 |
|---|---|---|---|---|---|---|---|---|---|---|---|---|
| 3 | $R$ | — | | | | | | | | | | |
| 5 | $R$ | — | — | $R$ | | | | | | | | |
| 7 | $R$ | $R$ | — | $R$ | — | — | | | | | | |
| 11 | $R$ | — | $R$ | $R$ | $R$ | — | — | — | $R$ | — | | |
| 13 | $R$ | — | $R$ | $R$ | — | — | — | — | $R$ | $R$ | — | $R$ |

We note that precisely half of the considered integers are quadratic residues in each case. This is proved for all odd primes $p$ in Theorem 1 below.

One might also see that the entries read the same from left to right as from right to left when $p \equiv 1 \pmod 4$. When $p \equiv 3 \pmod 4$, there is antisymmetry; that is, if one reverses each entry symbol as one reads from left to right, one obtains the entries from right to left. Each of these statements is true for general odd positive primes $p$; a proof is outlined in Problems 37 and 38 of Section 5.4.

Carl Friedrich Gauss derived a number of results on quadratic residues in his 1801 ***Disquisitiones Arithmeticae***. We state the most famous as Theorem 2 below but first note that among these results is the proof that, for a fixed odd positive prime $p$, the product of two quadratic residues or of two nonresidues is a quadratic residue while the product of a quadratic residue and a nonresidue is a nonresidue. We shall obtain this as a special case using the following results.

***Theorem 1*** ***Squares of Invertibles in*** $D$

*Let $V$ be the multiplicative group of invertibles of an integral domain $D$ whose characteristic is not 2. Let $W$ be the set of squares of elements of $V$. Then $W$ is a subgroup in $V$ and has index 2 when $V$ is finite.*

***Proof***

Let $\theta$ be the mapping from $V$ to itself with $\theta(v) = v^2$. Since $D$ is commutative,

$$\theta(ab) = abab = aabb = a^2b^2 = \theta(a)\theta(b);$$

that is, $\theta$ is a group homomorphism. It then follows from Theorem 2(c) of Section 3.3 that the image set $W$ is a subgroup in the codomain $V$.

Next we find the kernel $K$ of $\theta$. We note that $k$ is in $K$ if and only if $k^2 = 1$, which is equivalent to

$$(k - 1)(k + 1) = 0.$$

Since $D$ has no 0-divisors, this implies that $k$ is 1 or $-1$. Also, 1 and $-1$ are distinct since the characteristic of $D$ is not 2. Hence $K = \{1, -1\}$.

If $a^2$ is in $W$, the complete inverse image $\theta^{-1}(a^2)$ is the coset $aK = \{a, -a\}$ and the collection of all these cosets partitions $V$ into $r$ pairs, where $r = \text{ord } W$. Therefore, $\text{ord } V = 2r$ and $W$ has index 2 in $V$.

If $p$ is an odd prime, $\mathbf{Z_p}$ is a finite integral domain with characteristic different from 2. Then Theorem 1 tells us that the set $R$ of squares in $\mathbf{V_p}$ is a subgroup with index 2 in $\mathbf{V_p}$. If $N$ denotes the other coset of $R$ in $\mathbf{V_p}$, that is, the coset of nonsquares, then $\mathbf{V_p}/R = \{R, N\}$ has the multiplication table

| | $R$ | $N$ |
|---|---|---|
| $R$ | $R$ | $N$ |
| $N$ | $N$ | $R$ |

(2)

Translating these facts into congruence language, we see that half of the integers in $\{1, 2, \ldots, p-1\}$ are quadratic residues and that table (2) sums up concisely in group theoretic terms the content of Gauss' theorem on products of quadratic residues and nonresidues.

The following definition will help us state a famous and deep property of quadratic residues which was first discovered independently by Euler and Legendre in 1785 but not proved by them. Gauss discovered it in 1795, proved it in several ways, and published a proof in his *Disquisitiones Arithmeticae* of 1801.

### *Notation Legendre Symbol*

*If $p$ is an odd prime and* $\gcd(n, p) = 1$, *the Legendre symbol* $\left(\frac{n}{p}\right)$ *is defined as follows:*

$$\left(\frac{n}{p}\right) = 1 \text{ if } n \text{ is a quadratic residue modulo } p,$$

$$\left(\frac{n}{p}\right) = -1 \text{ if } n \text{ is a quadratic nonresidue modulo } p.$$

### *Theorem 2 Law of Quadratic Reciprocity*

*If $p$ and $q$ are distinct odd positive primes in* $\mathbf{Z}$,

$$\left(\frac{p}{q}\right)\left(\frac{q}{p}\right) = (-1)^{(p-1)(q-1)/4}.$$

Since all the known proofs are somewhat involved, we do not present a proof but instead refer the reader to one of the standard works on number theory listed in the bibliography. This law is useful in determining the quadratic residues and nonresidues for large primes.

## *Problems*

1. Find the quadratic residues in $\{1, 2, 3, \ldots, p-1\}$ modulo:
   (a) $p = 17$; (b) $p = 19$.
2. Find the quadratic residues modulo 23 in $\{1, 2, \ldots, 22\}$.
3. Show that the Law of Quadratic Reciprocity can be restated as follows: If $p$ and $q$ are distinct odd positive primes, then

$$\left(\frac{p}{q}\right) = \left(\frac{q}{p}\right)(-1)^{(p-1)(q-1)/4}.$$

4. Let $p$ and $q$ be distinct odd positive primes. Show that the Law of Quadratic Reciprocity is equivalent to the two following statements:
   (a) $\left(\frac{p}{q}\right) = \left(\frac{q}{p}\right)$ if either $p \equiv 1 \pmod 4$ or $q \equiv 1 \pmod 4$.
   (b) $\left(\frac{p}{q}\right) = -\left(\frac{q}{p}\right)$ if $p \equiv 3 \equiv q \pmod 4$.
5. Let $p$ be an odd positive prime with $p \neq 5$. Use the preceding problem to show that 5 is a quadratic residue modulo $p$ if and only if $p$ is a quadratic residue modulo 5—that is, if and only if $p \equiv \pm 1 \pmod 5$.
6. Let $p$ be an odd prime with $p > 3$. Find $\left(\frac{p}{3}\right)$ and $\left(\frac{3}{p}\right)$ under each of the following assumptions:
   (a) $p \equiv 1 \pmod{12}$, (b) $p \equiv 5 \pmod{12}$, (c) $p \equiv 7 \pmod{12}$, (d) $p \equiv 11 \pmod{12}$.
7. Let $p$ be an odd positive prime. Use Problem 38 of Section 5.4 to show that:
   (a) $\left(\frac{-1}{p}\right) = 1$ and $\left(\frac{-n}{p}\right) = \left(\frac{n}{p}\right)$ if $p \equiv 1 \pmod 4$.
   (b) $\left(\frac{-1}{p}\right) = -1$ and $\left(\frac{-n}{p}\right) = -\left(\frac{n}{p}\right)$ if $p \equiv 3 \pmod 4$.
8. Use Problems 6 and 7 to find $\left(\frac{-3}{p}\right)$ under each of the following assumptions; here $p$ is a prime:
   (a) $p \equiv 1 \pmod 6$, (b) $p \equiv 5 \pmod 6$.
9. Let $p$ be an odd prime. Then it is a theorem of Euler that $\gcd(p, n) = 1$ implies that

$$\left(\frac{n}{p}\right) \equiv n^{(p-1)/2} \pmod p.$$

Use this result to show the following:

(i) $\left(\frac{p-1}{p}\right) = \left(\frac{-1}{p}\right) = (-1)^{(p-1)/2}$.

(ii) $p - 1$ is a quadratic residue modulo $p$ if and only if $p \equiv 1 \pmod 4$.

(iii) If $m$ is an integer greater than 1 and $p|(m^2 + 1)$, then $p \equiv 1 \pmod 4$.

---

## *Review Problems*

1. Find all $n$ in $\{0, 1, 2, \ldots, 105\}$ such that $n$ is congruent to either 1 or $-1$ modulo $p$ for every $p$ in $\{2, 3, 5, 7\}$.

2. Find all $n$ in $\{0, 1, 2, \ldots, 71\}$ such that $n$ is congruent to either 1 or $-1$ modulo $p$ for both $p = 11$ and $p = 13$.

*3. Find the smallest integer $n > 1$ with

$$n \equiv \pm 1 \pmod p \text{ for } p \in \{2, 3, 5, 7, 11\}.$$

*4. Find the smallest integer $n > 1$ with

$$n \equiv \pm 1 \pmod p \text{ for } p \in \{2, 3, 5, 7, 11, 13, 17\}.$$

5. Let $R$ be a commutative ring. Prove that every $\alpha$ with $\deg \alpha > 0$ in $R[x]$ has an irreducible divisor in $R[x]$.

6. Let $p, a_1, a_2, \ldots, a_n$ be integers with $p$ a prime. Also, let $p|(a_1 a_2 \cdots a_n)$. Prove that $p|a_i$ for at least one $i$ in $\{1, 2, \ldots, n\}$ (and thus prove Lemma 2 of Section 1.7).

7. Let $D$ be an integral domain with prime characteristic $p$. Prove that

$$(x + y)^{p^n} = x^{p^n} + y^{p^n}$$

for all $x$ and $y$ in $D$ and all nonnegative integers $n$.

8. Let $D$ be an integral domain with prime characteristic $p$. Prove that

$$(x_1 + x_2 + \cdots + x_m)^{p^n} = (x_1)^{p^n} + (x_2)^{p^n} + \cdots + (x_m)^{p^n}$$

for all integers $m \geq 1$ and $n \geq 0$ and all $x_i$ in $D$.

9. Let $u_0, u_1, u_2, \ldots,$ be the sequence with $u_0 = u_1 = u_2 = 0$, $u_3 = 1$, and $u_n = 3u_{n-1} + 6u_{n-2} - 3u_{n-3} - u_{n-4}$ for $n > 3$. For $n > 1$, prove that:
(a) $u_2 + u_4 + \cdots + u_{2n} = (5u_{2n} - 3u_{2n-1} - u_{2n-2})/4$.
(b) $u_3 + u_5 + \cdots + u_{2n+1} = (5u_{2n+1} - 3u_{2n} - u_{2n-1} - 1)/4$.

10. Let $F_0, F_1, F_2, \ldots$ be the Fibonacci sequence with $F_0 = 0$, $F_1 = 1$, and $F_n = F_{n-1} + F_{n-2}$

for $n > 1$. Let $v_n = F_n F_{n-1} F_{n-2}/2$. Show that $v_n$ equals the $u_n$ of the preceding problem for $n > 1$.

11. Let $\phi$ be the Euler $\phi$-function. Find all $a$ in $\{1, 2, \ldots, 25\}$, such that $\phi(p^n) = a$ for some positive prime $p$ and positive integer $n$.
12. Find all the positive integers $n$ such that $\phi(n) = 96$.
13. Explain why $\phi(n)$ is an even integer for $n > 2$.
14. Let $a$ be a positive integer. Explain why $\{n : \phi(n) = a\}$ is a finite set of positive integers.
15. For $n$ in $\mathbf{Z}^+ = \{1, 2, 3, \ldots\}$, let $\sigma(n)$ be the sum of the positive integral divisors of $n$. Find all $n$ in $\mathbf{Z}^+$ with $\sigma(n) = 18$.
16. Is there a maximum value of $\sigma(n)$? Explain.
17. Show that $\sigma(n) = 2n - 1$ for an infinite number of $n$'s in $\mathbf{Z}^+$.
18. Let $p$ be a prime in $\mathbf{Z}^+$ and $u_n = p^n$. Show that
$$\sigma(u_n) = u_n + \sigma(u_{n-1}) \text{ for } n \in \mathbf{Z}^+.$$
19. For $n$ in $\mathbf{Z}^+$, let $\tau(n)$ be the number of $d$'s in $\mathbf{Z}^+$ with $d \mid n$. Find the smallest $n$ in $\mathbf{Z}^+$ with $\tau(n) = 14$.
20. Characterize the positive integers $n$ such that $\tau(n) = 2$.
21. Characterize the positive integers $n$ such that $\tau(n) = 3$.
22. For $n$ in $\mathbf{Z}^+$, let $A(n)$ be the number of ordered pairs $(x, y)$ of positive integers such that
$$\frac{1}{x} + \frac{1}{y} = \frac{1}{n}.$$
Tabulate $A(n)$ and $\tau(n^2)$ for $n = 1, 2, 3, 4, 5, 6$, and 7.
23. Find $p(9)$, that is, the number of unordered partitions of 9.
24. Find $p(10)$.
25. Let $p$ and $q$ be odd positive primes in $\mathbf{Z}$ with $q \equiv 1 \pmod 4$ and $q \not\equiv 1 \pmod 8$. Show that $p$ is a quadratic residue modulo $q$ if and only if $q$ is a quadratic residue modulo $p$.
26. For each of the following, find the three smallest odd positive primes $p$ in $\mathbf{Z}$ satisfying the condition:
    (i) $\left(\frac{p}{19}\right)\left(\frac{19}{p}\right) = 1.$
    (ii) $\left(\frac{p}{19}\right)\left(\frac{19}{p}\right) = -1.$

27. Let $I = (r)$ and $J = (s)$ be principal ideals in **Z**. Show that there exist $x$ in $I$ and $y$ in $J$ with $x + y = 1$ if and only if $r$ and $s$ are relatively prime.

28. Let $I = (r)$ and $J = (s)$ be principal ideals in **Z** and let $\gcd(r, s) = 1$. Prove that the intersection

$$(a + I) \cap (b + J)$$

of a coset of $I$ and a coset of $J$ is nonempty and is a coset of $I \cap J$.

29. Let $p$ be an odd positive prime in **Z**. Prove without group theory the following results that are established, using group theory, after Theorem 1 of Section 7.4:
    (i) Modulo $p$, the product of two quadratic residues or of two quadratic nonresidues is a quadratic residue.
    (ii) Modulo $p$, the product of a quadratic residue and a quadratic nonresidue is a quadratic nonresidue.

## *Supplementary and Challenging Problems*

1. Let $U$ be a commutative ring with unity. We say that ideals $I$ and $J$ in $U$ are ***relatively prime*** if there exist an $h$ in $I$ and a $k$ in $J$ such that $h + k = 1$. Given that $I$ and $J$ are relatively prime ideals in $U$, prove that the intersection

$$(a + I) \cap (b + J),$$

of any cosets of $I$ and $J$, respectively, is nonempty and is a coset of $I \cap J$. Also explain why this result is a generalization of Theorem 2 of Section 7.1.

2. State and prove a generalization of the Chinese Remainder Theorem that deals with a set $\{I_1, I_2, \ldots, I_r\}$ of $r$ pairwise relatively prime ideals in $U$. (See the preceding problem.)

3. Let $\alpha\colon \mathbf{Z}^+ \to N = \{0, 1, 2, \ldots\}$ be a mapping with the three following properties:
    (1) $\alpha(1) = 0$;
    (2) $\alpha(p) = 1$ if $p$ is prime;
    (3) $\alpha(mn) = n\alpha(m) + m\alpha(n)$ for all $m$ and $n$ in $\mathbf{Z}^+$.
    Prove that one and only one such $\alpha$ exists and express $\alpha(n)/n$ in terms of the standard factorization $n = (p_1)^{e_1}(p_2)^{e_2} \cdots (p_r)^{e_r}$.

4. Prove that a prime $p > 3$ is expressible as $p = m^2 + 3n^2$, with $m$ and $n$ in **Z**, if and only if $p \equiv 1 \pmod 6$.

5. Let $F$ be a finite field with $m$ elements. Prove that:
    (i) If $d > 0$ and $d \mid (m - 1)$, there are $\phi(d)$ elements with order $d$ in the multiplicative group $V$ of nonzero elements of $F$.
    (ii) $V$ is cyclic.

6. Let $p$ be an odd positive prime in $\mathbf{Z}$ and let

$$1 + \frac{1}{2} + \frac{1}{3} + \cdots + \frac{1}{p-1} = \frac{r}{s},$$

with $r$ and $s$ in $\mathbf{Z}$. Prove the following:
(i) $p \mid r$.
(ii) If $p > 3$, then $p^2 \mid r$.

7. For all positive integers $n$, let $A(n)$ be the number of ordered pairs $(x, y)$ of positive integers such that

$$\frac{1}{x} + \frac{1}{y} = \frac{1}{n}.$$

Prove that $A(n) = \tau(n^2)$.

8. For all positive integers $n$, let $B(n)$ be the number of ordered pairs $(x, y)$ of positive integers such that $\operatorname{lcm}[x, y] = n$.
(i) Prove that $B(n)$ is a multiplicative function; that is, $B(rs) = B(r)B(s)$ whenever $(r, s) = 1$.
(ii) Give a formula for $B(p^m)$ where $p$ and $m$ are positive integers with $p$ a prime.

9. State and solve the analogue of Problem 8 dealing with ordered triples $(x, y, z)$ of positive integers having $n$ as their least common multiple.

10. Use Sylvester's Inclusion-Exclusion Principle to find the one millionth term in the sequence

$$2, 3, 5, 6, 7, 10, \ldots, 26, 28, 29, 30, 31, 33, \ldots$$

that results when the perfect squares, cubes, and fifth powers are removed from the positive integers $1, 2, 3, \ldots$.

11. Let $\theta$ be a permutation on $\{1, 2, \ldots, n\}$. Show that

$$n \prod_{i=1}^{n} [i - \theta(i)] = n[1 - \theta(1)][2 - \theta(2)] \cdots [n - \theta(n)]$$

is always an even integer.

12. Let $\alpha$ and $\beta$ be permutations in $\mathbf{S}_{400}$. For $1 \le i \le 400$, let $c_i$ be the product $\alpha(i)\beta(i)$ in $\mathbf{Z}$. Prove that

$$c_1 c_2 \cdots c_{400} \equiv 1 \pmod{401}.$$

13. Let $T = \{1, 51, 2551, 1252551, 6251252551, \ldots\}$ consist of all the numbers formed by placing the digits of $5^n, 5^{n-1}, \ldots, 5^2, 5, 1$ side by side, where $n$ is a nonnegative integer. Find all the integers $t$ in $T$ such that $t = x^2 + y^2 + z^2$ with $x$, $y$, and $z$ in $\mathbf{Z}$.

14. Characterize the positive integers $n$ that are not expressible as the sum, $a + (a + 1) + \cdots + (a + b)$, of two or more consecutive positive integers.

15. Let $a \equiv b \pmod{p}$, with $p$ a prime. For all $n$ in $\mathbf{Z}^+$, prove that

$$a^{p^n} \equiv b^{p^n} \pmod{p^{n+1}}.$$

16. Let $a_1 = 2$ and $a_{n+1} = a_n^2 - a_n + 1$ for $n \in \mathbf{Z}^+$. Prove that $a_s$ and $a_t$ are relatively prime whenever $s \neq t$.

17. Let $a$, $b$, $c$, and $d$ be distinct integers and let

$$\begin{aligned}\alpha = x^4 &- (a + b + c + d)x^3 + (ab + ac + ad + bc + bd + cd)x^2\\ &- (bcd + acd + abd + abc)x + (abcd - 49).\end{aligned}$$

Show that the only possibility for an integral zero of $\alpha$ is $(a + b + c + d)/4$.

18. Prove that among any 10 consecutive integers there is at least one that is relatively prime to each of the others.

19. Show that for every positive integer $n$ there is a positive integer $m$ such that $(\sqrt{2} - 1)^n = \sqrt{m + 1} - \sqrt{m}$.

20. Show that, for each positive integer $n$, there is an integer $a_n$ with both

$$0 < a_n - (\sqrt{3} + 1)^{2n} < 1 \quad \text{and} \quad 2^{n+1} \mid a_n.$$

21. Let $n$ be an integer such that

$$n^2 = 10^d c_d + 10^{d-1} c_{d-1} + \cdots + 10c_1 + c_0,$$

with each $c_j$ in $\{1, 4, 9\}$. What are the possible values of the tens digit $c_1$ of $n^2$?

22. For each of the following, find all solutions in $\mathbf{Z}$:
    (i) $x^2 + 3 = u^2$.
    (ii) $4y^2 + 4y + 4 = u^2$.
    (iii) $y^2 + y + 1 = v^2$.
    (iv) $z^2 - z + 1 = v^2$.

23. Find all triples of integers $x$, $y$, and $z$ such that

$$x = 3y^2 + 1 \quad \text{and} \quad x^2 + x + 1 = 3z^2.$$

24. Find all solutions in integers $x$ and $y$ of $x^3 - 1 = y^2$.

25. Find all solutions in integers $x$ and $y$ of $x^3 + 1 = y^2$.

26. Prove that the product of four consecutive positive integers is never a perfect square or cube.

# 8

# *Algebraic Coding Theory*

In a modern technological society there are many occasions for sending information from one place to another. These messages can be sent by radio, over wires, by microwave transmitters, and through the mails, among other means, and the messages may be in speech or other sounds, in typed words, or in pictures. The senders and receivers may be human beings or computers.

Originally codes were designed to protect the privacy of the messages should they fall into the hands of unauthorized persons. Now they have additional purposes, such as increasing the convenience of the transmitting process and increasing the likelihood that the intended message will be attainable when the transmission is not perfect.

There is always a possibility of interference in the transmission due to adverse atmospheric conditions, nearby electronic devices, faulty equipment, and so on; such interference we will call ***noise***. Sometimes the sender has no way of knowing whether the message was received properly, or has limited opportunities for repeating the message; an example is the transmission of pictures of Saturn and its satellites from the Voyager spacecraft. Clearly, NASA, telephone companies, large computer firms, and many others need methods for increasing the probability that the desired content of messages will be received despite noise.

Figure 8.1 is a simplified diagram depicting a message going from the source to its destination via an encoder, a transmission channel that is subject to noise, and a decoder. If the code is good enough, errors introduced by noise in the channel will with high probability be detected and corrected.

***Figure 8.1***

The channel can be anything from the space between Saturn and Earth to a magnetic tape used for computer storage. Messages may be in various forms, but a very common form, especially where computers are involved, is a sequence of 0's and 1's, that is, a sequence of elements of $\mathbf{Z}_2 = \{0, 1\}$. Such a sequence is called a ***binary string***; its ***length*** is the number of digits in the string. We assume that when a string goes through the channel, a string of equal length is received but that some of the original digits may be turned into their opposites by the noise. We also assume that the probability of sending a 0 and receiving a 1 is the same as that of sending a 1 and receiving a 0. Such a channel is called a ***binary symmetric channel.*** The word binary indicates that only 0's and 1's are used, and the word symmetric conveys the fact that the two possible errors are equally likely.

Let $p$ be the probability that a specific digit will be received incorrectly. As is always true of probabilities, $0 \le p \le 1$. The cases $p = 0$ and $\frac{1}{2} < p \le 1$ are uninteresting, since $p = 0$ implies no noise whatever and $p > \frac{1}{2}$ should motivate us to read a received 0 as a 1 and a received 1 as a 0. In most applications $p$ will be quite small, though in radio transmission from the far reaches of the solar system, $p$ might be near $\frac{1}{2}$.

## 8.1 Parity Codes

Since there are 26 letters in our alphabet and 32 binary strings of length 5, one can convert a message written in English into binary form by replacing each $a$ with 00000, each $b$ with 00001, ..., each $z$ with 11001. The other 5-digit binaries would be available as punctuation marks. As a means of detecting mistransmissions that change only one digit of a 5-digit binary $\alpha = a_1 a_2 a_3 a_4 a_5$, we encode $\alpha$ as the 6-digit binary codeword $a_1 a_2 a_3 a_4 a_5 a_6$ where $a_6 = 0$ if $a_1 + a_2 + a_3 + a_4 + a_5$ is even, and $a_6 = 1$ if the sum of the digits of $\alpha$ is odd. To send the message "OK" we translate "O" into 01110 and "K" into 01010; these are encoded as 011101 and 010100, respectively. In this code, all 6-digit codewords $a_1 a_2 a_3 a_4 a_5 a_6$ are such that the sum of the six digits is even (since adding a zero to an even sum does not change the sum and adding a one to an odd sum makes the new sum even).

Our message "OK" is sent as 011101010100. If received without error, it is broken into 6-digit binaries as 011101/010100. Since each of these is a codeword, each is decoded by dropping the sixth digit, giving 01110/01010 and this is translated back as "OK."

If the message is received as 011001000110, it is broken up as 011001/000110. The first of these is an obvious error since 011001 has an odd sum of digits and hence is not possible for a codeword. The second is also wrong but is possible as a codeword; it would be decoded by dropping the last digit. The decoded message would be error/00011 and this would be translated back as "error D."

Thus this code detects (but does not correct) mistransmissions with an odd number of errors in a block of six digits and does not detect mistransmissions with an even number of errors in such a block.

## 8.2 Block Codes

Here we generalize on the example of the previous section. Let $W_n$ consist of all binary strings $x_1x_2 \cdots x_n$ (with each $x_i$ in $\{0, 1\}$). Let $s$ be fixed and assume that the messages to be sent are made up of a number of strings from $W_s$; such a string is called a ***word***. Before transmission, each word will be encoded by being replaced with a string, called a ***codeword***, in a fixed $W_t$. It is necessary for the error detection and correction aspects of the code that $t > s$. We also want distinct words to be represented by distinct codewords; hence encoding is done by an injective mapping $E$: $W_s \to W_t$. Let $E(W_s)$ denote the set of all codewords—that is, the image set of the mapping $E$. Since $t > s$, $E(W_s)$ is a proper subset of the codomain $W_t$. This allows the code to be designed so that the most likely mistransmissions will not be codewords; such errors are trivial to detect and in some cases are easy to correct.

Decoding is done by a mapping $D$ with domain $W_t$ and with the codomain either $W_s$ or $W_s \cup \{\text{error}\}$. The decoding function $D$ must be such that $D(\beta) = \alpha$ whenever $\beta$ is the codeword with $E(\alpha) = \beta$. If $\beta$ is not a codeword, $D(\beta)$ can be $D(\beta')$ when $\beta'$ is the unique codeword which in some sense is nearest to $\beta$. If $\beta$ is not a codeword and there is no clear way to tell what codeword was intended, $D(\beta)$ should be "error."

An ordered pair $[E, D]$, where $E$: $W_s \to W_t$ is such an encoding function and $D$ is such a decoding function, is called an ***(s, t) block code***. For example, the parity code of Section 8.1 is a (5, 6) block code. Analogously, for every positive integer $s$, there is a ***parity code*** that is an $(s, s + 1)$ block code. Parity codes are usually called "parity check codes" but we drop the word "check" to avoid confusion with the "parity check matrices" of Section 8.4. Parity codes detect certain errors but do not correct any mistransmissions. We next describe a code that corrects many errors.

***Example 1***

The ***triple repetition code*** is the $(s, 3s)$ block code with the encoding function $E$: $W_s \to W_{3s}$ given by

$$E(x_1x_2 \cdots x_s) = x_1x_2 \cdots x_s x_1x_2 \cdots x_s x_1x_2 \cdots x_s$$

and the decoding function $D$: $W_{3s} \to W_s$ given by

$$D(y_1y_2 \cdots y_{3s}) = z_1z_2 \cdots z_s$$

where for $i = 1, 2, \ldots, s$ the digit $z_i$ is chosen as the digit that appears at least twice in the ordered triple

$$T_i = (y_i, y_{s+i}, y_{2s+i}).$$

Since 0 and 1 are the only binary digits, one of them must appear at least twice in each $T_i$. When $s = 5$, a message 00110 is encoded as the codeword 001100011000110. If received correctly this is easily decoded as 00110. If the codeword is received incorrectly as 001100111000111, it is decoded by breaking it into blocks as 00110/01110/00111, using $T_1 = (0, 0, 0)$, $T_2 = (0, 1, 0)$, $T_3 = (1, 1, 1)$, $T_4 = (1, 1, 1)$, $T_5 = (0, 0, 1)$ to deduce that $z_1 = 0 = z_2$, $z_3 = 1 = z_4$, and $z_5 = 0$, and thus deciding that the original message was 00110. We note that this code corrects all mistransmissions involving at most one wrong digit in any triple $T_i$ but does not correct errors in two or three of the digits of a $T_i$.

## Problems for Sections 8.1 and 8.2

1. Let $[E, D]$ be the parity code with $s = 4$. Find: (a) $E(1110)$; (b) $E(1010)$; (c) $D(11111)$; (d) $D(11011)$.
2. Let $[E, D]$ be the parity code with $s = 5$. Find: (a) $E(00000)$; (b) $E(00001)$; (c) $D(110110)$; (d) $D(100110)$.
3. Using the parity code with $s = 6$, if the intended codeword 1101100 is mistransmitted as 1001000, would the decoder automatically detect an error? Explain.
4. If 1101100 is mistransmitted as 1001100, would the parity decoder detect an error? Explain.
5. (a) How many binary strings are there in $W_5$?
   (b) What percentage of these are codewords in the parity code with $s = 4$?
6. What is the probability that a randomly chosen string in $W_6$ will be a codeword in the parity code with $s = 5$?
7. The $(s, s + 1)$ block parity code detects mistransmissions with one wrong digit. Does it detect a codeword's mistransmission with:
   (a) two wrong digits; (b) three wrong digits; (c) four wrong digits?
8. Give a necessary and sufficient condition on the number of wrong digits for the $(s, s + 1)$ block parity code to detect the mistransmission of a codeword.
9. Let $[E, D]$ be the triple repetition (4, 12) block code. Find:
   (a) $E(1110)$; (b) $D(011010101111)$; (c) $D(101010101111)$.
10. Let $[E, D]$ be the triple repetition (5, 15) block code. Find:
    (a) $E(10001)$; (b) $D(010100101001010)$; (c) $D(011100110000110)$.
11. For the (3, 9) triple repetition code, give an example of a word $\alpha$, codeword $\beta$, and received string $\beta'$ such that:
    (a) $E(\alpha) = \beta$, $\beta' \neq \beta$, $D(\beta') = \alpha$.
    (b) $E(\alpha) = \beta$, $\beta' \neq \beta$, $D(\beta') \neq \alpha$.

12. Do Problem 11 for the (2, 6) triple repetition code.
13. Define $E$ and $D$ for a double repetition $(s, 2s)$ block code.
14. Define $E$ and $D$ for a quadruple repetition $(s, 4s)$ block code such that $D$ corrects some mistransmissions.
15. Give an advantage and a disadvantage for the $(s, 3s)$ triple repetition code over the $(s, 2s)$ double repetition code.
16. Give an advantage and a disadvantage for the $(s, 5s)$ quintuple repetition code over the $(s, 4s)$ quadruple repetition code.

## 8.3 Additive Groups of Binary Strings

The following definition enables one to use group theory in designing codes. Later, we will define multiplication and then be able to use ring theory.

### Definition 1 Addition in $W_n$

*The* **sum** *of binary strings* $x_1x_2 \cdots x_n$ *and* $y_1y_2 \cdots y_n$ *is the binary string* $z_1z_2 \cdots z_n$ *with* $z_i = 0$ *if* $x_i = y_i$ *and* $z_i = 1$ *if* $x_i \neq y_i$.

Clearly, in this definition $z_i = x_i + y_i \pmod 2$ for each $i$ and thus $W_n$ is isomorphic to the direct product of $n$ identical cyclic additive groups $[1] = \{0, 1\}$. Hence $W_n$ is an additive group.

### Theorem 1 Identity and Inverses

*The identity in* $W_n$ *is the string* $e_1e_2 \cdots e_n$ *with each* $e_i = 0$. *Every string in* $W_n$ *is its own negative (that is, additive inverse).*

The proof follows readily from Definition 1.

### Definition 2 Weight of a Binary String

*The* **weight** *of an* $\alpha = a_1a_2 \cdots a_n$ *in* $W_n$ *is the number of values of* $i$ *in* $\{1, 2, \ldots, n\}$ *for which* $a_i = 1$; wgt $\alpha$ *designates this weight.*

For example, if $\alpha = 01110$ one has wgt $\alpha = 3$.

***Definition 3 Distance in $W_n$***

*Let $\alpha = a_1a_2 \cdots a_n$ and $\beta = b_1b_2 \cdots b_n$ be in $W_n$. Then the number of values of $i$ in $\{1, 2, \ldots, n\}$ for which $a_i \neq b_i$ is called the* **distance** *between $\alpha$ and $\beta$ and is denoted by $d(\alpha, \beta)$.*

Weight and distance are also called ***Hamming weight*** and ***Hamming distance***, respectively, since these concepts were first used by R. W. Hamming, the originator of a family of codes discussed below.

As an example, the distance between $\alpha = 00111$ and $\beta = 01001$ is 3. One may note that $\alpha + \beta = 01110$ and thus that $d(\alpha, \beta) = \text{wgt}(\alpha + \beta)$ in this case. The following result states that this alternative method of calculating a distance works in all cases.

***Theorem 2 Distance as a Weight***

*For $\alpha$ and $\beta$ in $W_n$, $d(\alpha, \beta) = wgt(\alpha + \beta)$.*

The proof is left to the reader as Problem 9 below.

Since a string is its own negative, one also has $d(\alpha, \beta) = \text{wgt}(\alpha - \beta) = \text{wgt}(\beta - \alpha)$.

The distance function for ordered pairs of points in space satisfies

$$d(P, Q) + d(Q, R) \geq d(P, R),$$

which is called the ***Triangle Inequality***. The following result asserts that this also holds for distances between strings in $W_n$.

***Theorem 3 Triangle Inequality***

*$d(\alpha, \beta) + d(\beta, \gamma) \geq d(\alpha, \gamma)$ for all strings $\alpha, \beta, \gamma$ in $W_n$.*

***Proof***

Let $\alpha = a_1a_2 \cdots a_n$, $\beta = b_1b_2 \cdots b_n$ and $\gamma = c_1c_2 \cdots c_n$. For each $i$ such that $a_i \neq c_i$, one must have either $a_i \neq b_i$ or $b_i \neq c_i$. Hence $\text{wgt}(\alpha + \beta) + \text{wgt}(\beta + \gamma) \geq \text{wgt}(\alpha + \gamma)$. Using Theorem 2, this translates into the desired inequality.

***Definition 4 Transmission Error Pattern, Number of Errors***

*If $\alpha = a_1a_2 \cdots a_t$ is the codeword sent and $\beta = b_1b_2 \cdots b_t$ is the received binary string, the* **error pattern** *is the string $p_1p_2 \cdots p_t$ with $p_i = 0$ if $a_i = b_i$, and $p_i = 1$ if $a_i \neq b_i$. The* **number of errors** *is the number of values of $i$ in $\{1, 2, \ldots, t\}$ for which $a_i \neq b_i$.*

***Theorem 4 Error Formulas***

*Let $\alpha$, $\beta$, and $\pi$ be the intended codeword, received string, and error pattern, respectively. Then $\pi = \alpha + \beta$, and the number of errors is given by wgt $\pi$ or by $d(\alpha, \beta)$.*

***Proof***

Using Definitions 1 and 4, one sees that the location of the 1's in $\alpha + \beta$ gives the positions of the incorrectly received digits and hence $\text{wgt}(\alpha + \beta)$ is the number of erroneous digits. From Theorem 2, $\text{wgt}(\alpha + \beta) = d(\alpha, \beta)$.

***Example 1***

Let $\alpha = 00101001$ be an intended codeword in $W_8$ and $\beta = 00100101$ be the received string. Then the error pattern $\pi = \alpha + \beta = 00001100$. The location of the 1's in $\pi$ shows that the wrong digits in $\beta = b_1b_2 \cdots b_8$ are $b_5$ and $b_6$. Also the number of wrong digits is $\text{wgt}\, \pi = 2 = d(\alpha, \beta)$. We now consider three possibilities concerning the code being used. If the code is such that $\beta$ is also a codeword, the errors cannot be detected (except from the context). If the code is such that the distance between any two codewords is at least 3, then $\beta$ cannot be a codeword and the receiver knows that the transmission was incorrect. If the code is such that the distance between any two codewords is at least 5, then for any codeword $\gamma$ different from $\alpha$, the Triangle Inequality implies

$$d(\alpha, \beta) + d(\beta, \gamma) \geq d(\alpha, \gamma) \geq 5$$

and hence $d(\beta, \gamma) \geq 5 - d(\alpha, \beta) = 5 - 2 = 3$. Then $d(\beta, \gamma) \geq 3$ tells one that $\alpha$ is the nearest codeword to $\beta$ and it is reasonable to replace the erroneous $\beta$ with $\alpha$.

This example motivates the following general result:

***Theorem 5 Conditions for Detecting or Correcting Errors***

*For a code to detect all errors in $k$ or fewer digits, it is necessary and sufficient that the distance between any pair of codewords be at least*

*$k + 1$. For a code to correct all errors in $k$ or fewer digits, it is necessary and sufficient that the distance between any pair of codewords be at least $2k + 1$.*

The proof is omitted since its details are illustrated in Example 1.

### Problems

1. For $\alpha = 1010101$ and $\beta = 1100100$, find:
   (a) $\alpha + \beta$; (b) wgt $\alpha$; (c) wgt $\beta$; (d) $\text{wgt}(\alpha + \beta)$; (e) $d(\alpha, \beta)$.
2. Do the parts of Problem 1 for $\alpha = 00001111$ and $\beta = 11110101$.
3. If a codeword 1010101 is received as 1100100, what is the error pattern?
4. If a codeword 00001111 is received as 11110101, what is the error pattern?
5. (a) Find strings $\alpha, \beta, \gamma$ in $W_3$ such that $d(\alpha, \beta) = d(\beta, \gamma) = d(\gamma, \alpha) = 2$.
   (b) Do there exist four strings in $W_3$ such that the distance between any two of the four strings is at least 2?
6. What is the largest $m$ for which there exist $m$ strings $\alpha_1, \alpha_2, \ldots, \alpha_m$ in $W_4$ with $d(\alpha_i, \alpha_j) \geq 3$ for $i \neq j$?
7. Suppose that the only messages one wishes to convey are "yes," "no," and "not ready to answer." What is the smallest $n$ that allows one to transmit codewords in $W_n$ for these messages and have the code detect all mistransmissions having just one wrong digit?
8. What is the largest number of messages one can transmit with codewords from $W_3$ if the code is to correct all transmissions with one wrong digit?
9. Prove that $d(\alpha, \beta) = \text{wgt}(\alpha + \beta)$ and thus prove Theorem 2.
10. Let us define the ***product*** of strings $\alpha = a_1 a_2 \cdots a_n$ and $\beta = b_1 b_2 \cdots b_n$ to be the string $\alpha\beta = c_1 c_2 \cdots c_n$ with $c_i = 1$ when $a_i = 1 = b_i$, and $c_i = 0$ when either $a_i$ or $b_i$ is zero. Show that
$$\text{wgt}(\alpha + \beta) = \text{wgt}\,\alpha + \text{wgt}\,\beta - 2\,\text{wgt}(\alpha\beta).$$
11. Do the operations of addition as in Definition 1 and multiplication as in Problem 10 make $W_n$ into a commutative ring? Explain.

## 8.4 Codes Generated by Matrices

Many codes are described most easily in the language and notation of matrix theory. Below we use matrices over the ring $\mathbf{Z}_2 = \{0, 1\}$ to characterize the $(m, m + 1)$ block parity code, the $(m, 3m)$ block triple repetition code, and other block codes. First we abstract some material from Section 4.7.

We recall that the ***product*** $AB$ of matrices $A = (a_{ij})_{r \times s}$ and $B = (b_{ij})_{s \times t}$ is the matrix $C = (c_{ij})_{r \times t}$ with

$$c_{ij} = a_{i1}b_{1j} + a_{i2}b_{2j} + \cdots + a_{is}b_{sj};\ 1 \le i \le r,\ 1 \le j \le t.$$

The $m \times n$ ***zero matrix*** $0_{m \times n}$ has all entries equal to zero. The $m \times m$ ***identity matrix*** is the square matrix $I_m = (\delta_{ij})_{m \times m}$, where

$$\delta_{ij} = \begin{cases} 1 & \text{if } i = j, \text{ and} \\ 0 & \text{if } i \ne j. \end{cases}$$

The ***transpose*** of $A = (a_{ij})_{m \times n}$ is $A^T = (a'_{ij})_{n \times m}$, where $a'_{ij} = a_{ji}$.

A binary string $a_1a_2 \cdots a_m$ can be thought of as the $1 \times m$ matrix $\alpha = (a_1\ a_2 \cdots a_m)$. If $G = (g_{ij})_{m \times n}$, then the matrix product $\alpha G$ is a $1 \times n$ matrix $\beta = (b_1\ b_2 \cdots b_n)$. This means that one way to obtain an encoding function $E: W_m \to W_n$ is to find a suitable $m \times n$ matrix $G$ over $\mathbf{Z}_2$ and let $E(\alpha) = \alpha G$. We note that matrix multiplication modulo 2 is not difficult and is easily programmable.

***Definition 1 Matrix Code, Generator Matrix***

*An $(m, n)$ block code $[E, D]$ is an $m \times n$* **matrix code** *if there exists an $m \times n$ matrix $G$ over $\mathbf{Z}_2$ such that the codeword $E(\alpha)$ for a word $\alpha$ is the matrix product $\alpha G$; this $G$ is the* **generator matrix** *for the code.*

If $A = (a_{ij})_{m \times r}$, $B = (b_{ij})_{m \times s}$, and $C = (c_{ij})_{m \times t}$, let $(A{:}B{:}C)$ denote the $m \times (r + s + t)$ matrix $P = (p_{ij})$ with $p_{ij} = a_{ij}$ for $1 \le j \le r$, $p_{i,r+j} = b_{ij}$ for $1 \le j \le s$, and $p_{i,r+s+j} = c_{ij}$ for $1 \le j \le t$. Similarly, one can define $(A{:}B)$, $(A{:}B{:}C{:}D)$, and so on. For example,

$$(I_4 : I_4 : I_4) = \begin{pmatrix} 1 & 0 & 0 & 0 & 1 & 0 & 0 & 0 & 1 & 0 & 0 & 0 \\ 0 & 1 & 0 & 0 & 0 & 1 & 0 & 0 & 0 & 1 & 0 & 0 \\ 0 & 0 & 1 & 0 & 0 & 0 & 1 & 0 & 0 & 0 & 1 & 0 \\ 0 & 0 & 0 & 1 & 0 & 0 & 0 & 1 & 0 & 0 & 0 & 1 \end{pmatrix}.$$

Let $G$ be this $4 \times 12$ matrix and let $\alpha = (a_1\ a_2\ a_3\ a_4)$. Then the matrix product $\alpha G$ is easily seen to be

$$\beta = (a_1\quad a_2\quad a_3\quad a_4\quad a_1\quad a_2\quad a_3\quad a_4\quad a_1\quad a_2\quad a_3\quad a_4)$$

Thus $(I_4 : I_4 : I_4)$ is the generator matrix for the (4, 12) block triple repetition code. In general, the $(m, 3m)$ block triple repetition code has $(I_m : I_m : I_m)$ as its generator matrix.

Similarly, the $(m, m+1)$ block parity code is a matrix code. For example, the generator of the (3, 4) parity code is

$$G = \begin{pmatrix} 1 & 0 & 0 & 1 \\ 0 & 1 & 0 & 1 \\ 0 & 0 & 1 & 1 \end{pmatrix} = (I_3 : F), \text{ with } F = (1 \quad 1 \quad 1)^T,$$

since for this $G$ and $\alpha = (a_1 \; a_2 \; a_3)$, one has $\alpha G = (b_1 \; b_2 \; b_3 \; b_4)$, where $b_i = a_i$ for $1 \le i \le 3$ and $b_4 = a_1 + a_2 + a_3$ (with the addition performed in $\mathbf{Z}_2$, that is, modulo 2).

***Definition 2 Systematic Code***

*An $(m, n)$ block code $[E, D]$ is a* **systematic code** *if $E : a_1 a_2 \cdots a_m \mapsto b_1 b_2 \cdots b_n$ has $b_i = a_i$ for $i = 1, 2, \ldots, m$.*

In such a code, the first $m$ digits of the codeword $E(\alpha)$ are the digits of the word $\alpha$, and the extra $n - m$ digits serve to detect and correct errors. The digits $b_i$ with $1 \le i \le m$ are called ***information digits*** and the $b_i$ with $m < i \le n$ are ***check digits***.

***Theorem 1 Generator Matrix of a Systematic Code***

*An $(m, n)$ block matrix code is systematic if and only if its generator matrix is of the form $(I_m : F)$, with $F$ an $m \times (n - m)$ matrix.*

The proof is left to the reader as Problem 20 of this section.

***Notation***

*The symbol* $\begin{pmatrix} A \\ \because \\ B \end{pmatrix}$ *denotes* $(A^T : B^T)^T$.

*Example 1*

Let the generator matrix for a (3, 7) block code $[E, D]$ be

$$G = \begin{pmatrix} 1 & 0 & 0 & 0 & 1 & 1 & 1 \\ 0 & 1 & 0 & 1 & 0 & 1 & 1 \\ 0 & 0 & 1 & 1 & 1 & 0 & 1 \end{pmatrix} = (I_3 : F).$$

Then the word $\alpha = (a_1 \ a_2 \ a_3)$ has as its codeword $\alpha G = \beta = (b_1 \ b_2 \ b_3 \ b_4 \ b_5 \ b_6 \ b_7)$, where the rule for matrix multiplication leads to $b_i = a_i$ for $1 \le i \le 3$, $b_4 = a_2 + a_3$, $b_5 = a_1 + a_3$, $b_6 = a_1 + a_2$, and $b_7 = a_1 + a_2 + a_3$. In the last four of these equations, we replace the $a$'s with their equal $b$'s and get

$$b_4 = b_2 + b_3, b_5 = b_1 + b_3, b_6 = b_1 + b_2, \ b_7 = b_1 + b_2 + b_3.$$

Since each $b$ is its own negative in $\mathbf{Z_2}$, one can transpose (modulo 2) and thus obtain the system

$$\begin{array}{llllllll} & b_2 + b_3 + b_4 & & & & = 0 \\ b_1 & \quad + b_3 & & + b_5 & & = 0 \\ b_1 + b_2 & & & & + b_6 & = 0 \\ b_1 + b_2 + b_3 & & & & & + b_7 = 0 \end{array} \tag{1}$$

Using the rule for matrix multiplication, one sees that this system of four equations is equivalent to the single matrix equation

$$\beta H = 0_{1 \times 4} \tag{2}$$

where

$$H = \begin{pmatrix} 0 & 1 & 1 & 1 \\ 1 & 0 & 1 & 1 \\ 1 & 1 & 0 & 1 \\ 1 & 0 & 0 & 0 \\ 0 & 1 & 0 & 0 \\ 0 & 0 & 1 & 0 \\ 0 & 0 & 0 & 1 \end{pmatrix} = \begin{pmatrix} F \\ \cdot\cdot \\ I_4 \end{pmatrix}$$

is the transpose $(F^T : I_4)^T$ of the matrix of coefficients of the system (1). To illustrate the importance of the formula $\beta H = 0$ of (2), let $\alpha$ stand for the word 100 in $W_3$ or the equivalent matrix $(1 \ 0 \ 0)$. Then its codeword is $\beta = \alpha G = 1000111$. One finds that $\beta H = 0000$, as (2) promises for all codewords $\beta$. Now let $\beta$ be received with an error (due to noise) in the 3rd digit as $\gamma = 1010111$. This $\gamma$ can be written as $\beta + \delta$, with $\delta = 0010000$. Then

$$\gamma H = (\beta + \delta)H = \beta H + \delta H = 0 + \delta H = (0 \ \ 0 \ \ 1 \ \ 0 \ \ 0 \ \ 0 \ \ 0)H.$$

Using the rule for matrix multiplication, one can easily see that $\delta H$ is the 3rd row (1 1 0 1) of $H$. In general, if $\gamma$ is $\beta$ with only its $i$th digit changed, $\gamma H$ turns out to be the $i$th row of $H$. Since the seven rows of $H$ are all different from one another and no row of $H$ consists entirely of zeros, this enables us to correct all 1-digit errors and to detect some multiple errors. Decoding, with its error detecting and correcting aspects, is discussed below in a more general setting.

Example 1 motivates us to introduce some terminology.

***Definition 3 Check Matrix***

*Let $G = (I_m : F)$ be the generator matrix of a systematic code and $H = (F^T : I_{n-m})^T = \begin{pmatrix} F \\ \cdot\cdot \\ I_{n-m} \end{pmatrix}$. If no two rows of $H$ are identical and no row of $H$ has all of its entries zero, we say that $G$ is* **checkable** *and call $H$ the* **check matrix** *for $G$.*

The check matrix $H$, when it exists, is usually called the "parity check matrix." We drop the word "parity" to avoid confusion with "parity codes" (which are generally called "parity check codes").

We insist that $H$ have distinct rows so that we can tell which digit in $\gamma$ is wrong when a single error occurs. We do not allow $H$ to have a row of all zeroes so that $\gamma H$ with $\gamma$ having a single error should not be the same $0_{1\times(n-m)}$ as $\beta H$ with $\beta$ a codeword. One can easily choose the matrix $F$ in the generator $G = (I_m : F)$ so that $H = \begin{pmatrix} F \\ \cdot\cdot \\ I_{n-m} \end{pmatrix}$ has these properties and hence is the check matrix for $G$.

***Definition 4 Syndrome***

*If $H$ is the check matrix for the generator $G$ of an $(m, n)$ systematic code and $\gamma$ is a string in $W_n$, $\gamma H$ is the* **syndrome** *for $\gamma$.*

We now generalize on Example 1. Let the generator matrix for an $m \times n$ systematic code be a checkable matrix $G = (I_m : F)$, where $F = (f_{kj})_{m\times(n-m)}$. Then $H = \begin{pmatrix} F \\ \cdot\cdot \\ I_{n-m} \end{pmatrix}$ has the properties of Definition 3 and thus is the check matrix for $G$. Let $V_j$ be the $j$th column of $G$. Let $\alpha$ be the word $a_1a_2 \cdots a_m$ or its equivalent $1 \times m$ matrix $(a_1\ a_2\ \cdots\ a_m)$ and let $\alpha G = \beta = b_1b_2 \cdots b_n = (b_1\ b_2\ \cdots\ b_n)$.

Then the dot product $\alpha \cdot V_j = b_j$ for $1 \le j \le n$. Since the first $m$ columns of $G$ are from $I_m$, we see that $b_j = a_j$ for $1 \le j \le m$. Using the $V_j$ with $j > m$, we have

$$a_1 f_{1j} + a_2 f_{2j} + \cdots + a_m f_{mj} = b_{m+j}. \tag{3}$$

Replacing each $a_i$ with $b_i$ and transposing $b_{m+j}$ (modulo 2), one has

$$b_1 f_{1j} + b_2 f_{2j} + \cdots + b_m f_{mj} + b_{m+j} = 0 \qquad (1 \le j \le n - m). \tag{4}$$

The system of these $n - m$ equations is equivalent to the single matrix equation $\beta H = 0_{1 \times (n-m)}$, where $H = \begin{pmatrix} F \\ \cdot\cdot \\ I_{n-m} \end{pmatrix}$. Let $\delta_k$ be a string $d_1 d_2 \cdots d_n$ with $d_i = 0$ for all but the single choice $k$ for $i$. Then it follows from the rule for matrix multiplication that $\delta_k H$ is the $k$th row of $H$. The mistransmission of a codeword $\beta = b_1 b_2 \cdots b_n$ with only $b_k$ incorrect is the string $\gamma = \beta + \delta_k$. For this $\gamma$, the syndrome

$$\gamma H = (\beta + \delta_k)H = \beta H + \delta_k H = 0 + \delta_k H = \delta_k H$$

is thus the $k$th row of $H$. Since $H$ has the properties of Definition 3, we can use this fact to correct single errors and to detect some multiple errors, as we see next.

Let $\gamma = c_1 c_2 \cdots c_n$ be a general string in $W_n$. The ***syndrome decoding*** function $D$ is given by the following rules:

(a) If the syndrome $\gamma H$ for $\gamma$ is the zero matrix $0_{1 \times (n-m)}$ or is one of the last $n - m$ rows of $H$, $D(\gamma)$ is the string $c_1 c_2 \cdots c_m$ formed from the first $m$ digits of $\gamma$.

(b) If the syndrome $\gamma H$ equals the $i$th row of $H$ with $1 \le i \le m$, $D(\gamma)$ is $c_1 c_2 \cdots c_m$ with its $i$th digit changed.

(c) If $H$ is not as in (a) or (b), $D(\gamma) =$ "error."

Case (b) and part of case (a) indicate how single errors are corrected. Case (c) shows how some multiple errors are detected. In Problems 17 and 18 below, we will see that some multiple errors are neither detected nor corrected.

***Theorem 2 Zero Syndrome***

*Let $H$ be the check matrix for an $m \times n$ matrix code. Then the syndrome $\gamma H = 0_{1 \times (n-m)}$ if and only if $\gamma$ is a codeword.*

We have already proved that $\gamma H = 0$ when $\gamma$ is a codeword. The converse requires matrix theory beyond this text and is omitted.

### Problems

1. Write the generator matrix $G$ for the (4, 5) block parity code.
2. Write the generator matrix $G$ for the (2, 6) block triple repetition code.
3. Is the $(m, m + 1)$ block parity code a systematic code? Explain.
4. Is the $(m, 3m)$ triple repetition code systematic? Explain.
5. (a) Does the (2, 3) block parity code have a check matrix? Explain.
   (b) Does the (2, 6) block triple repetition code have a check matrix? Explain.
6. (a) Does the (3, 4) block parity code have a check matrix? Explain.
   (b) Does the (3, 9) block triple repetition code have a check matrix? Explain.
7. Let $Q_m$ be the $m \times m$ matrix $(q_{ij})$ with $q_{ij} = 0$ if $i = j$ and with $q_{ij} = 1$ if $i \neq j$. Does $G = (I_3 : Q_3)$ have a check matrix $H$? If so, find $H$.
8. Let $Q_m$ be as in Problem 7. Does $(I_4 : Q_4)$ have a check matrix $H$? If so, find $H$.
9. Find the generator matrix $G$ whose check matrix is

$$H = \begin{pmatrix} 1 & 1 & 0 & 0 \\ 1 & 0 & 1 & 1 \\ 0 & 1 & 1 & 0 \\ 0 & 1 & 0 & 1 \\ 1 & 0 & 0 & 0 \\ 0 & 1 & 0 & 0 \\ 0 & 0 & 1 & 0 \\ 0 & 0 & 0 & 1 \end{pmatrix}$$

10. Find the generator matrix $G$ whose check matrix is

$$H = \begin{pmatrix} 1 & 1 & 0 \\ 0 & 1 & 1 \\ 1 & 0 & 1 \\ 1 & 0 & 0 \\ 0 & 1 & 0 \\ 0 & 0 & 1 \end{pmatrix}.$$

11. For the code of Problem 9, decode each of the following strings after finding its syndrome:
    (a) 10000100; (b) 10000000; (c) 10010100; (d) 00000000.
12. For the code of Problem 10, decode each of the following strings after finding its syndrome:
    (a) 111000; (b) 011000; (c) 111100; (d) 000000.

13. What is the relationship between the number of digits in a syndrome and the number of check digits in a systematic code?

14. Use the fact that a checkable $m \times n$ matrix code has an $n \times (n - m)$ check matrix to find $m$ and $n$ so that the check matrix is $7 \times 3$.

15. Explain why:
    (a) a check matrix cannot be $n \times 2$ with $n > 3$.
    (b) a check matrix cannot be $n \times 3$ with $n > 7$.

16. (a) What is the largest $n$ for which there is an $n \times 4$ check matrix?
    (b) What is the largest $n$ for which there is an $n \times m$ check matrix? [Here $m$ is fixed.]

17. For the code of Problem 9, let $\alpha = (1\ 1\ 0\ 0)$ and its codeword $\alpha G$ be $\beta = (1\ 1\ 0\ 0\ 0\ 1\ 1\ 1)$. If noise changes $\beta$ into $\gamma = (1\ 1\ 1\ 1\ 1\ 1\ 1\ 1)$, would the decoding scheme detect or correct this triple error? Explain.

18. Construct an example of an undetectable error for the code of Problem 10.

19. Let $G = (I_m : F)$ be an $m \times n$ matrix over $\mathbf{Z}_2$ and $E: W_m \to W_n$ have the matrix product $\alpha G$ as $E(\alpha)$. Explain why $E$ is an injective mapping and hence is an encoding function.

20. Prove that an $(m, n)$ block matrix code is systematic if and only if its generator matrix $G$ is of the form $(I_m : F)$, with $F$ an $m \times (n - m)$ matrix (and thus prove Theorem 1).

21. Let $G$ be an $m \times n$ matrix over $\mathbf{Z}_2$. Show that the mapping $E: W_m \to W_n$ with $E(\alpha) = \alpha G$ is an additive group homomorphism.

22. Let $[E, D]$ be an $(m, n)$ block matrix code. Explain why $E$ is an additive group isomorphism from $W_m$ into $W_n$.

## 8.5 Group Codes and Coset Decoding

Here we use the concepts of subgroups and cosets to describe some efficient codes.

### Theorem 1 Matrix Encoding Isomorphism

*Let $E$ be the encoding function and $B$ be the set of codewords for an $m \times n$ matrix code. Then $E$ is an additive group isomorphism from $W_m$ into $W_n$, and the image set $B$ of the isomorphism is an additive subgroup in $W_n$.*

***Proof***

Let the generator be $G$. Using distributivity of matrix multiplication,

$$E(\alpha + \alpha') = (\alpha + \alpha')G = \alpha G + \alpha' G = E(\alpha) + E(\alpha').$$

Hence $E$ is a homomorphism. Since $E$ (as an encoding function) is injective, $E$ is an isomorphism from $W_m$ into $W_n$. The image set $B$ of $E$ (that is, the set of all codewords) is a subgroup in the codomain $W_n$ by Theorem 2(c) of Section 3.3.

We retain the notations of Theorem 1. Also let $2^m = k$ and $2^{n-m} = h$. Then there are $k$ words $\alpha_1, \ldots, \alpha_k$ in $W_m$ and so $k$ codewords $\beta_1, \ldots, \beta_k$ in $B$, with $\beta_i = E(\alpha_i)$ for $1 \leq i \leq k$. The additive subgroup $B$ has index $2^n/2^m = h$ in $W_n$. Hence, as in Lagrange's Theorem, there are $h$ cosets

$$\gamma_1 + B, \gamma_2 + B, \ldots, \gamma_h + B$$

of $B$ in $W_n$, and these cosets form a partition of the set $W_n$. In the representation $\gamma_i + B$ for one of these cosets, $\gamma_i$ could be any string in the coset. We choose the $\gamma_i$ in some specific way (to be described later). Then the $hk$ strings in $W_n$ are expressible uniquely in the form

$$\gamma_i + \beta_j; 1 \leq i \leq h, 1 \leq j \leq k.$$

A natural method of decoding is obtained by assuming that a received string $\gamma = \gamma_i + \beta_j$ in $W_n$ is the result of the mistransmission of the codeword $\beta_j$. In other words, we assume that $\gamma_i$ is the error pattern. This motivates us to define the decoding function $D$ so that $D(\gamma_i + \beta_j) = D(\beta_j)$. The original word $\alpha_j$ can then be found easily from a table of the encoding function as the $\alpha_j$ with $E(\alpha_j) = \beta_j = \alpha_j G$, where $G$ is the generator matrix. This decoding scheme is effective only if the $\gamma_i$ have small weights and hence the $\gamma_i$ are probable as error patterns.

***Definition 1 Coset Leaders, Coset Decoding***

*If strings $\gamma_1, \gamma_2, \ldots, \gamma_h$ exist in $W_n$ such that the cosets $\gamma_1 + B, \gamma_2 + B, \ldots, \gamma_h + B$ ($B$ being the set of codewords) partition $W_n$ and each $\gamma_i$ has minimal weight for the strings of its coset, $\gamma_i$ is the* **coset leader** *for any string in its coset. The* **coset decoding** *function $D$ then uses the following procedures to decode a string $\gamma$ in $W_n$:*

(a) *Find the coset leader $\gamma_i$ for $\gamma$.*

(b) *Subtract $\gamma_i$ from $\gamma$ to obtain the probable intended codeword $\beta$.*

(c) *Let $D(\gamma)$ be the word $\alpha$ such that $\alpha G = \beta$.*

***Example 1***

Let

$$G = \begin{pmatrix} 1 & 0 & 0 & 0 & 1 & 1 \\ 0 & 1 & 0 & 1 & 0 & 1 \\ 0 & 0 & 1 & 1 & 1 & 0 \end{pmatrix}, \qquad H = \begin{pmatrix} 0 & 1 & 1 \\ 1 & 0 & 1 \\ 1 & 1 & 0 \\ 1 & 0 & 0 \\ 0 & 1 & 0 \\ 0 & 0 & 1 \end{pmatrix}.$$

Then $G$ is the generator for a $3 \times 6$ matrix code and one finds that $H$ is its check matrix. The following table gives all the words $\alpha_j$ in $W_3$ and their codewords $\beta_j = \alpha_j G$ in $W_6$:

| $j$ | 1 | 2 | 3 | 4 | 5 | 6 | 7 | 8 |
|---|---|---|---|---|---|---|---|---|
| $\alpha_j$ | 000 | 001 | 010 | 011 | 100 | 101 | 110 | 111 |
| $\beta_j = \alpha_j G$ | 000000 | 001110 | 010101 | 011011 | 100011 | 101101 | 110110 | 111000 |

It follows from Theorem 1 that the set of codewords

$$B = \{000000, 001110, 010101, 011011, 100011, 101101, 110110, 111000\}$$

is an additive subgroup in $W_6$. $B$ has order $2^3 = 8$ and index $2^6/2^3 = 8$ in $W_6$. We next describe an array of the 64 strings of $W_6$ in 8 rows each having 8 entries.

Let $\gamma_1 = \beta_1 = 000000$. We place the codewords $\beta_1, \beta_2, \ldots, \beta_8$ of $B$ in the first row. We choose $\gamma_2$ as a string of least weight among the strings of $W_6$ which are not in $B$; to be specific, we let $\gamma_2 = 100000$. Then we place the strings $\gamma_2 + \beta_1, \gamma_2 + \beta_2, \ldots, \gamma_2 + \beta_8$ of the coset $\gamma_2 + B$ in the second row of the array. Next we let $\gamma_3 = 010000$, since that is one of the strings of least weight among the strings of $W_6$ which are not in $B$ nor in the coset $\gamma_2 + B$. We continue in this manner and obtain the following array (called Slepian's ***Standard Array*** for the code):

***Table 8.1 Standard Array***

Coset leaders $\gamma_i$ are in the first column; codewords $\beta_j$ are in the first row.

| | | | | | | | |
|---|---|---|---|---|---|---|---|
| 000000 | 100011 | 010101 | 001110 | 110110 | 101101 | 011011 | 111000 |
| 100000 | 000011 | 110101 | 101110 | 010110 | 001101 | 111011 | 011000 |
| 010000 | 110011 | 000101 | 011110 | 100110 | 111101 | 001011 | 101000 |
| 001000 | 101011 | 011101 | 000110 | 111110 | 100101 | 010011 | 110000 |
| 000100 | 100111 | 010001 | 001010 | 110010 | 101001 | 011111 | 111100 |
| 000010 | 100001 | 010111 | 001100 | 110100 | 101111 | 011001 | 111010 |
| 000001 | 100010 | 010100 | 001111 | 110111 | 101100 | 011010 | 111001 |
| 100100 | 000111 | 110001 | 101010 | 010010 | 001001 | 111111 | 011100 |

The elements in the leftmost column are the coset leaders $\gamma_i$. We note that in all rows but the last, the coset leader is the string in the coset (i.e., row) of least weight. In the last row there are three strings of minimal weight 2; one of these three was chosen arbitrarily as the coset leader.

A received string $\gamma$ is decoded by finding $\gamma$ in the table and decoding it as if it were the codeword $\beta_j$ at the top of the column containing $\gamma$. Thus 010001 would be assumed to be a mistransmission of the codeword 010101, and 000111 would be assumed to be a mistransmission of 100011; so the original message words would be assumed to be 010 and 100, respectively. (Since it is a systematic $3 \times 6$ matrix code, $D(\beta)$ is just the first three digits of $\beta$.)

It should be noted that a string $\gamma$ in the table is the sum $\gamma_i + \beta_j$ of its coset leader $\gamma_i$ and the codeword $\beta_j$ at the top of the column. In decoding, we assumed that $\gamma_i$, which equals $\gamma - \beta_j$, is the error pattern of the codeword $\beta_j$, and so we let $D(\gamma) = D(\beta_j)$.

**Theorem 2 Efficiency of Coset Decoding**

*Let $\beta$ and $\beta'$ be codewords and $\gamma$ be a coset leader. Then $d(\gamma + \beta, \beta) \leq d(\gamma + \beta, \beta')$.*

**Proof**

Since the set $B$ of codewords is an additive group, $\beta + \beta'$ is in $B$. Hence $\gamma$ and $\gamma + \beta + \beta'$ are in the same coset of $B$. By virtue of the method of selecting coset leaders, wgt $\gamma \leq \text{wgt}(\gamma + \beta + \beta')$. Then Theorem 2 of Section 8.3 implies, as desired, that

$$d(\gamma + \beta, \beta) = \text{wgt } \gamma \leq \text{wgt}(\gamma + \beta + \beta') = d(\gamma + \beta, \beta').$$

If $\gamma + \beta$ is a received string, with $\gamma$ a coset leader and $\beta$ a codeword, Theorem 2 tells us that $\beta$ is at least as likely to be the intended codeword as any other codeword $\beta'$.

**Theorem 3 Syndrome Decoding**

*Let $H$ be the check matrix for an $m \times n$ matrix code and $B$ be the subgroup of codewords in $W_n$. Then $\gamma$ and $\gamma'$ are in the same coset of $B$ if and only if they have equal syndromes.*

**Proof**

Let $\gamma$ and $\gamma'$ be strings in $W_n$. Then $\gamma$ and $\gamma'$ are in the same coset of $B$ if and only if $\gamma - \gamma'$ is in $B$. By Theorem 3 of Section 8.4, $\gamma - \gamma'$ is in $B$ if and only if its syndrome $(\gamma - \gamma')H = 0_{1 \times (n-m)}$. Finally, $(\gamma - \gamma')H = 0$ if and only if $\gamma H = \gamma' H$.

In the following example, we show how Theorem 3 helps us to decode using a simpler table than the standard array.

*Example 2*

Here we continue with the code of Example 1. To make it possible to decode without consulting the standard array, we introduce the table:

| coset leader $\gamma$ | 000000 | 100000 | 010000 | 001000 | 000100 | 000010 | 000001 | 100100 |
|---|---|---|---|---|---|---|---|---|
| syndrome $\gamma H$ | 000 | 011 | 101 | 110 | 100 | 010 | 001 | 111 |

The coset leaders were taken from the first column of the standard array and their syndromes calculated easily by matrix multiplication. To find $D(101100)$—that is, to decode the string $\gamma = 101100$—we calculate its syndrome $\gamma H$ and find it to be 001. Since the table of this example shows that 001 is the syndrome of 000001, Theorem 3 tells us that 101100 and 000001 are in the same coset of $B$. Hence 000001 is the coset leader for the coset containing $\gamma$. In coset decoding, we assume that this coset leader is the error pattern and thus that the intended codeword is $101100 + 000001 = 101101$. Looking for the codeword $\beta = 101101$ in the small table of codewords in Example 1, we see that $\beta$ is the codeword for the message $\alpha = 101$. (Since the code is a systematic $3 \times 6$ matrix code, $\alpha$ is found most easily as the first 3 digits of $\beta$.)

## Problems

1. For the code of Examples 1 and 2, find $D(110100)$ using the standard array. Check using the syndrome decoding of Example 2.
2. Do as in Problem 1 for $D(000101)$.
3. Let $H = \begin{pmatrix} 1 & 1 \\ 1 & 0 \\ 0 & 1 \end{pmatrix}$ be the check matrix for a systematic code generated by an $m \times n$ matrix $G$. Give the following:
   (a) $m$ and $n$;
   (b) the matrix $G$;
   (c) $h = 2^{n-m}$ and $k = 2^m$;
   (d) the set $B$ of the $k$ codewords;
   (e) a set of $h$ coset leaders;
   (f) the syndrome of each coset leader;
   (g) $D(111)$ using coset decoding;
   (h) $D(101)$ using the syndrome decoding method of Section 8.4.

4. Do analogues of Parts (a) through (f) of Problem 3 using

$$H = \begin{pmatrix} 1 & 1 & 0 \\ 0 & 1 & 1 \\ 1 & 0 & 0 \\ 0 & 1 & 0 \\ 0 & 0 & 1 \end{pmatrix}$$

as the check matrix.
   (g) Find $D(11101)$ using coset decoding.
   (h) Find $D(11111)$ using the syndrome decoding method of Section 8.4.
5. Let $B$ be the set of codewords for an $m \times n$ matrix code. Prove that either all the codewords in $B$ have even weight or half have even weight and half have odd weight.
6. Let $B$ be the additive group of codewords in Example 1.
   (a) Give the orders of the elements of $B$. (Order is as in Definition 3 of Section 2.7.)
   (b) Tabulate an isomorphism from $B$ onto the direct product $\mathbf{Z}_2 \times \mathbf{Z}_2 \times \mathbf{Z}_2$ of three cyclic groups $\{0, 1\}$ of order 2.

## 8.6 Hamming Codes

First we introduce a measure of the cost (in transmission effort per digit of the message) of a block code.

### Definition 1 Rate of a Block Code

*The* **rate** *of an* $(m, n)$ *block code is the ratio* $n/m$.

We see that the least expensive code (with certain desirable error detecting and correcting properties) is the one whose rate is closest to 1.

Now we let $H$ be a check matrix with $t$ columns, and we seek the number $n$ of rows in $H$ so that the code will have minimal rate. One sees easily that the generator $G$ is an $(n - t) \times n$ matrix and hence the rate is $n/(n - t)$. For fixed $t$, this is least when $n$ is as large as possible. The number of distinct $1 \times t$ rows with entries in $\{0, 1\}$ is $2^t$. Since a check matrix cannot have repeated rows or have a row of all zeros, the largest $n$ is $2^t - 1$. For this $n$, there are $n - t = m = 2^t - 1 - t$ message digits and $t$ check digits.

***Definition 2 Hamming Code***

*A* **Hamming Code** *for a given* $t > 1$ *is a* $(2^t - 1 - t,\ 2^t - 1)$ *block code that corrects* 1*-digit errors.*

We note that $(2^t - 1 - t, 2^t - 1)$ for $t = 2, 3, 4, \ldots$ is (1, 3), (4, 7), (11, 15), and so on. Let $H_2$ and $H_3$ be the transposes of the matrices.

$$H_2^T = \begin{pmatrix} 1 & 1 & 0 \\ 1 & 0 & 1 \end{pmatrix}, \qquad H_3^T = \begin{pmatrix} 1 & 1 & 1 & 0 & 1 & 0 & 0 \\ 1 & 1 & 0 & 1 & 0 & 1 & 0 \\ 1 & 0 & 1 & 1 & 0 & 0 & 1 \end{pmatrix}. \tag{1}$$

Then $H_2$ and $H_3$ are check matrices for (1, 3) and (4, 7) block codes, respectively. Since both correct 1-digit errors using syndrome decoding, they are Hamming Codes. There are Hamming Codes with analogous check matrices $H_t$ for all $t > 1$. For a particular $t$, the Hamming Code check matrix is unique except for permutation of its rows, and such permutation does not change the nature of the code.

If an $(m, n)$ block code corrects all error patterns of weight $g$ or less and has minimal rate among such codes, it is called a ***perfect*** $g$-error correcting code. The Hamming Codes, developed by R. W. Hamming in 1950, are perfect 1-error correcting codes. There are more sophisticated perfect codes that are multiple error correcting. Such codes were created independently by A. Hoquenghem in 1959 and by R. C. Bose and D. V. Ray-Chanduri in 1960; they are called ***BCH codes***.

### *Problems*

1. Find the generator matrix $G$ for the (4, 7) Hamming Code, that is, the $G$ with the $H_3$ of (1) as its check matrix.
2. Find $G$ for the (1, 3) Hamming Code, with the $H_2$ of (1) as check matrix.
3. Using the (4, 7) Hamming Code, encode:
   (a) 1001; (b) 0110; (c) 1000; (d) 0111.
4. Using the (4, 7) Hamming Code, encode:
   (a) 0101; (b) 0100; (c) 1110; (d) 1111.
5. Construct a standard array for the (4, 7) Hamming Code.
6. Construct a standard array for the (1, 3) Hamming Code.
7. Construct a coset leader–syndrome table for the (4, 7) Hamming Code.
8. Construct a coset leader–syndrome table for the (1, 3) Hamming Code.

9. Decode the following, using the array of Problem 5 and checking with the table of Problem 7:
   (a) 1101010; (b) 1111000; (c) 1100011.
10. Do as in Problem 9 for the following:
   (a) 0001110; (b) 1111110; (c) 1000000.

## 8.7 Modular Codes, Trapdoor Functions

Now we consider codes designed to maintain privacy of the messages rather than to combat transmission errors. Here elements of the multiplicative group $V_m$ of invertibles in $\mathbf{Z_m} = \{\overline{0}, \overline{1}, \ldots, \overline{m-1}\}$, for a suitable $m$, are used for message words and codewords instead of the binary strings of previous sections.

The modulus $m$ is chosen to be a product $pq$ of distinct positive primes (which should be very large in real life encryption). For $m = pq$, let $f = \phi(m) = (p-1)(q-1)$, where $\phi$ is the Euler $\phi$-function. Let $e$ be in $\mathbf{Z}^+$ and be relatively prime to $f$. Then there exist $d$ and $g$ in $\mathbf{Z}$ with $ed - fg = 1$. [See Corollary (b) to Theorem 2 of Section 1.4.] Thus $ed = fg + 1$ and for $\bar{x}$ in $V_m$ one has

$$\overline{x}^{ed} = \overline{x}^{fg+1} = (\overline{x}^f)^g\overline{x} = \overline{1}^g\overline{x} = \overline{x}$$

since $\overline{x}^f = \overline{x}^{\phi(m)} = \overline{1}$ in $V_m$ by Euler's Theorem (Theorem 1 of Section 4.3). The equation $\overline{x}^{ed} = \overline{x}$ for all $\overline{x}$ in $V_m$ means that the mappings

$$E: \overline{x} \mapsto \overline{x}^e, \qquad D: \overline{x} \mapsto \overline{x}^d, \tag{1}$$

from $V_m$ to itself, are inverses of each other. Hence these mappings are bijections. In particular, $E$ is injective and so is available as an encoding function.

### *Definition 1 Modular Code*

*The ordered pair $[E, D]$, with $E$ and $D$ as in* (1), *is the* **modular code** *with* **modulus** *$m$,* **encoding exponent** *$e$,* **decoding exponent** *$d$, and* **modular factors** *$p$ and $q$.*

Such codes are important in encryption for the following reasons. If the modular prime factors $p$ and $q$ are large enough (say, each with about 100 digits), we can publish the values of the modulus $m$ and the encoding exponent $e$ and thus make it possible for anyone to send us a coded message while keeping $p$, $q$, and $d$ secret so that only authorized persons can decode. This is based on the practical impossibility of factoring a very large number $m$ even when one knows that it is the product of two primes. To create one's own modular code, one needs two large primes, but modern electronic computers can obtain sufficient numbers of such primes.

We now illustrate the principles of modular codes using a case with fairly small modular factors; this code could be "cracked" rather easily.

***Example 1***

We announce that anyone wishing to send us a message should use the elements $\overline{2}, \overline{3}, \ldots, \overline{27}$ of the group $V_{899}$ for the letters $a, b, \ldots, z$ and that the message should be encoded by replacing each element $\overline{x}$ by $\overline{x}^{11}$. Thus the letter $O$ would be represented by the word $\overline{16}$ and encoded as $\overline{16}^{11}$. We use the fact that $\overline{900} = \overline{1}$ in $V_{899}$ to calculate $\overline{16}^{11}$ as follows:

$$\overline{16}^2 = \overline{256},\ \overline{16}^4 = \overline{256}^2 = \overline{65536} = \overline{72 \cdot 900 + 736} = \overline{72 + 736} = \overline{808},$$

$$\overline{16}^8 = \overline{808}^2 = \overline{652864} = \overline{725 \cdot 900 + 364} = \overline{725 + 364} = \overline{1089} = \overline{190},$$

$$\overline{16}^{10} = \overline{16}^8 \cdot \overline{16}^2 = \overline{190 \cdot 256} = \overline{48640} = \overline{54 \cdot 900 + 40} = \overline{54 + 40} = \overline{94},$$

$$\overline{16}^{11} = \overline{16}^{10} \cdot \overline{16} = \overline{94 \cdot 16} = \overline{1504} = \overline{900 + 604} = \overline{605}.$$

This tells us that the codeword to be sent is $\overline{605}$.

We as receivers of the codeword would know that the modular factors are 29 and 31. Hence, 2, 3, ..., 27 are all relatively prime to $m = 29 \cdot 31$ and the "words" $\overline{2}, \overline{3}, \ldots, \overline{27}$ are all in $V_m$. We would also know that $\phi(m) = 28 \cdot 30 = 840 = f$, and we could use the euclidean algorithm to obtain the equation $11 \cdot 611 - 840 \cdot 8 = 1$. This would tell us that $d = 611$ and that the decoding function is $D\colon \overline{x} \mapsto \overline{x}^{611}$.

Even for the relatively small $p$, $q$, and $e$ of Example 1, encoding and decoding are tedious. For very large $p$ and $q$, these calculations are still practical with computers; but factoring $m$ and thus being able to find the decoding function are not possible in any reasonable amount of time with present-day techniques and computers. As computing equipment gets faster, one can merely increase the size of $p$ and $q$ to maintain the qualities of the code.

An encoding function $E$, that can be described exactly while keeping its inverse $D$ secret, is called a ***trapdoor function***. The first practical construction of trapdoor functions was by R. L. Rivest, L. Shamir, and L. M. Adelman in their paper "A method for obtaining digital signatures and public-key cryptosystems," ***Communications of the Association for Computing Machinery***, 21 (1978), 120–126.

The application of this section depends on the lack of a practical computer algorithm for factoring a number $m$ that is known to be the product of two large primes. If such an algorithm were developed, the modular codes would cease to have their desirable "trapdoor" property and would have to be replaced by codes using other trapdoor functions. Information on such codes is given by M. E. Hellman in the paper "An overview of public key cryptography" *IEEE Communications Society Magazine* 16 (1978), 24–32. It is very probable that algebra and number theory will continue to be important in such applications.

### Problems

1. Use the encoding function $E: x \mapsto x^3$ to encode the following elements of $V_{15}$: (a) $\bar{2}$; (b) $\bar{4}$; (c) $\bar{7}$.
2. Use the encoding function $E: x \mapsto x^5$ to encode the following elements of $V_{21}$: (a) $\bar{2}$; (b) $\bar{4}$; (c) $\bar{5}$.
3. Let $E: \bar{x} \mapsto \bar{x}^{13}$ be the encoding function on $V_{77}$. Find the decoding exponent $d$ for the decoding function $D: \bar{x} \mapsto \bar{x}^d$.
4. Let $E: \bar{x} \mapsto \bar{x}^7$ be the encoding function on $V_{143}$. Find the decoding exponent $d$.
5. Which numbers in $\{2, 3, \ldots, 76\}$ are possible as encoding exponents $e$ or decoding exponents $d$ for a modular code using the elements of $V_{77}$?
6. How many numbers in $\{2, 3, \ldots, 142\}$ are possible as encoding exponents $e$ for a modular code on $V_{143}$?
7. In Example 1, five multiplications (most of them followed by reductions modulo $m$) were required to calculate $\bar{x}^{11}$ in $V_{899}$. How many multiplications are needed to find $\bar{x}^{101}$ in $V_{1133}$?
8. How many multiplications are needed to find $\bar{x}^{10247}$ in $V_{643063}$?

*9. Crack the code on $V_{1133}$ with $E(\bar{x}) = \bar{x}^{101}$; that is, find $D$.

*10. Crack the code on $V_{643063}$ with $E(\bar{x}) = \bar{x}^{10247}$.

11. If one knew that the prime factors $p$ and $q$ of 643063 are greater than 700, could one use an element of $V_{643063}$ to represent an ordered pair of letters in a modular code? Explain.
12. How large would the modular factors $p$ and $q$ have to be to allow elements $\bar{2}, \bar{3}, \ldots, \bar{r}$ of $V_m$ to represent ordered triples of letters in a modular code?

### Review Problems

1. In the (5, 6) block parity code, suppose that the received string is 011001. Assuming a 1-digit error, what are all the possibilities for the message word in $W_5$?
2. Do as in Problem 1 for the (4, 5) parity code and received string 01000.

3. For the (2, 6) triple repetition code, find:
   (a) $E(01)$; (b) $D(110110)$; (c) $D(100010)$.
4. For the (3, 9) triple repetition code find:
   (a) $E(101)$; (b) $D(011010110)$; (c) $D(111001011)$.
5. What minimum distance between codewords is required to be able to detect all error patterns of weight 4 or less?
6. Let the minimum distance between codewords be 5. For what weights of error patterns would the errors be detected?
7. Do Problem 5 with "detect" changed to "correct."
8. Do Problem 6 with "detect" changed to "correct."
9. How many strings are there in $W_9$?
10. How many strings are there in $W_{10}$?
11. Use the standard array (Table 8.1) of Example 1 in Section 8.5 to decode 110000.
12. Do as in Problem 11 to decode 011010.
13. Use the coset leader–syndrome table in Example 2 of Section 8.5 to decode 010111.
14. Do as in Problem 13 to decode 001101.
15. If the check matrix for a Hamming Code has 31 rows, how many binary digits are there in a message word? In a codeword?
16. Do Problem 15 with 31 replaced by 63.
17. Write a check matrix for a (11, 15) Hamming Code.
18. Find the generator matrix for the code of Problem 17.
19. Why would a modular code on $V_{35}$ be easy to crack?
20. Crack the modular code $E: \bar{x} \mapsto \bar{x}^5$ on $V_{221}$, that is, find $D$.

### *Supplementary and Challenging Problems*

1. If one permutes the rows of a check matrix $H$, what does that do to the digits of a syndrome $\gamma H$?
2. If the columns of the generator matrix $G$ for a systematic code are permuted to form a matrix $G'$ that is no longer of the form $(I_m : F_{n-m})$, could one still find a "check matrix" $H'$ for which "syndromes" $\gamma H'$ would enable us to correct 1-digit errors? Explain.
3. Let $H$ be the transpose of the matrix

$$H^T = \begin{pmatrix} 1 & 0 & 1 & 0 & 1 & 0 & 1 \\ 0 & 1 & 1 & 0 & 0 & 1 & 1 \\ 0 & 0 & 0 & 1 & 1 & 1 & 1 \end{pmatrix}.$$

   Find a generator $G$ for a matrix code such that $\beta H = 0_{1 \times 3}$ for all codewords $\beta$.

4. Let $H$ be the check matrix for an $m \times n$ generator $G$. Prove that $GH = 0_{m \times (n-m)}$.
5. Let $r$ and $s$ be in $\mathbf{Z}^+$ with $2^{r-1} \leq s < 2^r$. Show that $x^s$ can be calculated with no more than $2(r-1)$ multiplications.
6. Describe a practical method for computer search for a prime with 50 digits.
7. In $\mathbf{Z}_2[x]$ let $\pi$ be a prime polynomial of degree $d$ and $\gamma$ be relatively prime to $\pi$. Let $\mathrm{E}\colon a_1a_2 \cdots a_d \mapsto b_1b_2 \cdots b_d$ be the mapping from $W_d$ to itself with

$$(a_1 + a_2 x + \cdots + a_d x^{d-1}) - (b_1 + b_2 x + \cdots + b_d x^{d-1})$$

in the principal ideal $(\pi)$. Is E bijective? If so, describe its inverse function.
8. Is Problem 7 the basis for a code? If so, is it error detecting or does it serve for encryption?

# *Annotated Bibliography*

### *General*

Albert, A. Adrian, ed. *Studies in Modern Algebra* (MAA Studies in Mathematics, vol. 2). Buffalo, N.Y.: The Mathematical Association of America; Englewood Cliffs, N.J.: Prentice-Hall, 1963.

A collection of essays on recent advances in algebra by MacLane, Bruck, C. W. Curtis, Kleinfeld, and Paige.

Aleksandrov, A. D., A. N. Kolmogorov, and M. A. Lavrent'ev, eds. *Mathematics: Its Content, Methods, and Meaning*, 2nd edition (3 volumes). Cambridge, Mass.: MIT Press, 1969. (Paperback available.)

An excellent general view of mathematics by a group of Russian mathematicians. The section on groups is good for its discussion of symmetry and the applications of group theory to other parts of mathematics, for example, topology.

Bourbaki, N. *Éléments de mathématiques.* Paris: Hermann, 1942–1958.

Although the *Éléments* covers a wide range of mathematical ideas, the material on algebra has been particularly important in influencing terminology and notation. For experienced readers.

Courant, Richard, and Herbert Robbins. *What Is Mathematics?* Oxford: Oxford University Press, 1941.

A beautifully written, lucid account of some of the most interesting and important problems of mathematics, designed for the reader without much mathematical background. Excellent sections on number theory and on constructibility.

Hardy, G. H. *Ramanujan: Twelve Lectures on Subjects Suggested by His Life and Work.* Cambridge: University Press, 1940.

A great twentieth century mathematician develops some topics associated with his remarkable collaborator.

Herstein, I. N. *Topics in Algebra*, 2nd edition. Lexington, Mass.: Xerox College Publishing, 1975.

A popular intermediate level text in abstract algebra.

Jacobson, Nathan. *Basic Algebra I.* San Francisco: W. H. Freeman, 1974.

An important text that will strike some readers as something beyond the basic level.

Jacobson, Nathan. *Lectures in Abstract Algebra* (3 volumes). New York: Springer-Verlag, 1976, 1975, 1976.

A classic treatment of abstract algebra written at a fairly sophisticated level.

Lange, Serge. *Algebra.* Reading, Mass.: Addison-Wesley, 1965.
A widely used graduate level text. Very terse.

MacLane, Saunders, and Garrett Birkhoff. *Algebra.* New York: Macmillan, 1967.
A modern successor to the authors' early and pace-setting classic *A Survey of Modern Algebra.* Fairly sophisticated. It uses very up-to-date notation and terminology.

van der Waerden, B. L. *Modern Algebra* (2 volumes). New York: Ungar, 1949, 1950.
Originally published in German, this great classic was one of the first texts in abstract algebra and was used to introduce the subject into graduate curricula in the United States in the 1930's.


## *Group Theory*


Aschbacher, Michael. "The Classification of the Finite Simple Groups," *Mathematical Intelligencer, 3*: 2 (1981), 59–64.
An expository article on the history of the classification theorem for finite simple groups.

Gorenstein, Daniel. *Finite Groups.* New York: Harper & Row, 1968.
An advanced treatise on group theory.

Hall, Marshall, Jr. *The Theory of Groups.* New York: Macmillan, 1959.
An advanced, modern treatment of group theory. A good reference.

Rotman, Joseph. *Theory of Groups,* 2nd edition. Boston: Allyn & Bacon, 1973.
A standard graduate text in group theory.

Schattschneider, Doris. "The Plane Symmetry Groups: Their Recognition and Notation." *American Mathematical Monthly 85* (1978), 439–450.
A charming article on transformations that leave invariant designs or patterns in the plane. It includes many patterns and a discussion of connections with the art work of M. C. Escher.

Thompson, J. G., and W. Feit. "Solvability of Groups of Odd Order." *Pacific Journal of Mathematics, 13* (1963), 775–1029.
This is the now-famous paper containing the results mentioned in the biographical note on Burnside at the close of Section 2.13.


## *Rings*


Burton, David M. *A First Course in Rings and Ideals.* Reading, Mass.: Addison-Wesley, 1970.
An advanced undergraduate or beginning graduate text in rings and ideals, containing a great deal of material on the subject beyond the present text.

Divinsky, N. J. *Rings and Radicals*. Toronto: University of Toronto Press, 1965.
A graduate level treatment of rings, including fairly recent developments with an emphasis on the theory of radicals (Jacobson, Brown-McCoy, Levitski, and others).

McCoy, Neal H. *Rings and Ideals* (Carus Monograph No. 8). Washington, D.C.: Mathematical Association of America, 1948.
A classic work on rings in the well-known series of monographs written for the college-level audience.

McCoy, Neal H. *The Theory of Rings*. New York: Macmillan, 1964. (Paperback.)
A text for students with basic knowledge of abstract algebra. It contains some results which are comparatively recent in the development of the subject.

### Fields

Artin, Emil. *Galois Theory* (Notre Dame Mathematical Lecture No. 2), 2nd edition. Notre Dame, Ind.: University of Notre Dame Press, 1944.
A classic view of Galois theory by a great teacher and mathematician.

Lieber, Lillian R. *Galois and the Theory of Groups: A Bright Star in Mathesis*. Brooklyn, N.Y.: The Galois Institute of Mathematics and Art, 1956. (Presently out of print.)
An unconventional free-verse treatment of Galois theory which conveys the principal ideas of the subject, if not the details. Amusing illustrations are provided by the author's husband.

Pollard, Harry, and H. G. Diamond. *The Theory of Algebraic Numbers* (Carus Monograph No. 9). Washington, D.C.: The Mathematical Association of America, 1975.
An elementary treatment of algebraic number theory in the well known Carus series of books aimed at undergraduates.

### Theory of Numbers

Grosswald, Emil. *Topics from the Theory of Numbers*. New York: Macmillan, 1966.
Many of the usual topics of elementary number theory as well as an introduction to algebraic number theory and analytic number theory. Excellent treatment of the theory of partitions. Some work in abstract algebra included. Unusual for its treatment of many topics not included in other elementary texts.

Hardy, G. H., and E. M. Wright. *An Introduction to the Theory of Numbers*, 4th edition. Oxford: Clarendon Press, 1960.
The greatest classic of them all—at least in the theory of numbers and in the English language. The footnotes are almost as exciting as the text and the text is superb.

Khinchin, Aleksander Y. *Three Pearls of Number Theory.* Rochester, N.Y.: Graylock Press, 1952.

An elegant treatment of three proofs, by youthful mathematicians, of conjectures in number theory that resisted the efforts of mature mathematicians for some time.

LeVeque, W. J. *Studies in Number Theory* (MAA Studies in Mathematics, vol. 6). Washington, D.C.: The Mathematical Association of America; Englewood Cliffs, N.J.: Prentice-Hall, 1969.

A collection of essays on developments in number theory by well-known mathematicians: LeVeque, Lewis, J. Robinson, D. H. Lehmer, P. T. Bateman, and H. G. Diamond.

LeVeque, W. J. *Topics in Number Theory* (2 volumes). Reading, Mass.: Addison-Wesley, 1956.

A collection of topics in number theory, some not often found in easily accessible sources.

Niven, Ivan, and Herbert S. Zuckerman. *An Introduction to the Theory of Numbers*, 4th edition. New York: Wiley, 1980.

An important text with some unusual topics and many challenging, interesting problems.

Rademacher, Hans. *Lectures on Elementary Number Theory.* New York: Blaisdell, 1964. (Presently out of print.)

A beautifully written collection of classic theorems in number theory with comparatively straightforward proofs. A good reference for solutions of some of the famous and difficult problems of number theory.

Sierpinski, Waclaw. *250 Problems in Elementary Number Theory.* New York: American Elsevier, 1970.

A beautiful set of problems by a great mathematician. Some of the problems have been solved only in recent years.

Uspensky, J. V., and M. A. Heaslet. *Elementary Number Theory.* New York: McGraw-Hill, 1939.

This remains, in spite of its age, a standard work in number theory. It contains many derivations which show the real reasons why various formulas work rather than clever tricks which work but do not aid one's intuition. Many nontrivial problems.

## *Theory of Equations*

Uspensky, J. V. *Theory of Equations.* New York: McGraw-Hill, 1948. (Available in paperback.)

A very good text on the classical topics of the theory of equations, including methods for solving the cubic and quartic equations and methods of calculating or approximating roots of higher degree equations. A discussion of symmetric functions.

## *History of Mathematics*

Bell, Eric Temple. *The Development of Mathematics*, 2nd edition. New York: McGraw-Hill, 1945.

More or less chronological accounts of the growth and development of important areas of mathematics by one of the most entertaining and popular of mathematical writers.

Bell, Eric Temple. *Men of Mathematics*. New York: Simon and Schuster, 1961. (Available in paperback.)

An entertaining, anecdotal account of the lives of prominent mathematicians. For a popular treatment of the lives of Galois, Abel, and others this is highly recommended.

Boyer, Carl. *A History of Mathematics*. New York: Wiley, 1968.

A recent history, one of the few popular sources to cover results (in a limited manner) of the twentieth century.

Cajori, Florian. *A History of Mathematics*, 2nd edition. New York: Macmillan, 1919. (Presently out of print.)

An excellent, detailed history of mathematics by a recognized authority. It is, unfortunately, quite old; for twentieth century mathematics and twentieth century discoveries concerning earlier results, one must go elsewhere. Nevertheless, it remains a very valuable source on the subject.

Dickson, Leonard Eugene. *History of the Theory of Numbers* (Vols. I, II, and III). New York: Chelsea, 1952.

A comprehensive history of the literature on the Theory of Numbers up to approximately 1920.

Eves, Howard. *An Introduction to the History of Mathematics*, rev. ed. New York: Holt, Rinehart and Winston, 1964.

A popular, well written history, widely used as a textbook.

Eves, Howard. *In Mathematical Circles* (Vols. I and II). *Mathematical Circles Revisited; Mathematical Circles Squared*. Boston: Prindle, Weber & Schmidt, 1969, 1971, 1972.

A very entertaining collection of anecdotes and historical accounts of mathematics and mathematicians. Delightful reading for the layman as well as the professional.

Gleason, A. M., R. E. Greenwood, and L. M. Kelly. *The William Lowell Putnam Mathematical Competition, Problems and Solutions: 1938–1964*. Washington, D.C.: The Mathematical Association of America, 1980.

The problems of the first 25 competitions with solutions and references to related research in mathematics.

Infeld, Leopold. *Whom the Gods Love: The Story of Évariste Galois*. Reston, Va.: National Council of Teachers of Mathematics, 1978.

A biography of Galois by one of his modern admirers.

Kline, Morris. *Mathematical Thought from Ancient to Modern Times.* New York: Oxford University Press, 1972.

A detailed, up-to-date history of mathematics. If one is to own one history of mathematics, this is it.

Kramer, Edna. *The Nature and Growth of Modern Mathematics.* New York: Hawthorn, 1970.

A general book on the development of mathematical ideas. Excellent for biographical data on some twentieth century mathematicians. The accounts of the growth of algebra are easy to read and interesting. Also of interest is the chapter on Klein and his *Erlanger Programm.*

Newman, James R. *World of Mathematics* (4 volumes). New York: Simon and Schuster, 1956. (Available in paperback.)

A popular collection of excerpts on mathematics from many sources. A good set for reference as well as reading for pleasure.

Nový, Luboš. *Origins of Modern Algebra.* Leyden: Noordhof International Publishing, 1973.

This is a detailed history of abstract algebra between 1770, the year of Lagrange's important work, and 1870, the end of a very active period in the development of algebra in England.

Ore, Oystein. *Niels Henrik Abel, Mathematician Extraordinary.* Minneapolis, Minn.: University of Minnesota Press, 1957. (Presently out of print.)

A biography of Abel by a leading twentieth century mathematician.

Ore, Oystein. *Number Theory and Its History.* New York: McGraw-Hill, 1948.

An excellent, easy-to-read book on number theory with historical accounts and portraits of mathematicians.

Smith, David E. *A Source Book in Mathematics* (2 volumes). New York: Dover, 1959. (Paperback.)

Here one can find the original papers on the cubic, quartic, and quintic equations by Cardano, Ferrari (Cardano), Abel, and Galois, translated into English.

Struik, Dirk J. *A Concise History of Mathematics.* New York: Dover, 1948. (Paperback.)

A good introduction to the history of mathematics, but very concise. There is nothing on twentieth century developments.

Struik, Dirk J., ed. *A Source Book in Mathematics, 1200–1800.* Cambridge, Mass.: Harvard University Press, 1969.

A collection of original papers by great mathematicians up to 1800 along with comments and background by the editor. Much of abstract algebra is missing here because of the 1800 terminal date. However, it is a good source if one wishes to see excerpts from the original publications on the solution of equations by Cardano and Lagrange, for example.

Wussing, Hans. *Die Genesis des abstrakten Gruppenbegriffes.* Berlin: VEB Deutscher Verlag der Wissenschaften, 1969.

An excellent description of the work of Lagrange, Ruffini, Cauchy, Galois, Kronecker, Jordan, Klein, Netto, Hölder, and others in group theory and the subjects in which group theory grew. In German.

### *Applications*

Beckenbach, Edwin F., ed. *Applied Combinatorial Mathematics.* New York: Wiley, 1964.

A collection of papers of which one is especially interesting in the present context: "Polya's Theory of Counting," by N. G. de Bruijn. Group theory is used to develop a fundamental theorem in combinatorial mathematics.

Birkhoff, Garrett, and Thomas C. Bartee. *Modern Applied Algebra.* New York: McGraw-Hill, 1970.

Applications of abstract algebra to problems of computer design, programming languages, coding and decoding, mathematical linguistics, radar.

Cornwell, J. F. *Group Theory and Electronic Energy Bonds in Solids.* Amsterdam: North-Holland, 1969.

Basic group theory with matrix representations. Group theory in quantum mechanics and other branches of physics.

Cotton, F. Albert. *Chemical Applications of Group Theory.* New York: Interscience, 1963.

A book written by a chemist attempting to make clear to chemists applications of group theory to that subject, in particular symmetry groups and their relation to problems of molecular symmetry, representation of groups, along with other topics. The knowledge of chemistry required may discourage mathematicians.

Dornhoff, Larry L., and Franz E. Hohn. *Applied Modern Algebra.* New York: Macmillan, 1978.

An accessible treatment of a wide range of algebraic topics with applications, including Boolean algebra and coding theory. Some topics are not strictly algebraic. For example, there are discussions of some topics from graph theory (including the four-color problem and the traveling salesperson problem) and from combinatorics (the Polya enumeration theorem).

Dyson, Freeman J. "Mathematics in the Physical Sciences." In *The Mathematical Sciences. A Collection of Essays.* Cambridge, Mass.: MIT Press, 1969.

An easy-to-read essay on the uses of group theory in the discovery of elementary particles in physics.

Dyson, Freeman J. *Symmetry Groups in Nuclear and Particle Physics.* New York: W. A. Benjamin, 1966. (Cloth or paperback.)

A lecture-note and reprint volume on applications of group theory to nuclear and particle physics. Primarily for physicists.

Gell-Mann, Murray, and Yuval Ne'eman. *The Eightfold Way*. New York: W. A. Benjamin, 1965.

Applications of group theory to the present day study of elementary particles.

Gill, Arthur. *Applied Algebra for the Computer Sciences*. Englewood Cliffs, N.J.: Prentice-Hall, 1976.

The algebra needed for a variety of areas of computer science, including some Boolean algebra and switching theory, and a short chapter on coding, as well as sections on networks, finite-state machines, automata, and so on.

Hall, Lowell H. *Group Theory and Symmetry in Chemistry*. New York: McGraw-Hill, 1969.

An upper division–graduate level text for students of chemistry, introducing the elements of group theory with applications to the symmetry description of molecular structure. Some matrix algebra is included as well as representation theory. Fascinating computer-drawn stereoscopic illustrations are included along with a special viewer.

Hohn, Franz E. *Applied Boolean Algebra*, 2nd edition. New York: Macmillan, 1966.

This is a simple, easy-to-read introduction to the applications of Boolean algebra.

Kohavi, Zvi. *Switching and Finite Automata Theory*, 2nd edition. New York: McGraw-Hill, 1978.

A popular text by an engineer. It starts with error-correcting codes, moves on to lattices and Boolean algebra with applications to switching, and then to a number of more advanced topics.

Levinson, Norman. "Coding Theory: A Counterexample to G. H. Hardy's Conception of Applied Mathematics." *American Mathematical Monthly*, 77 (1970), 249–258.

Applications of finite fields and number theory to coding.

Liu, C. L. *Introduction to Applied Combinatorial Mathematics*. New York: McGraw-Hill, 1968.

An up-to-date extensive first course in combinatorics.

Loebl, Ernest M., ed. *Group Theory and Its Applications*. New York: Academic Press, 1968.

A collection of essays on applications of group theory to quantum mechanics, solid state physics, atomic spectroscopy, nuclear theory, elementary particle theory, and relativity.

Lomont, John S. *Applications of Finite Groups*. New York: Academic Press, 1959.

Applications of group theory to mathematical physics, chemistry, and crystallography.

Macgillavry, Caroline H. *Symmetry Aspects of M. C. Escher's Periodic Drawings*. Utrecht: International Union of Crystallography, 1965.

An interesting exposition of the relationship between groups, symmetry, and a very popular artist. Handsomely illustrated with color and black-and-white illustrations of Escher's work.

MacWilliams, F. J., and N. J. A. Sloane. *The Theory of Error-Correcting Codes.* Amsterdam: North-Holland Publishing Co., 1978.

An encyclopedic treatment of error-correcting codes by authors who have made major contributions to the field. Although the book covers very sophisticated topics, the early chapters are easy to read.

Marcus, Mitchell P. *Switching Circuits for Engineers*, 2nd edition. Englewood Cliffs, N.J.: Prentice-Hall, 1967.

An engineer's view of Boolean algebra and switching theory, with a chapter on codes.

Meijer, Paul H. E., and Edmond Bauer. *Group Theory, The Applications to Quantum Mechanics.* Amsterdam: North-Holland, 1962.

Much physics, not so much group theory. Some interesting material on rotational crystal symmetry.

Wigner, Eugene. *Group Theory and Its Applications to the Quantum Mechanics of Atomic Spectra.* New York: Academic Press, 1959.

An advanced treatise on the applications of group theory to physics by an eminent twentieth century physicist.

# *Answers, Hints, or Solutions for Most Odd-Numbered Problems*

## *Section 1.1, page 8*

**1.** *Hint:* See Example 2 of this section.
**3.** For all m in $\mathbf{Z}^+$, $L_{3m}$ is even and $L_{3m-2}$ and $L_{3m-1}$ are odd. The proof is similar to that of Example 1.
**5.** (ii) $T_n = (n+1)/2n$. Yes. Yes. **7.** (ii) $A_n = n^2$.
**9.** (i) $v_n = (-1/2)^{n-1}$.
(ii) *Hint:* Show that $v_{k+1} = -v_k/2$ and then use mathematical induction.
(iv) $u_n = (2/3)[1 - (-1/2)^n]$.
**11.** $B_n = (n+1)! - 1$. **13.** $a_n = F_{n+1}$.

## *Section 1.2, page 13*

**1.** Let $a$, $b$, $d$, $h$, and $k$ be integers with $d$ an integral divisor both of $a$ and $b$. Then $d$ is a divisor of $ha + kb$ in $\mathbf{Z}$.
**5.** (a) $-1, 1$. (b) All integers. (c) 4. (d) 6. (e) $\pm 1, \pm 2, \pm 3, \pm 6$.
**7.** (a) $0, \pm 3, \pm 6, \pm 9, \pm 12, \pm 15, \pm 18$.
(b) Yes; the set $\mathbf{Z}$, of all the integers, is one of the possibilities for the set $T$.

**9.** 0, 2, 4, 6. These are the only possibilities since $10x + 14y$ must be even. They are of this form since $0 = 10 \cdot 0 + 14 \cdot 0$, $2 = 10 \cdot 3 + 14(-2)$, $4 = 10 \cdot 6 + 14(-4)$, $6 = 10 \cdot 9 + 14(-6)$.
**13.** The only integral divisors of $p$, and hence of $-p$, are $\pm 1$ and $\pm p$.
**15.** $x = k$, $y = h - qk$. **17.** (a) 6. (b) 2. (c) 6. (d) 3.
**19.**

| $x$ | $-31$ | $-5$ | 5 | 7 | 9 | 11 | 21 | 47 |
|---|---|---|---|---|---|---|---|---|
| $y$ | $-6$ | $-8$ | $-18$ | $-44$ | 34 | 8 | $-2$ | $-4$ |

## *Sections 1.3 and 1.4, page 21*

**1.** (a) $q = 11$, $r = 16$. (b) $q = -12$, $r = 1$. (c) $q = 0 = r$.
**3.** (a) $d = 11$, $e = 7$, $f = 1$, $g = 1$. (b) 1, 2, 3, 4. (c) 1, 5. **5.** 1 or $-1$.
**7.** (i) 11. This follows from $(22, 55) = 11$ and Theorem 2(a).
(ii) $x = 3$, $y = -1$.
**9.** 1, 2, 3, 4, 6, 8, 9, 12, 18, 24, 36, 72. **11.** (i) 1, 19.
**21.** (a) By Theorem 2(e) of Section 1.3, $S$ consists of all the (integral) multiples of the smallest positive integer $t$ of $S$. Hence $t$ must be one of the positive common divisors 1, 2, 17, and 34 of 170 and $-102$. In each case, the multiples of $t$ include the multiples of 34.
(b) Yes. $S$ might be the set of all integers.

**23.** (iii) $x = 23$, $y = 21$ and $x = 13$, $y = 9$. **25.** *Hint:* See Problem 23(ii).

**27.** By Example 1 of Section 1.2, a common (integral) divisor of $n$ and $n + 1$ must be a divisor of their difference 1.

**29.** The positive integral divisors 1, 2, 5, and 10 of the difference 10 of $n$ and $n + 10$.

**31.** *Hint:* $n^2 + n + 1 = (n + 2)(n - 1) + 3$.

## *Sections 1.5 and 1.6, page 29*

**1.**

| | $(c, d)$ | $[c, d]$ | $(c, d)[c, d]$ | $cd$ |
|---|---|---|---|---|
| (a) | 4 | 24 | 96 | 96 |
| (b) | 7 | 35 | 245 | 245 |
| (c) | 1 | 42 | 42 | 42 |

**3.** $d = 907$, $x = 257$, and $y = 302$. **5.** 14.

**9.** The hypothesis implies that there are integers $m$, $h$, and $k$ with $bc = am$ and $1 = ah + bk$. Then $c = ach + bck = ach + amk = a(ch + mk)$. Hence $a|c$.

**11.** Let $m = [u, v]$. Then $m = au = bv$, with $a$ and $b$ in **Z**. Hence $tm = atu = btv$ is a common integral multiple of $tu$ and $tv$.

## *Section 1.7, page 33*

**1.** (i) $2^4 \cdot 3^2 \cdot 5 \cdot 7$. (ii) $2^8 \cdot 3^4 \cdot 5^2 \cdot 7^2$. (iii) $2^{12} \cdot 3^6 \cdot 5^3 \cdot 7^3$.

**3.** (i) 1, 2, 4, 8. (ii) 1, 2, 4, 8, 16. (iii) 1, 2, 4, 8, 16, 32. (iv) 1, 2, 4, 8, 16, 32, 64.

**7.** $(a + 1)(b + 1)(c + 1)$.

**11.** All the exponents $e_i$ must be even.

**21.** $(a, b) = 2^6 \cdot 19 \cdot 23^3$, $[a, b] = 2^{30} \cdot 3 \cdot 5^{21} \cdot 7^4 \cdot 11^2 \cdot 19^5 \cdot 23^7$.

**27.** $2^4$ (or 16).

## *Review Problems (Chapter 1), page 36*

**1.** (i) 97. (iii) $x = -15$, $y = 14$.

**3.** The gcd is a positive integral divisor of every linear combination of $n$ and $n + 4$ and hence of their difference 4.

**5.** (ii) 12. (iii) Yes. See Theorem 2(e) of Section 1.3.

## *Section 2.1, page 40*

**1.** $\begin{pmatrix} 1 & 2 \\ 1 & 2 \end{pmatrix}$, $\begin{pmatrix} 1 & 2 \\ 2 & 1 \end{pmatrix}$.

**3.** (a) $4! = 4 \cdot 3 \cdot 2 \cdot 1 = 24$. (b) $5! = 120$. **5.** $3! = 6$.

**7.** (i) $\gamma$: $1 \mapsto 3$, $2 \mapsto 2$, $3 \mapsto 1$. (ii) $\delta$: $1 \mapsto 1$, $2 \mapsto 3$, $3 \mapsto 2$.

**9.** $\begin{pmatrix} 1 & 6 & 3 & 2 & 4 & 5 \\ 6 & 3 & 1 & 4 & 2 & 5 \end{pmatrix}$.

**11.** (i) 1. (ii) 1, 3, 6. (iii) 2, 4.

**13.** $1 \mapsto 1, 2 \mapsto 2, 3 \mapsto 3, 4 \mapsto 4$; $1 \mapsto 2, 2 \mapsto 1, 3 \mapsto 3, 4 \mapsto 4$;
$1 \mapsto 3, 2 \mapsto 2, 3 \mapsto 1, 4 \mapsto 4$; $1 \mapsto 4, 2 \mapsto 2, 3 \mapsto 3, 4 \mapsto 1$;
$1 \mapsto 1, 2 \mapsto 3, 3 \mapsto 2, 4 \mapsto 4$; $1 \mapsto 1, 2 \mapsto 4, 3 \mapsto 3, 4 \mapsto 2$;
$1 \mapsto 1, 2 \mapsto 2, 3 \mapsto 4, 4 \mapsto 3$; $1 \mapsto 2, 2 \mapsto 1, 3 \mapsto 4, 4 \mapsto 3$;
$1 \mapsto 3, 2 \mapsto 4, 3 \mapsto 1, 4 \mapsto 2$; $1 \mapsto 4, 2 \mapsto 3, 3 \mapsto 2, 4 \mapsto 1$.

## *Section 2.2, page 47*

**1.** Each of $\varepsilon\varepsilon$ and $\theta\theta$ is given by $1 \mapsto 1, 2 \mapsto 2$ while each of $\varepsilon\theta$ and $\theta\varepsilon$ is given by $1 \mapsto 2, 2 \mapsto 1$.

**3.** (i) $\rho^2 = \begin{pmatrix} 1 & 2 & 3 \\ 3 & 1 & 2 \end{pmatrix}$. (ii) $\phi\rho = \begin{pmatrix} 1 & 2 & 3 \\ 2 & 1 & 3 \end{pmatrix} = \rho^2\phi$.

(iv)

| | $\varepsilon$ | $\rho$ | $\rho^2$ | $\phi$ | $\rho\phi$ | $\rho^2\phi$ |
|---|---|---|---|---|---|---|
| $\rho^2$ | $\rho^2$ | $\varepsilon$ | $\rho$ | $\rho^2\phi$ | $\phi$ | $\rho\phi$ |
| $\phi$ | $\phi$ | $\rho^2\phi$ | $\rho\phi$ | $\varepsilon$ | $\rho^2$ | $\rho$ |
| $\rho\phi$ | $\rho\phi$ | $\phi$ | $\rho^2\phi$ | $\rho$ | $\varepsilon$ | $\rho^2$ |
| $\rho^2\phi$ | $\rho^2\phi$ | $\rho\phi$ | $\phi$ | $\rho^2$ | $\rho$ | $\varepsilon$ |

**5.** (i) $\alpha^2$: $1 \mapsto 3, 2 \mapsto 2, 3 \mapsto 6, 4 \mapsto 4, 5 \mapsto 5, 6 \mapsto 1$.
(ii) $\alpha^2\alpha$: $1 \mapsto 1, 2 \mapsto 4, 3 \mapsto 3, 4 \mapsto 2, 5 \mapsto 5, 6 \mapsto 6$.
(iii) $\alpha\alpha^2 = \alpha^2\alpha$. [See part (ii).]
(iv) $\alpha^{-1}$: $1 \mapsto 3, 2 \mapsto 4, 3 \mapsto 6, 4 \mapsto 2, 5 \mapsto 5, 6 \mapsto 1$.

## *Section 2.3, page 54*

**1.** (a) $\theta^{-1} = \theta^3 = \theta^{101} = \theta$; $\theta^2 = \theta^4 = \theta^{100} = \varepsilon$, where $\varepsilon$: $1 \mapsto 1, 2 \mapsto 2$.
(b) $\theta^{-1}$: $1 \mapsto 3, 2 \mapsto 1, 3 \mapsto 2$; $\theta^2 = \theta^{-1}$; $\theta^3$: $1 \mapsto 1, 2 \mapsto 2, 3 \mapsto 3$; $\theta^4 = \theta$; $\theta^{100} = \theta$; $\theta^{101} = \theta^{-1}$.
(c) $\theta^{-1} = \theta = \theta^3 = \theta^{101}$; $\theta^2 = \theta^4 = \theta^{100} = \varepsilon$: $1 \mapsto 1, 2 \mapsto 2, 3 \mapsto 3, 4 \mapsto 4$.

**3.** $(\alpha\beta)^2$: $1 \mapsto 1, 2 \mapsto 2, 3 \mapsto 3$ while $\alpha^2\beta^2$: $1 \mapsto 3, 2 \mapsto 1, 3 \mapsto 2$.

**5.** Given that $ab = ba$, one has

$$(ab)^2 = (ab)(ab) = a(ba)b = a(ab)b = (aa)(bb) = a^2b^2.$$

**7.** $1 \cdot 1 = 1 = (-1)(-1)$, $1 \cdot (-1) = -1 = (-1) \cdot 1$; hence the set is closed under multiplication. Multiplication of any integers is associative. 1 is an identity. Each of 1 and $-1$ is its own inverse. Hence $\{1, -1\}$ is a group under the given operation.

**9.** (i) $\{1, \beta, \gamma\}$ is closed and associative under multiplication, 1 is an identity; 1 is its own inverse while $\beta$ and $\gamma$ are inverses of each other.
(ii) This is proved in Example 2.

**11.**

| | e | $b$ | $c$ | $d$ |
|---|---|---|---|---|
| e | e | $b$ | $c$ | $d$ |
| $b$ | $b$ | e | $d$ | $c$ |
| $c$ | $c$ | $d$ | e | $b$ |
| $d$ | $d$ | $c$ | $b$ | e |

**13.** Yes; this is seen from multiplications $\mathbf{ee} = \mathbf{e} = bb$ and $\mathbf{e}b = b = b\mathbf{e}$ in a group $\{\mathbf{e}, b\}$ with e as its identity. (See Problem 8.)

**15.** One choice is $\alpha$: $1 \mapsto 2$, $2 \mapsto 1$, $3 \mapsto 3$; $\beta$: $1 \mapsto 2$, $2 \mapsto 3$, $3 \mapsto 1$; and $\gamma$: $1 \mapsto 3$, $2 \mapsto 1$, $3 \mapsto 2$. Then $\alpha\beta$ and $\gamma\alpha$ have $1 \mapsto 3$, $3 \mapsto 1$, $2 \mapsto 2$ but $\beta \neq \gamma$.

**19.** $G$ is commutative if and only if the table is symmetric about the main diagonal (i.e., the one from the top left corner to the bottom right corner).

**21.** First we note that $a \neq a^2$ since $a = a^2$ would imply that $a\mathbf{e} = aa$ and then $\mathbf{e} = a$ using left cancellation. Hence $\mathbf{e}$, $a$, and $a^2$ are three distinct elements. Also, the given $a^3 = \mathbf{e}$ implies that

$$\cdots = a^{-6} = a^{-3} = \mathbf{e} = a^3 = a^6 = \cdots,$$
$$\cdots = a^{-5} = a^{-2} = a = a^4 = a^7 = \cdots,$$
$$\cdots = a^{-4} = a^{-1} = a^2 = a^5 = a^8 = \cdots.$$

All parts of the problem now follow.

**23.** If $xx \neq \mathbf{e}$, $x$ is not its own inverse. Such elements come in pairs $\{x, x^{-1}\}$ and hence there are an even number of them.

**25.** For all $a$ and $b$ in $G$, the hypothesis implies that $a^2 = b^2 = (ab)^2 = \mathbf{e}$. Hence $a^2b^2 = \mathbf{ee} = \mathbf{e} = (ab)^2$ and it follows from Problem 6 that $ab = ba$.

**27.** $bcda = a^{-1}(abcd)a = a^{-1}\mathbf{e}a = \mathbf{e}$.

**29.** An entry appears more than once in the row (and in the column) for c.

## *Section 2.4, page 62*

**1.** (a) (123), (132), (124), (142), (134), (143), (234), (243).
(b) (12)(34), (13)(24), (14)(23).

**3.**

| | $\gamma^{-1}$ | $\gamma^2$ | $\gamma^3$ | $\gamma^4$ | $\gamma^5$ | $\gamma^6$ |
|---|---|---|---|---|---|---|
| (a) | (132) | (132) | (1) | (123) | (132) | (1) |
| (b) | (1432) | (13)(24) | (1432) | (1) | (1234) | (13)(24) |
| (c) | (15432) | (13524) | (14253) | (15432) | (1) | (12345) |
| (d) | (165432) | (135)(246) | (14)(25)(36) | (153)(264) | (165432) | (1) |

**5.** (i) (123); (ii) (1234); (iii) (12345); (iv) (123456).

**7.** $(12)(13)(14)\cdots(1s)$ or $(1s)(2s)(3s)\cdots(\overline{s-1}\,s)$.

**9.** (i) $(acb)$; (ii) $(xyz) = (xzy)^2$; (iii) $(abc)^5 = (acb)$; (iv) $(xyz) = (xzy)^5$.

**11.** $(a_1a_3a_5\ldots a_{2r-1})(a_2a_4a_6\ldots a_{2r})$. **13.** $(12)(13) \neq (13)(12)$ in $\mathbf{S_n}$ for $n \geq 3$.

**15.** (i) (1) = (12)(12), (123) = (12)(13), (132) = (13)(12).
(ii) The only interesting cases are (123)(132) = (132)(123) = (1).
(iii) (1) is its own inverse while (123) and (132) are inverses of each other.

**17.** Yes, as $\gamma_1\gamma_2 \cdots \gamma_{2h}\delta_1\delta_2 \cdots \delta_{2k}$. The integer $2h + 2k$ is even.
**19.** Yes. $\alpha^{-1} = \gamma_{2h}^{-1}\gamma_{2h-1}^{-1} \cdots \gamma_1^{-1} = \gamma_{2h}\gamma_{2h-1} \cdots \gamma_1$, since a transposition is its own inverse.
**21.**

| | (1) | $\theta$ | $\theta^2$ | $\theta^3$ | $\theta^4$ |
|---|---|---|---|---|---|
| (1) | (1) | $\theta$ | $\theta^2$ | $\theta^3$ | $\theta^4$ |
| $\theta$ | $\theta$ | $\theta^2$ | $\theta^3$ | $\theta^4$ | (1) |
| $\theta^2$ | $\theta^2$ | $\theta^3$ | $\theta^4$ | (1) | $\theta$ |
| $\theta^3$ | $\theta^3$ | $\theta^4$ | (1) | $\theta$ | $\theta^2$ |
| $\theta^4$ | $\theta^4$ | (1) | $\theta$ | $\theta^2$ | $\theta^3$ |

**23.** 6. **27.** 12.
**29.** (i) $\theta$ cannot have a cycle $(abc \ldots)$ in its standard form since that would imply that $\theta^2$ sends $a$ to $c$ and hence $\theta^2 \neq (1)$.

## *Section 2.5, page 67*

**1.**

| | (1) | (123) | (132) |
|---|---|---|---|
| (1) | (1) | (123) | (132) |
| (123) | (123) | (132) | (1) |
| (132) | (132) | (1) | (123) |

The table shows that $A$ is closed under multiplication. Also the inverse of each element of $A$ is in $A$ since (1) is its own inverse while (123) and (132) are inverses of each other. Now Lemma 2 tells us that $A$ is a subgroup in $\mathbf{S}_3$.

**3.** $(123)(23) = (13)$, which is not in $D$. **5.** $(123)^{-1} = (132)$, which is not in $F$.
**7.** (iii) No; if $\beta$ is in a subgroup, so are $\beta^2$, $\beta^3$, and $\beta^4 = (1)$ by closure under multiplication.
(iv) $\{(1), \beta^2\}$ is a subgroup in $H$, and in $\mathbf{S}_4$, since it is closed under multiplication, and each of its elements is its own inverse.
**9.** (i) Every group of order 2 consists of the identity $\mathbf{e}$ and another element $b$ such that $b^2 = \mathbf{e}$. (See Problem 8 of Section 2.3.)
(ii) $\{(1), (23)\}$, $\{(1), (13)\}$, $\{(1), (12)\}$.
**11.** (i) This part is similar to Problem 7(iv).
(ii) $\{1, \mathbf{i}\}$ is not closed under multiplication, and hence is not a subgroup, since $\mathbf{ii} = -1$ is not in this set. Alternatively, $\mathbf{i}^{-1} = -\mathbf{i}$ is not in $\{1, \mathbf{i}\}$.
**13.** Yes; $a^ia^j = a^{i+j} = a^{j+i} = a^ja^i$ for $0 \leq i < s$ and $0 \leq j < s$. Commutativity in $G$ follows from commutativity of addition of integers.
**15.** (i) Yes; if $h$ and $k$ are in $H$ then $hk = kh$ since multiplication in $H$ is the same as in $G$ and $G$ is abelian.
(ii) No; $\mathbf{S}_3$ is not abelian even though its subgroup $\{(1)\}$ is.
**17.** The desired results follow from Lemma 2 or Theorem 1. **19.** *Hint:* Use Lemma 2.
**25.** (ii) When $G = \mathbf{S}_3$ and $s = 2$, $H = \{(1), (123), (132)\}$ is a subgroup in $\mathbf{S}_3$. (See Problem 1.)
(iii) When $G = \mathbf{S}_3$ and $s = 3$, $H = \{(1), (23), (13), (12)\}$ is not a subgroup in $\mathbf{S}_3$ since it is not closed under multiplication; for example, $(12)(13) = (123)$, which is not in $H$.
**27.** (i) $G$, $\{1, -1\}$, $\{1\}$. (ii) Yes.

## *Section 2.6, page 74*

**1.** (a) It is a group.
(b) It is not a group since it does not satisfy Axiom 2 (Identity).
(c) It is a group.
**3.** (a) It is not a group since it does not satisfy Axiom 3 (Inverses).
(b) It is not a group since it does not satisfy Axiom 1 (Closure).
(c) It is a group.
**5.** The positive integers.
**7.** Assume that there exists an additive group $G = \{z, b, c\}$ of order 3 with $z$ as the identity. Show that $b + c = z = c + b$, $2 \cdot b = c$, $2 \cdot c = b$, and $3 \cdot b = z = 3 \cdot c$ in such a group.
**9.** $-(-a) = a$, for every $a$ of an additive group.
**11.** $\begin{pmatrix} 2 & 4 & 4 \\ 8 & -12 & 6 \end{pmatrix}$.
**13.** (a) $6\sqrt{2}[\cos(3\pi/4) + \mathbf{i}\sin(3\pi/4)]$;
(b) $9[\cos(\pi/2) + \mathbf{i}\sin(\pi/2)]$;
(c) $5(\cos\pi + \mathbf{i}\sin\pi)$.
**15.** (a) $4\sqrt{3} + 4\mathbf{i}$; (b) $-1 - \mathbf{i}$.
**17.** $-2^{27} + 2^{27}\sqrt{3}\mathbf{i}$.

## *Section 2.7, page 82*

**1.** (i) 1; (ii) 2; (iii) 3; (iv) 4; (v) 5.
**3.** (1243); it has order 4.
**5.** (i) 2; (ii) 6; (iii) 4; (iv) 10.
**7.** (i) 6; (ii) 4; (iii) 20; (iv) 12; (v) 60; (vi) 60.
**9.** (i) $[(1)] = \{(1)\}$ has order 1; $[(23)] = \{(1), (23)\}$, $[(13)] = \{(1), (13)\}$, and $[(12)] = \{(1), (12)\}$ have order 2; and $[(123)] = [(132)] = \{(1), (123), (132)\}$ has order 3.
(ii) No element of $\mathbf{S}_3$ has order 6 (that is, generates $\mathbf{S}_3$).
**11.** (i) $\alpha$ has order 3, $\beta$ has order 12; (ii) $(-1 - \mathbf{i}\sqrt{3})/2$; (iii) $\beta, \beta^5, \beta^7, \beta^{11}$.
**13.** (i) 4; (ii) 4; (iii) 3; (iv) 12.
**15.** (a) $\{\mathbf{e}, a^{15}\}$; (b) $\{\mathbf{e}, a^{10}, a^{20}\}$; (c) $\{\mathbf{e}, a^6, a^{12}, a^{18}, a^{24}\}$;
(d) $\{\mathbf{e}, a^5, a^{10}, a^{15}, a^{20}, a^{25}\}$.
**17.** (a) $a^{15}$; (b) $a^{10}, a^{20}$; (c) $a^3, a^9, a^{21}, a^{27}$.
**19.** The order of $a$ is 1, 3, 5, or 15.
**21.** (i) 1; (ii) $m - 1$. **23.** (12345678), (14725836), (16385274), and (18765432).
**25.** Yes. See Problem 8 and Example 2 of Section 2.3.
**27.**

| | 0 | 1 | 2 | 3 | 4 |
|---|---|---|---|---|---|
| 0 | 0 | 1 | 2 | 3 | 4 |
| 1 | 1 | 2 | 3 | 4 | 0 |
| 2 | 2 | 3 | 4 | 0 | 1 |
| 3 | 3 | 4 | 0 | 1 | 2 |
| 4 | 4 | 0 | 1 | 2 | 3 |

**37.** (a) 1, 2; (b) 1, 3; (c) 1, 2, 3, 4; (d) 1, 5; (e) 1, 2, 3, 4, 5, 6; (f) 1, 3, 5, 7; (g) 1, 2, 4, 5, 7, 8.

**41.** In the additive group **Z**, $[1] = \{n : n = m \cdot 1, m \in \mathbf{Z}\}$ is **Z** as is $[-1] = \{n : n = m(-1), m \in \mathbf{Z}\}$. There are no other generators.

## *Section 2.8, page 88*

**1.** (a) $x_2^2 + x_3^2 - x_1^2$.
(b) $x_2^2 + x_1^2 - x_3^2$.
(c) $x_2^2 + x_3^2 + x_4^2$.

**3.** (1), (14).

**5.** (a) even; (b) odd; (c) even; (d) even.

**7.** $\gamma_i$ is a product of $s_i - 1$ transpositions; hence $\theta$ is a product of

$$(s_1 - 1) + (s_2 - 1) + \cdots + (s_r - 1) = s_1 + s_2 + \cdots + s_r - r$$

transpositions.

**13.** $\{(1), (123), (132)\}$.

**15.** We can let $\alpha = \gamma_1\gamma_2\cdots\gamma_h$ and $\beta = \delta_1\delta_2\cdots\delta_k$ with the $\gamma_i$ and $\delta_j$ transpositions. Then

$$\alpha^{-1}\beta^{-1}\alpha\beta = \gamma_h\cdots\gamma_1\delta_k\cdots\delta_1\gamma_1\cdots\gamma_h\delta_1\cdots\delta_k$$

is a product of $2(h + k)$ transpositions and so is even.

## *Section 2.9, page 93*

**1.** (i) $\rho^2 = (13)(24)$, $\rho^3 = (1432)$, $\rho\phi = (14)(23)$, $\rho^2\phi = (13)$, $\rho^3\phi = (12)(34)$ and these are the remaining elements of the octic group.
(ii) $\phi\rho = (12)(34) = \rho^3\phi$.

**3.** See the table on the inside back cover. **5.** $\{(1), (13)(24)\}$ or $\{\varepsilon, \rho^2\}$.

**7.** (i) $[(1234)] = \{(1), (1234), (13)(24), (1432)\}$.
(ii) $\{(1), (13), (24), (13)(24)\}$ and $\{(1), (12)(34), (13)(24), (14)(23)\}$.
(iii) $\{\varepsilon, \rho^2\}$, $\{\varepsilon, \phi\}$, $\{\varepsilon, \rho\phi\}$, $\{\varepsilon, \rho^2\phi\}$, $\{\varepsilon, \rho^3\phi\}$.

**11.** (i) Yes. It is a finite subset closed under multiplication.
(ii) (1), $\rho^2$, $\rho^3\phi$, $\rho\phi$.
(iii) Yes. $K = H$.

**13.** (i) $\phi\rho = (25)(34)(12345) = (12)(35)$, $\rho^{-1}\phi = (15432)(25)(34) = (12)(35)$.

**15.** (i) Assuming the vertices numbered as in the square, the permutations are (1), (12)(34), (13)(24), and (14)(23).
(ii) This table is easily obtained from that for the octic group.

## *Section 2.10, page 97*

**1.** (12)(34), (13)(24), (14)(23). **3.** No.
**5.** {(1), (123), (132)}, {(1), (124), (142)}, {(1), (134), (143)}, {(1), (234), (243)}.
**7.** $\mathbf{A_4}$.
**9.** (i)

| $\theta$ | $\{\alpha_1\theta, \alpha_5\theta, \alpha_9\theta\}$ |
|---|---|
| $\alpha_1, \alpha_5$, or $\alpha_9$ | $\{\alpha_1, \alpha_5, \alpha_9\}$ |
| $\alpha_2, \alpha_6$, or $\alpha_{10}$ | $\{\alpha_2, \alpha_6, \alpha_{10}\}$ |
| $\alpha_3, \alpha_7$, or $\alpha_{11}$ | $\{\alpha_3, \alpha_7, \alpha_{11}\}$ |
| $\alpha_4, \alpha_8$, or $\alpha_{12}$ | $\{\alpha_4, \alpha_8, \alpha_{12}\}$ |

(ii) 4.
**11.** (i)

| $\theta$ | $\{\alpha_1\theta, \alpha_2\theta, \alpha_3\theta, \alpha_4\theta\}$ |
|---|---|
| $\alpha_1, \alpha_2, \alpha_3$, or $\alpha_4$ | $\{\alpha_1, \alpha_2, \alpha_3, \alpha_4\}$ |
| $\alpha_5, \alpha_6, \alpha_7$, or $\alpha_8$ | $\{\alpha_5, \alpha_6, \alpha_7, \alpha_8\}$ |
| $\alpha_9, \alpha_{10}, \alpha_{11}$ or $\alpha_{12}$ | $\{\alpha_9, \alpha_{10}, \alpha_{11}, \alpha_{12}\}$ |

(ii) 3. (iii) *Hint:* Use Table 2.2.
**13.** 15; 20; 24. **15.** {1, 2, 3, 5}.
**17.**

| $\theta$ | $\alpha_3$ | $\alpha_4$ | $\alpha_5$ | $\alpha_6$ | $\alpha_7$ | $\alpha_8$ | $\alpha_9$ | $\alpha_{10}$ | $\alpha_{11}$ | $\alpha_{12}$ |
|---|---|---|---|---|---|---|---|---|---|---|
| $\theta^{-1}\alpha_2\theta$ | $\alpha_2$ | $\alpha_2$ | $\alpha_4$ | $\alpha_4$ | $\alpha_4$ | $\alpha_4$ | $\alpha_3$ | $\alpha_3$ | $\alpha_3$ | $\alpha_3$ |

**19.** $\{\alpha_1, \alpha_2, \alpha_3, \alpha_4\}$
**21.** (i) Yes.
(ii) No. *Hint:* Square (1), $(ab)$, $(abc)$, $(abcd)$, $(abcde)$, $(abcdef)$, $(ab)(cd)$, $(ab)(cd)(ef)$, $(ab)(cde)$, $(ab)(cdef)$, and $(abc)(def)$.
**23.** $H_0 = \mathbf{A_4}$; $H_1 = \{\alpha_1, \alpha_2, \alpha_3, \alpha_4\}$; $H_2, H_3, H_4$, and $H_5$ are (in any order) the subgroups $\{\alpha_1, \alpha_5, \alpha_9\}$, $\{\alpha_1, \alpha_6, \alpha_{11}\}$, $\{\alpha_1, \alpha_7, \alpha_{12}\}$, and $\{\alpha_1, \alpha_8, \alpha_{10}\}$, $H_6$, $H_7$, and $H_8$ are (in any order) the subgroups $\{\alpha_1, \alpha_2\}$, $\{\alpha_1, \alpha_3\}$, and $\{\alpha_1, \alpha_4\}$; and $H_9 = \{\alpha_1\}$.

## *Section 2.11, page 103*

**1.** (i) $H$ and $\{(13)(24), (14)(23)\} = \{\alpha_3, \alpha_4\}$.
(ii) $\alpha_5 H = \{\alpha_5, \alpha_6\}$ is different from $H\alpha_5 = \{\alpha_5, \alpha_8\}$.
(iii) $K$ is abelian; hence any subgroup in $K$ is normal in $K$.
**3.** (i) Yes. (ii) Yes. (iii) 3.
(iv) $N$, $\{\alpha_5, \alpha_6, \alpha_7, \alpha_8\}$, and $\{\alpha_9, \alpha_{10}, \alpha_{11}, \alpha_{12}\}$ are the left cosets and also the right cosets of $N$ in $\mathbf{A_4}$.
(v) Yes. $\theta N = N\theta$ for every $\theta$ in $\mathbf{A_4}$.
**5.** For every $h$ of a subgroup $H$ in the center of $G$ and every $g$ in $G$, $gh = hg$. Hence $gH = Hg$ and $H$ is normal in $G$.
**7.** One left (or right) coset consists of the even permutations and the other consists of the odd permutations.

**9.** (i) The only possible orders for elements of $G$ are 1, 2, and 4. If $G$ is not cyclic, each element $g$ has order 1 or 2 and so satisfies $g^2 = \mathbf{e}$.
(ii) This follows from (i) and Problem 11 of Section 2.3.
**15.** (i) $\{1, 2, 3, 4, 5\}$. (ii) 60.
**17.** (ii) Yes.

## *Section 2.12, page 111*

**1.** $\alpha_3 H \cdot \alpha_5 H = \{(123), (243), (142), (134)\}$.
**3.**

| | $N$ | $P$ | $Q$ |
|---|---|---|---|
| $N$ | $N$ | $P$ | $Q$ |
| $P$ | $P$ | $Q$ | $N$ |
| $Q$ | $Q$ | $N$ | $P$ |

**5.** (i) $\mathbf{Z}$ is abelian; so all of its subgroups are normal.
(ii) m. (iii) Every integer is in one of these four distinct cosets.
**7.** Let $K = \{(23), (13), (12)\}$. Then $\theta H = H = H\theta$ for $\theta$ in $H$ and $\theta H = K = H\theta$ for $\theta$ in $K$. Hence $\theta H = H\theta$ for all $\theta$ in $\mathbf{S}_3$.

| | $H$ | $K$ |
|---|---|---|
| $H$ | $H$ | $K$ |
| $K$ | $K$ | $H$ |

**9.** The solution is the same as for Problem 7 except that here the cosets are $H$ and $K = (-1)H$.
**13.** The order of $G/N$ is $p$, a prime, and all groups of prime order are cyclic.
**15.** (a) $\overline{2m\pi} = m \cdot \overline{2\pi} = m \cdot \bar{0} = \bar{0}$ for all integers $m$.
(b) For every real number $a$, there is an integer $n$ such that $n\pi < a \leq (n+1)\pi$. If $n$ is even, let $b = a - n\pi$; if $n$ is odd, let $b = a - (n+1)\pi$.
(c) $24 \cdot \overline{\pi/12} = \overline{2\pi} = \bar{0}$ and $m \cdot \overline{\pi/12} \neq \bar{0}$ for $0 < m < 24$.
(d) No positive integral multiple of 1 is a positive integral multiple of $2\pi$.
**17.** *Hint*: Show that $G/N = \{N, P\}$ with $N^2 = N = P^2$.
**19.** *Hint*: State the order of $b^2$ and use Problem 17.

## *Section 2.13, page 118*

**1.** (i) (1), (12)(34), (13)(24), (14)(23).
(ii) No. Theorem 1 tells us that a normal subgroup with index 3 contains all the cubes, that is, all the elements in part (i).
**3.** If a subgroup $H$ of order 6 existed in $\mathbf{A}_4$, it would have index 2 and so be normal. Then it would contain the nine squares in $\mathbf{A}_4$ (that is (1) and the eight 3-cycles). Since a group of order 6 cannot have 9 distinct elements, no such group exists.

**5.** (ii) The identity (1), the twenty-four 5-cycles, and the fifteen products of two disjoint 2-cycles are all the cubes in $\mathbf{A}_5$. (The twenty 3-cycles are not cubes.)
(iii) Theorem 1 tells us that a normal subgroup with index 3 would contain the 40 cubes of part (ii). But such a subgroup would only have $60/3 = 20$ elements.

**7.** (i) Since $(xyz) = (xzy)^2$, $(xyz)$ is in $N$ by Theorem 1.
(ii) Since $(abcde) = (adbec)^2$, $(abcde)$ is in $N$.
(iii) This follows from closure of $N$ and the hint.

**9.** Such a subgroup would have order $120/5 = 24$ but would also have to contain the 96 fifth powers in $\mathbf{S}_5$. [Every permutation with a standard form of the type (1), $(ab)$, $(abc)$, $(abcd)$, $(ab)(cd)$, $(ab)(cde)$, or $(abc)(de)$ is a fifth power.]

**13.** Let $G_0 = G$, $G_1 = \{\mathbf{e}, a^5, a^{10}, a^{15}, a^{20}\}$, and $G_2 = \{\mathbf{e}\}$. Since $G$ is cyclic, it is abelian. This helps us see that $G_{i+1}$ is a normal subgroup with prime index 5 in $G_i$ for $i = 0, 1$.

## *Section 2.14, page 121*

**1.** (a) The only subgroup of order 4 is $\{(1), (12)(34), (13)(24), (14)(23)\}$.
(b) $K = \{(34)^{-1}\theta(34) : \theta \in H\}$. Two other 3-subgroups are $[(134)]$ and $[(234)]$.

**5.** The octic group of symmetries of a square.

## *Review Problems (Chapter 2), page 122*

**1.** Proof by contradiction: $ab^{-1} = ac^{-1}$ implies that $b^{-1} = c^{-1}$ which in turn implies that $b = c$.

**3.** $\alpha = (1352)(46)$ and $\alpha^{-1} = (1253)(46)$ have order 4; $\beta = (163254)$ has order 6; and $\beta^{99} = \beta^3 = (12)(34)(56)$ has order 2.

**5.** (a) 20; the smallest positive integer $n$ with $(a^{36})^n = \mathbf{e}$ is 20.
(b) $\mathbf{e}, a^{16}, a^{32}, a^{48}, a^{64}$
(c) $\mathbf{e}, a^{10}, a^{20}, a^{30}, a^{40}, a^{50}, a^{60}, a^{70}$.
(d) $a^8, a^{24}, a^{56}, a^{72}$.

**7.** $\{\mathbf{e}\}$, $\{\mathbf{e}, a^3\}$, $\{\mathbf{e}, a^2, a^4\}$, $G$. Yes; each is finite and closed under the operation of $G$.

**9.** 181,440.

**11.** (a) Yes. Lagrange's Theorem and $a^{32} \neq \mathbf{e}$ imply that $a$ has order 64.

**19.** 420.

## *Section 3.1, page 132*

**1.** $\theta(n) = n + 1$.

**3.** (i) One such $\theta$ has $\theta(n) = n$ for all $n$ in $X$.
(ii) No. The element 4 of $Y$ is not the image under $\theta$ of any $x$ in $X$.

**5.** (i) No. $\theta(-1) = \theta(1)$ but $-1 \neq 1$, for example.
(ii) Yes. Every $x$ in $N$ is an image (of $x$ or $-x$).
(iii) $-5, 5$.

**7.** (i) No. (ii) No. **9.** $3^2 = 9$.

**13.** (i) Yes. See Lemma 1 of Section 2.12.

(ii)

| $\alpha$ | $\varepsilon$ | $\rho$ | $\rho^2$ | $\rho^3$ | $\phi$ | $\rho\phi$ | $\rho^2\phi$ | $\rho^3\phi$ |
|---|---|---|---|---|---|---|---|---|
| $f(\alpha)$ | $N$ | $B$ | $N$ | $B$ | $C$ | $D$ | $C$ | $D$ |

(iii) Yes. $N$ is the identity of $G/N$.

(iv) Yes. $f(\alpha)f(\alpha^{-1}) = f(\alpha\alpha^{-1}) = f(\varepsilon) = N$, the identity of $G/N$.

(v) $\varepsilon, \rho^2$.

**15.** The mapping $\gamma$ such that $\gamma(x) = z$ whenever $\alpha(x) = y$ and $\beta(y) = z$.

## *Section 3.2, page 140*

**1.** (i)

| $x$ | $\mathbf{e}$ | $a$ | $a^2$ | $a^3$ | $a^4$ |
|---|---|---|---|---|---|
| $\theta_1(x)$ | $\mathbf{e}'$ | $b$ | $b^2$ | $b^3$ | $b^4$ |
| $\theta_2(x)$ | $\mathbf{e}'$ | $b^2$ | $b^4$ | $b$ | $b^3$ |
| $\theta_3(x)$ | $\mathbf{e}'$ | $b^3$ | $b$ | $b^4$ | $b^2$ |
| $\theta_4(x)$ | $\mathbf{e}'$ | $b^4$ | $b^3$ | $b^2$ | $b$ |

(ii) An isomorphism $\theta$ from $[a]$ to $[b]$ is completely determined once $\theta(a)$ is chosen as one of the four generators of $[b]$.

**3.** $\theta_2$: $\mathbf{e} \mapsto \mathbf{e}'$, $b \mapsto u$, $c \mapsto w$, $d \mapsto v$.
$\theta_3$: $\mathbf{e} \mapsto \mathbf{e}'$, $b \mapsto v$, $c \mapsto u$, $d \mapsto w$.
$\theta_4$: $\mathbf{e} \mapsto \mathbf{e}'$, $b \mapsto v$, $c \mapsto w$, $d \mapsto u$.
$\theta_5$: $\mathbf{e} \mapsto \mathbf{e}'$, $b \mapsto w$, $c \mapsto u$, $d \mapsto v$.
$\theta_6$: $\mathbf{e} \mapsto \mathbf{e}'$, $b \mapsto w$, $c \mapsto v$, $d \mapsto u$.

**7.** Yes. *Hint*: Use $a'b' = (ab)' = (ba)' = b'a'$.

**9.**

| $\alpha$ | (1) | (123) | (132) | (23) | (13) | (12) |
|---|---|---|---|---|---|---|
| $f(\alpha)$ | (1) | (123) | (132) | (23) | (13) | (12) |
| $g(\alpha)$ | (1) | (123) | (132) | (13) | (12) | (23) |
| $h(\alpha)$ | (1) | (123) | (132) | (12) | (23) | (13) |

**13.** $\beta = (123)$. **19.** $\theta_1$ and $\theta_2$, with $\theta_1(x) = x$ and $\theta_2(x) = -x$ for all $x$ in $\mathbf{Z}$.

**25.** 4.

**31.** (b) $L$ is the conjugate of $H$ by $a^{-1}$.

## *Section 3.3, page 149*

**1.** (i)

| $g$ | $\mathbf{e}$ | $a$ | $a^2$ | $a^3$ | $a^4$ | $a^5$ | $a^6$ | $a^7$ | $a^8$ |
|---|---|---|---|---|---|---|---|---|---|
| $\theta(g)$ | (1) | (123) | (132) | (1) | (123) | (132) | (1) | (123) | (132) |

(ii) $\{\mathbf{e}, a^3, a^6\}$.

(iii) $\theta^{-1}[(123)] = \{a, a^4, a^7\}$, $\theta^{-1}[(132)] = \{a^2, a^5, a^8\}$.

**5.** (i)

| | $\bar{0}$ | $\bar{1}$ | $\bar{2}$ |
|---|---|---|---|
| $\bar{0}$ | $\bar{0}$ | $\bar{1}$ | $\bar{2}$ |
| $\bar{1}$ | $\bar{1}$ | $\bar{2}$ | $\bar{0}$ |
| $\bar{2}$ | $\bar{2}$ | $\bar{0}$ | $\bar{1}$ |

(ii)

| $x$ | $-5$ | $-4$ | $-3$ | $-2$ | $-1$ | $0$ | $1$ | $2$ | $3$ | $4$ | $5$ |
|---|---|---|---|---|---|---|---|---|---|---|---|
| $\theta(x)$ | $\bar{1}$ | $\bar{2}$ | $\bar{0}$ | $\bar{1}$ | $\bar{2}$ | $\bar{0}$ | $\bar{1}$ | $\bar{2}$ | $\bar{0}$ | $\bar{1}$ | $\bar{2}$ |

**7.** (i) $\theta(ab) = |ab| = |a| \cdot |b| = \theta(a)\theta(b)$ for all $a$ and $b$ in $\mathbf{R}^{\pm}$.
(ii) $\{-1, 1\}$. (iii) $\{-6, 6\}$.

**9.** (i) $\theta(a)\theta(b) = \mathbf{e}'\mathbf{e}' = \mathbf{e}' = \theta(ab)$, for all $a$ and $b$ in $G$.
(ii) $G$.

**15.** (ii) One example has $G = \mathbf{S}_3$ and $s = 2$.

**17.** (i) $\theta(m + n) = a^{m+n} = a^m a^n = \theta(m)\theta(n)$.
(ii) The set of integral multiples of $q$.
(iii) $\{0\}$.

**19.** (iii) $a$ and $b'$ must have the same order.

**23.** (c) $K$ consists of all $\alpha$ with $f(\alpha) = 1$.

**25.** Let $a$ have order 2, $G = [a] = \{\mathbf{e}, a\}$, and $G' = \{\mathbf{e}\}$.

## *Section 3.4, page 156*

**1.**

| $\theta$ | (1) | (123) | (132) | (23) | (13) | (12) |
|---|---|---|---|---|---|---|
| $f(\theta)$ | (1) | (123)(465) | (132)(456) | (14)(25)(36) | (15)(26)(34) | (16)(24)(35) |

**3.** 1, 2, 3, 4, 5, 6.

**5.** (i) $q_3$. (ii) $q_2, q_4, q_5, q_6, q_7, q_8$. (iii) $\{q_1, q_3\}$.
(iv) $\{q_1, q_2, q_3, q_4\}$, $\{q_1, q_5, q_3, q_7\}$, $\{q_1, q_6, q_3, q_8\}$.

**9.** $\{q_1, q_3\}$.

## *Section 3.5, page 162*

**3.** $\{\varepsilon, \rho^2, \phi, \rho^2\phi\}$.

**11.** $\{\varepsilon, \rho^2\}$.

**31.** If $a$ and $b$ are in $G$, then $a$ is in $G_i$ and $b$ is in $G_j$ for some $i$ and $j$ in $\mathbf{Z}^+$. If $j \le i$, $G_j \le G_i$ and we define the product of $a$ and $b$ in $G$ to be their product in $G_i$. If $j > i$, the product in $G$ is as in $G_j$.

**33.** $B = \{(1)\}$, $C$ and $D$ are $\{\alpha_5, \alpha_6, \alpha_7, \alpha_8\}$ and $\{\alpha_9, \alpha_{10}, \alpha_{11}, \alpha_{12}\}$, and $E = \{\alpha_2, \alpha_3, \alpha_4\}$.

**37.** $H$, $\{(123), (243), (142), (134)\}$, $\{(132), (143), (243), (124)\}$.

## *Section 3.6, page 167*

**1.** (i) (1, 1), (1, 3), (1, 4), (2, 1), (2, 3), (2, 4).
(ii) (1, 1), (3, 1), (4, 1), (1, 2), (3, 2), (4, 2).
(iii) No.

**3.** $mn$.

**5.** (i) The table is that of a cyclic group $[g]$ of order 6 with $(\mathbf{e}, \mathbf{e}')$ the identity, $(a, b) = g$, $(a^2, \mathbf{e}') = g^2$, $(\mathbf{e}, b) = g^3$, $(a, \mathbf{e}') = g^4$, and $(a^2, b) = g^5$.
(ii) Yes. See answer to part (i).
(iii) No. $G$ is cyclic while $\mathbf{S}_3$ is not. Hence it follows from Example 2 of Section 3.2 that $G$ and $\mathbf{S}_3$ are not isomorphic.
(iv)

| $x$ | $(\mathbf{e}, \mathbf{e}')$ | $(\mathbf{e}, b)$ | $(a, \mathbf{e}')$ | $(a, b)$ | $(a^2, \mathbf{e}')$ | $(a^2, b)$ |
|---|---|---|---|---|---|---|
| $h(x)$ | $\mathbf{e}$ | $\mathbf{e}$ | $a$ | $a$ | $a^2$ | $a^2$ |

**9.** Yes. Projection preserves the operations.

## *Section 3.7, page 172*

**11.** 16.
**13.** $M$ is reflexive and transitive but is not symmetric or a linear order relation.
**15.** $P$ is symmetric but is not reflexive, transitive, or a linear order relation.
**17.** Yes.
**23.** (ii) Let $S = \{1, 2\}$ and $L$ be the subset $\{(2, 2)\}$ of $S \times S$.

## *Section 3.8, page 175*

**1.** (2, 2), (2, 4), (2, 6), (2, 12), (3, 3), (3, 6), (3, 12), (4, 4), (4, 12), (6, 6), (6, 12), (12, 12).
**3.** There is no minimal element; 12 is the maximal element.
**5.** $m$ is the maximal element for $(S, \geq)$.
**7.** The relation of a poset is antisymmetric.

**11.** (i)

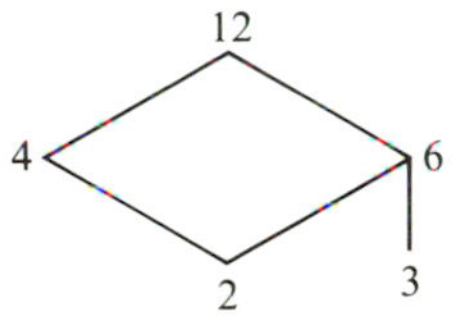


***Figure A.1***

(ii) One does not have segments for the ordered pairs $(a, a)$ implied by reflexiveness nor for the ordered pairs (2, 12) and (3, 12) implied by transitivity.
**13.** $M$ is reflexive, antisymmetric, and transitive and hence is a partial ordering. $M$ is not symmetric nor a linear ordering.

## *Section 3.9, page 178*

**1.** (a) 10010; $\bar{A} = \{s_2, s_3, s_5\}$ has 01101 as its binary numeral.
(b) $B = \{s^2, s^3, s^4\}$; $\bar{B} = \{s_1, s_5\}$.
(c) 00000. (d) $c_i = 1$ unless $a_i = 0 = b_i$. (e) Yes.
**3.** Yes. **5.** Yes. **7.** Yes, $S$. **9.** $3^n$. **11.** Yes.

**13.** (a) This is just a different notation for $(P(S), \subseteq)$, where $S$ has $n$ elements.
(b) $00 \cdots 0$. (c) $11 \cdots 1$.

**15.** $(T, \subseteq)$ is reflexive, antisymmetric, and transitive since $(P(S), \subseteq)$ has these properties and $T$ is a subcollection of $P(S)$.

## *Section 3.10, page 182*

**1.** (a) $3^3$. (b) $3^9$. (c) $3^6$. (d) $3^{3^n}$.

**3.** The identity under $*$ is 1. Only 1 has an inverse.

**5.** There is no identity and hence no possibility of inverses.

**7.** (a)

| $A$ | $\varnothing$ | $\{1\}$ | $\{2\}$ | $\{1, 2\}$ |
|---|---|---|---|---|
| $\bar{A}$ | $\{1, 2\}$ | $\{2\}$ | $\{1\}$ | $\varnothing$ |

(b)

| $\cup$ | $\varnothing$ | $\{1\}$ | $\{2\}$ | $\{1, 2\}$ |
|---|---|---|---|---|
| $\varnothing$ | $\varnothing$ | $\{1\}$ | $\{2\}$ | $\{1, 2\}$ |
| $\{1\}$ | $\{1\}$ | $\{1\}$ | $\{1, 2\}$ | $\{1, 2\}$ |
| $\{2\}$ | $\{2\}$ | $\{1, 2\}$ | $\{2\}$ | $\{1, 2\}$ |
| $\{1, 2\}$ | $\{1, 2\}$ | $\{1, 2\}$ | $\{1, 2\}$ | $\{1, 2\}$ |

(c) $\varnothing$.

**9.** (a)

| $\cap$ | 000 | 001 | 010 | 011 | 100 | 101 | 110 | 111 |
|---|---|---|---|---|---|---|---|---|
| 000 | 000 | 000 | 000 | 000 | 000 | 000 | 000 | 000 |
| 001 | 000 | 001 | 000 | 001 | 000 | 001 | 000 | 001 |
| 010 | 000 | 000 | 010 | 010 | 000 | 000 | 010 | 010 |
| 011 | 000 | 001 | 010 | 011 | 000 | 001 | 010 | 011 |
| 100 | 000 | 000 | 000 | 000 | 100 | 100 | 100 | 100 |
| 101 | 000 | 001 | 000 | 001 | 100 | 101 | 100 | 101 |
| 110 | 000 | 000 | 010 | 010 | 100 | 100 | 110 | 110 |
| 111 | 000 | 001 | 010 | 011 | 100 | 101 | 110 | 111 |

(b) 111.

**11.** Each statement holds for all subsets in $P(S)$.

**13.** (a) True. (b) True. (c) True.

**15.** (a) $\varnothing$. (b) $S$.

**17.** $c_i = \max(a_i, b_i)$ for $1 \leq i \leq n$.

## *Section 3.11, page 188*

**1.** (a) $(Z^+, +)$ is a semigroup. It is not a monoid nor a group.
(b) $(Z^+, \cdot)$ is a semigroup and is a monoid but is not a group.

**3.** No. No. No. **5.** (a) Yes. (b) Not unless $S = \varnothing$, since only $\varnothing$ has an inverse.

**7.** (a) $(\mathbf{Z}^+, +)$. (b) No; a monoid is a semigroup with an identity.

**9.** $\langle 2 \rangle = \{2, 4, 8, 16, 32, 64, 28, 56, 12, 24, 48, 96, 92, 84, 68, 36, 72, 44, 88, 76, 52\}$.

## *Section 3.12, page 192*

**5.**

| $\vee$ | $O$ | $I$ |
|---|---|---|
| $O$ | $O$ | $I$ |
| $I$ | $I$ | $I$ |

**7.** (a)

| $\vee$ | 1 | 3 | 5 | 15 |
|---|---|---|---|---|
| 1 | 1 | 3 | 5 | 15 |
| 3 | 3 | 3 | 15 | 15 |
| 5 | 5 | 15 | 5 | 15 |
| 15 | 15 | 15 | 15 | 15 |

| $\wedge$ | 1 | 3 | 5 | 15 |
|---|---|---|---|---|
| 1 | 1 | 1 | 1 | 1 |
| 3 | 1 | 3 | 1 | 3 |
| 5 | 1 | 1 | 5 | 5 |
| 15 | 1 | 3 | 5 | 15 |

| $x$ | 1 | 3 | 5 | 15 |
|---|---|---|---|---|
| $x'$ | 15 | 5 | 3 | 1 |

(b) Yes. (c) Yes. (d) Yes.

**9.** (i) 1 for $\vee$ and 8 for $\wedge$.
(ii) No.

**17.** $O$ is the minimal element and $I$ the maximal element.

**19.** $A \leq B$ if and only if $A$ is a subset of $B$.

**27.** Yes. The isomorphism is the mapping $f$ with $f(O) = O$ and $f(I) = I$.

**29.**

| $x$ | 000 | 001 | 010 | 100 | 011 | 101 | 110 | 111 |
|---|---|---|---|---|---|---|---|---|
| $f(x)$ | 1 | 3 | 5 | 7 | 15 | 21 | 35 | 105 |

## *Section 3.13, page 199*

**1.** Yes; $x \vee (y \wedge y') = x$.

**3.** No; $\alpha(0, 1, 1) = 0$ and $\beta(0, 1, 1) = 1$, for example.

**5.**

| $(x, y)$ | $x \Rightarrow y$ | $y'$ | $x'$ | $y' \Rightarrow x'$ |
|---|---|---|---|---|
| (0, 0) | 1 | 1 | 1 | 1 |
| (0, 1) | 1 | 0 | 1 | 1 |
| (1, 0) | 0 | 1 | 0 | 0 |
| (1, 1) | 1 | 0 | 0 | 1 |

**7.** (i) $\beta(0, 0) = 1 \vee 0 = 1$. (ii) $\alpha(0, 0) = 1$. (iii) true.

**9.** Yes; $\alpha(x, y) = \beta(x, y)$ for all $x$ and $y$ in $\{0, 1\}$.

**11.** $x \Leftrightarrow y'$ (or $x' \Leftrightarrow y$, among other possibilities).

**13.** $\alpha(0, 1, 1) = 0$; the check is not valid.

**15.** No; neither (i) nor (ii) of Example 4 is satisfied.

**17.** Let $R$, $S$, and $T$ denote $p|(ab)$, $p|a$, and $p|b$, respectively.

**19.** (a) Contingency; true for $(x, y) = (0, 0)$ and false for $(x, y) = (1, 0)$.
(b) Tautology; true for all $x$ and $y$.
(c) Contradiction; false for all $x$ and $y$.
(d) Tautology; always true.
(e) Contingency; true for $(x, y, z, w) = (1, 1, 1, 1)$, false for $(x, y, z, w) = (1, 0, 0, 0)$.

## *Section 3.14, page 204*

1.

| $x$ | e | $a$ | $a^2$ | $a^3$ | $a^4$ | $a^5$ | $a^6$ | $a^7$ | $a^8$ | $a^9$ | $a^{10}$ | $a^{11}$ |
|---|---|---|---|---|---|---|---|---|---|---|---|---|
| $\gamma(x)$ | e | $c$ | e | $c$ | e | $c$ | e | $c$ | e | $c$ | e | $c$ |

$\alpha$, $\beta$, and $\gamma$ are all homomorphisms.

## *Review Problems (Chapter 3), page 209*

1. (i)

| $x$ | e | $a$ | $a^2$ | $a^3$ |
|---|---|---|---|---|
| $\theta_1(x)$ | 1 | **i** | $-1$ | $-$**i** |
| $\theta_2(x)$ | 1 | $-$**i** | $-1$ | **i** |

(ii) The orders of elements must be preserved under an isomorphism.

3. (a) Using the fact that $f(\theta) = \rho^2$ for $\theta$ in $\{\rho, \rho^3\}$ and $f(\theta) = \varepsilon$ otherwise, one sees easily that $f$ is neither an injection nor a surjection. It is not a homomorphism since, for example, $f(\rho\phi) \neq f(\rho)f(\phi)$.
(b) $g$ is both injective and surjective (i.e., bijective) but is not a homomorphism.

9.

| $(x, y)$ | (e, e) | (e, $b$) | (e, $c$) | ($b$, e) | ($b$, $b$) | ($b$, $c$) | ($c$, e) | ($c$, $b$) | ($c$, $c$) |
|---|---|---|---|---|---|---|---|---|---|
| $(b, c)(x, y)$ | ($b$, $c$) | ($b$, e) | ($b$, $b$) | ($c$, $c$) | ($c$, e) | ($c$, $b$) | (e, $c$) | (e, e) | (e, $b$) |

11. 8 antisymmetric relations of which 2 are reflexive.
13. $D$ is reflexive, antisymmetric, and transitive and hence is a partial ordering. It is not symmetric.
15. (a) $2^4 = 16$. (b) $4^{16}$. (c) $16^4$ of which $16 \cdot 15 \cdot 14 \cdot 13$ are injective.
17. (a) $4^4$. (b) $4^{16}$. (c) $4^{10}$.
19. (a) It is a monoid with 1 as identity. It is not a group since 0 has no inverse.
(b) 102. (c) No, no, it has no identity.
21. $(P(S), \cup, \cap, \bar{\ }, \varnothing, S)$ with $S$ having 5 elements.
25. (a) Tautology; $\alpha(x, y) = 1$ for all $x$ and $y$.
(b) Tautology; $\beta(x, y, z) = 1$ for all $x$, $y$, and $z$.

## *Section 4.1, page 220*

1. (i)

| $+$ | 0 | $b$ |
|---|---|---|
| 0 | 0 | $b$ |
| $b$ | $b$ | 0 |

(ii)

| $\cdot$ | 0 | $b$ |
|---|---|---|
| 0 | 0 | 0 |
| $b$ | 0 | $b$ |

(iii)

| $\cdot$ | 0 | $b$ |
|---|---|---|
| 0 | 0 | 0 |
| $b$ | 0 | 0 |

(iv) In part (ii), $b$ is the unity. There is no unity in part (iii).

3. Yes. It meets the conditions in the definition of a ring.
13. (a) No. (b) No. (c) Yes. (d) Yes.
19. (ii) Yes.

**23.** Since $\{1, c, d\}$ is a multiplicative group, Example 2 of Section 2.3 helps us see that the table is as follows:

| | 0 | 1 | $c$ | $d$ |
|---|---|---|---|---|
| 0 | 0 | 0 | 0 | 0 |
| 1 | 0 | 1 | $c$ | $d$ |
| $c$ | 0 | $c$ | $d$ | 1 |
| $d$ | 0 | $d$ | 1 | $c$ |

**25.** *Hint*: Use distributivity and the fact that the additive group of $R$ is cyclic.

## *Section 4.2, page 234*

**5.** (ii) $\{\bar{0}, \bar{3}, \bar{6}, \bar{9}, \overline{12}\}$. **9.** $m = 0$.

**11.** (i) $\bar{0}$. (ii) Yes, $\bar{1}$. (iii) Yes. (iv) No.

**13.** (i) The integral multiples of 14.
(ii) The integers of the form $14n + 1$, with $n$ in $\mathbf{Z}$.

**15.** (a)

| $x$ | $\bar{0}$ | $\bar{1}$ | $\bar{2}$ | $\bar{3}$ | $\bar{4}$ | $\bar{5}$ |
|---|---|---|---|---|---|---|
| $-x$ | $\bar{0}$ | $\bar{5}$ | $\bar{4}$ | $\bar{3}$ | $\bar{2}$ | $\bar{1}$ |

(b)

| $x$ | $\bar{0}$ | $\bar{1}$ | $\bar{2}$ | $\bar{3}$ | $\bar{4}$ | $\bar{5}$ |
|---|---|---|---|---|---|---|
| order of $x$ | 1 | 6 | 3 | 2 | 3 | 6 |

(c) $\bar{0}$, $\bar{2}$, and $\bar{4}$. No. The element $\bar{1}$ is not of the form $\bar{2} \cdot \bar{a}$, with $\bar{a}$ in $\mathbf{Z}_6$.

(d) $\bar{a} = \bar{1} = \bar{b}$ and $\bar{a} = \bar{5} = \bar{b}$.

(e) $\{\bar{1}, \bar{5}\}$.

(f)

| $\bar{a}$ | $\bar{2}$ | $\bar{3}$ | $\bar{3}$ | $\bar{4}$ |
|---|---|---|---|---|
| $\bar{b}$ | $\bar{3}$ | $\bar{2}$ | $\bar{4}$ | $\bar{3}$ |

(g) $\{\bar{0}, \bar{1}, \bar{3}, \bar{4}\}$.

(h)

| $x$ | $\bar{0}$ | $\bar{1}$ | $\bar{2}$ | $\bar{3}$ | $\bar{4}$ | $\bar{5}$ |
|---|---|---|---|---|---|---|
| $y = z$ | $\bar{0}$ | $\bar{0}$ | $\bar{2}$ | $\bar{0}$ | $\bar{0}$ | $\bar{2}$ |

(i) $\bar{0}, \bar{1}, \bar{3}, \bar{4}$.

**21.** (a) $(\bar{2}) = \{\bar{0}, \bar{2}, \bar{4}, \bar{6}\}$. (b) Same as part (a).
(c) $(\bar{3}) = \{\bar{0}, \bar{1}, \bar{2}, \bar{3}, \bar{4}, \bar{5}, \bar{6}, \bar{7}\}$. (d) Same as part (c).

**23.** (ii) 6.

**25.** (a) $\bar{0}$, $\bar{1}$, $\bar{5}$, and $\bar{6}$ satisfy $x^2 = x$; $\bar{5}$ is the unity of $\{\bar{0}, \bar{5}\}$.
(c) $(\bar{0})$, $(\bar{1})$, $(\bar{6})$, $(\overline{10})$, $(\overline{15})$, $(\overline{16})$, $(\overline{25})$.

**27.** (i)

| $\bar{a}$ | $\bar{0}$ | $\bar{1}$ | $\bar{2}$ | $\bar{3}$ | $\bar{4}$ | $\bar{5}$ | $\bar{6}$ | $\bar{7}$ | $\bar{8}$ | $\bar{9}$ | $\overline{10}$ | $\overline{11}$ |
|---|---|---|---|---|---|---|---|---|---|---|---|---|
| $\overline{4a}$ | $\bar{0}$ | $\bar{4}$ | $\bar{2}$ | $\bar{0}$ | $\bar{4}$ | $\bar{2}$ | $\bar{0}$ | $\bar{4}$ | $\bar{2}$ | $\bar{0}$ | $\bar{4}$ | $\bar{2}$ |

**29.** This follows from Problem 34 of Section 2.7.

**31.** $\{\bar{0}\}$, $\{\bar{0}, \overline{12}\}$, $\{\bar{0}, \bar{8}, \overline{16}\}$, $\{\bar{0}, \bar{6}, \overline{12}, \overline{18}\}$, $\{\bar{0}, \bar{4}, \bar{8}, \overline{12}, \overline{16}, \overline{20}\}$, $\{\bar{0}, \bar{3}, \bar{6}, \bar{9}, \overline{12}, \overline{15}, \overline{18}, \overline{21}\}$, $\{\bar{0}, \bar{2}, \bar{4}, \bar{6}, \bar{8}, \overline{10}, \overline{12}, \overline{14}, \overline{16}, \overline{18}, \overline{20}, \overline{22}\}$, and $\mathbf{Z}_{24} = \{\bar{0}, \bar{1}, \bar{2}, \bar{3}, \ldots, \overline{23}\}$.

**33.** (ii) The mapping $\theta$ with $\theta(x) = \bar{0}$ for all $x$ in $\mathbf{Z}_{30}$.

**45.** Using Theorem 4 of Section 4.1, one shows that $S$ is a subring in $G$. But $S$ is not an ideal in $G$ since $(1 + 2\mathbf{i})(1 + \mathbf{i}) = -1 + 3\mathbf{i}$, for example, shows that $S$ is not closed under inside-outside multiplication.

## *Section 4.3, page 245*

**1.** (i)

| $x$ | $\overline{1}$ | $\overline{5}$ | $\overline{7}$ | $\overline{11}$ | $\overline{13}$ | $\overline{17}$ |
|---|---|---|---|---|---|---|
| order of $x$ | 1 | 6 | 3 | 6 | 3 | 2 |

(ii) Yes. $\overline{5}$ and $\overline{11}$ are generators.

**3.** (iii) $\overline{1}\cdot\overline{2}\cdot\overline{3}\cdot\overline{4}\cdot\overline{5}\cdot\overline{6} = -1$ in $\mathbf{V}_7$. (iv) $7|(6! + 1)$.

**11.**

| $m$ | 1 | 2 | 3 | 4 | 5 | 6 | 7 | 8 | 9 | 10 |
|---|---|---|---|---|---|---|---|---|---|---|
| $\phi(m)$ | 1 | 1 | 2 | 2 | 4 | 2 | 6 | 4 | 6 | 4 |

**13.** $\phi(3^n) = 2\cdot 3^{n-1}$. **15.** $x = 19$.

**17.** (ii) For all integers $a$ and $h$, the integers $a$ and 1 are relatively prime and $ah \equiv 1 \pmod 1$.

**19.** $\overline{1}, \overline{5}, \overline{7}, \overline{11}$.

**21.** (i) One example is $a = 3$, $b = 2$, $c = 4$, and $m = 6$.

**25.** (i) 4. (ii) 3. (iii) 1. (iv) 2. (v) 4. (vi) 2.

**27.** $x = 9$. Since $94 \equiv 3 \pmod{13}$, $y$ is also 9.

**35.**

| $m$ | 11 | 12 | 13 | 14 | 15 | 16 | 17 | 18 | 19 | 20 |
|---|---|---|---|---|---|---|---|---|---|---|
| $\phi(m)$ | 10 | 4 | 12 | 6 | 8 | 8 | 16 | 6 | 18 | 8 |

**41.** 0, 1, 4.

**49.** *Hint*: See Examples 4 and 5 of this section.

## *Section 4.4, page 254*

**1.** For the multiplication table, see the answer to Problem 23 of Section 4.1. The addition table is:

| $+$ | 0 | 1 | $c$ | $d$ |
|---|---|---|---|---|
| 0 | 0 | 1 | $c$ | $d$ |
| 1 | 1 | 0 | $d$ | $c$ |
| $c$ | $c$ | $d$ | 0 | 1 |
| $d$ | $d$ | $c$ | 1 | 0 |

**7.** (iii) $p$.

**19.** A subring $S$ in $D$ is a subdomain if and only if the unity of $D$ is in $S$.

**21.** *Hint:* Every abelian group of order 6 is cyclic and so has elements of order 1, 2, 3, and 6.

**27.** (a) 100. (b) 1.

**33.** No; $(0, 1)(1, 0) = (0, 0)$ shows that $D \times D'$ has 0-divisors.

## *Section 4.5, page 258*

**21.** 5/2 and 17/2.

**27.** (i) $\{x, y, z\} = \{2, 3, 5\}$. (ii) $\{x, y, z\} = \{2, 3, 5\}$.

**31.**

| $x$ | $-2$ | $-1$ | 2 |
|---|---|---|---|
| $y$ | $\pm 3$ | $\pm 2$ | $\pm 7$ |

### *Section 4.6, page 263*

**5.** (iii)

| + | 0 | 1 | $c$ |
|---|---|---|---|
| 0 | 0 | 1 | $c$ |
| 1 | 1 | $c$ | 0 |
| $c$ | $c$ | 0 | 1 |

| | 0 | 1 | $c$ |
|---|---|---|---|
| 0 | 0 | 0 | 0 |
| 1 | 0 | 1 | $c$ |
| $c$ | 0 | $c$ | 1 |

**9.** Only one. If $\theta$ is an automorphism of $\mathbf{Z}_{23}$, then $\theta(\bar{0}) = \bar{0}$, $\theta(\bar{1}) = \bar{1}$, $\theta(\bar{2}) = \theta(\bar{1} + \bar{1}) = \theta(\bar{1}) + \theta(\bar{1}) = \bar{1} + \bar{1} = \bar{2}$, and so on.

**15.** $a = \bar{3}$

**17.** $x = \overline{14}$, $\quad y = \bar{2}$.

**21.** *Hint:* See Problem 24 of Section 4.2.

**23.** No. $\theta(x)$ may be 0 for all $x$ in $F$.

### *Section 4.7, page 273*

**1.** The ring $M(\mathbf{Z}_2)$ (of $2 \times 2$ matrices with entries in $\mathbf{Z}_2$).

**3.** (a) $\begin{pmatrix} 0 & 0 \\ 0 & 0 \end{pmatrix}$. (b) Yes; see part (a).

**5.** Each entry is 0.

**7.** Let $a_{ij} = 1$ for each $i$ and $j$.

**9.** One choice is $A = \begin{pmatrix} 1 & 0 \\ 1 & 0 \end{pmatrix}$, $B = \begin{pmatrix} 0 & 0 \\ 0 & 0 \end{pmatrix}$, $C = \begin{pmatrix} 0 & 0 \\ 1 & 1 \end{pmatrix}$.

**11.** No; $A$ can be the zero matrix.

**13.** *Hint:* Use displays (7) and (8) to convert to form (6) and then use matrix multiplication.

### *Review Problems (Chapter 4), page 282*

**3.** $\bar{1}, \bar{5}, \overline{11}, \overline{13}, \overline{17}, \overline{19}, \overline{23}, \overline{25}, \overline{29}, \overline{31}, \overline{37}$, and $\overline{41}$.

**5.** 101, 107, 113, 131, 137, 149, 167, 173, 179, 191, and 197.

**7.** (i) $\bar{0}, \bar{1}, \bar{5}$, and $\overline{16}$, (ii) $\bar{5}$ is the unity of $(\bar{5}) = \{\bar{0}, \bar{5}, \overline{10}, \overline{15}\}$; also $\overline{16}$ is the unity of $(\bar{4}) = \{\bar{0}, \bar{4}, \bar{8}, \overline{12}, \overline{16}\}$.

**11.** $x = 40$.

**13.** (i)

| $v$ | $\bar{1}$ | $\bar{7}$ | $\overline{11}$ | $\overline{13}$ | $\overline{17}$ | $\overline{19}$ | $\overline{23}$ | $\overline{29}$ |
|---|---|---|---|---|---|---|---|---|
| $v^{-1}$ | $\bar{1}$ | $\overline{13}$ | $\overline{11}$ | $\bar{7}$ | $\overline{23}$ | $\overline{19}$ | $\overline{17}$ | $\overline{29}$ |

**17.** (i) 3. (ii) Since $943 \equiv 3 \pmod{10}$, the answer is the same as in (i).

**31.** (i)

| | $\bar{1}$ | $\bar{3}$ | $\bar{5}$ | $\bar{7}$ |
|---|---|---|---|---|
| $\bar{1}$ | $\bar{1}$ | $\bar{3}$ | $\bar{5}$ | $\bar{7}$ |
| $\bar{3}$ | $\bar{3}$ | $\bar{1}$ | $\bar{7}$ | $\bar{5}$ |
| $\bar{5}$ | $\bar{5}$ | $\bar{7}$ | $\bar{1}$ | $\bar{3}$ |
| $\bar{7}$ | $\bar{7}$ | $\bar{5}$ | $\bar{3}$ | $\bar{1}$ |

**35.** (ii) $\{\mathbf{V}_5, \mathbf{V}_{10}\}$ and $\{\mathbf{V}_8, \mathbf{V}_{12}\}$.

**41.** (i)

| $x$ | $\overline{1}$ | $\overline{3}$ | $\overline{5}$ | $\overline{9}$ | $\overline{11}$ | $\overline{13}$ | $\overline{15}$ | $\overline{17}$ | $\overline{19}$ | $\overline{23}$ | $\overline{25}$ | $\overline{27}$ |
|---|---|---|---|---|---|---|---|---|---|---|---|---|
| $x^2$ | $\overline{1}$ | $\overline{9}$ | $\overline{25}$ | $\overline{25}$ | $\overline{9}$ | $\overline{1}$ | $\overline{1}$ | $\overline{9}$ | $\overline{25}$ | $\overline{25}$ | $\overline{9}$ | $\overline{1}$ |

(ii) $K = \{\overline{1}, \overline{13}, \overline{15}, \overline{27}\}$.

(iii) Let $L = \{\overline{3}, \overline{11}, \overline{17}, \overline{25}\}$ and $M = \{\overline{5}, \overline{9}, \overline{19}, \overline{23}\}$. Then the natural map is given by:

| $x$ | $\overline{1}$ | $\overline{3}$ | $\overline{5}$ | $\overline{9}$ | $\overline{11}$ | $\overline{13}$ | $\overline{15}$ | $\overline{17}$ | $\overline{19}$ | $\overline{23}$ | $\overline{25}$ | $\overline{27}$ |
|---|---|---|---|---|---|---|---|---|---|---|---|---|
| $\alpha(x)$ | $K$ | $L$ | $M$ | $M$ | $L$ | $K$ | $K$ | $L$ | $M$ | $M$ | $L$ | $K$ |

(iv)

| | $K$ | $L$ | $M$ |
|---|---|---|---|
| $K$ | $K$ | $L$ | $M$ |
| $L$ | $L$ | $M$ | $K$ |
| $M$ | $M$ | $K$ | $L$ |

**43.**

| $x$ | $\overline{1}$ | $\overline{3}$ | $\overline{5}$ | $\overline{9}$ | $\overline{11}$ | $\overline{13}$ |
|---|---|---|---|---|---|---|
| $\theta_1(x)$ | $\overline{1}$ | $\overline{5}$ | $\overline{11}$ | $\overline{7}$ | $\overline{13}$ | $\overline{17}$ |
| $\theta_2(x)$ | $\overline{1}$ | $\overline{11}$ | $\overline{5}$ | $\overline{13}$ | $\overline{7}$ | $\overline{17}$ |

**45.** (b) Yes. $\mathbf{V}_{12}$ is isomorphic to $\mathbf{V}_8$.
(c) No. $\mathbf{V}_5$ and $\mathbf{V}_{10}$ are cyclic while a direct product of two groups of order 2 is not cyclic.

**55.** $p = 2$. **59.** (a) 138. (b) 1.

**63.** Four. The distinct ideals are $(\overline{0})$, $(\overline{1})$, $(\overline{3})$, and $(\overline{13})$.

**65.** Two. [*Hint:* Problems 28(c) and 26(i) of Section 4.2 may facilitate the proof.]

## *Section 5.1, page 295*

**3.** $a_1 = -4, a_0 = 13$. **11.** $S$ is the set of all $m + n\sqrt{2}$ with $m$ and $n$ in $\mathbf{Z}$.

**13.** (a) $(3 - \sqrt{2})^{-1} = (3/7) + (1/7)\sqrt{2}$

## *Section 5.2, page 303*

**1.** (ii) $x^3, x^3 + x^2, x^3 + x, x^3 + 1, x^3 + x^2 + x, x^3 + x^2 + x + 1$.
(iii) $x^3 + x^2 + 1, x^3 + x + 1$.

**3.** $100_3, 102_3, 110_3, 111_3, 120_3, 121_3$. **5.** $m^d$.

**11.** No. If $n > \deg \alpha$ and $n > \deg \beta$, the coefficient of $x^n$ is 0 in $\alpha$ and in $\beta$ and hence in their sum.

**15.** $x^2 = x \cdot x = (x + \overline{3})(x + \overline{6})$. **27.** (ii) $x$. **33.** (i) Yes. (ii) No. (iii) Yes.

## *Section 5.3, page 313*

**1.** (i) $\overline{0}, \overline{2}, \overline{4}, \overline{6}, \overline{8}, \overline{10}$; (ii) $\overline{1}, \overline{2}, \overline{5}, \overline{7}, \overline{10}, \overline{11}$; (iii) $\overline{2}, \overline{10}$; (iv) $\overline{2}x^3 + x, x^3 + \overline{2}x$.

**3.** (i) 4. (ii) 8. (iii) 16. (iv) 32. **5.** $\gamma = (1/2)x - (3/4)$, $\rho = -(9/4)x + (29/4)$.

**7.** $\beta = x^{n-1} + ax^{n-2} + \cdots + a^{n-2}x + a^{n-1}$. **15.** The principal ideal generated by $2x$.

**17.** Yes. $A$ is symmetric, reflexive, and transitive. **27.** 36.

### *Section 5.4, page 322*

**1.** 69. **9.** (ii) Four.
**17.** (a) 2. (b) 1. (c) 2. (d) 9. (e) 10. (f) 2.
**25.** (a) $r_1 + r_2 + \cdots + r_d = -a_{d-1}$.
(b) $r_1 r_2 \cdots r_d = (-1)^d a_0$.
**29.** $h_5 = (5ab^3c - 5a^2bc^2 - b^5)/a^5$.
**31.** (i) $\rho = 2x$.
(ii) $\alpha(\mathbf{i}) = 2\mathbf{i}$.
(iii) $\alpha(\mathbf{i}) = \rho(\mathbf{i})$ since $\alpha(\mathbf{i}) = \gamma(\mathbf{i})\beta(\mathbf{i}) + \rho(\mathbf{i})$ and $\beta(\mathbf{i}) = 0$.
**35.** (ii) $c = -1/3$.

### *Section 5.5, page 329*

**1.** The integer root is $-1$; the others are $-1 \pm \sqrt{3}$.
**3.** $1, -3, \mathbf{i}\sqrt{2}, -\mathbf{i}\sqrt{2}$.
**5.** (a) $5/3, (-3 \pm \sqrt{17})/2$. (b) $-1, 2/3, 5/4$.
(c) 1 is a root with multiplicity 3 and $-2$ is a simple root.
(d) $0, 2/3, -1/3, (1 \pm \mathbf{i}\sqrt{3})/6$. (e) $\pm\mathbf{i}, \pm\mathbf{i}\sqrt{2}$.
**7.** $\pm 1, \pm 3, \pm 7, \pm 21$.
**9.** Yes. In lowest terms, $35/21 = 5/3$; and this meets the conditions of Theorem 2.
**13.** Either $s = 0$ or both $s|a_1$ and $t|a_d$. **17.** (i) $\alpha = x^4 - 10x^2 + 1$.

### *Section 5.6, page 339*

**1.** $3(x-1)(x-2)(x-3)(x-5)$. **3.** $\bar{2}x^3 + x^2 + x + \bar{1}$.
**7.** $(n^4 - 2n^3 - n^2 + 2n)/4$. **9.** $[n(n+1)/2]^2$.
**11.** (i) $A_1 = 2, A_2 = 4, A_3 = 7, A_4 = 11$.
(iii) $(n^2 + n + 2)/2$.
**25.** $\beta = 1 + 6B_2 + 6B_3$.

### *Section 5.7, page 352*

**3.** $3 + (4 + \sqrt{7})\mathbf{i}, -3 + (-4 + \sqrt{7})\mathbf{i}$
**9.** (a) $D = -23$; one root is real and the others are complex conjugates.
(b) $D = 0$; there is a root with multiplicity greater than 1.
(c) $D = 392$; the roots are real and distinct.
**11.** $-(r+s)$, $-(wr + w^2s)$, and $-(w^2r + ws)$ where $r^3 = (-1 + \sqrt{5})/2$, $s = -1/r$, and $w = (-1 + \mathbf{i}\sqrt{3})/2$.
**15.** (a) $-1 \pm \mathbf{i}\sqrt{2}, 3 \pm \sqrt{6}$.
(b) $(5 \pm \sqrt{17})/2, (5 \pm \sqrt{29})/2$.
(c) The only root is $-1$, which has multiplicity 4.

### Section 5.8, page 358

**1.** $\delta = x^2 + x + \bar{1}$, $\xi = x$, $\eta = x + 1$.
**11.** (ii) Every polynomial in $K$ has $s$ as a zero.
**13.** *Hint:* See Section 4.8.

### Section 5.9, page 363

**5.** (a) $x = (\bar{2}x + \bar{3})(\bar{3}x + \bar{2})$.
(b) $x + \bar{2} = (\bar{2}x + \bar{1})(\bar{3}x + \bar{2})$.
(c) $x + \bar{3} = (\bar{2}x + \bar{3})(\bar{3}x + \bar{5})$.

### Section 5.10, page 366

**1.** (a) $[(17/5)/(x - 2)] + [(8/5)/(x + 3)]$.
(b) $[2/(x + 1)] - [3/(x + 1)^2] - [4/(x^2 + 4)]$.
**3.** $x + [\bar{4}/(x - \bar{1})] + [\bar{4}/(x + \bar{1})]$.
**7.** $[(x - r_1)(x - r_2)\cdots(x - r_n)]^{-1} = a_1(x - r_1)^{-1} + \cdots + a_n(x - r_n)^{-1}$ where $a_k = [(r_k - r_1)(r_k - r_2)(r_k - r_3)\cdots(r_k - r_{k-1})(r_k - r_{k+1})\cdots(r_k - r_n)]^{-1}$.
**11.** Let a ppd-term in **Q** be of the form $c/p^k$ with $p$ a positive prime in **Z**, $k$ a positive integer, and $c$ an integer satisfying $|c| < p$. Then every rational number is the sum of an integer and a finite number of ppd-terms in **Q**.

### Section 5.11, page 373

**1.** (iii) $3^3 = 27$.
**3.** Since $25 = 5^2$, we seek a prime polynomial $\pi$ of degree 2 in $\mathbf{Z}_5[x]$ and find that $x^2 + \bar{2}$ is such a $\pi$. Then Theorem 1 tells us that

$$\mathbf{Z}_5[x]/(x^2 + \bar{2})$$

is a field with 25 elements.
**11.** $x^3 - 6x - 6$.

### Section 5.12, page 377

**3.** (i) Two.
(ii) One. [There is only one cube root of 2, the real number $\sqrt[3]{2}$, in $\mathbf{Q}(\sqrt[3]{2})$.]
**5.** (i) 6. (ii) 2.

### *Review Problems (Chapter 5), page 378*

**5.** $x^5 - x$. **9.** (i) $\rho = 3x - 9$. (ii) $\rho = \bar{3}x + \bar{1}$.
**11.** (i) Eight. (ii) Two. **13.** $x^{14} - \bar{2}$.
**15.** The monic generator is $x^{17} - x$. **17.** The rational root is 5/2; the other roots are $\pm 3\mathbf{i}$.
**19.** $\bar{2}x^3 + x^2 + \bar{2}$. **21.** $\alpha = 5(x-2)^2 + (x-2)^3$. The multiplicity of the zero 2 is 2.
**23.** (ii) Let $r^3 = (-7 + \mathbf{i}\sqrt{1323})/2$, $s = 7/r$, and $w = (-1 + \mathbf{i}\sqrt{3})/2$. Also, let $u$ take on the values $-(r+s)$, $-(wr + w^2s)$, and $-(w^2r + ws)$. Then the zeros are given by $x = (u-1)/3$.
**25.** (ii) Yes. (iii) $x - r$. **27.** 406.
**29.** No; 99 is not a power of a prime in **Z**.
**33.** Yes. (For the proof, show that $\theta$ sends 1 to 1 and 1 + 1 to 1 + 1.)

### *Section 6.2, page 390*

**9.** Let $a = \sqrt{7}$, $b = \sqrt{3 + a}$, $c = \sqrt{3 - a}$. Then the four roots, $\pm b$ and $\pm c$, are in $\mathbf{Q}(a, b, c)$.

### *Section 7.1, page 405*

**1.** $8x \equiv 37 \pmod{141}$. **3.** 209 and 419.
**5.** 59 and 119. **7.** 548.
**11.** (i) $a = 21$, $r = 77$; (ii) $b = 42$, $s = 154$; (iii) $c = 24374$, $t = 30030$.

### *Section 7.2, page 407*

**5.** Let $a$ be an integer and let $X = \{a, a+1, a+2, \ldots\}$. If $S$ is a subset of $X$ such that $a$ is in $S$ and $a + k + 1$ is in $S$ whenever $a, a+1, a+2, \ldots, a+k$ are in $S$, then $S = X$.
**13.** $\binom{k+n-1}{n} = (k+n-1)!/n!(k-1)!$
**17.** $A_n = (n^2 + 11n + 12)/2$. **31.** *Hint:* See Problem 13.

### *Section 7.3, page 417*

**1.** (a) 16; (b) 1680; (c) 16. **3.** (a) 5, 8, 10, 12; (b) 7, 9, 14, 18.
**13.** (iii) 65, 104, 105, 112, 130, 140, 144, 156, 168, 180, 210. (There are 11 $n$'s.)
**25.** (ii) 1, 4, 9, 16.
**27.** $n = q^{p-1}$, with $q$ a positive prime.
**29.** $2^4 \cdot 3 = 48$.

**33.**

| $n$ | 1 | 2 | 3 | 4 | 5 | 6 | 7 | 8 |
|---|---|---|---|---|---|---|---|---|
| $p(n)$ | 1 | 2 | 3 | 5 | 7 | 11 | 15 | 22 |

### *Section 7.4, page 425*

**1.** (a) 1, 2, 4, 8, 9, 13, 15, 16.
(b) 1, 4, 5, 6, 7, 9, 11, 16, 17.

### *Review Problems (Chapter 7), page 426*

**1.** 1, 29, 41, 71. **3.** 461.

**11.**

| $a$ | 1 | 2 | 2 | 4 | 4 | 6 | 6 | 8 | 10 | 12 | 16 | 16 | 18 | 18 | 20 |
|---|---|---|---|---|---|---|---|---|---|---|---|---|---|---|---|
| $p$ | 2 | 2 | 3 | 2 | 5 | 3 | 7 | 2 | 11 | 13 | 2 | 17 | 3 | 19 | 5 |
| $n$ | 1 | 2 | 1 | 3 | 1 | 2 | 1 | 4 | 1 | 1 | 5 | 1 | 3 | 1 | 2 |

**15.** Only $n = 10$. **19.** $2^6 \cdot 3 = 192$.
**21.** The squares of positive primes in **Z**. **23.** 30.

### *Sections 8.1 and 8.2, page 434*

**1.** (a) 11101. (b) 10100. (c) error. (d) 1101.
**3.** No; 1001000 is also a codeword.
**5.** (a) $2^5 = 32$. (b) 50% (or 1/2).
**7.** (a) No. (b) Yes. (c) No.
**9.** (a) 111011101110. (b) 1110. (c) 1010.
**11.** (a) $\alpha = 000$, $\beta = 000000000$, $\beta' = 100000000$.
(b) $\alpha = 000$, $\beta = 000000000$, $\beta' = 100100000$.
**13.** $E(a_1a_2 \cdots a_s) = a_1a_2 \cdots a_s a_1a_2 \cdots a_s$, $D(b_1b_2 \cdots b_{2s})$ is $b_1b_2 \cdots b_s$ if $b_i = b_{s+i}$ for $1 \leq i \leq s$ and is "error" otherwise.
**15.** The triple repetition code corrects single-digit errors and the double repetition code merely detects such errors. The disadvantage is that the $(s, 3s)$ code requires more transmission time and effort.

### *Section 8.3, page 438*

**1.** (a) 0110001. (b) 4. (c) 3. (d) 3. (e) 3.
**3.** 0110001.
**5.** (a) 000, 110, 101. (b) Yes; for example 000, 110, 101, 011.
**7.** 3; one can use the codewords of the answer to Problem 5(a).
**11.** Yes; $W_n$ is made into the direct product $\mathbf{Z}_2 \times \mathbf{Z}_2 \times \cdots \times \mathbf{Z}_2$ of $n$ copies of $\mathbf{Z}_2$.

## *Section 8.4, page 444*

**1.** $G = (I_4 : F)$ with $F = (1\ \ 1\ \ 1\ \ 1)^T$.
**3.** Yes. Its generator $G = (I_m : F)$ with $F = (1\ \ 1\ \cdots\ 1)^T$, and the mapping $E: W_m \to W_{m+1}$ with $E(\alpha) = \alpha G$ is injective.
**5.** (a) No; $G = (I_2 : F)$ with $F = (1\ \ 1)^T$, and $H = (F^T : I_1)^T$ has repeated rows.
(b) Yes. Let $F = (I_2 : I_2)$. Then $G = (I_2 : F)$, and $H = (F^T : I_4)^T$ has no row of zeros nor does it have repeated rows.
**7.** Yes; $H = (Q_3 : I_3)^T$.
**9.** $$\begin{pmatrix} 1 & 0 & 0 & 0 & 1 & 1 & 0 & 0 \\ 0 & 1 & 0 & 0 & 1 & 0 & 1 & 1 \\ 0 & 0 & 1 & 0 & 0 & 1 & 1 & 0 \\ 0 & 0 & 0 & 1 & 0 & 1 & 0 & 1 \end{pmatrix}.$$
**11.** (a) $\gamma H = 1000$, $D(\gamma) = 1000$. (b) $\gamma H = 1100$, $D(\gamma) = 0100$.
(c) $\gamma H = 1101$, $D(\gamma) =$ error. (d) $\gamma H = 0000$, $D(\gamma) = 0000$.
**13.** They are equal.
**15.** (a) There are only 3 distinct 2-vectors with at least one nonzero entry.
(b) There are only 7 distinct 3-vectors with at least one nonzero entry.
**17.** No; $\gamma H = 1011$ and $D(\gamma) = 1011$, which is incorrect.
**19.** *Hint:* Use the systematic form of $G$.

## *Section 8.5, page 449*

**1.** 110.
**3.** (a) $m = 1$, $n = 3$. (b) $G = (1\ \ 1\ \ 1)$. (c) $h = 4$, $k = 2$. (d) $B = \{000, 111\}$.
(e) $\{000, 001, 010, 100\}$.
(f)

| $\gamma$ | 000 | 001 | 010 | 100 |
|---|---|---|---|---|
| $\gamma H$ | 00 | 01 | 10 | 11 |

(g) 1. (h) 1.

## *Section 8.6, page 451*

**1.** $$\begin{pmatrix} 1 & 0 & 0 & 0 & 1 & 1 & 1 \\ 0 & 1 & 0 & 0 & 1 & 1 & 0 \\ 0 & 0 & 1 & 0 & 1 & 0 & 1 \\ 0 & 0 & 0 & 1 & 0 & 1 & 1 \end{pmatrix}.$$
**3.** (a) 1001100. (b) 0110011. (c) 1000111. (d) 0111000.
**5.** The entry in the $i$th row and $j$th column is $\gamma_i + \beta_j$ where the coset leaders $\gamma_1, \gamma_2, \ldots, \gamma_8$ are

0000000, 1000000, 0100000, 0010000,
0001000, 0000100, 0000010, 0000001

and the codewords $\beta_1, \beta_2, \ldots, \beta_{16}$ are

0000000, 0001011, 0010101, 0011110,
0100110, 0101101, 0110011, 0111000,
1000111, 1001100, 1010010, 1011001,
1100001, 1101010, 1110100, 1111111.

**7.**

| $\gamma$ | 0000000 | 1000000 | 0100000 | 0010000 |
|---|---|---|---|---|
| $\gamma H$ | 000 | 111 | 110 | 101 |

| $\gamma$ | 0001000 | 0000100 | 0000010 | 0000001 |
|---|---|---|---|---|
| $\gamma H$ | 011 | 100 | 010 | 001 |

**9.** (a) 1101. (b) 0111. (c) 1100.

### *Section 8.7, page 454*

**1.** (a) $\overline{8}$. (b) $\overline{4}$. (c) $\overline{13}$. **3.** $d = 37$.
**5.** 7, 11, 13, 17, 19, 23, 29, 31, 37, 41, 43, 47, 49, 53, 59, 61, 67, 71, 73.
**7.** 9. **11.** Yes; one could use $\overline{2}, \overline{3}, \ldots, \overline{677}$ to represent the 676 ($= 26^2$) ordered pairs of letters.

### *Review Problems (Chapter 8), page 454*

**1.** 01100, 01101, 01110, 01000, 00100, 11100.
**3.** (a) 010101. (b) 11. (c) 10.
**5.** 5. **7.** 9. **9.** $2^9 = 512$. **11.** 111. **13.** 010.
**15.** 26 digits in words, 31 digits in codewords.
**17.** The transpose of

$$\begin{pmatrix} 1 & 0 & 0 & 0 & 1 & 1 & 1 & 0 & 0 & 0 & 1 & 1 & 1 & 0 & 1 \\ 0 & 1 & 0 & 0 & 1 & 0 & 0 & 1 & 1 & 0 & 1 & 1 & 0 & 1 & 1 \\ 0 & 0 & 1 & 0 & 0 & 1 & 0 & 1 & 0 & 1 & 1 & 0 & 1 & 1 & 1 \\ 0 & 0 & 0 & 1 & 0 & 0 & 1 & 0 & 1 & 1 & 0 & 1 & 1 & 1 & 1 \end{pmatrix}$$

**19.** $f = \phi(35) = 24$ and every $\bar{e}$ in $V_{24}$ is its own inverse. Hence the "secret" decoding function has to be the same as the announced encoding function. This could easily be discovered while encoding.

# *Index*